Thermodynamics and Statistical Mechanics of Small Systems

Thermodynamics and Statistical Mechanics of Small Systems

Special Issue Editors

Andrea Puglisi
Alessandro Sarracino
Angelo Vulpiani

MDPI • Basel • Beijing • Wuhan • Barcelona • Belgrade

MDPI

Special Issue Editors

Andrea Puglisi
Istituto dei Sistemi Complessi–CNR and Dipartimento di Fisica
Università degli studi di Roma "La Sapienza"
Italy

Alessandro Sarracino
Istituto dei Sistemi Complessi-CNR and
Dipartimento di Ingegneria, Università della Campania "Luigi Vanvitelli"
Italy

Angelo Vulpiani
Dipartimento di Fisica
Università degli studi di Roma "La Sapienza"
Italy

Editorial Office
MDPI
St. Alban-Anlage 66
Basel, Switzerland

This edition is a reprint of the Special Issue published online in the open access journal *Entropy* (ISSN 1099-4300) from 2017–2018 (available at:
http://www.mdpi.com/journal/entropy/special_issues/small_systems).

For citation purposes, cite each article independently as indicated on the article page online and as indicated below:

LastName, A.A.; LastName, B.B.; LastName, C.C. Article title. *Journal Name* **Year**, Article number, Page Range.

ISBN 978-3-03897-057-6 (Pbk)
ISBN 978-3-03897-058-3 (PDF)

Cover image courtesy of Andrea Baldassarri, Andrea Puglisi, Alessandro Saracino and Angelo Vulpiani.

Contents

Contents

About the Special Issue Editors

Andrea Puglisi, First Researcher, was born in Rome, Italy, on 11/6/1973, graduated from Rome Sapienza University in 1998 and took a Ph.D. in Physics in 2002, both under the supervision of Angelo Vulpiani. He has been the Marie-Curie fellow at Orsay, Paris (2003–2004) and a postdoctoral researcher at Rome Sapienza (2005–2008), and a researcher at CNR-ISC (2009–2018). He is now First Researcher at CNR-ISC, based at Sapienza University. His interest are granular materials (theory and experiments), non-equilibrium statistical mechanics and computational cognitive science. He has written 130 scientific papers on international journals and one book (Transport and Fluctuations in Granular Fluids, Springer 2014).

Alessandro Sarracino, Researcher, was born in Naples, Italy, on 22/12/1981. He obtained his Laurea degree in Physics from the University of Naples "Federico II" in 2005 and his PhD in Physics from the University of Salerno in 2009. He worked in Rome at the CNR Institute for Complex Systems since 2010–2018. He also spent two years in Paris at the University Pierre et Marie Curie in 2014–15. His research interests are in the field of non-equilibrium statistical mechanics, fluctuation-dissipation relations and non-linear responses, with applications to aging systems, granular matter and Brownian motors. He is the author of more than 40 scientific papers.

Angelo Vulpiani, Vulpiani, Full Professor of Theoretical Physics, was born in Borgorose (Rieti), Italy, on 08/08/1954, graduated from Rome University in 1977, supervised by Gianni Jona-Lasinio. He has been a CNR Fellow (1978–1981), Assistant Professor at Rome University (1981–1988), Associate Professor at the University of L'Aquila(1988–1991) and then at the University of Rome (1991–2000). At present he is Full Professor of Theoretical Physics in the Physics Department of the University of Rome "Sapienza", and is a Fellow of the Institute of Physics. He was a visiting fellow at several research institutes and universities in France, Belgium, Sweden, Denmark, and the United States. His scientific interests include chaos and complexity in dynamic systems, statistical mechanics of nonequilibrium and disordered systems, developed turbulence, and phenomena of transport, diffusion and foundations of physics. He has written about 250 scientific papers on international journals and nine books.

entropy

MDPI

Editorial

Thermodynamics and Statistical Mechanics of Small Systems

Andrea Puglisi [1,*], Alessandro Sarracino [1] and Angelo Vulpiani [2,3]

[1] Consiglio Nazionale delle Ricerche (CNR), Istituto dei Sistemi Complessi (ISC), c/o Dipartimento di Fisica, Universita' Sapienza Roma, p.le A. Moro 2, 00185 Roma, Italy; ale.sarracino@gmail.com

[2] Dipartimento di Fisica, Università degli studi di Roma "La Sapienza", Piazzale A. Moro 5, 00185 Roma, Italy; Angelo.Vulpiani@roma1.infn.it

[3] Centro Interdisciplinare "B. Segre", Accademia dei Lincei, 00100 Roma, Italy

* Correspondence: andrea.puglisi@roma1.infn.it

Received: 15 May 2018; Accepted: 21 May 2018; Published: 23 May 2018

Keywords: statistical mechanics; small systems; stochastic thermodynamics; non-equilibrium fluctuations; large deviations

---

A challenging frontier in modern statistical physics is concerned with systems with a small number of degrees of freedom, far from the thermodynamic limit. Beyond the general interest in the foundation of statistical mechanics, the relevance of this subject is due to the recent increase of resolution in the observation and in the manipulation of biological and man-made objects at micro- and nano-scales. The peculiar feature of small systems is the role played by fluctuations, which cannot be neglected and are responsible for many non-trivial behaviors. The study of fluctuations of thermodynamic quantities, such as energy or entropy, goes back to Einstein, Onsager, and Kubo; more recently, interest in this matter has grown with the establishment of new fluctuation–dissipation relations, which hold even in non-linear regimes, and of the so-called stochastic thermodynamics. Such a turning point has received a great impulse from the study of systems that are far from thermodynamic equilibrium, due to very long relaxation times, as in disordered systems, or due to the presence of external forcing and dissipation, as in granular or active matter. Applications of the thermodynamics and statistical mechanics of small systems range from molecular biology to micro-mechanics, including, among others, models of nano-transport, of Brownian motors, and of (living or artificial) self-propelled organisms.

The Contributions

In this special issue, we collect 20 contributions, spanning the above mentioned subjects. In particular, the main addressed topics are as follows:

1. entropy production and stochastic thermodynamics (8);
2. heat transport and entropy in nonlinear chains and long-range systems (4);
3. granular and other dissipative systems (2);
4. phase transitions and large deviations in probabilistic models (2);
5. coarse-graining techniques (2);
6. ferromagnetic models (2).

Topic (1): Entropy Production and Stochastic Thermodynamics

In [1], the stochastic energetics approach is applied to a Stirling engine model. Such an engine is realized by coupling a passive and an active system. The latter is modeled through a non-Gaussian noise with persistency: this choice allows the authors to study two possible origins of discrepancy with respect to thermodynamic engines (coupled to equilibrium systems), finding that persistency is more important.

In [2], the interesting problem of optimizing the energetic cost of maintaining a non-equilibrium state is discussed. The important result concerns the existence of a minimum bounding such a cost, which is expressed as an information-theoretic measure of distinguishability between the target non-equilibrium state and the underlying equilibrium distribution.

In [3], the possibility of a Clausius relation for active systems is investigated on the basis of the so-called Active Ornstein-Uhlenbeck Particles (AOUP) model. It is shown that a mapping from the AOUP model to an underdamped model with non-uniform viscosity and temperature can shed light on this question, but induces an ambiguity in the determination of the parity under time-reversal of some forces in the system, leaving at least two possible definitions of entropy production. One of the two possible choices leads to an entropy production that is consistent with detailed balance in the system and can be expressed in a Clausius-like fashion.

In [4], the problem of moving a system, in a finite time, from an equilibrium state to a different one is studied. The question here is when such a transformation occurs optimally in the sense of producing a minimum amount of entropy. The answer is a set of constraints on the possible protocols, particularly on their time-derivatives. Some interesting examples related to recent experiments are discussed.

In [5], the author revisits the classical problem of a system in contact with a collection of harmonic oscillators initially in thermal equilibrium, which may represent a thermal bath. The new ingredient is a time-dependent system-bath coupling which is shown to lead to an additional harmonic force acting on the system. The consequences for heat and work functionals of stochastic thermodynamics, in classical as well as quantum systems, are also worked out.

In [6], a magnetic quantum thermal engine is considered where the energy levels are degenerate. The analytical expression of the relation between the magnetic field and temperature along the adiabatic process is calculated, including the efficiency as a function of the compression ratio.

In [7], two interacting stochastic systems are considered, under the point of view of information exchange. An information landscape and an information flux are defined and seen to influence different aspects of the systems' dynamics. Connections with the entropy production of non-equilibrium thermodynamics are investigated.

In [8], a stochastic model of nanopore is investigated in the framework of information dynamics, focusing on the local and specific entropy rates computed in simulations. Those metrics are put in relation with the fluctuations of the current in the nanopore.

Topic (2): Heat Transport and Entropy in Nonlinear Chains and Long-Range Systems

In [9], transport of mass and energy through a discrete nonlinear Schrödinger chain in contact with a heat reservoir and a pure dissipator is considered. Depending on the heat bath temperature, two interesting regimes are observed, featuring a non-monotonous shape of the temperature profiles across the chain (at low temperature) and a spontaneous emergence of discrete breathers (at high temperature), whose statistics can be described in terms of large deviations.

In [10], a multi-partitioned piston model coupled with two thermal baths at different temperatures is discussed in the framework of kinetic theory, obtaining the values of the main thermodynamic quantities characterizing the stationary non-equilibrium states: a good agreement with Fourier's law in the thermodynamic limit is obtained.

In [11], the effect of both short- and long-range interparticle interactions in the Nosé–Hoover dynamics of many-body Hamiltonian systems is investigated. It was found that the equilibrium properties of the system coincides with that within the canonical ensemble; however, in the case with only long-range interactions, the momentum distribution relaxes to its Gaussian form in equilibrium over a scale that diverges with the system size. This study brings to the fore the crucial role that interactions play in deciding the equivalence between Nosé–Hoover and canonical equilibrium.

In [12], the relation between Kolmogorov entropy and the largest Lyapunov exponent in systems with long-range interactions is studied. In particular, Lyapunov spectra for Coulombic and gravitational versions of the one-dimensional systems of parallel sheets with periodic boundary conditions are computed, showing that the largest Lyapunov exponent can be viewed as a precursor of the transition that becomes more pronounced as the system size increases.

Topic (3): Granular and Other Dissipative Systems

In [13], the interesting issue of Kovacs-like memory effects, which characterize some athermal dissipative systems whose stationary states cannot be completely characterized by macroscopic variables such as pressure, volume, and temperature, is addressed. Within the linear response regime, it is proved that the observed non-monotonic relaxation is consistent with the monotonic decay of the non-equilibrium entropy.

In [14], the transport properties of a low-density granular gas immersed in an active fluid, modeled as a non-uniform stochastic thermostat, are investigated. Navier–Stokes hydrodynamic equations can describe the steady flow in the system, even for high inelasticity.

Topic (4): Phase Transitions and Large Deviations in Probabilistic Models

In [15], an example of the condensation of fluctuations is considered in probabilistic models with power-law distributions. This kind of phenomenon occurs when there is a critical threshold above which the fluctuation in the system is fed by just one degree of freedom. The paper focuses on the evaluation of the participation ratio as a generic indicator of condensation.

In [16], a stochastic logistic model with multiplicative noise, which shows a transition for sufficiently strong noise, is studied. Such a transition between different solutions is analyzed in terms of entropy and information length.

Topic (5): Coarse-Graining Techniques

In [17], a systematic coarse-graining methodology to treat many particle molecular systems using cluster expansion techniques is discussed. This allows for the building of effective Hamiltonians with interaction potentials with two, three, (and more) body interactions.

In [18], spatial block analysis as a method of efficient extrapolation of thermodynamic quantities from finite-size computer simulations of a large variety of physical systems is reviewed. Such a method provides promising results for simple liquids and liquid mixtures.

Topic (6): Ferromagnetic Models

In [19], the authors focus on the Ising model with a small number of sites, comparing numerical results about the magnetization (in 1d, 2d, and 3d and without periodic boundary conditions) with past results for the thermodynamic limit.

In [20], the spinor analysis is applied to exactly evaluate spin–spin correlation functions in the 2d rectangular Ising model. The author shows the different results (even in terms of short- or long-range order) emerging from different boundary conditions.

Acknowledgments: We express our thanks to the authors of the above contributions, and to the journal Entropy and MDPI for their support during this work.

Conflicts of Interest: The authors declare no conflict of interest.

References

1. Zakine, R.; Solon, A.; Gingrich, T.; van Wijland, F. Stochastic Stirling Engine Operating in Contact with Active Baths. *Entropy* **2017**, *19*, 193. [CrossRef]
2. Horowitz, J.; England, J. Information-Theoretic Bound on the Entropy Production to Maintain a Classical Nonequilibrium Distribution Using Ancillary Control. *Entropy* **2017**, *19*, 333. [CrossRef]
3. Puglisi, A.; Marini Bettolo Marconi, U. Clausius Relation for Active Particles: What Can We Learn from Fluctuations. *Entropy* **2017**, *19*, 356. [CrossRef]

4. Muratore-Ginanneschi, P.; Schwieger, K. An Application of Pontryagin's Principle to Brownian Particle Engineered Equilibration. *Entropy* **2017**, *19*, 379. [CrossRef]
5. Aurell, E. On Work and Heat in Time-Dependent Strong Coupling. *Entropy* **2017**, *19*, 595. [CrossRef]
6. Peña, F.; González, A.; Nunez, A.; Orellana, P.; Rojas, R.; Vargas, P. Magnetic Engine for the Single-Particle Landau Problem. *Entropy* **2017**, *19*, 639. [CrossRef]
7. Zeng, Q.; Wang, J. Information Landscape and Flux, Mutual Information Rate Decomposition and Connections to Entropy Production. *Entropy* **2017**, *19*, 678. [CrossRef]
8. Gilpin, C.; Darmon, D.; Siwy, Z.; Martens, C. Information Dynamics of a Nonlinear Stochastic Nanopore System. *Entropy* **2018**, *20*, 221. [CrossRef]
9. Iubini, S.; Lepri, S.; Livi, R.; Oppo, G.; Politi, A. A Chain, a Bath, a Sink, and a Wall. *Entropy* **2017**, *19*, 445. [CrossRef]
10. Caprini, L.; Cerino, L.; Sarracino, A.; Vulpiani, A. Fourier's Law in a Generalized Piston Model. *Entropy* **2017**, *19*, 350. [CrossRef]
11. Gupta, S.; Ruffo, S. Equilibration in the Nosé–Hoover Isokinetic Ensemble: Effect of Inter-Particle Interactions. *Entropy* **2017**, *19*, 544. [CrossRef]
12. Kumar, P.; Miller, B. Lyapunov Spectra of Coulombic and Gravitational Periodic Systems. *Entropy* **2017**, *19*, 238. [CrossRef]
13. Plata, C.; Prados, A. Kovacs-Like Memory Effect in Athermal Systems: Linear Response Analysis. *Entropy* **2017**, *19*, 539. [CrossRef]
14. Vega Reyes, F.; Lasanta, A. Hydrodynamics of a Granular Gas in a Heterogeneous Environment. *Entropy* **2017**, *19*, 536. [CrossRef]
15. Gradenigo, G.; Bertin, E. Participation Ratio for Constraint-Driven Condensation with Superextensive Mass. *Entropy* **2017**, *19*, 517. [CrossRef]
16. Kim, E.; Tenkès, L.; Hollerbach, R.; Radulescu, O. Far-From-Equilibrium Time Evolution between Two Gamma Distributions. *Entropy* **2017**, *19*, 511. [CrossRef]
17. Tsourtis, A.; Harmandaris, V.; Tsagkarogiannis, D. Parameterization of Coarse-Grained Molecular Interactions through Potential of Mean Force Calculations and Cluster Expansion Techniques. *Entropy* **2017**, *19*, 395. [CrossRef]
18. Heidari, M.; Kremer, K.; Potestio, R.; Cortes-Huerto, R. Fluctuations, Finite-Size Effects and the Thermodynamic Limit in Computer Simulations: Revisiting the Spatial Block Analysis Method. *Entropy* **2018**, *20*, 222. [CrossRef]
19. Vogel, E.; Vargas, P.; Saravia, G.; Valdes, J.; Ramirez-Pastor, A.; Centres, P. Thermodynamics of Small Magnetic Particles. *Entropy* **2017**, *19*, 499. [CrossRef]
20. Mei, T. Exact Expressions of Spin-Spin Correlation Functions of the Two-Dimensional Rectangular Ising Model on a Finite Lattice. *Entropy* **2018**, *20*, 277. [CrossRef]

Article

Stochastic Stirling Engine Operating in Contact with Active Baths

Ruben Zakine [1], Alexandre Solon [2], Todd Gingrich [2] and Frédéric van Wijland [1],*

[1] Laboratoire Matière et Systèmes Complexes (MSC), Université Paris Diderot, Sorbonne Paris Cité, UMR 7057 CNRS, 75205 Paris, France; ruben.zakine@univ-paris-diderot.fr

[2] Department of Physics, Massachusetts Institute of Technology, Cambridge, MA 02139, USA; solon@mit.edu (A.S.); toddging@mit.edu (T.G.)

* Correspondence: fvw@univ-paris-diderot.fr; Tel.: +33-1-5727-6254

Academic Editors: Andrea Puglisi, Alessandro Sarracino, Angelo Vulpiani, Eliodoro Chiavazzo and Kevin H. Knuth
Received: 17 March 2017; Accepted: 21 April 2017; Published: 27 April 2017

Abstract: A Stirling engine made of a colloidal particle in contact with a nonequilibrium bath is considered and analyzed with the tools of stochastic energetics. We model the bath by non Gaussian persistent noise acting on the colloidal particle. Depending on the chosen definition of an isothermal transformation in this nonequilibrium setting, we find that either the energetics of the engine parallels that of its equilibrium counterpart or, in the simplest case, that it ends up being less efficient. Persistence, more than non-Gaussian effects, are responsible for this result.

Keywords: active matter; stochastic energetics; Stirling engine

1. Introduction

Every well-educated physicist has heard of Carnot or Stirling cycles. In equilibrium thermodynamics of macroscopic systems (such as a gas enclosed in some container), a cycle is a periodic sequence of transformations the system is subjected to, with a view, as far as engines are considered, extracting work from the system. For a Carnot cycle, this is the well-known adiabatic-isothermal-adiabatic-isothermal sequence, while, for a Stirling cycle, the adiabatic transformations are replaced with isochoric ones. The analysis of small, microscopic or nanoscopic systems, such as a colloidal particle in some solvent, in contrast with the nineteenth century fluid systems, poses theoretical and experimental challenges. The former have been overcome by the advent of stochastic energetics at the end of the nineties [1]. Stochastic energetics (or stochastic thermodynamics) encompass a series of concepts and methods that allow one to define work, heat, dissipation, energy, etc. at an instantaneous and fluctuating level. By taking averages, one usually recovers (with often no need to consider the limit of macroscopic systems) the standard principles of thermodynamics. The gain, however, is enormous in that stochastic energetics also allows one to quantify fluctuations, which may not be negligible for small-scale systems. An excellent review on the latest developments of stochastic thermodynamics is that by Seifert [2], while the earlier Schmiedl and Seifert paper [3] focuses specifically on the analysis of stochastic engines. Experimental realizations pose challenges of their own. These are concerned with the control of small-size objects (often by means of optical tweezers), coupled to the need to control other parameters of the experiment. The bath temperature is one of them. Another one is the optical trap stiffness that can be seen as playing a role analogous to the volume of the container enclosing the gas in the macroscopic version. The conjugate parameter (analogous to the pressure) is the particle position (squared). The first colloidal-made engines were concerned with a Stirling cycle [4,5], in which a sequence of transformations by which the bath temperature and the trap stiffness were varied was applied to the colloidal particle. This is no place to discuss what an adiabatic transformation actually means at the level of a colloidal particle in a solvent, suffice it to say that this has very recently been defined [6] and put to work in an actual Carnot cycle [7]. A lot remains to be done at the experimental level and theoretical level alike, but it is fair to say that things are pretty well-understood as far as the theoretical framework is concerned. However, a somewhat unexpected generalization of these cycles seen as transformations between equilibrium states has recently been put forward by Krishnamurthy et al. [8]. The generalization, in the spirit of the seminal work of Wu and Libchaber [9], consists of replacing the equilibrium bath by an active bath containing living bacteria in a stationary yet nonequilibrium state. The sequence of transformations thus occurs between nonequilibrium steady-states instead of between equilibrium ones. Due to the nonequilibrium nature of the bacterial bath, there is no way to define a *bona fide* temperature. There are, however, several ways to define an energy scale expressing the level of energetic activity of the bath (all of which reduce, under equilibrium conditions, to the physical temperature). The proposal of [8] is to use the colloid's position fluctuations via $T_{act} = \frac{k}{2}\langle x^2 \rangle$ (where k is the trap stiffness). Another possibility would have been the following: in the absence of any confining force, the colloidal particle will eventually diffuse away from its initial position, so that we might then expect $\langle (x(t) - x(0))^2 \rangle = \frac{2T}{\gamma} t$, where T is yet another acceptable active temperature (this would be the asymptotic slope in Figure 2 of [9]). One might be inclined, somewhat subjectively, to view T as better expressing the intrinsic properties of the bath, while T_{act} must result from a balance between the bath and some external force. We will come back to that point at a later stage.

The purpose of this work is to analyze the results of [8] in the light of a specific modeling of the bacterial bath. We argue that, in the presence of the nonequilibrium bath, the Stirling engine efficiency depends on whether T or T_{act} is held fixed during the isothermal transformation, a distinction which does not apply to equilibrium baths for which T and T_{act} are equal. Our modeling relies on a single hypothesis: the bath enters the colloid's motion only through an extra noise term, and the noise statistics alone encode for the effect of the bath. Inspired by the suggestion of [8] that non-Gaussian statistics are essential, we will propose that the noise to which the colloid is subjected may have itself non-Gaussian statistics (recent advances of stochastic energetics for non-Gaussian but white processes [10,11] have taught us how to manipulate such signals) and possibly possess persistence properties. We will begin by a reminder of the properties of the stochastic Stirling engine between equilibrium reservoirs. We will then consider the extension to nonequilibrium bath and see how equilibrium results are *not* affected by choosing isothermal processes based on T_{act}. Then, we will adopt a definition of active temperature based on the colloid's diffusion constant and show that energy balance considerations are deeply modified and that the persistence of the noise is of key importance.

2. Stirling Cycle between Equilibrium States: A Quick Review

2.1. Modeling the Motion of a Colloidal Particle

The standard description of the dynamics of a colloidal particle in a solvent rests on a Langevin equation governing the evolution of the particle's position $x(t)$. In the overdamped limit relevant to the description of a micron-sized particle, this Langevin equation reads

$$\gamma \frac{\mathrm{d}x}{\mathrm{d}t} = -\partial_x V + \gamma \eta, \tag{1}$$

where γ is the friction coefficient characterizing the viscous drag of the particle in the solvent (this is the inverse mobility). The external potential V depends on the particle's position x and an external control parameter of the potential (like the stiffness of the harmonic trap). Finally, η, which, with the chosen normalization, has the dimension of a velocity, stands for a Gaussian white noise with correlations $\langle \eta(t)\eta(t') \rangle = \frac{2T}{\gamma} \delta(t - t')$. Under those conditions, where the dissipation kernel exactly matches the noise correlator, as prescribed by Kubo [12], the colloidal particle is in equilibrium (provided, of course, the external potential is not time dependent). In experimental setups, the potential is harmonic and the particle's motion is tracked in two-dimensional space, $\mathbf{r} = (x, y)$ and $V(x, k) = \frac{k}{2}(x^2 + y^2)$. We will stick to a one-dimensional description for notational simplicity. In the nonequilibrium setting we want to describe here, we shall encapsulate the effects of the interactions of the colloidal particle with its nonequilibrium environment into a single ingredient, namely the noise statistics. However, there is no reason to expect the noise resulting from the interactions of the colloidal particle with the bacteria bath to be either Gaussian or white. We postpone the analysis of such active noises to the next section and now proceed with a reminder of [3,4].

2.2. Energetics of the Stirling Cycle

In this subsection, we briefly review the results presented in [4,5]. This serves as a way to set notations straight and to define the quantities of interest. A Stirling cycle $ABCDA$ is made of the following sequence of states in the stiffness-temperature space (k, T):

$$A : (k_2, T_2) \overset{\text{isothermal}}{\longrightarrow} B : (k_1, T_2) \overset{\text{isochoric}}{\longrightarrow} C : (k_1, T_1) \overset{\text{isothermal}}{\longrightarrow} D : (k_2, T_1) \overset{\text{isochoric}}{\longrightarrow} A, \qquad (2)$$

where the terminology *isochoric* of course refers to an iso-stiffness transformation. This cycle is sketched in Figure 1.

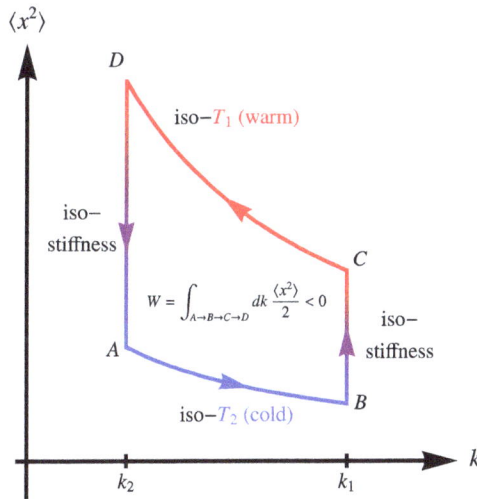

Figure 1. Schematic diagram of the Stirling cycle in stiffness-position space. Unlike its thermodynamic counterpart, the cycle is run counter-clockwise but is nevertheless an engine cycle.

We will denote by $a = k_1 / k_2 > 1$ the stiffness ratio (a large value of k is analogous to a more compressed state). The warm source is at T_1 while the cold source is at T_2 ($T_1 > T_2$). The instantaneous fluctuating energy of the particle is $V(x, k) = \frac{k}{2}x^2$. The work done on the colloid along a protocol driving it from state i to state f is $W = \int_i^f dt \frac{dk}{dt} \frac{\partial V}{\partial k} = \int_i^f dk \frac{1}{2}x^2$. The heat received by the colloid during the same step is given by the integral of the entropy production along the given protocol:

$$Q = - \int_i^f dt\, T\sigma, \qquad (3)$$

where $\sigma = T^{-1}\dot{x}(\gamma\dot{x} - \gamma\eta) = -T^{-1}k\dot{x}x$ is also the rate of work performed by the particle on the bath, and thus Q is the work performed by the bath on the particle. Altogether, we thus have $Q = \int_i^f k x dx$. If $p_{\text{eq}}(x) = e^{-kx^2/2T}/\sqrt{2\pi T/k}$ is the equilibrium distribution, then, up to a constant $S = -\int dx p_{eq}(x) \ln p_{eq}(x) = -\frac{1}{2} \ln \frac{k}{2\pi T} + \frac{1}{2}$ is the equilibrium entropy and $\langle Q \rangle = \int_i^f T dS$, with $dS = -\frac{1}{2}\frac{dk}{k} + \frac{1}{2}\frac{dT}{T}$. This is consistent with $Q = \int_i^f k x dx = [\frac{kx^2}{2}]_i^f - \int_i^f dk \frac{x^2}{2}$, which is a promotion of the first law $V_f - V_i = W + Q$ to stochastic energies. Using $\langle x^2 \rangle = T/k$, it is a simple exercise to determine the average heat received by the system during each step, $\langle Q_{AB} \rangle = -\frac{1}{2}T_2 \ln a < 0$, $\langle Q_{BC} \rangle = \frac{1}{2}(T_1 - T_2) > 0$, $\langle Q_{CD} \rangle = \frac{1}{2}T_1 \ln a > 0$ and $\langle Q_{DA} \rangle = -\frac{1}{2}(T_1 - T_2) < 0$. Correspondingly, $\langle W_{AB} \rangle = \frac{T_2}{2} \ln a$, $\langle W_{BC} \rangle = 0$, $\langle W_{CD} \rangle = -\frac{T_1}{2} \ln a$ and $\langle W_{DA} \rangle = 0$. The total average work received by the colloid is $\langle W \rangle = \langle W_{AB} + W_{CD} \rangle = -\frac{1}{2}(T_1 - T_2) \ln a < 0$. This means that the engine provides some work on average. Given that $Q_1 = Q_{BC} + Q_{CD}$ and $Q_2 = Q_{AB} + Q_{DA}$ are the heat effectively received by the colloid and the heat effectively given by the colloid to the bath, respectively, we define $\mathcal{E} = \frac{|\langle W \rangle|}{\langle Q_1 \rangle}$ as the engine's efficiency. The result is

$$\mathcal{E} = \frac{\langle Q_1 + Q_2 \rangle}{\langle Q_1 \rangle} = \frac{(T_1 - T_2) \ln a}{T_1 - T_2 + T_1 \ln a}. \tag{4}$$

If a perfect regenerator was used during the isochoric cooling $D \to A$, then the energy given out during this isochoric cooling could be used for the heating during the isochoric heating $B \to C$. Then, the heat received by the colloid would reduce to $Q_1 = Q_{CD}$ and the efficiency would become $\mathcal{E}_C = \frac{(T_1 - T_2) \ln a}{T_1 \ln a} = 1 - \frac{T_2}{T_1}$ (this Carnot efficiency is of course an upper bound for $\mathcal{E} = \frac{\mathcal{E}_C \ln a}{\mathcal{E}_C + \ln a}$ as given in Equation (4)). Again, these results can all be found in [4]. We have now set the stage for the purpose of this work, which is to re-examine each of these steps when the colloidal particle is in contact with nonequilibrium baths just as was carried out experimentally in [8].

3. Engine Operating between Nonequilibrium Baths

3.1. Modified Langevin Equation

We stick to our hypothesis that the effects of the bacterial bath can be entirely encoded into a single random process, so that now the colloid's position evolves according to

$$\gamma \dot{x} = -kx + \gamma \eta_{act}, \tag{5}$$

where the active noise η_{act} is a characteristic feature of the bacterial bath. Assuming this random signal inherits its properties from the bacteria making up the bath, we may expect that not only will the noise display non-Gaussian statistics, but it will also exhibit persistence properties captured by some memory kernel in the noise correlations. One way to substantiate our hypothesis on a more mathematical basis is to view the colloidal probe as interacting with the bacteria via some potential and then integrate out the degrees of freedom of the bacteria. Adapting the Vernon and Feynman approach [13] to this classical and nonequilibrium context can be seen to give rise to a dissipation kernel that depends on the bacteria-colloid interactions only, while the noise correlations (which are built from the noise felt by the bacteria themselves) in addition involve the persistence time of the bacteria. Furthermore, non-Gaussian statistics of the effective noise felt by the colloid follows directly from the non-Gaussian statistics of the noise felt by each individual active bacterium. Our approximation thus retains exactly these two features and forgets about further equilibrium-like memory effects. Among existing models, we may cite Run-and-Tumble noise, Active Brownian noise (see [14] for a review), Active Ornstein–Uhlenbeck noise [15] or even white yet non-Gaussian [10]. Following [8], we define a first active temperature T_{act} by the steady-state value of x^2: $T_{act} \equiv k\langle x^2 \rangle$. However, we introduce another active temperature that we denote by T by means of the colloid's mean-square displacement in the absence of a confining force, namely, at $k = 0$, we expect that

$$\langle (x(t) - x(0))^2 \rangle = \frac{2T}{\gamma} t \tag{6}$$

at large times, so that $T = \frac{\gamma}{2t} \int_0^t dt_1 dt_2 \langle \eta_{act}(t_1) \eta_{act}(t_2) \rangle$. We stress that neither T nor T_{act} are bona fide temperatures. They merely are energy scales reflecting how the bath injects energy into the colloids. The simple fact that these temperatures may not be equal out of equilibrium highlights that these temperatures cannot be endowed with any thermodynamics meaning.

3.2. The Energetics Is Not Altered If We Use Iso-T_{act} Steps

Using the definition of the work $W_{i \to f} = \int_i^f dk \frac{x^2}{2}$, we see that $\langle W_{i \to f} \rangle = \int_i^f dk \frac{T_{act}}{2k}$, which leads to the exact same expressions for the work as found in the previous section, up to the replacement of the equilibrium temperature with T_{act}. Similarly, following [2], we base our analysis on the fact that the heat is given by the work exerted by the bath on the colloid, namely, $Q_{i \to f} = \int_i^f dt \dot{x}(-\gamma \dot{x} + \gamma \eta_{act})$, which again simplifies into $Q_{i \to f} = \int_i^f dt \dot{x}(kx)$ and thus $Q_{i \to f} = [\frac{kx^2}{2}]_i^f - \int_i^f dk \frac{x^2}{2}$. After taking averages, we are back onto the expression found in equilibrium, again up to the replacement of temperatures by the corresponding T_{act}s. Hence, within that set of definitions and within our modeling, a quasistatic engine operating between nonequilibrium baths cannot outperform an equilibrium one. In light of the experiments of [8], this leaves us with a puzzle that we will address in the discussion section. In the following section, we suggest that perhaps another definition of the active isothermal process might lead to more striking differences with respect to an equilibrium engine.

4. Energetics Using the Diffusion Constant as an Active Temperature

In this section, we re-examine the Stirling engine operating between nonequilibrium baths using the temperature T defined in Equation (6) via the diffusion constant of an unconstrained particle. An isothermal process will now be understood as a process at constant T. Physically, this requirement is arguably more natural than processes at constant T_{act}. Indeed, T is an intrinsic measure of the activity of the bath, which can usually be tuned easily by the experimentalist, while T_{act} results from a balance between the bath's activity and a given external potential.

This new definition immediately requires us to adopt specific models for the active noise η_{act} because the explicit dependence of $\langle x^2 \rangle$ on T and k is now of crucial importance. We examine successively the case in which η_{act} is a non-Gaussian but white noise, and then a persistent noise with two-point correlations exponentially decreasing in time, a case that encompasses Ornstein–Uhlenbeck noise, Run-and-Tumble or Active Brownian noise.

4.1. A Bath with White but Non-Gaussian Statistics

Let's now assume that the active nature of the bacterial bath only surfaces through the non-Gaussian statistics of the noise η_{nG} appearing in the Langevin equation,

$$\gamma \dot{x} = -\partial_x V + \gamma \eta_{nG}, \tag{7}$$

while memory effects can be ignored in a first approximation. A non-Gaussian white noise $\eta_{nG}(t)$ can be formed by compounding Poisson point processes with random and independent amplitudes [16]. In practice, a realization of the noise over a time interval $[0, \mathcal{T}]$ is generated by first drawing a number of points, n, from a Poisson distribution with mean $\nu \mathcal{T}$. Then, a collection of times t_i, with $i = 1, \ldots, n$, are drawn uniformly in $[0, \mathcal{T}]$. To each t_i is associated a jump amplitude c_i, where the c_is are independent but identically distributed random variables with distribution $p(c)$. The non-Gaussian, white noise is constructed as the composition of these random-amplitude Poisson jumps:

$$\eta_{nG}(t) = \sum_i c_i \delta(t - t_i). \tag{8}$$

The generating functional of $\eta_{nG}(t)$ is $\langle e^{\int dt \, j(t) \eta_{nG}(t)} \rangle = e^{\nu \int dt (\langle e^{c \, j(t)} \rangle_p - 1)}$, where the p index denotes an average with respect to c and $j(t)$ is the field conjugate to η_{nG}. The two parameters defining

Entropy **2017**, *19*, 193

the noise statistics are the hitting frequency ν and the full jump distribution p. The Gaussian white noise limit is recovered as $\nu \to \infty$ and $\langle c^2 \rangle_p \to 0$ while $\nu \langle c^2 \rangle_p$ remains fixed. The noise has cumulants

$$\langle \eta_{nG}(t_1) \ldots \eta_{nG}(t_n) \rangle_{\text{cumulant}} = \nu \langle c^n \rangle_p \delta(t_1 - t_2) \ldots \delta(t_{n-1} - t_n). \tag{9}$$

We denote by $T/\gamma = \nu \langle c^2 \rangle_p / 2$ so that $\langle \eta_{nG}(t) \eta_{nG}(t') \rangle = (2T/\gamma)\delta(t - t')$ and T matches the definition given in Equation (6). It is possible to show that, in the case of the non-Gaussian white noise, this T is actually identical to our prior definition $T_{\text{act}} = \langle x^2 \rangle / k$. To prove this, we start from the master equation for the probability that $x(t)$ takes the value x at time t, $P(x, t)$, which reads

$$\partial_t P(x, t) = \gamma^{-1} \partial_x (kx P(x, t)) + \nu \int dc \, p(c) \left(P(x - c, t) - P(x, t) \right), \tag{10}$$

which we multiply by x^2 and integrate over x. This directly leads to

$$\frac{d}{dt} \langle x^2 \rangle = -\frac{2k}{\gamma} \langle x^2 \rangle + \nu \langle c^2 \rangle_p. \tag{11}$$

Hence, it results that in steady-state $k \langle x^2 \rangle = T$, independently of the non-Gaussian noise specifics. This is an extension of equipartition to a nonequilibrium context. An identical equipartition holds for an underdamped Langevin equation with non-Gaussian noise, for which $m\dot{v} = -\gamma v - V' + \gamma \eta_{nG}$ leads to $m \frac{d\langle v^2 \rangle}{dt} = -\gamma \langle v^2 \rangle - \frac{d}{dt} \langle V \rangle + \frac{2\gamma T}{m}$. This indeed forces $\langle mv^2/2 \rangle = T/2$ in the steady-state, irrespective of the white noise statistics. In a similar vein, one can also see that $\frac{d}{dt} \langle xv + \frac{\gamma}{m} \frac{x^2}{2} \rangle = \langle v^2 \rangle - \frac{1}{m} \langle xV' \rangle$, which leads to $\langle xV' \rangle = m \langle v^2 \rangle = T$ in the nonequilibrium steady-state. It immediately follows that the average works and heats will be unchanged with respect to the equilibrium discussion of Section 2.2. Note, however, that, in line with [2,10], the heat is not given anymore by the entropy production $\delta Q = -T\sigma dt$ as in Equation (3). It would be an interesting task to try and evaluate the corresponding σ, which is beyond the scope of the present discussion. (This might be feasible in an expansion in the jump size a at fixed active T. Such an expansion around a Gaussian white noise is admittedly questionable in view of Pawula's theorem [17] but can be controlled when manipulated with care [18]). Finally, that equipartition holds does not preclude strong non-Gaussian effects to show up in the colloid's position pdf. For instance, choosing $p(c) = e^{-|c|/a}/(2a)$ (with $\langle c^2 \rangle_p = 2a^2$) for the distribution of jumps allows us to find the stationary state distribution $P_{ss}(x)$. Indeed, with this choice of jump statistics [19] for $\alpha = \frac{T}{2ka^2} - \frac{1}{2}$ positive, we have $P_{ss}(x) = C|x/a|^\alpha K_\alpha(|x|/a)$ and $C = 2^{-\alpha} a^{-1}/(\sqrt{\pi} \Gamma(1/2 + \alpha))$, and thus $\langle x^2 \rangle = \frac{T}{k}$ can be explicitly verified. Note P_{ss} exhibits a cusp at the origin for $0 \le \alpha \le 1/2$, namely, for $ka^2 < T < 2ka^2$.

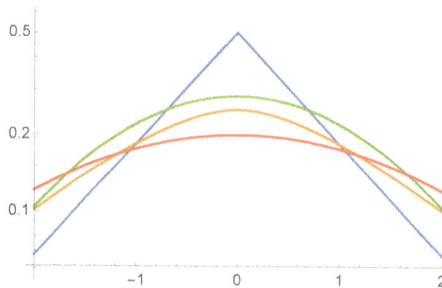

Figure 2. Log of the probability of the colloid's position as a function of position (for a unit a), in equilibrium with Gaussian statistics (red at $T/k = 2$, green at $T/k = 4$) or out of equilibrium as given by P_{ss} (blue at $T/k = 2$, orange at $T/k = 4$).

It comes as no surprise that the position statistics in the steady-state are strongly non-Gaussian as illustrated in Figure 2. However, it simply turns out that these non-Gaussian fluctuations do not interfere with the energy balance of the Stirling engine (Appendix B shows explicitly that the value of the kurtosis of the position distribution is uncorrelated from the efficiency). We now turn our attention to an active noise displaying some persistence properties with Gaussian statistics.

4.2. A Bath with a Persistent Noise

We now address more realistic modelings of the noise produced by the bacterial bath, in the form of a stochastic force imparted on the colloid that captures the persistent motion of an active particle. Such persistent noise arises from three common classes of active dynamics: Run-and-Tumble particles, active Brownian particles, and active Ornstein–Uhlenbeck motion. All three classes exhibit noise correlations that decay exponentially with a characteristic time τ:

$$\langle \eta_P(t) \eta_P(t') \rangle = \frac{2T}{\gamma} \times \frac{e^{-\frac{|t-t'|}{\tau}}}{2\tau}. \tag{12}$$

The prefactor T in Equation (12) matches our definition for T from Equation (6). We show in the Appendix A how the different models give rise to Equation (12) and relate T and τ to the microscopic parameters of the dynamics. Here, we adopt a unified description of the three different models by analyzing the impact of their shared noise correlator, Equation (12). Note that we restrict our discussion to one space dimension only for simplicity. In higher dimensions, the correlator of each component of the (vectorial) noise is still given by Equation (12), and, by symmetry, our results trivially generalize to a spherically harmonic potential.

If we interpret the isothermal transformations of Figure 1 as iso-T processes (as opposed to iso-T_{act}), the energetics of our Stirling cycle now differ from the equilibrium analysis of Section 2.2. During an iso-T protocol, $\langle x^2 \rangle$ does not trace an isotherm with the form $\langle x^2 \rangle \propto k^{-1}$. The new form of the isotherm depends only on the two-point correlator $\langle \eta_P(t) \eta_P(t') \rangle$ and not on higher-order correlations, allowing us to simultaneously treat all three types of active motion. Indeed, in Fourier space, Equation (7) reads

$$(k + i\gamma\omega)\tilde{x}(\omega) = \gamma \tilde{\eta}_P(\omega), \tag{13}$$

with the Fourier transform defined as $\tilde{f}(\omega) = \int_{-\infty}^{+\infty} f(t) e^{-i\omega t} dt$. One can then show that in steady state

$$\langle x^2 \rangle = \int_{-\infty}^{+\infty} \frac{d\omega}{2\pi} \frac{\gamma^2 \langle \tilde{\eta}_P(\omega) \tilde{\eta}_P(-\omega) \rangle}{k^2 + \omega^2 \gamma^2}. \tag{14}$$

For the noise correlator Equation (12), $\langle \tilde{\eta}_P(\omega) \tilde{\eta}_P(-\omega) \rangle = 2T/(\gamma(1 + \tau^2 \omega^2))$ so that we obtain

$$\langle x^2 \rangle = \frac{T}{k(1 + k\tau/\gamma)}. \tag{15}$$

In the notation of Section 3.1, $T_{act} \equiv k\langle x^2 \rangle = T/(1 + \Omega\tau)$, where we have defined the frequency $\Omega \equiv k/\gamma$.

Let us proceed, then, with the cycle Equation (2) in which we consider isothermal processes at fixed T. The average work has the expression $\langle W_{i \to f} \rangle = \frac{1}{2} \int_i^f dk \frac{T}{k(1+\Omega\tau)}$, which is zero along an isochoric protocol, but which now reads $\langle W_{i \to f} \rangle = \frac{T}{2} [\ln \frac{k}{1+\Omega\tau}]_i^f$ along an iso-T protocol. Similarly, the average heat reads $\langle Q_{i \to f} \rangle = [\frac{T/2}{1+\Omega\tau}]_i^f - \frac{1}{2} \int_i^f dk \frac{T}{k(1+\Omega\tau)}$. Putting everything together, we find

$$\langle Q_{AB} \rangle = \frac{T_2}{2} \left[\frac{1}{1 + \Omega_1 \tau} - \frac{1}{1 + \Omega_2 \tau} \right] - \frac{T_2}{2} \ln \left[a \frac{1 + \Omega_2 \tau}{1 + \Omega_1 \tau} \right],$$

$$\langle Q_{BC} \rangle = \frac{(T_1 - T_2)/2}{1 + \Omega_1 \tau} > 0,$$

$$\langle Q_{CD} \rangle = \frac{T_1}{2} \left[\frac{1}{1 + \Omega_2 \tau} - \frac{1}{1 + \Omega_1 \tau} \right] + \frac{T_1}{2} \ln \left[a \frac{1 + \Omega_2 \tau}{1 + \Omega_1 \tau} \right],$$

$$\langle Q_{DA} \rangle = -\frac{(T_1 - T_2)/2}{1 + \Omega_2 \tau} < 0,$$

(16)

while the average works are given by

$$\langle W_{AB} \rangle = \frac{T_2}{2} \ln a \frac{1 + \Omega_2 \tau}{1 + \Omega_1 \tau}, \quad \langle W_{CD} \rangle = -\frac{T_1}{2} \ln a \frac{1 + \Omega_2 \tau}{1 + \Omega_1 \tau}, \quad \langle W_{BC} \rangle = \langle W_{DA} \rangle = 0,$$

(17)

and thus $\mathcal{E} = \frac{-\langle W \rangle}{\langle Q_{BC} + Q_{CD} \rangle}$ in the $T_1 \gg T_2$ limit is

$$\mathcal{E}_{\text{sat}} = \frac{\ln a \frac{1 + \Omega_2 \tau}{1 + \Omega_1 \tau}}{\ln a \frac{1 + \Omega_2 \tau}{1 + \Omega_1 \tau} + \frac{1}{1 + \Omega_2 \tau}}.$$

(18)

In the limit of small correlation time ($\Omega_1 \tau \ll 1$), we find that

$$\mathcal{E}_{\text{sat}} \simeq \frac{\ln a}{1 + \ln a} - \Omega_2 \tau \frac{a - 1 - \ln a}{(1 + \ln a)^2} + \mathcal{O}(\tau^2).$$

(19)

In Equation (19), the correction to $\mathcal{E}_{\text{sat}}$ actually remains negative at arbitrary values of τ: the efficiency saturates to a *lower* value due to the persistent properties of the noise when compared to equilibrium (no persistence). The cost of maintaining nonequilibrium steady-state is not paid off by an improved efficiency! While the available work has increased, the required energy to operate the engine has increased by an even larger amount.

Let us stress here that the generality of these results, which depend only on the two-point correlator of the noise but not on higher-order statistics, can seem rather surprising because the behavior of active Ornstein–Uhlenbeck, Run-and-Tumble and Active Brownian particles in an harmonic trap are all qualitatively different. The case of an active Ornstein–Uhlenbeck noise in a quadratic potential is special in that the colloidal particle actually *is* in equilibrium [15]. The equilibrium distribution is $p_{\text{eq}}(x) \sim e^{-\frac{k(1 + \Omega \tau)}{2T} x^2}$ (whether and how this extends beyond a quadratic potential was discussed in [20]), from which $T_{\text{act}} = T/(1 + \Omega \tau)$ is readily extracted. On the contrary, Active Brownian and Run-and-Tumble particles have a richer (nonequilibrium) physics. In particular, if the particle is persistent on a time scale larger than Ω^{-1}, the steady-state distribution is not peaked around $x = 0$ (the particle spends most of its time on the edge of the trap) [14]. It is therefore surprising that these differences do not affect the thermodynamics of our engine.

4.3. A Bath Described by a More General Langevin Equation

A more general description of the effect of the bath on the colloidal particle includes memory effect in the dissipation as well, as discussed by Berthier and Kurchan [21] in a different context. One obtains a Langevin equation of the form

$$\gamma \int_{-\infty}^{t} dt' K_{\text{diss}}(t - t') \dot{x}(t') = -kx + \gamma \eta; \quad \langle \eta(t) \eta(t') \rangle = \frac{2T}{\gamma} K_{\text{inj}}(|t - t'|)$$

(20)

with generic injection and dissipation kernels K_{inj} and K_{diss}. Equilibrium is achieved on condition that $K_{\text{inj}}(\omega)$ and $\text{Re} K_{\text{diss}}(\omega)$ are equal. The ratio

$$T_{\text{eff}}(\omega) = T \frac{K_{\text{inj}}(\omega)}{\text{Re} K_{\text{diss}}(\omega)} \qquad (21)$$

tells us about the mismatch between injection and dissipation. In the cases considered previously, with $K_{\text{diss}}(\omega) = 1$, we ended up with $T_{\text{eff}}(\omega) = \frac{T}{1+(\omega\tau)^2}$. The characteristic frequency of the relaxation within the harmonic well being Ω, we a posteriori understand that in the regime where persistence matters, namely, when $\Omega\tau \gg 1$, $T_{\text{eff}}(\omega > \Omega) \to 0$, and thus the position spectrum will be cut-off beyond $\omega = \Omega$. It is thus no surprise that eventually $\langle x^2 \rangle < T/k$, or $T_{\text{act}} < T$. Of course, in the presence of a more complicated dissipation kernel K_{diss}, the latter inequality can be challenged. Should one devise a dissipation kernel such that $T_{\text{act}} > T$, one might reasonably suspect that the thermodynamic efficiency could exceed that of the equilibrium Stirling engine.

Let us therefore consider how the energetics of the cycle in Figure 1 are impacted by the relationship between T_{act} and T. In particular, we repeat the analysis of Section 4.2 by assuming $T_{\text{act}} = Tf(k)$ for a generic function f. The derivation proceeds exactly as before and we obtain for the heat and work on each segment

$$\langle W_{AB} \rangle = \frac{T_2}{2}\left[G(k_1) - G(k_2)\right]; \quad \langle W_{CD} \rangle = -\frac{T_1}{2}\left[G(k_1) - G(k_2)\right]; \quad \langle W_{BC} \rangle = \langle W_{DA} \rangle = 0; \qquad (22)$$

$$\langle Q_{AB} \rangle = -\frac{T_2}{2}\left[G(k_1) - G(k_2)\right] + \frac{T_2}{2}\left[f(k_1) - f(k_2)\right]; \quad \langle Q_{BC} \rangle = \frac{(T_1 - T_2)}{2}f(k_1); \qquad (23)$$

$$\langle Q_{CD} \rangle = \frac{T_1}{2}\left[G(k_1) - G(k_2)\right] - \frac{T_1}{2}\left[f(k_1) - f(k_2)\right]; \quad \langle Q_{DA} \rangle = \frac{(T_2 - T_1)}{2}f(k_2); \qquad (24)$$

where $G(k)$ is defined such that $G' = f/k$. The maximum efficiency in the limit $T_1 \gg T_2$ then reads

$$\mathcal{E}_{\text{sat}} = \frac{G(k_1) - G(k_2)}{f(k_1) + G(k_1) - G(k_2)}. \qquad (25)$$

This leads to the simple criterion that, in this limit, the cycle outperforms an equilibrium Stirling engine if and only if

$$\int_{k_2}^{k_1} \frac{f(k)}{kf(k_1)}\,\mathrm{d}k > \int_{k_2}^{k_1} \frac{\mathrm{d}k}{k}. \qquad (26)$$

In particular, if $f(k)$ is an increasing function of k in the range $[k_2; k_1]$, the active engine outperforms the equilibrium one. This could correspond to a physical situation in which energy injection happens at a particular, finite length scale. As an example, a semi-flexible filament immersed in a bath of Active Brownian particles is excited at a characteristic length scale [22] and could thus be a candidate to realize such an efficient engine.

5. Discussion: Back to Experiments

We have assumed all along that the tagged colloidal particle is subjected to a noise that inherits its properties from those of the bath while the rest of its dynamics are unchanged. That the effect of the nonequilibrium bath can be encoded in a single random signal as an extra force does not seem to be an outrageous hypothesis, though, given the size of the bacteria used in [8], comparable to that of the colloidal particle, perhaps hydrodynamic effects should be taken into account, as well as further memory effects (in the dissipation kernel and in noise correlations). Within that framework, we might anticipate that bacteria in a constant temperature medium maintain a fixed activity, and thus the statistical properties of the random forces should remain unchanged throughout the isothermal transformation. However, because the random forces do not emerge from equilibrium thermal fluctuations, there is not a unique way to characterize which aspect of the random forces remains constant. We have shown that, with a definition of the isothermal process based on a iso-"potential energy", we see no reason for the equilibrium results to be altered in any way. Coming back to [8], aside from the limiting efficiency in the high $T_1 \gg T_2$ limit, our theoretical observation is altogether rather consistent with the experiments.

It is interesting to consider how the nonequilibrium nature of the baths could result in an engine efficiency that differs from the equilibrium efficiency. To this end, we have suggested an alternative definition of an isothermal process in which the active temperature is defined through the diffusion constant of a particle without any external potential. With this definition, in stark contrast to the iso-"potential energy" definition, the efficiency of a Stirling engine takes a dramatically different form that involves the persistence time of the noise produced by the bacteria. Interestingly, we are able to pinpoint memory effects as being responsible for the deviation from equilibrium-like efficiencies. Non-Gaussian statistics alone is not a sufficient ingredient (we have shown equipartition to hold in the limiting non-Gaussian but white scenario).

From an experimental standpoint, achieving an isothermal process defined through the diffusion constant certainly involves working at constant ambient temperature. However, it also involves working at constant activity for the bacteria. The latter are eventually propelled by molecular motors which consume ATP (which should therefore be in sufficient amounts). In other words, food for the bacteria must remain at a constant level (and the bacteria population as well). The former, namely the ambient temperature, controls the chemical kinetics behind the active processes actually propelling the bacteria. Lowering the ambient temperature will affect the kinetics of the bacteria and will lead to a decreased diffusion constant. Whether and how the persistence time of the bacteria's motion is affected is, in our view, an interesting and challenging question beyond the scope of the present work. An alternative way to vary the diffusion constant might be to vary food concentration.

We hope the suggestion to use our alternative active temperature will trigger further experiments along the lines of [8].

Acknowledgments: The authors thank Julien Tailleur and Jordan Horowitz for many discussions. Alexandre Solon and Todd Gingrich acknowledge funding from the Betty and Gordon Moore Foundation.

Author Contributions: All authors contributed equally to this work. All authors have read and approved the final manuscript.

Conflicts of Interest: The authors declare no conflict of interest.

Appendix A. Active Particle Dynamics

For completeness, we define here Active Brownian and Run-and-Tumble particles (hereafter, ABPs and RTPs). In both cases, the noise entering the Langevin equation Equation (1) is written as a force of constant magnitude f_0 in a fluctuating direction $\mathbf{u}$, a unit vector. In arbitrary dimension (ABPs are only defined in $d \geq 2$),

$$\gamma\dot{\mathbf{r}} = -\nabla V + \gamma f_0 \mathbf{u}. \tag{A1}$$

For ABPs, the direction $\mathbf{u}$ undergoes rotational diffusion, while, for RTPs, a new direction is picked uniformly at a constant rate α. Let us show that, in both cases, each component of the noise has correlations given by Equation (12). We focus here on the 2d case. The derivation follows in the same way in higher dimensions (and $d = 1$ for RTPs).

In 2d, $\mathbf{u}$ is parametrized by an angle θ, $\mathbf{u} = (\cos\theta, \sin\theta)$. For ABPs, the Fokker–Planck equation associated with the evolution of the angle reads

$$\partial_t \mathcal{P}_t(\theta) = D_r \frac{\partial^2 \mathcal{P}(\theta)}{\partial\theta^2} \tag{A2}$$

with D_r, the rotational diffusion coefficient. This gives for the x-component of $u_x = \cos\theta$

$$\partial_t \langle\cos\theta\cos\theta_0\rangle = D_r \int d\theta \cos\theta\cos\theta_0 \frac{\partial^2 \mathcal{P}(\theta)}{\partial\theta^2} = -D_r\langle\cos\theta\cos\theta_0\rangle, \tag{A3}$$

where the last equality follows from integrating by parts. We thus have $\langle \cos\theta(t+t_0)\cos\theta(t_0)\rangle = \frac{1}{2}e^{-D_r t}$ so that, in the notations of Equation (12), $\tau = D_r^{-1}$ and $T = f_0^2/(2D_r)$.

One gets a similar result for RTPs which obey the Master equation

$$\partial_t \mathcal{P}_t(\theta) = -\alpha\mathcal{P}(\theta) + \alpha\int\frac{d\theta'}{2\pi}\mathcal{P}(\theta'), \tag{A4}$$

with α being the tumble rate. This leads in the same way to $\tau = \alpha^{-1}$ and $T = f_0^2/(2\alpha)$.

Appendix B. Is Kurtosis Related to Efficiency?

One may want to quantify deviations to the Gaussian distribution for the position [8]. We define $\mu_n(X)$ the n-th moment of a random variable X and we can compute renormalized kurtosis κ defined as follows: $\kappa_x = \frac{\mu_4(x)}{3\mu_2(x)^2} - 1 = \frac{\langle x^4\rangle}{3\langle x^2\rangle^2} - 1$ if $\langle x\rangle = 0$. Let us focus on the ABP case assuming that the derivation of the fourth moment for RTPs is similar. The Fokker–Planck equation in the 2d case for ABPs writes:

$$\partial_t \mathcal{P}(\mathbf{r},\theta) = \frac{1}{\gamma}\nabla\cdot(\mathcal{P}(\mathbf{r},\theta)\nabla V) - f_0\mathbf{u}(\theta)\cdot\nabla\mathcal{P}(\mathbf{r},\theta) + D_r\frac{\partial^2}{\partial\theta^2}\mathcal{P}(\mathbf{r},\theta). \tag{A5}$$

We take a quadratic potential $V = \frac{1}{2}kr^2$. In the stationary regime, multiplying the two members by x^4 and performing integration with respect to $\mathbf{r}$ and θ gives:

$$\langle x^4\rangle = f_0\frac{\gamma}{k}\langle x^3\cos\theta\rangle = \frac{f_0}{\Omega}\langle x^3\cos\theta\rangle. \tag{A6}$$

Similarly, we obtain:

$$\langle x^3\cos\theta\rangle = \frac{3f_0}{D_r+3\Omega}\langle x^2\cos^2\theta\rangle; \quad \langle x^2\cos^2\theta\rangle = \frac{1}{2D_r+\Omega}\left(D_r\langle x^2\rangle + f_0\langle x\cos^3\theta\rangle\right); \tag{A7}$$

$$\langle x\cos^3\theta\rangle = \frac{1}{9D_r+\Omega}\left(\frac{3}{8}f_0 + 6D_r\langle x\cos\theta\rangle\right); \quad \langle x^2\rangle = \frac{f_0}{\Omega}\langle x\cos\theta\rangle; \quad \langle x\cos\theta\rangle = \frac{f_0}{2(D_r+\Omega)}. \tag{A8}$$

Using Equations (A6)–(A8), we get:

$$\kappa_x = -\frac{\Omega(7D_r+3\Omega)}{2(2D_r+\Omega)(D_r+3\Omega)} < 0. \tag{A9}$$

We might wonder whether the kurtosis can give indications on the efficiency of the stochastic Stirling engine. We can also compute kurtosis of x for RTPs in 1d as we know the distribution [23]. For this case, we have $\kappa_x = -\frac{2\Omega}{\alpha+3\Omega}$ with α being the tumbling rate. Here, $\kappa_x < 0$ and we have proved in Section 4.2 that efficiency was still lower than the efficiency of the equilibrium case. This result should be compared to the kurtosis for x that satisfies the steady state distribution of Section 4.1, where the noise is white and non-Gaussian. For $P_{ss}(x) = C|x/a|^s K_s(|x|/a)$ and $C = 2^{-s}a^{-1}/(\sqrt{\pi}\Gamma(1/2+s))$, kurtosis $\kappa_x = 2/(1+2s)$ is strictly positive and the maximum efficiency is still the equilibrium one. Hence, the kurtosis of the position distribution does not indicate how the efficiency relates to that of an equilibrium Stirling engine.

References

1. Sekimoto, K. *Stochastic Energetics*; Lecture Notes in Physics; Springer: Berlin/Heidelberg, Germany, 2010.
2. Seifert, U. Stochastic thermodynamics, fluctuation theorems and molecular machines. *Rep. Prog. Phys.* **2012**, *75*, 126001.
3. Schmiedl, T.; Seifert, U. Efficiency at maximum power: An analytically solvable model for stochastic heat engines. *Eur. Lett.* **2008**, *81*, 20003.

4. Blickle, V.; Bechinger, C. Realization of a micrometre-sized stochastic heat engine. *Nat. Phys.* **2011**, *8*, 143–146.

5. Horowitz, J.M.; Parrondo, J.M.R. Thermodynamics: A Stirling effort. *Nat. Phys.* **2012**, *8*, 108–109.

6. Martínez, I.A.; Roldán, E.; Dinis, L.; Petrov, D.; Rica, R.A. Adiabatic Processes Realized with a Trapped Brownian Particle. *Phys. Rev. Lett.* **2015**, *114*, 120601.

7. Martinez, I.A.; Roldan, E.; Dinis, L.; Petrov, D.; Parrondo, J.M.R.; Rica, R.A. Brownian Carnot engine. *Nat. Phys.* **2016**, *12*, 67–70.

8. Krishnamurthy, S.; Ghosh, S.; Chatterji, D.; Ganapathy, R.; Sood, A.K. A micrometre-sized heat engine operating between bacterial reservoirs. *Nat. Phys.* **2016**, *12*, 1134–1138.

9. Wu, X.L.; Libchaber, A. Particle Diffusion in a Quasi-Two-Dimensional Bacterial Bath. *Phys. Rev. Lett.* **2000**, *84*, 3017–3020.

10. Kanazawa, K.; Sagawa, T.; Hayakawa, H. Stochastic Energetics for Non-Gaussian Processes. *Phys. Rev. Lett.* **2012**, *108*, 210601.

11. Kanazawa, K.; Sagawa, T.; Hayakawa, H. Heat conduction induced by non-Gaussian athermal fluctuations. *Phys. Rev. E* **2013**, *87*, 052124.

12. Kubo, R. The fluctuation-dissipation theorem. *Rep. Prog. Phys.* **1966**, *29*, 255.

13. Feynman, R.P.; Vernon, F.L. The theory of a general quantum system interacting with a linear dissipative system. *Ann. Phys.* **1963**, *24*, 118–173.

14. Solon, A.; Cates, M.; Tailleur, J. Active brownian particles and run-and-tumble particles: A comparative study. *Eur. Phys. J. Spec. Top.* **2015**, *224*, 1231–1262.

15. Szamel, G. Self-propelled particle in an external potential: Existence of an effective temperature. *Phys. Rev. E* **2014**, *90*, 012111.

16. Van Kampen, N. *Stochastic Processes in Physics and Chemistry*; Elsevier: Amsterdam, The Netherlands, 1992; Volume 1.

17. Pawula, R. Generalizations and extensions of the Fokker–Planck–Kolmogorov equations. *IEEE Trans. Inf. Theory* **1967**, *13*, 33–41.

18. Popescu, D.M.; Lipan, O. A Kramers-Moyal Approach to the Analysis of Third-Order Noise with Applications in Option Valuation. *PLoS ONE* **2015**, *10*, e0116752.

19. Fodor, E.H.; Hayakawa, J.T.; van Wijland, F. What is the role of non Gaussian noise in assemblies of self-propelled active particles? in preparation.

20. Fodor, É.; Nardini, C.; Cates, M.E.; Tailleur, J.; Visco, P.; van Wijland, F. How far from equilibrium is active matter? *Phys. Rev. Lett.* **2016**, *117*, 038103.

21. Berthier, L.; Kurchan, J. Non-equilibrium glass transitions in driven and active matter. *Nat. Phys.* **2013**, *9*, 310–314.

22. Nikola, N.; Solon, A.P.; Kafri, Y.; Kardar, M.; Tailleur, J.; Voituriez, R. Active Particles with Soft and Curved Walls: Equation of State, Ratchets, and Instabilities. *Phys. Rev. Lett.* **2016**, *117*, 098001.

23. Tailleur, J.; Cates, M.E. Sedimentation, trapping, and rectification of dilute bacteria. *Eur. Lett.* **2009**, *86*, 60002.

MDPI

Article

Information-Theoretic Bound on the Entropy Production to Maintain a Classical Nonequilibrium Distribution Using Ancillary Control

Jordan M. Horowitz * and Jeremy L. England

Physics of Living Systems Group, Department of Physics, Massachusetts Institute of Technology,
400 Technology Square, Cambridge, MA 02139, USA; jengland@mit.edu
* Correspondence: jhorowit@mit.edu; Tel.: +1-617-452-2904

Received: 23 March 2017; Accepted: 1 July 2017; Published: 4 July 2017

Abstract: There are many functional contexts where it is desirable to maintain a mesoscopic system in a nonequilibrium state. However, such control requires an inherent energy dissipation. In this article, we unify and extend a number of works on the minimum energetic cost to maintain a mesoscopic system in a prescribed nonequilibrium distribution using ancillary control. For a variety of control mechanisms, we find that the minimum amount of energy dissipation necessary can be cast as an information-theoretic measure of distinguishability between the target nonequilibrium state and the underlying equilibrium distribution. This work offers quantitative insight into the intuitive idea that more energy is needed to maintain a system farther from equilibrium.

Keywords: nonequilibrium thermodynamics; dissipation; relative entropy

---

1. Introduction

Small systems are continually bombarded by noise from their surroundings. Sometimes this noise is helpful; thermal and chemical fluctuations are the fuel that power biological molecular motors [1]. More often, though, noise is a nuisance. Fluctuations in gene expression or transcription can lead to errors in downstream macromolecules, like RNA, that can be detrimental to a cell's function [2]. Noise can also interfere with the functioning of artificial mesoscopic devices, such as micromechanical [3] and nanomechanical resonators [4,5]. In all these situations, an ancillary control mechanism can be employed to suppress fluctuations. This can take the form of a kinetic proofreading scheme [2] or the addition of an auxiliary control device that employs active feedback, as in a Maxwell's demon [6–12].

No matter the control mechanism, the effect is to force the system into a statistical state distinct from its noisy equilibrium, where it will inevitably dissipate energy. Thus maintaining a system away from equilibrium comes with an energetic cost. Attempts at predicting the properties of such nonequilibrium states by minimizing the energy dissipation have a long history, starting with Prigogine and coworkers [13] within linear irreversible thermodynamics (see also [14]). However, it seems no such general variational principle exists beyond the linear regime [14–16]. As such, our goal in this work is not to characterize the nonequilibrium state through a thermodynamic variational principle. Instead, we aim to characterize the energetic requirement to hold an originally equilibrium system in a prescribed out-of-equilibrium state using an additional external control system that does not alter the original system's properties. Indeed, previously in Refs. [17,18], we showed that for specific classes of externally imposed controls, this minimum energetic cost could be expressed simply in terms of the systems underlying equilibrium dynamics. In this Article, we expand this program to include new control mechanisms, and in the process offer a unifying perspective on these previous results. In particular, we demonstrate that for various control mechanisms the minimum entropy production (or dissipation) to keep a mesoscopic system in a specified nonequilibrium distribution can be expressed as a time derivative of the relative entropy between the target distribution and the uncontrolled equilibrium Boltzmann distribution. This information-theoretic characterization quantitatively characterizes the intuitive notion that the farther a system is from equilibrium the more energy must be dissipated to maintain it.

2. Setup

We have in mind a small mesoscopic system making random transitions among a set of discrete mesostates, or configurations, $i = 1, \ldots, N$, each with (free) energy E_i. We can visualize this dyanmics occurring on a graph (or network) like in Figure 1, where each configuration is assigned a vertex (or node), and possible transitions are represented by edges (or links).

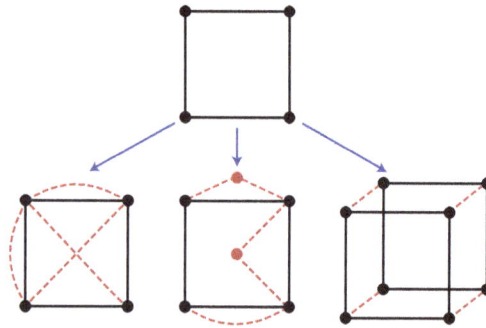

Figure 1. Illustration of three types of control: (**Top**) Graph representation of the system's configuration space without control. Mesoscopic configurations are represented as vertices (or nodes) with edges signifying allowed transitions; (**Bottom**) Control is implemented by adding additional edges (red dashed) or nodes (red dots) in order to drive the system into a nonequilibrium distribution. From Left to Right: Edge control, Node control, and Auxiliary control.

The dynamical evolution is modeled as a Markov jump process on our graph according to transition rates W_{ij} from $j \to i$, with $W_{ij} \neq 0$ only when $W_{ji} \neq 0$, so that every transition has a reverse. As such, the system's time-dependent probability density $p_i(t)$ evolves according to the Master Equation [19]

$$\partial_t p_i(t) = \sum_{j \neq i} W_{ij} p_j(t) - W_{ji} p_i(t), \tag{1}$$

which constitutes a probability conservation equation with probability currents

$$J_{ij}(p) = W_{ij} p_j(t) - W_{ji} p_i(t). \tag{2}$$

Now, in the absence of any external control, we assume that our system relaxes to a thermal equilibrium steady state at inverse temperature $\beta = 1/k_B T$, given by the Boltzmann distribution $p_i^{eq} = e^{\beta(F^{eq} - E_i)}$ with equilibrium free energy $F^{eq} = -k_B T \ln \sum_i e^{-\beta E_i}$. To guarantee this, we impose detailed balance on the transition rates [19],

$$W_{ij} p_j^{eq} = W_{ji} p_i^{eq}. \tag{3}$$

In equilibrium, each transition is counter-balanced by its reverse. Our goal is then to maintain the system in a predetermined target nonequilibrium steady state $p^* \neq p^{eq}$, and to characterize the minimum dissipation necessary.

3. Minimum Dissipation to Suppress Fluctuations

We are interested in pushing and holding our system into a statistical state distinct from the equilibrium Boltzmann distribution. One could imagine a variety of schemes to accomplish this goal. Perhaps the simplest, is if we have complete control to vary the system's energy function $\{E_i\}$. We could then hold the system in the statistical state p^* by shifting all the energy levels to $E_i^* = -k_B T \ln p_i^*$, thereby making the target state the new equilibrium state: $p_i^* = e^{-\beta E^*}$. After an initial transient relaxation, the system would then remain in p^* indefinitely as it is in equilibrium. While there is a one time energetic cost to vary the energy levels equal to the free energy difference [20], the system can be held in p^* for free. Implementing such a protocol, however, generically requires very fine control over all the individual energies, which often is prohibitive [20]. As a result, there are a number of situations where this is not possible or not desirable. For example, nature does not utilize this control mechanism; in cells, where the free energies of molecules are fixed, noise reduction is implemented by coupling together various driven chemical reactions that constantly burn energy [2,21,22]. Whereas fluctuations in quantum mesoscopic devices are often suppressed by coupling an auxiliary device that continually and coherently extracts noise through feedback [4,23–25].

Motivated by this observation, we analyze control mechanisms where we cannot alter the internal energies $\{E_i\}$. Instead, the statistical state of our system is manipulated by introducing additional pathways. In particular, the scenarios we address, depicted in Figure 1, are: (i) Edge control—additional driven transitions (or edges) are added with transition rates $\{M_{lk}\}$, which model the coupling of additional thermodynamic reservoirs; (ii) Node control—additional configurations are incorporated and coupled to the original network through driven transitions with rates $\{M_{lk}\}$, allowing for ancillary intermediate configurations, such as in dissipative catalysis; (iii) Auxiliary control—an entirely new system is coupled to the controlled system, as in feedback control; and finally (iv) Chemical control—where new chemical reactions are included. Though ostensibly a special case of edge control, it adds new complications due to the possibility of breaking conservation laws. The addition of such controllers alters the system's dynamics leading to a modified Master Equation (cf. (1))

$$\partial_t p_i(t) = \sum_{j \neq i} W_{ij} p_j(t) - W_{ji} p_i(t) + \sum_{j \neq i} M_{ij} p_j(t) - M_{ji} p_i(t)$$
$$= \sum_{j \neq i} J_{ij}(p) + J_{ij}^M(p). \tag{4}$$

Yet no matter which control mechanism is employed, we assume the net effect is to push our target system into the nonequilibrium steady state p^*. While designing such control is generically a challenging problem, we take it as a given and instead focus on the minimum cost.

Our only assumption is that the additional control transition rates satisfy a local detailed balance relation connecting them to the entropy flow into the thermodynamic reservoir that mediates the transition [26,27]:

$$\ln \frac{M_{kl}}{M_{lk}} = \Delta s_{kl}^e. \tag{5}$$

For example, if we implement control by coupling a thermal reservoir at a different inverse temperature β', then we require $\ln(M_{kl}/M_{lk}) = \beta'(E_l - E_k)$, which is proportional to the heat flow into the environment.

The local detailed balance relation implies that our super-system, composed of the system of interest and the controller, with rates $\{\mathcal{W}_{ij}^\kappa\} = \{W_{ij}, M_{ij}\}$, where κ specifies an uncontrolled or controlled transition, satisfies the second law of thermodynamics: Namely, the (irreversible) entropy production is positive

$$\dot{S}_i = k_B \sum_{i>j,\kappa} J_{ij}^\kappa(p) \ln \frac{\mathcal{W}_{ij}^\kappa p_j}{\mathcal{W}_{ji}^\kappa p_i} \geq 0, \tag{6}$$

which is typically split into the derivative of the Shannon entropy

$$\partial_t S = -k_B \partial_t \sum_i p_i \ln p_i = k_B \sum_{i>j,\kappa} J_{ij}^\kappa(p) \ln \frac{p_j}{p_i}, \tag{7}$$

and the entropy flow

$$\dot{S}_e = k_B \sum_{i>j,\alpha} J_{ij}^\kappa(p) \ln \frac{\mathcal{W}_{ij}^\kappa}{\mathcal{W}_{ji}^\kappa}. \tag{8}$$

Our goal will be to bound the entropy production (or dissipation) over all controls that fix the steady state distribution to be p^*. Due to the local detailed balance relations (3) and (5), we can always connect the dissipation to the underlying energetics. Thus, our bound on the entropy production can always be reframed as a minimum energetic cost.

3.1. Edge Control

We begin our investigation with the edge control scheme, where we add a collection of additional edges to the graph, corresponding to new transitions mediated by additional thermal or chemical reservoirs. This analysis was originally carried out in [17]. We briefly review it here, as this control scheme is the simplest and all the following developments will build on it.

In this scenario the super-system produces entropy in the controlled steady state p^* at a rate

$$\dot{S}_i = k_B \sum_{i>j} J_{ij}(p^*) \ln \frac{W_{ij} p_j^*}{W_{ji} p_i^*} + k_B \sum_{k>l} J_{kl}^M(p^*) \ln \frac{M_{kl} p_l^*}{M_{lk} p_k^*}. \tag{9}$$

We now wish to bound this sum solely in terms of properties of the system's environment as codified by the $\{W_{ij}\}$ and the target distribution p^*.

To this end, we observe that not only is the total entropy production positive, but link by link the entropy production is positive, $J_{kl} \ln(M_{kl} p_l / M_{lk} p_k) \geq 0$ [12], which follows readily from the inequality $(x - y) \ln(x/y) \geq 0$. Thus, each control edge only contributes additional dissipation, implying that the only unavoidable dissipation occurs along the system's original links:

$$\dot{S}_i \geq \dot{S}_{\min} = k_B \sum_{i>j} J_{ij}(p^*) \ln \frac{W_{ij} p_j^*}{W_{ji} p_i^*} \geq 0. \tag{10}$$

No matter how control is implemented, the system will inevitable make jumps along the original links, and those will on average dissipate free energy into the environment when the system is held in the target state p^*.

We now offer some physical insight into the meaning of (10). To this end, we recognize that $\dot{S}_{\min}$ is the entropy production rate of the equilibrium dynamics when the statistical state is the target state p^*. In other words, it represents the instantaneous entropy production we would observe if we turned off the control and allowed p^* to begin to relax to equilibrium. An enlightening reformulation of this observation is offered by recalling the intimate connection between the time derivative of the relative entropy, $D(f||g) = \sum_i f_i \ln(f_i/g_i)$ [28], and the entropy production rate:

$$-k_B \partial_t D(p(t)||p^{eq}) = \sum_{i>j} J_{ij}(p) \ln \frac{W_{ij} p_j}{W_{ji} p_i}, \tag{11}$$

which is a direct consequence of detailed balance (3). As such, we immediately recognize that the minimum dissipation (10) can be equivalently formulated as

$$\dot{S}_i \geq \dot{S}_{\min} = -k_B \partial_t^{eq} D(p^*||p^{eq}), \tag{12}$$

where the derivative ∂_t^{eq} should be understood to operate on p^* as if it were evolving under the uncontrolled equilibrium dynamics. As the relative entropy is an information-theoretic measure of distinguishability [28], Equation (12) quantifies precisely the intuitive fact that it costs more to control a system the farther it is from equilibrium.

We note that this analysis immediately offers the condition under which we saturate the minimum. As our bound originates in setting aside the extraneous entropy production due to the control transitions, we immediately find as a consequence that this additional entropy production is zero when the control transitions operate thermodynamically reversibly. This requires them to operate much faster than the system dynamics, so that at any instant the system is locally detailed balanced with respect to the control transitions on a link-by-link basis. In other words, the optimal transition rates must verify

$$M_{kl}^* p_l^* = M_{lk}^* p_k^*. \tag{13}$$

Indeed, this implies the optimal dissipation on each driven link (cf. (5)) should be

$$\Delta s_{kl}^* = \ln \frac{M_{kl}^*}{M_{lk}^*} = \ln \frac{p_k^*}{p_l^*}. \tag{14}$$

3.2. Node Control

More than a fundamental result, the preceding analysis outlines an approach for characterizing the minimum dissipation to hold a system out of equilibrium. We now carry out this analysis again in a new scenario, but allow the addition of C extra nodes in the network and edges connecting them (Figure 1).

When we add additional nodes, the system plus controller will have $\alpha = 1, \ldots, N + C$ configurations, with a steady-state distribution ρ_α^{ss} over the super-system. Now, in this case control will be successful when in the resulting steady state the relative likelihood of the N original states are in the target distribution $p_i^* = \rho_i^{ss}/\mathcal{P}$, where $\mathcal{P} = \sum_{i=1}^N \rho_i^{ss}$. Unfortunately, this is not sufficient to fix the dissipation rate on the original set of links, as the currents are left undetermined. Indeed, it is possible to have the subset of N system nodes in the target distribution p^*, but have $\mathcal{P}$ small; leading to small currents on the uncontrolled links $J_{ij}(\rho^{ss}) = \mathcal{P}J_{ij}(p^*)$ and negligible dissipation (cf. (6)). Thus to arrive at a sensible bound we must also fix the probability currents $J_{ij}(\rho^{ss}) = \mathcal{P}J_{ij}(p^*)$ on the original uncontrolled links, or equivalently $\mathcal{P}$, the total probability to be in the original configurations. In effect, we are maintaining the function of the system, as the currents represent different possible tasks for the system, e.g., they are the rate of production of a molecule or the rate at which heat flux is converted into useful work.

With this setup, the minimum entropy production rate is again given by the entropy production on the original undriven links in the global steady state

$$\dot{S}_i \geq \dot{S}_{min} = k_B \sum_{i>j} J_{ij}(\rho^{ss}) \ln \frac{W_{ij}\rho_j^{ss}}{W_{ji}\rho_i^{ss}}. \tag{15}$$

To make this an expression that only depends on p^* and $\mathcal{P}$, we substitute $\rho_i^{ss} = p_i^*\mathcal{P}$ to find

$$\dot{S}_{min} = k_B \mathcal{P} \sum_{i>j} J_{ij}(p^*) \ln \frac{W_{ij}p_j^*}{W_{ji}p_i^*}$$

$$= -k_B \mathcal{P} \partial_t^{eq} D(p^* || p^{eq}). \tag{16}$$

Again, the minimum dissipation is dictated by how different the target state is from equilibrium, but here weighted by the total probability $\mathcal{P}$, which fixes the system's currents. Similarly, optimality is reached when the additional edges that connect the control nodes are very fast, minimizing their contribution to the entropy production.

3.3. Auxiliary Control

Another scheme for control is the addition of an entirely new system, which we call the auxiliary. This scheme was original analyzed in [18] for quantum mesoscopic devices modeled by a Markovian quantum Master equation. In an effort to unify various results, we recapitulate this argument here, translated into classical language.

We now amend our state space with the addition of an auxiliary control system with states $\alpha = 1, \ldots, C$, so that each configuration of the super-system is labeled by the pair (i, α). Transition rates of the original system are assumed unaltered, but the new transitions between auxiliary states $\{M_{\alpha\gamma}^i\}$ must depend on the system state in order to implement the feedback control. Such a structure is called bipartite [10,29–31], and is captured in the graph structure by the absence of diagonal links where the system and auxiliary transition simultaneously (Figure 1). Here, control is successful when the steady-state distribution $\rho_{i\alpha}^{ss}$ has a marginal distribution on the system that is the target distribution $\sum_{\alpha=1}^{C} \rho_{i\alpha}^{ss} = p_i^*$.

Again we can bound the total entropy production in the system plus auxiliary with the entropy production on just the system links

$$\dot{S}_i \geq k_B \sum_{(i,\alpha) > (j,\gamma)} J_{ij}(\rho^{ss}) \ln \frac{W_{ij} \rho_{j\gamma}^{ss}}{W_{ji} \rho_{i\alpha}^{ss}}. \tag{17}$$

At this point the lower bound still depends on the full distribution ρ^{ss} over the entire super-system. However, if we coarse-grain over the auxiliary, we can use the monotonicity of the relative entropy under coarse-graining [28], to weaken the bound to

$$\dot{S}_i \geq \dot{S}_{min} = k_B \sum_{i>j} J_{ij}(p^*) \ln \frac{W_{ij} p_j^*}{W_{ji} p_i^*}$$

$$= -k_B \partial_t^{eq} D(p^* || p^{eq}). \tag{18}$$

This result was originally derived in the context of quantum mesoscopoic devices [18]. Here we have reframed it in classical language.

3.4. Controlling Chemical Reaction Networks

As a final scenario, we turn to the control of a chemical reaction network. This scenario adds an additional complication: the incorporation of additional control reactions can break an underlying conservation law of the equilibrium dynamics [32,33]. For example, adding a chemostat that exchanges matter breaks the conservation of particle number or mass. This observation requires a slight modification of (12).

To set the stage, consider a chemical reaction network with configurations specified by the vector of chemical species number $\mathbf{X} = \{X_1, \ldots, X_S\}$. Transitions then correspond to chemical reactions that change the value of $\mathbf{X}$ subject to system-specific constraints; an illustrative example of which is pictured in Figure 2. For simplicity, we take the only constraint to be particle number $\mathcal{N} = \sum_{i=1}^{S} X_i$. In which case, the equilibrium steady state is a Poisson distribution constrained to the manifold of fixed particle number, $p_{\mathcal{N}}^{eq}(\mathbf{X}) = \prod_i \frac{\hat{x}_i^{X_i}}{X_i!} e^{-\hat{x}_i} \delta(\sum_j X_j - \mathcal{N})$ [19]; and detailed balance respects the constraints as well:

$$W_{\mathbf{X}'\mathbf{X}} p_{\mathcal{N}}^{eq}(\mathbf{X}) = W_{\mathbf{X}\mathbf{X}'} p_{\mathcal{N}}^{eq}(\mathbf{X}'). \tag{19}$$

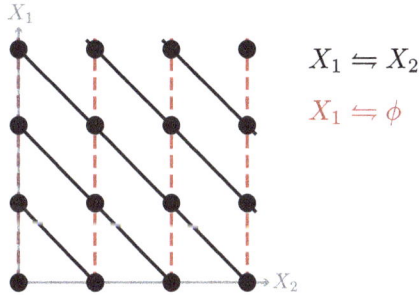

Figure 2. Illustration of chemical control: Two species X_1 and X_2 interconvert through a single chemical reaction $X_1 \rightleftharpoons X_2$ that conserves particle number, depicted as solid black lines. As a result, the dynamical evolution of this chemical reaction network is restricted to a single diagonal subspace of the network of states. Control can be implemented by chemostating one of the species, say X_1, by allowing X_1 molecules to be added and subtracted from the reaction volume through the reaction $X_1 \rightleftharpoons \phi$, depicted as red dashed lines. As this reaction breaks the particle number conservation law, it extends the possible configurations the system can dynamically explore.

Now, for control we add new reactions that maintain the system in a fixed target distribution $p^*(\mathbf{X})$ over chemical space, which may not respect our particle number constraint, i.e., it may have support on configurations $\mathbf{X}$ that have different numbers of total particle number (cf. Figure 2). To make this explicit, we split the target distribution $p^*(\mathbf{X}) = p^*_{\mathcal{N}}(\mathbf{X})\mathcal{P}_{\mathcal{N}}$ into two controlloable pieces: the conditional probability given the total particle number, $p^*_{\mathcal{N}}(\mathbf{X})$ and the probability to have $\mathcal{N}$ particles, $\mathcal{P}_{\mathcal{N}}$. With this splitting, the minimum dissipation to maintain p^* again is only due to the entropy produced in the original reactions

$$
\begin{aligned}
\dot{S}_{\min} &= k_{\mathrm{B}} \sum_{\mathbf{X}' > \mathbf{X}} J_{\mathbf{X}'\mathbf{X}}(p^*) \ln \frac{W_{\mathbf{X}'\mathbf{X}} p^*(\mathbf{X})}{W_{\mathbf{X}\mathbf{X}'} p^*(\mathbf{X}')} \\
&= k_{\mathrm{B}} \sum_{\mathcal{N}} \mathcal{P}_{\mathcal{N}} \sum_{\mathbf{X}' > \mathbf{X}} J_{\mathbf{X}'\mathbf{X}}(p^*_{\mathcal{N}}) \ln \frac{W_{\mathbf{X}'\mathbf{X}} p^*_{\mathcal{N}}(\mathbf{X})}{W_{\mathbf{X}\mathbf{X}'} p^*_{\mathcal{N}}(\mathbf{X}')},
\end{aligned}
\tag{20}
$$

An information-theoretic interpretation is provided by recalling that the equilibrium transitions $W_{\mathbf{X}'\mathbf{X}}$ conserve particle number, to find

$$
\dot{S}_{\min} = -k_{\mathrm{B}} \sum_{\mathcal{N}} \mathcal{P}_{\mathcal{N}} \partial_t^{\mathrm{eq}} D(p^*_{\mathcal{N}} || p^{\mathrm{eq}}_{\mathcal{N}}).
\tag{21}
$$

In a chemical system, the minimum dissipation depends only on how different the target distribution is from the equilibrium distribution on the conserved sectors, whereas shifting only the number distribution $\mathcal{P}_{\mathcal{N}}$ can in principle be accomplished for free. This conclusion should remain true when there are additional conservation laws as well. Note that when the target distribution conserves particle number, $\mathcal{P}_{\mathcal{N}} = 1$ and we recover (12).

4. Discussion

We investigated the minimum entropy production or free energy dissipation to maintain a system in a target nonequilibrium distribution using ancillary control. We found that in a variety of scenarios this minimum cost can be formulated using the information-theoretic relative entropy as a measure of how distinguishable the target nonequilibrium state is from equilibrium. Our analysis further revealed that the minimum is reached when the driven control transitions operated reversibly.

As in previous analyses of nonequilibrium thermodynamics, the relative entropy [34–39] appeared as a key tool in characterizing dissipation. In these previous works, however, the relative entropy compared the true evolution of the system to the underlying stationary state. By contrast, here we find that when using external control the cost is characterized by the time-variation of the relative entropy under a fictitious uncontrolled equilibrium dynamics, evaluated against the unperturbed equilibrium state.

Looking ahead, we note that while we had in mind throughout the paper autonomous control, nonautonomous control through reversible hidden pumps offer an intriguing alternative to saturate our energetic bound [40]. Additionally, we have focused on the average or typical behavior of the state of our system, but a number recent predictions, collectively known as thermodynamic uncertainty relations, relate the dissipation to fluctuations in currents or flows [41–45]. Such work suggests that it would be intriguing to understand how our lower bound is modified, when one wants to use external control to constrain not just the typical state, but fluctuations as well. Our hope is that the approach developed here offers the possibility of quantifying the minimum energetic cost of nonequilibrium states in other more general scenarios.

Acknowledgments: This work was supported by the Gordon and Betty Moore Foundation through Grant GBMF4343. Jeremey L. England further acknowledges the Cabot family for their generous support of Massachusetts Institute of Technology (MIT).

Author Contributions: Jordan M. Horowitz and Jeremey L. England conceived of the project and wrote the paper; Jordan M. Horowitz carried out the project. Both authors have read and approved the final manuscript.

Conflicts of Interest: The authors declare no conflict of interest.

References

1. Parrondo, J.M.R.; De Cisneros, B.J. Energetics of Brownian motors: A review. *Appl. Phys. A* **2002**, *75*, 179–191.
2. Bialek, W. *Biophysics: Searching for Principles*; Princeton University Press: Princeton, NJ, USA, 2012.
3. Li, T.; Kheifets, S.; Raizen, M. Millikelvin cooling of an optically trapped microsphere in vacuum. *Nat. Phys.* **2011**, *7*, 527–530.
4. Tian, L. Ground state cooling of a nanomechanical resonator via parametric linear coupling. *Phys. Rev. B* **2009**, *79*, 193407.
5. Palomaki, T.A.; Harlow, J.W.; Teufel, J.D.; Simmonds, R.W.; Lehnert, K.W. Coherent state transfer between itinerant microwave fields and a mechanical oscillator. *Nature* **2013**, *495*, 210–214.
6. Horowitz, J.M.; Sagawa, T.; Parrondo, J.M.R. Imitating chemical motors with optimal information motors. *Phys. Rev. Lett.* **2013**, *111*, 010602.
7. Munakata, T.; Rosinberg, M.L. Entropy production and fluctuation theorems under feedback control: The molecular refrigerator model revisited. *J. Stat. Mech.* **2012**, *2012*, P05010.
8. Parrondo, J.M.R.; Horowitz, J.M.; Sagawa, T. Thermodynamics of information. *Nat. Phys.* **2015**, *11*, 131–139.
9. Sandberg, H.; Delvenne, J.C.; Newton, N.J.; Mitter, S.K. Maximum work extraction and implementation costs for nonequilibrium Maxwell's demons. *Phys. Rev. E* **2014**, *90*, 042119.
10. Horowitz, J.M.; Esposito, M. Thermodynamics with continuous information flow. *Phys. Rev. X* **2014**, *4*, 031015.
11. Horowitz, J.M.; Sandberg, H. Second-law-like inequalities with information and their interpretations. *New J. Phys.* **2014**, *15*, 125007.
12. Shiraishi, N.; Sagawa, T. Fluctuation theorem for Partially-masked nonequilibrium dynamics. *Phys. Rev. E* **2015**, *91*, 012130.
13. Kondepudi, D.; Prigogine, I. *Modern Thermodynamics: From Heat Engines to Dissipative Structures*, 2nd ed.; John Wiley & Sons, Ltd.: Chichester, UK, 2014.
14. Maes, C.; Netočný, K. Minimum entropy production principle from a dynamical fluctuation law. *J. Math. Phys.* **2007**, *48*, 053306.
15. Bruers, S.; Maes, C.; Netočný, K. On the validity of entropy production principles for linear electrical circuits. *J. Stat. Phys.* **2007**, *129*, 725–740.

16. Polettini, M.; Esposito, M. Nonconvexity of the relative entropy for Markov dynamics: A Fisher information approach. *Phys. Rev. E* **2013**, *88*, 012112.

17. Horowitz, J.M.; Zhou, K.; England, J.L. Minimum energetic cost to maintain a target nonequilibrium state. *Phys. Rev. E* **2017**, *95*, 042102.

18. Horowitz, J.M.; Jacobs, K. Energy cost of controlling mesoscopic quantum systems. *Phys. Rev. Lett.* **2015**, *115*, 130501.

19. Van Kampen, N.G. *Stochastic Processes in Physics and Chemistry*, 3rd ed.; Elsevier Ltd.: New York, NY, USA, 2007.

20. Esposito, M.; Van den Broeck, C. Second law and Landauer principle far from equilibrium. *Europhys. Lett.* **2011**, *95*, 40004.

21. Sartori, P.; Granger, L.; Lee, C.; Horowitz, J. Thermodynamic costs of information processing in sensory adaptation. *PLoS Comput. Biol.* **2014**, *10*, e1003974.

22. Sartori, P.; Piglotti, S. Thermodynamics of error correction. *Phys. Rev. X* **2015**, *5*, 041039.

23. Horowitz, J.M.; Jacobs, K. Quantum effects improve the energy efficiency of feedback control. *Phys. Rev. E* **2014**, *89*, 042134.

24. Hamerly, R.; Mabuchi, H. Advantages of coherent feedback for cooling quantum oscillators. *Phys. Rev. Lett.* **2012**, *109*, 173602.

25. Schliesser, A.; Rivière, R.; Anetsberger, G.; Arcizet, O.; Kippenberg, T.J. Resolved-sideband cooling of a micromechanical oscillator. *Nat. Phys.* **2008**, *5*, 415–419.

26. Esposito, M.; Van den Broeck, C. Three faces of the second law. I. Master equation formulation. *Phys. Rev. E* **2010**, *82*, 011143.

27. Seifert, U. Stochastic thermodynamics, fluctuation theorems, and moleculer machines. *Rep. Prog. Phys.* **2012**, *75*, 126001.

28. Cover, T.M.; Thomas, J.A. *Elements of Information Theory*, 2nd ed.; Wiley-Interscience: Flatbush, NY, USA, 2006.

29. Barato, A.C.; Hartich, D.; Seifert, U. Rate of mutual information between coarse-grained non-Markovian variables. *J. Stat. Phys.* **2013**, *153*, 460–478.

30. Diana, G.; Esposito, M. Mutual entropy production in bipartite systems. *J. Stat. Mech. Theor. Exp.* **2014**, *2014*, P04010.

31. Hartich, D.; Barato, A.C.; Seifert, U. Stochastic thermodynamics of bipartite systems: Transfer entropy inequalities and a Maxwell's demon interpretation. *J. Stat. Mech.* **2014**, *2014*, P02016.

32. Polettini, M.; Esposito, M. Irreversible thermodynamics of open chemical networks. I. Emergent cycles and broken conservation laws. *J. Chem. Phys.* **2014**, *141*, 024117.

33. Rao, R.; Esposito, M. Nonequilibrium thermodynamics of chemical reaction networks: Wisdom from stochastic thermodynamics. *Phys. Rev. X* **2016**, *6*, 041064.

34. Kawai, R.; Parrondo, J.M.R.; Van den Broeck, C. Dissipation: The Phase-Space Perspective. *Phys. Rev. Lett.* **2007**, *98*, 080602.

35. Parrondo, J.M.R.; Van den Broeck, C.; Kawai, R. Entropy production and the arrow of time. *New J. Phys.* **2009**, *11*, 073008.

36. Spohn, H.; Lebowitz, J.L. Irreversible thermodynamics for quantum systems weakly coupled to thermal reservoirs. In *Advances in Chemical Physics: For Ilya Prigogine*; Rice, S.A., Ed.; John Wiley & Sons: Hoboken, NJ, USA, 1978; Volume 38.

37. Ge, H.; Qian, H. Physical origins of entropy produciton, free energy dissipation, and their mathematical representations. *Phys. Rev. E* **2010**, *81*, 051133.

38. Gaveau, B.; Schulman, L.S. A general framework for non-equilibrium phenomena: the master equation and its formal consequences. *Phys. Lett. A* **1997**, *229*, 347–353.

39. Procaccia, I.; Levine, R.D. Potential work: A statistical-mechanical approach to systems in disequilibrium. *J. Chem. Phys.* **1976**, *65*, 3357.

40. Esposito, M.; Parrondo, J.M.R. Stochastic thermodynamics of hidden pumps. *Phys. Rev. E* **2015**, *91*, 052114.

41. Barato, A.C.; Seifert, U. Thermodynamic uncertainty relation for biomolecular processes. *Phys. Rev. Lett.* **2015**, *114*, 158101.

42. Pietzonka, P.; Barato, A.C.; Seifert, U. Affinity-and topology-dependent bound on current fluctuations. *J. Phys. A Math. Theor.* **2016**, *49*, 34LT01.

43. Pietzonka, P.; Barato, A.C.; Seifert, U. Universal bounds on current fluctuations. *Phys. Rev. E* **2016**, *93*, 052145.

Entropy **2017**, *19*, 333

44. Gingrich, T.; Horowitz, J.M.; Perunov, N.; England, J.L. Dissipation bounds all steady-state current fluctuations. *Phys. Rev. Lett.* **2016**, *116*, 120601.
45. Gingrich, T.R.; Rotskoff, G.M.; Horowitz, J.M. Inferring dissipation from current fluctuations. *J. Phys. A Math. Theor.* **2017**, *50*, 184004.

![entropy](entropy logo)

MDPI

Article

Clausius Relation for Active Particles: What Can We Learn from Fluctuations

Andrea Puglisi [1],*and Umberto Marini Bettolo Marconi [2,3]

[1] CNR ISC and Dipartimento di Fisica, Sapienza Università di Roma, p.le A. Moro 2, 00185 Roma, Italy
[2] Scuola di Scienze e Tecnologie, Università di Camerino, Via Madonna delle Carceri, 62032 Camerino, Italy; umberto.marinibettolo@unicam.it
[3] Istituto Nazionale di Fisica Nucleare (INFN), Perugia 06123, Italy
*   Correspondence: andrea.puglisi@roma1.infn.it; Tel.: +39-06-4991-3491

Received: 12 June 2017; Accepted: 12 July 2017; Published: 13 July 2017

Abstract: Many kinds of active particles, such as bacteria or active colloids, move in a thermostatted fluid by means of self-propulsion. Energy injected by such a non-equilibrium force is eventually dissipated as heat in the thermostat. Since thermal fluctuations are much faster and weaker than self-propulsion forces, they are often neglected, blurring the identification of dissipated heat in theoretical models. For the same reason, some freedom—or arbitrariness—appears when defining entropy production. Recently three different recipes to define heat and entropy production have been proposed for the same model where the role of self-propulsion is played by a Gaussian coloured noise. Here we compare and discuss the relation between such proposals and their physical meaning. One of these proposals takes into account the heat exchanged with a non-equilibrium active bath: such an "active heat" satisfies the original Clausius relation and can be experimentally verified.

Keywords: active particles; entropy production; Clausius relation

1. Introduction

Active particle systems have recently attracted the increasing interest of scientists of different disciplines since they sit at the intersection of biology, chemistry and physics [1,2]. A central feature of these materials is that its elementary constituents convert energy from the environment via metabolic or chemical reactions into direct motion but also dissipate energy producing heat by friction in order to move inside a surrounding solvent [3]. Therefore, the complex behavior of active particles can only be described by the tools of non equilibrium statistical physics, such as kinetic theory, statistical mechanics of non-equilibrium processes and stochastic thermodynamics [4]. A fascinating question, which naturally comes to our mind, is how thermodynamics shapes biological functions in living organisms [5] such as motility and self-propulsion and in particular which is the entropy production associated with their non equilibrium steady states [6]. This question requires a notion of heat generated by self-propulsion and dissipated in the thermostatted solvent.

Within careful calorimetric experiments, one is able to measure the heat dissipated into the solvent by a microbial colony [7]. Such total heat depends upon many biological functions which are not included in active models, but one could devise smart experiments (e.g., by varying motility without changing other functions) in order to assess the fraction of heat strictly generated by self-propulsion. Observing the associated fluctuations is perhaps a much harder task, if not impossible. However, before encountering experimental limitations, one finds limits in the theory.

At our mesoscopic level, the definition of heat has to be framed within stochastic thermodynamics [8–11]. A problem, however, arises when thermal fluctuations are discarded: such an approximation is adopted in many models of active particles, since temperature is negligible with respect to the energy associated with both self-propulsion and external forces. In active models, some noise is retained to describe the non-deterministic nature of the self-propulsion force, but it usually acts on time-scales and energy scales much larger than molecular agitation of the solvent. This "athermal" nature of active particle models is similar to that in granular models [12] or in models of macroscopic friction [13]. While it is very useful—sometimes even inevitable—for the purpose of analytic calculations or numerical simulations, it leads to a mismatch between entropy production and heat [14]. Basically, the relation between total entropy production and dissipated heat loses its similitude with the original Clausius form and involves additional terms. For this reason, it does not provide a clear constraint on heat divided by temperature, as it occurs for the Clausius relation in macroscopic thermodynamics. Such a problem has already been noticed in some models of active particles [15], and in systems with feedback [16–18].

Recently it has been shown that the above fallacy is bypassed in a model of active particles where Gaussian colored noise plays the role of self-propulsion [19–21]. Even if thermal fluctuations are neglected, a notion of "coarse-grained heat" can be introduced, together with a spatial-dependent effective temperature, such that the original Clausius relation is fully recovered. The crucial point here is that both such a "coarse-grained heat" and the effective temperature can be measured in experiments and therefore a test of this active Clausius relation can be attempted. In the last year other two proposals have appeared in the literature [22,23], devoted to define entropy production and heat in the same identical model. The purpose of the present paper is to discuss the connection between those different Clausius relations. Apart from this comparison, two novelties are present here with respect to [21]: (1) a single trajectory level of description is adopted, while in [21] an ensemble average had been considered; (2) a generalization of [21] to more than one dimension and interacting particles is presented. In order to simplify the discussion, the main discussion is focused on the 1d case with a time-independent potential, which is sufficient to show the main difference between the definitions of heat and entropy production in the three works considered [21–23]. The generalisation of the proposal in [21] to multi-dimensional cases with a time-dependent potential is also discussed at the end.

In Section 2, we review the basic facts of Clausius relation in macroscopic thermodynamics and in stochastic (or mesoscopic) thermodynamics. In Section 3, we introduce the active model with Gaussian colored noise with thermal fluctuations, where a "microscopic Clausius relation" is trivially satisfied, and then show what happens at the coarse-grained level, when inertia and thermal fluctuations are neglected. The three recipes appeared in [21–23] to connect entropy production, dissipated heat and temperature are reviewed and compared.

2. Heat and Entropy Production: From Macroscopic to Stochastic Thermodynamics

Here we revise a few elementary facts of thermodynamics, at the macroscopic level and at the mesoscopic one. The macroscopic level is the one presented in thermodynamic textbooks, where there are no fluctuations: we denote quantities at this level with capital letters. The mesoscopic level is the topic of an intense research exploded roughly in the last two decades, and is dominated by fluctuations: we denote quantities at this level with small letters. Averaging out the fluctuations of the mesoscopic level (an operation which—in general—is automatically obtained in the limit of a very large number of constituents) brings back the results of the macroscopic one. Through the whole paper, we set the Boltzmann constant $k_B = 1$.

2.1. Macroscopic Level

For a large system one may measure heat, for instance with a calorimeter, and call it δQ: we consider it positive when going from the bath to the system. The second principle of thermodynamics guarantees that in a transformation where entropy S changes by a quantity dS there is a non-negative entropy production

$$\delta \Sigma = dS - \frac{\delta Q}{T} \geq 0, \tag{1}$$

where we used a shorthand notation where—for instance $\delta\Sigma$ and δQ are non-exact differentials, while dS is an exact differential. In a "quasi-static" transformation the equal sign holds, i.e., $\delta\Sigma = 0$.

In the absence of a direct way of measuring the entropy of a system (e.g., if quasi-static transformations are not available), other relations derived from Equation (1) and involving measurable observables are useful. For instance from Equation (1) it follows

$$dS \geq \frac{\delta Q}{T} \tag{2}$$

and therefore the existence of a "minimum work" that can be extracted, also called difference of free energy dF:

$$TdS \geq dE - \delta W \rightarrow \delta W \geq dE - TdS = dF. \tag{3}$$

Another consequence that is derived way from Equation (1) is to consider a cyclical transformation ($dS = 0$), where it implies

$$\oint \frac{\delta Q(t)}{T(t)} \leq 0, \tag{4}$$

which is the celebrated Clausius relation. This can be tested in experiments and is the founding principle of the theory of heat engines, efficiency, etc.

It is important to underline that if—hypothetically—the total entropy production was something different from $dS - \delta Q/T$, i.e., if

$$\delta\Sigma = dS - \frac{\delta Q}{T} + \Sigma_{an} \geq 0, \tag{5}$$

then all the above relations, including the Clausius relation, would not hold anymore due to the presence of an "anomalous" entropy production term, Σ_{an}. However it is quite difficult to imagine Equation (5) in macroscopic thermodynamics, since the very definition of macroscopic entropy production is the difference between dS and $\delta Q/T$ [6]. On the contrary, equations similar to (5) have appeared in the literature in a stochastic thermodynamic treatment of systems with feedback and model of self-propelled particles [15–18,23].

2.2. Mesoscopic Level

When a small system is considered, a stochastic description is necessary in order to incorporate fluctuations. At thermodynamic equilibrium the stochastic evolution must be consistent with micro-reversibility. More precisely, the couple "system plus thermostat" describes all degrees of freedom of the world and therefore it has to satisfy an exact symmetry under time-reversal: when the heat bath is replaced by an effective stochastic bath force, time-reversal is mapped into the equivalence of probabilities of a trajectory and its time-reversal, which coincides with detailed balance if the process is Markovian.

When an external, non-conservative, force is applied to the system, one may expect that the stochastic bath force is not changed (for instance if the bath is very large and is weakly affected by the external force). This amounts to say that the non-equilibrium model contains the sum of two forces which both concur to change the energy of the system: the external force does work, the bath force brings heat [11]. A notion of entropy production rate σ of a trajectory $\omega(t)$, for Markovian stochastic systems, has been introduced in [8] and revisited in [9]. It can be summarized as

$$\int_0^t \delta\sigma(t') = \ln \frac{\text{prob}[\{\omega(t')\}_0^t]}{\text{prob}[\{\overline{\omega}(t-t')\}_0^t]} = \ln \frac{p[\omega(0)]}{p[\overline{\omega}(t)]} + \ln \frac{\text{prob}[\{\omega(t')\}_0^t|\omega(0)]}{\text{prob}[\{\overline{\omega}(t-t')\}_0^t|\overline{\omega}(t)]} \tag{6}$$

$$= \int_0^t ds + \int_0^t \delta s_m, \tag{7}$$

where $\overline{\omega}$ is the time-reversal of the phase-space variables (typically positions are unchanged and velocities are reflected), $s(t) = -\ln p[\omega(t)]$ is the microscopic Gibbs entropy in the point $\omega(t)$ in phase space and δs_m is the so-called entropy production of the surrounding medium [9]. In the rest of the paper we consider, for simplicity, the infinitesimal version of Equation (6), i.e., $\delta\sigma = ds + \delta s_m$. An average over noise realizations and initial conditions is expected to give back the macroscopic quantities, i.e., $\delta\Sigma = \langle\delta\sigma\rangle$ and $dS = \langle ds\rangle$, such that Equation (1) implies $\langle\delta s_m\rangle = -\delta Q/T$. Indeed in many models at constant temperature, one has $\delta s_m = -\delta q/T$ with δq the heat injected by the bath force, which satisfies $\langle\delta q\rangle = \delta Q$. The total entropy production $\int_0^t \delta\sigma(t')$ satisfies the Fluctuation-Relation at any time $t > 0$ and this guarantees that $\delta\Sigma = \langle\delta\sigma\rangle$ is non-negative [8]. In a stationary state $\langle\delta s_m\rangle \geq 0$ then follows.

As a useful example, let us consider the evolution of a colloidal particle of mass m, position and velocity $x(t), u(t)$ in one dimension, under the action of an external potential $\phi(x)$ and of a non-conservative external force $f_{nc}(t)$.

$$dx(t) = u(t)dt \tag{8a}$$

$$mdu(t) = -\gamma u(t)dt + \sqrt{2\gamma T}dW(t) - \phi'[x(t)]dt + f_{nc}(t)dt, \tag{8b}$$

with $dW(t)$ the Wiener infinitesimal increment (with variance dt). Defining energy as $e = mu^2/2 + \phi(x)$, it is easy to see that heat (going from the bath into the system) reads

$$\delta q = de - \delta w = u \circ [-\gamma u dt + \sqrt{2\gamma T}dW(t)] \tag{9}$$

where we have defined the work $\delta w = u f_{nc}dt$, and $\circ$ denotes products which must be integrated according to the Stratonovich rule.

For this model, it is possible to compute the conditional probability appearing in (6) and therefore compute δs_m. The result depends upon the parity of f_{nc} under time-reversal [24]. In simple cases, for instance when magnetic fields are not involved [25], such a force is assigned even parity under time-reversal. In this case, one gets (see Appendix A)

$$\delta\sigma = ds - \frac{u \circ [-\gamma u dt + \sqrt{2\gamma T}dW]}{T} = ds - \frac{\delta q}{T}, \tag{10}$$

which is ≥ 0 on average, leading to the usual Clausius relation.

On the contrary if f_{nc} is odd, for instance if the coarse-graining has delivered a force which is proportional to odd powers of the velocity of external bodies [14,26], or if magnetic fields are involved [25], the relation (10) does not hold anymore. In such cases, things seem to improve when the so-called *conjugated* dynamics is considered, by changing the sign of odd external non-conservative forces when computing the probability of inverse paths appearing in the denominator of Equation (6) [18,24,26–29]: basically this amounts to change the parity of the force and get back the result in Equation (10). The problem of such an artificial prescription, however, is that the conjugated dynamics cannot be realized in experiments and therefore an empirical evaluation (i.e., without a detailed knowledge of the equation of motions) of the conjugated probability is not available, neither it is possible to experimentally observe the associated fluctuation relation.

3. Active Particles: The Coarse-Grained Heat and Clausius Relation

The analogy between stochastic and macroscopic thermodynamics, Equation (10), rests upon two main ingredients: (1) the heat bath must be modeled as a stochastic force which—if non-conservative forces are removed—satisfies detailed balance with respect to the equilibrium probability distribution ($\delta\sigma \equiv 0$) and (2) the non-conservative forces are even under time-reversal, a fact which is expected to be realized when the microscopic forces are not velocity-dependent (e.g., there are no Lorentz forces) and the coarse-graining does not change or mix their parity. Many models of active particles abandon such basic facts (in particular detailed balance [4]), with the aim of describing the relevant variables (such as positions or orientations of the micro-swimmers) which evolve on scales much slower than those affected by thermal agitation. An interesting example of model of active swimmers where this procedure can be analyzed is one where self-propulsion takes the form of an Ornstein-Uhlenbeck process: its non-zero correlation time represents the persistence of motion due to activity. Active particles of biophysical interest, for instance bacteria or sperms, display such a finite time decay of autocorrelation, but of course can exhibit more complicate form of the decay, i.e., not necessarily an exponential. Several authors have discussed the fair comparison, at some level (for instance comparing the spatial correlations in the presence of external potentials or the phase separation effects induced by activity), between other models or real experiments and the Ornstein-Uhlenbeck model considered here, see for instance [30–32].

Here we introduce the model at a space-time scale fine enough to describe the real velocity u of the particle and thermal fluctuations:

$$dx(t) = u(t)dt \tag{11a}$$

$$mdu(t) = -\gamma u(t)dt + \sqrt{2\gamma T_b}dW(t) + f_a(t)dt - \phi'[x(t)]dt, \tag{11b}$$

where T_b is the environmental (solvent) temperature and the active force satisfies

$$df_a(t) = -\frac{f_a(t)}{\tau}dt + \frac{\gamma\sqrt{2D_a}}{\tau}dW_2(t), \tag{12}$$

with dW_2 another (independent) Wiener increment with variance dt. Here we consider for simplicity the 1-particle case in one dimension, with a potential $\phi(x)$ which does not depend upon time. Later we generalize some of the results to many interacting particles and with a time-dependent potential.

Note that, when $\phi = 0$, $\langle x^2 \rangle \sim 2(D_a + T_b/\gamma)t$ for large times. Based upon such a bare diffusivity, the "active bath temperature" $T_a = \gamma D_a$ is usually defined. In [21] a "mass-less" active temperature $T_\tau = D_a/\tau$ was defined. In this paper we show that it is not necessary, if an "effective mass" $\mu = \gamma\tau$ is used, as in [23]. Of course there is no thermostat at temperature T_a, such a temperature is only useful to define a relevant energy scale.

Since active micro-swimmers are usually dispersed in viscous liquids, it is much more common to find the overdamped version of the model [19], which describes the position of the particle on a time-scale slower than the relaxation time due to inertia:

$$dx(t) = \frac{\sqrt{2\gamma T_b}dW(t) + f_a(t)dt - \phi'[x(t)]dt}{\gamma}. \tag{13}$$

3.1. Heat Dissipation into the Solvent

Interpreting f_a as an external force derived—through the coarse-graining of the full microscopic dynamics—from forces which do not depend upon velocities, it is reasonable to consider it even. According to the recipe of stochastic thermodynamics discussed above, Equation (6) applied to Equation (11) or Equation (13), see Appendix A, one gets Equation (10), with

$$\delta q_b = u \circ [-\gamma u dt + \sqrt{2\gamma T_b}dW(t)], \tag{14}$$

which is the heat absorbed from the reservoir, satisfying in the steady state the Clausius relation at constant temperature, i.e.,

$$\delta Q_b = \langle \delta q_b \rangle \leq 0. \tag{15}$$

The interpretation is obvious, the active force $f_a(t)$ acts as an external non-conservative force and transfers energy in the system which is dissipated into the bath. This can be measured by ordinary calorimetry in the solvent [33]. As discussed above, such a measurement is in principle very difficult in experiments with living micro-swimmers, since released heat is affected by many other non-equilibrium biological functions. A promising direction could be the use of artificial active particles [3].

3.2. Removing the Solvent from the Description

Since T_b is orders of magnitude smaller than active temperatures, it is very useful—also for computational purposes—to remove it from Equation (13), keeping only

$$\dot{x} = \frac{f_a(t) - \phi'(x)}{\gamma}. \tag{16}$$

At this point, an important ingredient of the bath force (its noise) has disappeared and the basic recipe of stochastic thermodynamics cannot be applied straightforwardly. Still, it is useful to find a measure of "distance from equilibrium" and relate it to parameters and observable quantities. Considering that f_a is random, one is tempted to consider $-\gamma\dot{x} + f_a(t)$ as an effective bath and define a heat as $\dot{x} \circ [-\gamma\dot{x} + f_a(t)]$. However, the random force $f_a(t)$ is non-Markovian and therefore the standard recipe of stochastic thermodynamics brings in complications [34–36].

The simplest way to get rid of the non-Markovian character of the noise is to time-derive Equation (16), obtaining

$$dx(t) = u(t)dt \tag{17a}$$

$$\mu du(t) = -\gamma u(t)dt + \sqrt{2\gamma T_a}dW(t) - \phi'[x(t)]dt - \tau\phi''[x(t)]u(t)dt \tag{17b}$$

$$= -\gamma\Gamma(x)u(t)dt + \gamma\sqrt{2D_a}dW(t) - \phi'[x(t)]dt \tag{17c}$$

where we have introduced the effective mass $\mu = \gamma\tau$ and the space-dependent viscosity correction $\Gamma(x) = 1 + \frac{\tau}{\gamma}\phi''(x)$.

As highlighted by the two versions in Equations (17b)–(17c), the evolution of the effective velocity u is affected by the conservative force $-\phi'(x)$ and by an additional force that can be interpreted in two different ways: (1) an equilibrium bath at temperature T_a plus a non conservative force $f_{nc} = -\tau\phi''(x)u$ which is *odd* under time-reversal; or (2) a non-equilibrium bath with space-dependent viscosity modulated according to the function $\Gamma(x)$. In the next two subsections, we see the consequences of such different interpretations, which change both the definition of entropy production as well as of heat.

3.2.1. Equilibrium Bath with a Non-Conservative Force: Conjugated Entropy Production

This interpretation is considered in [23]. The authors propose to define heat as the energy injected by the force $-\gamma udt + \gamma\sqrt{2D_a}dW$, as if it were an equilibrium bath

$$\delta q_1 = u \circ (-\gamma udt + \gamma\sqrt{2\gamma T_a}dW). \tag{18}$$

To derive the entropy production, the authors consider the formula (6) with the probability of the time-reversed path (which appears in the denominator) computed according to a dynamics where the force $f_{nc}(t)$ is replaced by $-f_{nc}(t)$, as discussed at the end of Section 2.2. This idea is justified by the authors by showing that such a change of sign is necessary in order to make invariant under time-reversal the dynamics without the bath. However such an argument is not really compelling. The terms $-\gamma udt + \gamma\sqrt{2\gamma T_a}dW$ do not correspond to any well-defined part of the physical system which could be identified as an equilibrium bath: the first term is the viscous damping due to the solvent, the second term is the fluctuating part of the derivative of the self-propulsion. It is a mathematical coincidence that together they form a Ornstein-Uhlenbeck process of the same form of equilibrium bath forces. In our opinion, it is quite arbitrary detaching them from Equation (17) and there is no reason why the rest of the equation (once those terms are removed) should satisfy the invariance under time-reversal.

According to the "conjugated" prescription, one gets for the case of a single particle considered here (see Appendix A)

$$\delta\Sigma = ds - \frac{\delta q_1}{T_a} + \frac{\tau^2}{2T_a}(du)^2\phi'', \tag{19}$$

where $(du)^2 \approx 2T_a dt/(\gamma\tau^2)$. The average can be written as

$$\delta\Sigma = dS - \frac{\delta Q_1}{T_a} + \frac{dt}{\gamma}\langle\phi''\rangle \geq 0, \tag{20}$$

with $\delta Q_1 = \langle\delta q_1\rangle$.

A peculiarity of this recipe is that it gives a non-zero average entropy production also for the harmonic case $\phi(x) \sim x^2$. Since in the harmonic case Equation (17) satisfies detailed balance, it is unclear if such a peculiarity is an advantage or not. Moreover, as already discussed, entropy production computed with the conjugated reversed dynamics is not accessible in experiments. Most importantly, in our opinion Equation (20) hardly deserves the name "Clausius relation", as it does not give the same important information about the sign of the average heat.

3.2.2. Equilibrium Bath with a Non-Conservative Force: Standard Entropy Production

In [22] the authors consider Formula (6) (without conjugation for the reversed dynamics) applied to the dynamics in Equation (17c). However all terms giving exact deterministic differentials are thrown away, leading to an approximate formula (see Appendix A)

$$\delta\Sigma \approx \frac{\tau^2 u\phi'' \circ du}{T_a}. \tag{21}$$

In the steady state the neglected terms have zero average, and indeed only the average formula is reported in [22]. Fluctuations and large deviations functions, however, may keep the memory of those terms [37–39].

Another difficulty of formula (21) is its connection with heat. In the end of their paper, the authors manage to show that Equation (16) can be mapped exactly into a generalized Langevin equation with memory. This equation can be broken into a viscoelastic bath at equilibrium at temperature T plus a non-conservative force. Within such a description, the average of the entropy production rate in Equation (21) can be written as $\mathcal{J}/T$, where $\mathcal{J}$ is the heat flux dissipated into the bath. A simple formula for such a "viscoelastic" heat or its—local or global—average is not given in [22]. Most importantly, some terms of fluctuations of entropy production are neglected which could be relevant for large deviation functions and the validity of the Fluctuation Relation [37–39].

3.2.3. Non-Equilibrium Bath

If the standard recipe of stochastic thermodynamics, Equation (6), is used without neglecting any term, one gets (see Appendix A)

$$\delta\sigma = ds - \frac{\delta q_2}{\theta(x)}, \tag{22}$$

with the "active bath heat" defined as

$$\delta q_2 = u \circ df_{ab}, \tag{23}$$

the "active bath force" as

$$df_{ab}(t) = -\gamma\Gamma[x(t)]u(t)dt + \gamma\sqrt{2D_a}dW(t) \tag{24}$$

$$= -\gamma\Gamma[x(t)]u(t)dt + \sqrt{2\gamma\Gamma[x(t)]\theta(x)}dW(t) \tag{25}$$

and the "local active temperature" $\theta(x) = T_a/\Gamma(x)$. It is clear that—as in the general formulation, Equation (9)—δq_2 corresponds to the variation of the total energy $e = \frac{\mu u^2}{2} + \phi(x)$ due to the bath force. The interpretation of $\theta(x)$ as a local active temperature is supported by the observation that a local Maxwellian with temperature $\theta(x)$ is an approximate solution for the local velocity distribution, with "small" violations of detailed balance, see [21] for details. We underline that $\theta(x)$ is immediately accessible in experiments: indeed the external potential $\phi(x)$ is directly controlled by the experimentalist (for instance by means of optical fields). The parameters γ, τ and D_a can be measured by independent measurements with single particles in the fluid without potential.

Averaging Equation (22), one gets

$$\delta\Sigma = dS - \left\langle \frac{\delta q_2(x)}{\theta(x)} \right\rangle \geq 0, \tag{26}$$

which in the steady state ($dS = 0$) is identical to the Clausius relation [21]. Interestingly, the local average $\mathring{q}(x)$ of the dissipated heat flux reads

$$\mathring{q}(x) = \gamma\Gamma(x) \left[\frac{\theta(x)}{\mu} n(x) - \int dv u^2 p(x,u) \right], \tag{27}$$

where $n(x) = \int dv p(x,u)$. This is an additional argument in favour of the simplicity and consistency of the picture discussed in the present section: the "active heat" is exactly proportional to the difference between the local active temperature $\theta(x)$ and the empirical temperature $\langle u^2 \rangle_x$. The empirical temperature is "attracted" by the local active temperature but the non-uniformity of such a temperature prevents full relaxation: the mismatch is a source of flowing heat. Equation (27) shows a straightforward way to measure such "active heat". Indeed such a measurement only amounts to measure $n(x)$ and $\langle u^2 \rangle_x$ (for instance by means of a fast camera attached to a microscope), while all other variables are parameters of the experimental setup, controlled by the experimentalist. Once one has measured the local heat $\mathring{q}(x)$ an experimental verification of the Clausius relation, Equation (26), is immediately accessible, since

$$\left\langle \frac{\delta q_2(x)}{\theta(x)} \right\rangle = \int dx \frac{\mathring{q}(x)}{\theta(x)} \tag{28}$$

We note that when the potential does not depend upon time, as in all our equations up to this point, the active heat δq_2 has zero average. Nevertheless, the average entropy production $\delta\Sigma$ has non-zero average, apart from the harmonic case $\phi(x) \sim x^2$ which is a special case where $\theta(x)$ is uniform [21].

When more particles are involved, a (local and time-dependent) diagonalisation procedure can always set back the problem in the case of a single particle. The multi-particles and multi-dimensional version of Equation (17) reads

$$\mu du_i = -\gamma\Gamma_{ij}(\mathbf{r})u_j dt - \partial_i\phi(\mathbf{r})dt + \gamma\sqrt{2D_a}dW_i, \tag{29}$$

with $\Gamma_{ij} = \delta_{ij} + \frac{\tau}{\gamma}\partial_j\partial_i\phi$ and indexes running over all particles and all Cartesian components and the Einstein summation convention is assumed. The potential ϕ includes both external and internal forces. Since the matrix $\Gamma_{ij}(\mathbf{r})$ is symmetric, an orthogonal matrix $P_{ij}(\mathbf{r})$ always exists such that $P\Gamma P^T = D$ with $D_{ij}(\mathbf{r}) = \lambda_i(\mathbf{r})\delta_{ij}$. By defining the rotated coordinates $\mathbf{R} = P\mathbf{r}$ and velocities $\mathbf{U} = P\mathbf{u}$, and recalling that the gradient rotates as a vector and the rotation of the vector of independent white noises gives again a vector of independent white noises, it is straightforward to get the formula:

$$\mu dU_i = -\gamma\lambda_i(\mathbf{R})U_i dt - \partial_{R_i}\phi + \gamma\sqrt{2D_a}dW_i. \tag{30}$$

Computation of the entropy production leads, therefore, to

$$\delta\sigma(t) = ds(t) - \sum_i \frac{\delta q_{2,i}(t)}{\theta_i[\mathbf{R}(t)]}, \tag{31}$$

with $\theta_i(\mathbf{R}) = T_a/\lambda_i(\mathbf{R})$ the i-th component of the local active temperature and

$$\delta q_{2,i} = U_i \circ [-\gamma\lambda_i(\mathbf{R})U_i dt + \gamma\sqrt{2D_a}dW]. \tag{32}$$

Notice that Equation (31) generalizes the Clausius relation to a system with different temperatures θ_i.

As an example, in the case of an active particle moving in a plane a subject to a central potential $\phi(r) = \phi(\mathbf{r})$, we have the following Cartesian representation of the matrix $D_{ij}(r) = D_r(r)\hat{r}_i\hat{r}_j + D_t(r)(\delta_{1j} - \hat{r}_i\hat{r}_j)$ with $D_r(r) = 1 + \frac{\tau}{\gamma}\phi''(r)$ and $D_t(r) = 1 + \frac{\tau}{\gamma}\phi'(r)/r$. The two temperatures are $\theta_r(r) = 1/D_r(r)$ and $\theta_t(r) = 1/D_t(r)$.

3.3. Time-Dependent Potential

When an external transformation is considered, i.e., a time-dependent potential $\phi(x,t)$ is taken into account, Equation (17c) is replaced with

$$\mu du(t) = -\gamma\Gamma(x)u(t)dt + \gamma\sqrt{2D_a}dW(t) - \partial_x\phi[x(t),t]dt - \tau\partial_t\partial_x\phi[x(t),t]dt \tag{33}$$

Also, in this case, we get (see Appendix A) the validity of the mesoscopic Clausius relation Equation (22) with the active heat Equation (23). Time-dependent potentials are at the basis, for instance of realizations of heat engines [40].

Very Slow Transformations

Imagine a very slow transformation from a $\phi(x,t_1)$ to a new $\phi(x,t_2)$: this means transforming the original non-equilibrium steady state ("NESS", at $t < t_1$) to a new non-equilibrium steady state (for $t \gg t_2$). As discussed above, in the initial and final NESS there is heat going steadily to the bath, even without the transformation. Therefore for very slow transformations $\Delta Q \to -\infty$ and the Clausius relation becomes useless. For this reason, Oono-Paniconi [41], then Hatano-Sasa [42], Bertini et al. [43] and Maes [44] have found expressions for the so-called "excess heat", i.e., heat which is released for the sole purpose of the transformation: this heat is obtained removing the "housekeeping heat" (necessary for the steady states) from the total ΔQ. All the mentioned proposals have been given for overdamped systems, where certain symmetries are more clear but also less general. Active particles have some kind of inertia or persistence which cannot be disregarded and therefore do not comply with such an assumption. It would be interesting to see the above simple ideas applied to the model in Equation (16) with a slow transformation of the potential.

4. Conclusions

In this paper, we have reviewed and compared three different prescriptions to extend the Clausius formula to active systems, i.e. to particles able to self-propel by means of metabolic processes or chemical reactions and to dissipate energy by a frictional mechanism. Those relations between the heat dissipation, entropy and work appear at the mesoscopic level, where fluctuations are taken into account by means of a stochastic description, but the contribution to these fluctuations coming from the molecular bath is neglected, leaving a certain freedom in defining heat and entropy production.

While all three methods are admissible and do not contradict any general principle, our point of view gives indications that only one of these prescriptions can be considered as a stochastic version of the original Clausius heat theorem, that is Equation (22) with "active heat" defined in (23).

The stochastic "active" version of Clausius formula we have derived coincides with the one recently presented by using ensemble averaged quantities [21].

We conclude offering a physical interpretation of our finding. In this model there is a level of coarse-graining which is sufficiently *isolated* from the fastest (neglected) degrees of freedom: at such a level of description the system behaves as a microscopic heat engine in contact with a continuum spectrum of temperatures, i.e., a particle which transfers heats between different positions of space in an environment where each position is thermostatted at a different temperature $\theta(x)$. Basically there are regions (where $\theta(x) > \langle u^2 \rangle_x$) where heat enters the system and other regions where heat leaves the system.

Acknowledgments: We warmly akcnowledge useful discussions and communications with Luca Cerino, Alessandro Sarracino and Frederic van Wijland.

Author Contributions: A.P. and U.M.B.M. equally contributed to the work reported.

Conflicts of Interest: The authors declare no conflicts of interest.

Appendix A. Entropy Production

We consider here a generalization of the dynamics in Equation (8) with $f_{nc}(x, u, t)$ representing any kind of time-dependent term: it can be external or internal (that is function also of system's degrees of freedom), odd or even under time-reversal, and we consider both the normal or the conjugated dynamics for the probability of the time-reversed path. In particular, we assume $dx = udt$ and

$$mdu = d\alpha(x, u, t) = -\gamma u(t)dt + \sqrt{2\gamma T}dW(t) - \phi'[x(t)]dt + f_{nc}(x, u, t)dt \tag{A1}$$

$$mdu = d\alpha^*(x, u, t) = -\gamma u(t)dt + \sqrt{2\gamma T}dW(t) - \phi'[x(t)]dt + \overline{f}_{nc}(x, u, t)dt \tag{A2}$$

to generate the dynamics of the forward and reversed paths, respectively. The Wiener increments $dW(t)$ have variance dt. When the standard entropy production is computed, the dynamics is the same, i.e., $\overline{f}_{nc}(x, u, t) = f_{nc}(x, u, t)$. On the contrary for the conjugated entropy production $\overline{f}_{nc}(x, u, t) = -f_{nc}(x, u, t)$.

Following Equation (6) (factorized for the Markovian dynamics), the infinitesimal entropy discharged into the surrounding medium reads

$$\delta s_m(t) = \ln \frac{\exp\{-[mdu_t - d\alpha(x_t, u_t, t)]^2/(4\gamma Tdt)\}}{\exp\{-[mdu_t - d\alpha^*(x_{t+dt}, -u_{t+dt}, t+dt)]^2/(4\gamma Tdt)\}} \tag{A3}$$

$$= -\frac{1}{4\gamma Tdt}\left\{[mdu_t + \gamma u_t dt + \phi'(x_t)dt - f_{nc}(x_t, u_t, t)dt]^2 \right.$$
$$\left. -[mdu_t - \gamma u_{t+dt}dt + \phi'(x_{t+dt})dt - \overline{f}_{nc}(x_{t+dt}, -u_{t+dt}, t+dt)dt]^2\right\} \tag{A4}$$

$$= -\frac{1}{\gamma T}\left[m\gamma du_t \circ u_t + \gamma u_t \phi'(x_t)dt - \gamma \frac{u_t f_{nc}(x_t, u_t, t) + u_{t+dt}\overline{f}_{nc}(x_{t+dt}, -u_{t+dt}, t+dt)}{2}dt \right.$$
$$- mdu\frac{f_{nc}(x_t, u_t, t) - \overline{f}_{nc}(x_{t+dt}, -u_{t+dt}, t+dt)}{2}$$
$$\left. -\frac{\phi'(x_t)f_{nc}(x_t, u_t, t) - \phi'(x_{t+dt})\overline{f}_{nc}(x_{t+dt}, -u_{t+dt}, t+dt)}{2}dt\right]. \tag{A5}$$

In the first passage we have used the fact that the time-reversal of du_t is $-u_t - (-u_{t+dt}) = du_t$. In the second passage we have neglected terms which goes to zero faster than dt and we have replaced $du_t(u_t + u_{t+dt})/2$ with $du_t \circ u_t$.

The cases considered in this paper are the following:

- The standard case in Equation (8) where $f_{nc}(t)$ is even and external (i.e., it does not depend upon x, v): $\bar{f}_{nc}(t) = f_{nc}(t)$. In this case, the last two terms in Equation (A5) become of higher order in dt and one gets:

$$\delta s_m(t) = -\frac{1}{T} u_t \circ [mdu_t + \phi'(x_t)dt - f_{nc}(t)dt] \tag{A6}$$

which immediately gives Equation (10).

- The case considered in [23], where $f_{nc}(x, u, t) = -\tau u \phi''(x)$ and ("conjugated entropy production") $\bar{f}_{nc}(x, u, t) = \tau u \phi''(x)$. In this case the last term becomes of higher order in dt, while the term du^2 cannot be discarded (as it contains $dW^2 \sim dt$), and therefore one gets

$$\delta s_m(t) = -\frac{1}{T} \left\{ u_t \circ [mdu_t + \phi'(x_t)dt - f_{nc}(x_t, v_t, t)dt] + \frac{\tau^2}{2} du_t^2 \phi''(x_t) \right\}, \tag{A7}$$

(where we have used $m \equiv \mu = \gamma\tau$), that is Equation (19).

- The case considered in [21,22], where $f_{nc}(x, u, t) = -\tau u \phi''(x)$ and (according to the standard definition of stochastic entropy production) $\bar{f}_{nc}(x, u, t) = -\tau u \phi''(x)$. In this case the third term in Equation (A5) is of higher order in dt. All the other terms must be kept, giving

$$\delta s_m(t) = -\frac{1}{T} \left(1 + \frac{\tau}{\gamma} \phi''(x_t) \right) u_t \circ [mdu_t + \phi'(x_t)dt]. \tag{A8}$$

If no terms are neglected, it gives exactly Equation (22).

- If in Equation (A8) the exact differentials ($u_t \circ du_t = du_t^2/2$, $u_t \phi'(x_t)dt = d\phi(x_t)$ and $u_t \phi'(x_t)\phi''(x_t)dt = d[\phi'(x_t)]^2/2$) are removed, then only one terms remains:

$$\delta s_m(t) \approx -\frac{1}{T} \tau^2 \phi''(x_t) u_t \circ du_t, \tag{A9}$$

i.e., Equation (21).

- If a time-dependent potential is considered, then a second non-conservative force appears $f_{nc,2}(x, t) = -\tau \partial_x \partial_t \phi(x, t)$. We stress that the dependence upon time of $\phi(x, t)$ is external, i.e., (keeping the standard recipe of stochastic thermodynamics) the probability of the reversed dynamics is generated by the same equation, that is no change of sign attributed to ∂_t. Basically we have $\bar{f}_{nc,2}(x, t) = -\tau \partial_x \partial_t \phi(x, t)$. Introducing $f_{nc,2}$ in Equation (A3) leads to the appearance of two new addends in the brackets [...] of Equation (A5): one totally new term $-\tau u_t \partial^2 \phi(x_t, t) f_{nc,2}(x_t, t)$ coming from the product $f_{nc} f_{nc,2}$; one surviving term $-\gamma u_t f_{nc,2}(x_t, t)$ in the third addend. No new terms appear inside the fourth and fifth addend. In conclusion one gets

$$\delta s_m(t) = -\frac{1}{T} \left(1 + \frac{\tau}{\gamma} \partial_x^2 \phi(x_t, t) \right) u_t \circ [mdu_t + \phi'(x_t)dt - f_{nc,2}(x_t, t)], \tag{A10}$$

which gives again Equation (22).

References

1. Ramaswamy, S. The Mechanics and Statistics of Active Matter. *Annu. Rev. Condens. Matter Phys.* **2010**, *1*, 323–345.
2. Marchetti, M.C.; Joanny, J.F.; Ramaswamy, S.; Liverpool, T.B.; Prost, J.; Rao, M.; Simha, R.A. Hydrodynamics of soft active matter. *Rev. Mod. Phys.* **2013**, *85*, 1143.
3. Bechinger, C.; Leonardo, R.D.; Löwen, H.; Reichhardt, C.; Volpe, G.; Volpe, G. Active particles in complex and crowded environments. *Rev. Mod. Phys.* **2016**, *88*, 045006.
4. Cates, M. Diffusive transport without detailed balance in motile bacteria: Does microbiology need statistical physics? *Rep. Prog. Phys.* **2012**, *75*, 042601.
5. England, J.L. Statistical physics of self-replication. *J. Chem. Phys.* **2013**, *139*, 121923.

6. De Groot, S.R.; Mazur, P. *Non-Equilibrium Thermodynamics*; Dover Publications: New York, NY, USA, 1984.
7. Gustafsson, L. Microbiological calorimetry. *Thermochim. Acta* **1991**, *193*, 145–171.
8. Lebowitz, J.L.; Spohn, H. A Gallavotti-Cohen-Type Symmetry in the Large Deviation Functional for Stochastic Dynamics. *J. Stat. Phys.* **1999**, *95*, 333–365.
9. Seifert, U. Entropy production along a stochastic trajectory and an integral fluctuation theorem. *Phys. Rev. Lett.* **2005**, *95*, 040602.
10. Marconi, U.M.B.; Puglisi, A.; Rondoni, L.; Vulpiani, A. Fluctuation-dissipation: Response theory in statistical physics. *Phys. Rep.* **2008**, *461*, 111–195.
11. Sekimoto, K. *Stochastic Energetics*; Springer: Berlin, Germany, 2010.
12. Puglisi, A.; Visco, P.; Barrat, A.; Trizac, E.; van Wijland, F. Fluctuations of Internal Energy Flow in a Vibrated Granular Gas. *Phys. Rev. Lett.* **2005**, *95*, 110202.
13. Baule, A.; Touchette, H.; Cohen, E.G.D. Stick-slip motion of solids with dry friction subject to random vibrations and an external field. *Nonlinearity* **2011**, *24*, 351–372.
14. Cerino, L.; Puglisi, A. Entropy production for velocity-dependent macroscopic forces: The problem of dissipation without fluctuations. *Europhys. Lett.* **2015**, *111*, 40012.
15. Ganguly, C.; Chaudhuri, D. Stochastic thermodynamics of active Brownian particles. *Phys. Rev. E* **2013**, *88*, 032102.
16. Kim, K.H.; Qian, H. Entropy Production of Brownian Macromolecules with Inertia. *Phys. Rev. Lett.* **2004**, *93*, 120602.
17. Munakata, T.; Rosinberg, M.L. Entropy production and fluctuation theorems under feedback control: The molecular refrigerator model revisited. *J. Stat. Mech.* **2012**, *2012*, P05010.
18. Munakata, T.; Rosinberg, M.L. Entropy Production and Fluctuation Theorems for Langevin Processes under Continuous Non-Markovian Feedback Control. *Phys. Rev. Lett.* **2014**, *112*, 180601.
19. Maggi, C.; Marconi, U.M.B.; Gnan, N.; Leonardo, R.D. Multidimensional stationary probability distribution for interacting active particles. *Sci. Rep.* **2015**, *5*, 10742.
20. Marconi, U.M.B.; Gnan, N.; Paoluzzi, M.; Maggi, C.; Leonardo, R.D. Velocity distribution in active particles systems. *Sci. Rep.* **2016**, *6*, 23297.
21. Marconi, U.M.B.; Puglisi, A.; Maggi, C. Heat, temperature and Clausius inequality in a model for active brownian particles. *Sci. Rep.* **2017**, *7*, 46496.
22. Fodor, E.; Nardini, C.; Cates, M.E.; Tailleur, J.; Visco, P.; van Wijland, F. How Far from Equilibrium Is Active Matter? *Phys. Rev. Lett.* **2016**, *117*, 038103.
23. Mandal, D.; Klymko, K.; DeWeese, M.R. Entropy production and fluctuation theorems for active matter. *arXiv* **2017**, arxiv:1704.02313.
24. Chetrite, R.; Gawedzki, K. Fluctuation Relations for Diffusion Processes. *Commun. Math. Phys.* **2008**, *282*, 469–518.
25. Pradhan, P.; Seifert, U. Nonexistence of classical diamagnetism and nonequilibrium fluctuation theorems for charged particles on a curved surface. *Europhys. Lett.* **2010**, *89*, 37001.
26. Kwon, C.; Yeo, J.; Lee, H.K.; Park, H. Unconventional entropy production in the presence of momentum-dependent forces. *J. Korean Phys. Soc.* **2016**, *68*, 633.
27. Chernyak, V.Y.; Chertkov, M.; Jarzynski, C. Path-integral analysis of fluctuation theorems for general Langevin processes. *J. Stat. Mech. Theory Exp.* **2006**, *2006*, P08001.
28. Spinney, R.E.; Ford, I.J. Nonequilibrium Thermodynamics of Stochastic Systems with Odd and Even Variables. *Phys. Rev. Lett.* **2012**, *108*, 170603.
29. Chun, H.; Noh, J.D. Microscopic theory for the time irreversibility and the entropy production. *arXiv* **2017**, arxiv:1706.01691.
30. Maggi, C.; Paoluzzi, M.; Pellicciotta, N.; Lepore, A.; Angelani, L.; Leonardo, R.D. Generalized Energy Equipartition in Harmonic Oscillators Driven by Active Baths. *Phys. Rev. Lett.* **2014**, *113*, 238303.
31. Szamel, G. Self-propelled particle in an external potential: Existence of an effective temperature. *Phys. Rev. E* **2014**, *90*, 012111.
32. Farage, T.F.F.; Krinninger, P.; Brader, J.M. Effective interactions in active Brownian suspensions. *Phys. Rev. E* **2015**, *91*, 042310.
33. Speck, T. Stochastic Thermodynamics for active matter. *Europhys. Lett.* **2016**, *114*, 30006.

34. Zamponi, F.; Bonetto, F.; Cugliandolo, L.F.; Kurchan, J. A fluctuation theorem for non-equilibrium relaxational systems driven by external forces. *J. Stat. Mech.* **2005**, *2005*, P09013.
35. Speck, T.; Seifert, U. Jarzynski Relation, Fluctuation Theorems, and Stochastic Thermodynamics for Non-Markovian Processes. *J. Stat. Mech.* **2007**, *2007*, L09002.
36. Crisanti, A.; Puglisi, A.; Villamaina, D. Non-equilibrium and information: The role of cross-correlations. *Phys. Rev. E* **2012**, *85*, 061127.
37. Visco, P. Work fluctuations for a Brownian particle between two thermostats. *J. Stat. Mech.* **2006**, *2006*, P06006.
38. Puglisi, A.; Rondoni, L.; Vulpiani, A. Relevance of initial and final conditions for the fluctuation relation in Markov processes. *J. Stat. Mech.* **2006**, *2006*, P08010.
39. Bonetto, F.; Gallavotti, G.; Giuliani, A.; Zamponi, F. Chaotic hypothesis, fluctuation theorem, singularities. *J. Stat. Phys.* **2006**, *123*, 39, doi:10.1007/s10955-006-9047-5.
40. Zakine, R.; Solon, A.; Gingrich, T.; van Wijland, F. Stochastic Stirling Engine Operating in Contact with Active Baths. *Entropy* **2017**, *19*, 193.
41. Oono, Y.; Paniconi, M. Steady state thermodynamics. *Prog. Theor. Phys. Suppl.* **1998**, *130*, 29–44.
42. Hatano, T.; Sasa, S. Steady-state thermodynamics of Langevin systems. *Phys. Rev. Lett.* **2001**, *86*, 3463–3466.
43. Bertini, L.; Sole, A.D.; Gabrielli, D.; Jona-Lasinio, G.; Landim, C. Macroscopic fluctuation theory. *Rev. Mod. Phys.* **2015**, *87*, 593–636.
44. Maes, C.; Netocný, K. A Nonequilibrium Extension of the Clausius Heat Theorem. *J. Stat. Phys.* **2014**, *154*, 188–203.

entropy

MDPI

Article

An Application of Pontryagin's Principle to Brownian Particle Engineered Equilibration

Paolo Muratore-Ginanneschi [1,*] and Kay Schwieger [2]

[1]   Department of Mathematics and Statistics, University of Helsinki, P.O. Box 68, FIN-00014 Helsinki, Finland
[2]   iteratec GmbH, Zettachring 6, 70567 Stuttgart, Germany; kay.schwieger@gmail.com
*   Correspondence: paolo.muratore-ginanneschi@helsinki.fi; Tel.: +358-2-94151482

Received: 3 July 2017; Accepted: 20 July 2017; Published: 24 July 2017

Abstract: We present a stylized model of controlled equilibration of a small system in a fluctuating environment. We derive the optimal control equations steering in finite-time the system between two equilibrium states. The corresponding thermodynamic transition is optimal in the sense that it occurs at minimum entropy if the set of admissible controls is restricted by certain bounds on the time derivatives of the protocols. We apply our equations to the engineered equilibration of an optical trap considered in a recent proof of principle experiment. We also analyze an elementary model of nucleation previously considered by Landauer to discuss the thermodynamic cost of one bit of information erasure. We expect our model to be a useful benchmark for experiment design as it exhibits the same integrability properties of well-known models of optimal mass transport by a compressible velocity field.

Keywords: fluctuation phenomena; random processes; noise and Brownian motion; nonequilibrium and irreversible thermodynamics; control theory; stochastic processes

PACS: 05.40.-a; 05.70.Ln; 02.30.Yy; 02.50.Ey

1. Introduction

An increasing number of applications in micro- and sub-micro-scale physics calls for the development of general techniques for engineered finite-time equilibration of systems operating in a thermally-fluctuating environment. Possible concrete examples are the design of nano-thermal engines [1,2] or of micro-mechanical oscillators used for high precision timing or sensing of mass and forces [3].

A recent experiment [4] exhibited the feasibility of driving a micro-system between two equilibria over a control time several orders of magnitude faster than the natural equilibration time. The system was a colloidal micro-sphere trapped in an optical potential. There is consensus that non-equilibrium thermodynamics (see, e.g., [5]) of optically-trapped micron-sized beads is well captured by Langevin–Smoluchowski equations [6]. In particular, the authors of [4] took care of showing that it is accurate to conceptualize the outcome of their experiment as the evolution of a Gaussian probability density according to a controlled Langevin–Smoluchowski dynamics with gradient drift and constant diffusion coefficient. Finite time equilibration means that at the end of the control horizon, the probability density is the solution of the stationary Fokker–Planck equation. The experimental demonstration consisted of a compression of the confining potential. In such a case, the protocol steering the equilibration process is specified by the choice of the time evolution of the stiffness of the quadratic potential whose gradient yields the drift in the Langevin–Smoluchowski equation. As a result, the set of admissible controls is infinite. The selection of the control in [4] was then based on simplicity of implementation considerations.

A compelling question is whether and how the selection of the protocol may stem from a notion of optimal efficiency. A natural indicator of efficiency in finite-time thermodynamics is entropy production. Transitions occurring at minimum entropy production set a lower bound in the Clausius inequality. Optimal control of these transitions is, thus, equivalent to a refinement of the second law of thermodynamics in the form of an equality.

In the Langevin–Smoluchowski framework, entropy production optimal control takes a particularly simple form if states at the end of the transition are specified by sufficiently regular probability densities [7]. Namely, the problem admits an exact mapping into the well-known Monge–Kantorovich optimal mass transport [8]. This feature is particularly useful because the dynamics of the Monge–Kantorovich problem is exactly solvable. Mass transport occurs along free-streaming Lagrangian particle trajectories. These trajectories satisfy boundary conditions determined by the map, called the Lagrangian map, transforming into each other the data of the problem, the initial and the final probability densities. Rigorous mathematical results [9–11] preside over the existence, qualitative properties and reconstruction algorithms for the Lagrangian map.

The aforementioned results cannot be directly applied to optimal protocols for engineered equilibration. Optimal protocols in finite-time unavoidably attain minimum entropy by leaving the end probability densities out of equilibrium. The qualitative reason is that optimization is carried over the set of drifts sufficiently smooth to mimic all controllable degrees of freedom of the micro-system. Controllable degrees of freedom are defined as those varying over typical time scales much slower than the time scales of Brownian forces [12]. The set of admissible protocols defined in this way is too large for optimal engineered equilibration. The set of admissible controls for equilibration must take into account also extra constraints coming from the characteristic time scales of the forces acting on the system. From the experimental slant, we expect these restrictions to be strongly contingent on the nature and configuration of peripherals in the laboratory setup. From the theoretical point of view, the self-consistence of Langevin–Smoluchowski modeling imposes a general restriction. The time variation of drift fields controlling the dynamics must be slow in comparison to Brownian and inertial forces.

In the present contribution, we propose a refinement of the entropy production optimal control adapted to engineered equilibration. We do this by restricting the set of admissible controls to those satisfying a non-holonomic constraint on accelerations. The constraint relates the bound on admissible accelerations to the path-wise displacement of the system degrees of freedom across the control horizon. Such displacement is a deterministic quantity, intrinsically stemming from the boundary conditions inasmuch as we determine it from the Lagrangian map.

This choice of the constraint has several inherent advantages. It yields an intuitive hold on the realizability of the optimal process. It also preserves the integrability properties of the optimal control problem specifying the lower bound to the second law. This is so because the constraint allows us to maintain protocols within the admissible set by exerting on them uniform accelerating or decelerating forces. On the technical side, the optimal control problem can be handled by a direct application of the Pontryagin maximum principle [13]. For the same reasons as for the refinement of the second law [7], the resulting optimal control is of the deterministic type. This circumstance yields a technical simplification, but it is not a necessary condition in view of extensions of our approach. We will return to this point in the conclusions.

The structure of the paper is as follows. In Section 2 we briefly review the Langevin–Smoluchowski approach to non-equilibrium thermodynamics [14]. This section can be skipped by readers familiar with the topic. In Section 3, we introduce the problem of optimizing the entropy production. In particular we explain its relation with the Schrödinger diffusion problem [15,16]. This relation, already pointed out in [17], has recently attracted the attention of mathematicians and probabilists interested in rigorous application of variational principles in hydrodynamics [18]. In Section 4, we formulate the Pontryagin principle for our problem. Our main result follows in Section 5, where we solve in explicit form the optimal protocols. Sections 6 and 7 are devoted to applications.

In Section 6, we revisit the theoretical model of the experiment [4], the primary motivation of our work. In Section 7, we apply our results to a stylized model of controlled nucleation obtained by manipulating a double-well potential. Landauer and Bennett availed themselves of this model to discuss the existence of an intrinsic thermodynamic cost of computing [19,20]. Optimal control of this model has motivated in more recent years several theoretical [21] and experimental works [22–24].

Finally, in Section 8, we compare the optimal control we found with those of [25]. This reference applied a regularization technique coming from instanton calculus [26] to give a precise meaning to otherwise ill-defined problems in non-equilibrium thermodynamics, where terminal cost seems to depend on the control rather than being a given function of the final state of the system.

In the conclusions, we discuss possible extensions of the present work. The style of the presentation is meant to be discursive, but relies on notions in between non-equilibrium physics, optimal control theory and probability theory. For this reason, we include in the Appendices some auxiliary information as a service to the interested reader.

2. Kinematics and Thermodynamics of the Model

We consider a physical process in a d-dimensional Euclidean space ($\mathbb{R}^d$) modeled by a Langevin–Smoluchowski dynamics:

$$d\xi_t = -\partial_{\xi_t} U(\xi_t, t)\, dt + \sqrt{\frac{2}{\beta}}\, d\omega_t \tag{1}$$

The stochastic differential $d\omega_t$ stands here for the increment of a standard d-dimensional Wiener process at time t [6]. $U\colon \mathbb{R}^d \otimes \mathbb{R} \mapsto \mathbb{R}$ denotes a smooth scalar potential, and β^{-1} is a constant sharing the same canonical dimensions as U. We also suppose that the initial state of the system is specified by a smooth probability density:

$$P(q \leq \xi_{t_i} < q + dq) = p_i(q) d^d q \tag{2}$$

Under rather general hypotheses, the Langevin–Smoluchowski Equation (1) can be derived as the scaling limit of the overdamped non-equilibrium dynamics of a classical system weakly coupled to a heat bath [27]. The Wiener process in (1) thus embodies thermal fluctuations of order β^{-1}. The fundamental simplification entailed by (1) is the possibility to establish a framework of elementary relations linking the dynamical to the statistical levels of description of a non-equilibrium process [14,28]. In fact, the kinematics of (1) ensures that for any time-autonomous, confining potential, the dynamics tends to a unique Boltzmann equilibrium state.

$$p_{eq}(q) \propto \exp\left(-\beta U(q)\right)$$

Building on the foregoing observations [14], we may then identify U over a finite-time horizon with the internal energy of the system. The differential of U:

$$dU(\xi_t, t) = dt\, \partial_t U(\xi_t, t) + d\xi_t^{1/2} \partial_{\xi_t} U(\xi_t, t) \tag{3}$$

yields the energy balance in the presence of thermal fluctuations due to interactions with the environment. We use the notation $\overset{1/2}{\cdot}$ for the Stratonovich differential [6]. From (3), we recover the first law of thermodynamics by averaging over the realizations of the Wiener process. In particular, we interpret:

$$\mathcal{W} = E \int_{t_o}^{t_f} dt\, \partial_t U(\xi_t, t) \tag{4}$$

as the average work done on the system. Correspondingly,

$$Q = -\mathrm{E} \int_{t_0}^{t_f} \mathrm{d}\xi_t^{1/2} \partial_{\xi_t} U(\xi_t, t) \tag{5}$$

is the average heat discarded by the system into the heat bath, and therefore:

$$W - Q = \mathrm{E} \left(U(\xi_{t_i}, t_f) - U(\xi_{t_f}, t_f) \right) \tag{6}$$

is the embodiment of the first law.

The kinematics of stochastic processes [29] allows us also to write a meaningful expression for the second law of thermodynamics. The expectation value of a Stratonovich differential is in general amenable to the form:

$$Q = -\mathrm{E} \int_{t_i}^{t_f} \mathrm{d}t \, (v \cdot \partial_{\xi_t} U)(\xi_t, t) \tag{7}$$

where:

$$v(q, t) = -\partial_q \left(U(q, t) + \frac{1}{\beta} \ln p(q, t) \right) \tag{8}$$

is the current velocity. For a potential drift, the current velocity vanishes identically at equilibrium. As is well known from stochastic mechanics [30,31], the current velocity permits couching the Fokker–Planck equation into the form of a deterministic mass transport equation (see also Appendix B). Hence, upon observing that:

$$\mathrm{E} \int_{t_i}^{t_f} \mathrm{d}t \, (v \cdot \partial_{\xi_t} \ln p)(\xi_{t_f}, t_f) = \mathrm{E} \int_{t_i}^{t_f} \mathrm{d}t \left(\partial_t + v_t \cdot \partial_{\xi_t} \right) \ln p(\xi_{t_f}, t_f) = \mathrm{E} \ln \frac{p(\xi_{t_f}, t_f)}{p(\xi_{t_i}, t_i)} \tag{9}$$

we can recast (7) into the form:

$$Q_T = Q - \frac{1}{\beta} \mathrm{E} \ln \frac{p(\xi_{t_f}, t_f)}{p(\xi_{t_i}, t_i)} = \mathrm{E} \int_{t_i}^{t_f} \mathrm{d}t \, \| v(\xi_t, t) \|^2 \tag{10}$$

which we interpret as the second law of thermodynamics (see, e.g., [32]). Namely, if we define $\mathcal{E} = \beta \, Q_T$ as the total entropy change in $[t_i \, t_f]$, (10) states that the sum of the entropy generated by heat released into the environment plus the change of the Gibbs–Shannon entropy of the system is positive definite and vanishes only at equilibrium. The second law in the form (10) immediately implies a bound on the average work done on the system. To evince this fact, we avail ourselves of the equality:

$$W = \mathrm{E} \left(U(\xi_{t_f}, t_f) - U(\xi_{t_i}, t_i) + \frac{1}{\beta} \ln \frac{p(\xi_{t_f}, t_f)}{p(\xi_{t_i}, t_i)} \right) + Q_T \tag{11}$$

and define the current velocity potential:

$$F(q, t) = U(q, t) + \frac{1}{\beta} \ln p(q, t)$$

We then obtain:

$$W = \mathrm{E} \left(F(\xi_{t_f}, t_f) - F(\xi_{t_i}, t_i) \right) + Q_T \geq \mathrm{E} \left(F(\xi_{t_f}, t_f) - F(\xi_{t_i}, t_i) \right)$$

In equilibrium thermodynamics, the Helmholtz free energy is defined as the difference:

$$\mathcal{F} = \mathcal{U} - \beta^{-1}\mathcal{S}$$

between the internal energy $\mathcal{U}$ and entropy $\mathcal{S}$ of a system at temperature β^{-1}. This relation admits a non-equilibrium extension by noticing that the information content [33] of the system probability density:

$$S(q,t) = -\ln \mathrm{p}(q,t)$$

weighs the contribution of individual realizations of (1) to the Gibbs–Shannon entropy. We refer to [29] for the kinematic and thermodynamic interpretation of the information content as the osmotic potential. We also emphasize that the notions above can be given an intrinsic meaning using the framework of stochastic differential geometry [17,31]. Finally, it is worth noticing that the above relations can be regarded as a special case of macroscopic fluctuation theory [34].

3. Non-Equilibrium Thermodynamics and Schrödinger Diffusion

We are interested in thermodynamic transitions between an initial state (2) at time t_i and a pre-assigned final state at time t_f also specified by a smooth probability density:

$$\mathrm{P}(q \leq \xi_{t_f} < q + \mathrm{d}q) = \mathrm{p}_f(q)\mathrm{d}^d q \tag{12}$$

We also suppose that the cumulative distribution functions of (2) and (12) are related by a Lagrangian map $\boldsymbol{\ell} \colon \mathbb{R}^d \mapsto \mathbb{R}^d$ such that:

$$\mathrm{P}(\xi_{t_i} < q) = \mathrm{P}(\xi_{t_f} < \boldsymbol{\ell}(q)) \tag{13}$$

According to the Langevin–Smoluchowski dynamics (1), the evolution of probability densities obeys a Fokker–Planck equation, a first order in time partial differential equation. As a consequence, the price we pay to steer transitions between assigned states is to regard the drift in (1) not as an assigned quantity, but as a control. A priori, a control is only implicitly characterized by the set of conditions that make it admissible. Informally speaking, admissible controls are all those drifts steering the process $\{\xi_t, t \in [t_i, t_f]\}$ between the assigned end states (2) and (12) while ensuring that at any time $t \in [t_i, t_f]$, the Langevin–Smoluchowski dynamics remains well defined.

Schrödinger [15] considered already in 1931 the problem of controlling a diffusion process between assigned states. His work was motivated by the quest for a statistical interpretation of quantum mechanics. In modern language [35,36], the problem can be rephrased as follows. Given (2) and (12) and a reference diffusion process, determine the diffusion process interpolating between (2) and (12) while minimizing the value of its Kullback–Leibler divergence (relative entropy) [37] with respect to the reference process. A standard application (Appendix A) of the Girsanov formula [6] shows that the Kullback–Leibler divergence of (1) with respect to the Wiener process is:

$$\mathcal{K}(\mathrm{P} \parallel \mathrm{P}_\omega) = \frac{\beta}{2}\,\mathrm{E}\int_{t_i}^{t_f} \mathrm{d}t \,\|\partial_{\xi_t} U(\xi_t,t)\|^2 \tag{14}$$

P and P_ω denote respectively the measures of the process solution of (1) with drift $-\partial_q U(q,t)$ and of the Wiener process ω. The expectation value on the right-hand side is with respect to P as elsewhere in the text. A now well-established result in optimal control theory (see, e.g., [35,36]) is that the optimal value of the drift satisfies a backward Burgers equation with the terminal condition specified by the solution of the Beurling–Jamison integral equations. We refer to [35,36] for further details. What interests us here is to emphasize the analogy with the problem of minimizing the entropy production $\mathcal{E}$ in a transition between assigned states.

Several observations are in order at this stage.

The first observation is that also (10) can be directly interpreted as a Kullback–Leibler divergence between two probability measures. Namely, we can write (Appendix A):

$$\mathcal{K}(\mathrm{P} \parallel \mathrm{P}_R) = \frac{\beta}{2}\,\mathrm{E}\int_{t_i}^{t_f} \mathrm{d}t\,\|v(\xi_t, t)\|^2 \tag{15}$$

for P_R the path-space measure of the process:

$$\mathrm{d}\xi_t = \partial_{\xi_t} U(\xi_t, t)\,\mathrm{d}t + \sqrt{\frac{2}{\beta}}\,\mathrm{d}\omega_t \tag{16}$$

evolving backward in time from the final condition (12) [38,39].

The second observation has more far reaching consequences for optimal control. The entropy production depends on the drift of (1) exclusively through the current velocity (8). Hence, we can treat the current velocity itself as the natural control quantity for (15). This fact entails major simplifications [7]. The current velocity can be thought of as deterministic rather than as a stochastic velocity field (see [29] and Appendix B). Thus, we can couch the optimal control of (15) into the problem of minimizing the kinetic energy of a classical particle traveling from an initial position q at time t_i and a final position $\ell(q)$ at time t_f specified by the Lagrangian map ℓ (13). In other words, entropy production minimization in the Langevin–Smoluchowski framework is equivalent to solving a classical optimal transport problem [8].

The third observation comes as a consequence of the second one. The optimal value of the entropy production is equal to the Wasserstein distance [40] between the initial and final probability measures of the system; see [41] for details. This fact yields a simple characterization of the Landauer bound and permits a fully-explicit analysis of the thermodynamics of stylized isochoric micro-engines (see [42] and the references therein).

Finally, the construction of Schrödinger diffusions via optimal control of (14) corresponds to a viscous regularization of the optimal control equations occasioned by the Schrödinger diffusion problem (15).

4. Pontryagin's Principle for Bounded Accelerations

An important qualitative feature of the solution of the optimal control of the entropy production is that the system starts from (2) and reaches (12) with non-vanishing current velocity. This means that the entropy production attains a minimum value when the end-states of the transition are out-of-equilibrium. We refer to this lower bound as the refinement of the second law.

Engineered equilibration transitions are, however, subject to at least two further types of constraints not taken into account in the derivation of the refined second law. The first type of constraint is on the set of admissible controls. For example, admissible controls cannot vary in an arbitrary manner: the fastest time scale in the Langevin–Smoluchowski dynamics is set by the Wiener process. The second type is that end-states are at equilibrium. In mathematical terms, this means that the current velocity must vanish identically at t_i and t_f.

We formalize a deterministic control problem modeling these constraints. Our goal is to minimize the functional:

$$\mathcal{E} = \int_{t_i}^{t_f} \mathrm{d}t\,\beta\,\|v_t\|^2 \tag{17}$$

over the set of trajectories generated for any given choice of the measurable control α_t by the differential equation:

$$\dot{\chi}_t = v_t \tag{18a}$$

$$\dot{v}_t = \alpha_t \tag{18b}$$

satisfying the boundary conditions:

$$\chi_{t_i} = q \qquad \& \qquad \chi_{t_f} = \ell(q) \tag{19}$$

We dub the dynamical variable χ_t the running Lagrangian map as it describes the evolution of the Lagrangian map within the control horizon. We restrict the set of admissible controls $\mathbb{A} = \{\alpha_t, t \in [t_i, t_f]\}$ to those enforcing equilibration at the boundaries of the control horizon:

$$v_{t_i} = 0 \qquad \& \qquad v_{t_f} = 0 \tag{20}$$

whilst satisfying the bound:

$$|\alpha_t^{(i)}| \le \frac{K^{(i)}(q)}{(t_f - t_i)^2} \qquad \forall t \in [t_i, t_f] \qquad \forall i = 1, \ldots, d \tag{21}$$

We suppose that the $K^{(i)}(q) > 0\, i = 1, \ldots, d$ are strictly positive functions of the initial data q of the form:

$$K^{(i)}(q) \propto |\ell^{(i)}(q) - q^{(i)}| \tag{22}$$

The constraint is non-holonomic inasmuch as it depends on the initial data of a trajectory. The proportionality (22) relates the bound on acceleration to the Lagrangian displacement needed to satisfy the control problem. Finally, we emphasize that the rate of change v_t of the running Lagrangian map is related to the current velocity (8) by a standard change of hydrodynamic coordinates from Lagrangian to Eulerian, which we write explicitly in formula (33) below.

We resort to the Pontryagin principle [13] to find normal extremals of (17). We defer the statement of the Pontryagin principle, as well as the discussion of abnormal extremals to Appendix C. We proceed in two steps. We first avail ourselves of Lagrange multipliers to define the effective cost functional:

$$\mathcal{A} = \int_{t_i}^{t_f} dt \left(\beta \parallel v_t \parallel^2 + \eta_t \cdot (\dot{\chi}_t - v_t) + \theta_t \cdot (\dot{v}_t - \alpha_t) \right)$$

subject to the boundary conditions (19) and (20). Then, we couch the cost functional into an explicit Hamiltonian form:

$$\mathcal{A} = \int_{t_i}^{t_f} dt \left(\eta_t \cdot \dot{\chi}_t + \theta_t \cdot \dot{v}_t - H(\chi_t, v_t, \eta_t, \theta_t, \alpha_t) \right) \tag{23}$$

with:

$$H(\chi_t, v_t, \eta_t, \theta_t, \alpha_t) = \eta_t \cdot v_t + \theta_t \cdot \alpha_t - \beta \parallel v_t \parallel^2$$

Pontryagin's principle yields a rigorous proof of the intuition that extremals of the optimal control equations correspond to stationary curves of the action (23) with Hamiltonian:

$$H_\star(\chi_t, v_t, \eta_t, \theta_t) = \max_{\alpha \in \mathbb{A}} H(\chi_t, v_t, \eta_t, \theta_t, \alpha_t) = \eta_t \cdot v_t + \frac{\sum_{i=1}^{d} K^{(i)} |\theta_t^{(i)}|}{(t_f - t_i)^2} - \beta \parallel v_t \parallel^2$$

In view of the boundary conditions (19), (20), extremals satisfy the Hamilton system of equations formed by (18a) and:

$$\dot{v}_t^{(i)} = \partial_{\theta_t} H_\star = \frac{K^{(i)}}{(t_f - t_i)^2} \, \text{sgn} \, \theta_t^{(i)} \tag{24a}$$

$$\dot{\eta}_t = -\partial_{\chi_t} H_\star = 0 \tag{24b}$$

$$\dot{\theta}_t = -\partial_{v_t} H_\star = -\eta_t + 2\beta v_t \tag{24c}$$

In writing (24a), we adopt the convention:

$$\text{sgn} \, 0 = 0$$

5. Explicit Solution in the 1*d* Case

The extremal Equations (18a) and (24) are time-autonomous and do not couple distinct vector components. It is therefore not too restrictive to focus on the $d = 1$ case in the time horizon $[0, T]$.

The Hamilton equations are compatible with two behaviors: a "push-region" where the running Lagrangian map variable evolves with constant acceleration:

$$\ddot{\chi}_t = \frac{K}{T^2} \, \text{sgn} \, \theta_t \qquad \& \qquad \theta_t \neq 0$$

and a "no-action" region specified by the conditions:

$$\theta_t = 0 \qquad \& \qquad -\eta_\star + 2\beta v_\star = 0 \tag{25}$$

where χ_t follows a free streaming trajectory:

$$\dot{\chi}_t = v_\star$$

We call switching times the values of t corresponding to the boundary values of a no-action region. Switching times correspond to discontinuities of the acceleration α_t. Drawing from the intuition offered by the solution of the unbounded acceleration case, we compose push and no-action regions to construct a single solution trajectory satisfying the boundary conditions. If we surmise that during the control horizon, only two switching times occur, we obtain:

$$v_t = \begin{cases} \dfrac{K}{T^2} t \, \text{sgn} \, \theta_0 & t \in [0, t_1) \\[2mm] \dfrac{K \, t_1}{T^2} \, \text{sgn} \, \theta_0 & t \in [t_1, t_2] \\[2mm] \dfrac{K}{T^2} \left(t_1 \, \text{sgn} \, \theta_0 + (t - t_2) \, \text{sgn} \, \theta_T \right) & t \in (t_2, T] \end{cases} \tag{26}$$

which implies:

$$\theta_t = \begin{cases} \theta_0 - \dfrac{\beta \, K \, t \, (2 \, t_1 - t)}{T^2} \, \text{sgn} \, \theta_0 & t \in [0, t_1) \\[2mm] 0 & t \in [t_1, t_2] \\[2mm] \dfrac{K \, (t - t_2)^2}{T^2} \, \text{sgn} \, \theta_T & t \in (t_2, T] \end{cases} \tag{27}$$

The self-consistence of the solution fixes the initial data in (27):

$$\theta_0 = \frac{\beta K t_1^2}{T^2} \, \text{sgn} \, \theta_0$$

whilst the requirement of vanishing velocity at $t = T$ determines the relation between the switching times:

$$t_2 = T + \frac{\text{sgn} \, \theta_0}{\text{sgn} \, \theta_T} t_1$$

Self-consistence then dictates:

$$\text{sgn} \, \theta_{t_f} = -\, \text{sgn} \, \theta_{t_0}$$

We are now ready to glean the information we unraveled by solving (24), to write the solution of (18a):

$$\chi_t = q + \begin{cases} \dfrac{K t^2}{2 T^2} \, \text{sgn} \, \theta_0 & t \in [0, t_1) \\[2mm] \dfrac{K t_1 \, (2 t - t_1)}{2 T^2} \, \text{sgn} \, \theta_0 & t \in [t_1, T - t_1] \\[2mm] K \dfrac{2 t_1 \, (T - t_1) - (T - t)^2}{2 T^2} \, \text{sgn} \, \theta_0 & t \in (T - t_1, T] \end{cases} \tag{28}$$

The terminal condition on χ_t fixes the values of t_1 and $\text{sgn} \, \theta_{t_0}$:

$$\ell(q) = q + \frac{K(q) \, t_1 \, (T - t_1)}{T^2} \, \text{sgn} \, \theta_{t_0}$$

The equation for t_1 is well posed only if:

$$\text{sgn} \, \theta_{t_0} = \text{sgn} \left(\ell(q) - q \right) \tag{29}$$

The only admissible solution is then of the form:

$$t_1 = \frac{T}{2} \left(1 - \sqrt{1 - 4\delta} \right) \tag{30}$$

The switching time is independent of q in view of (22). It is realizable as long as:

$$\delta = \frac{|\ell(q) - q|}{K(q)} \leq \frac{1}{4} \qquad \forall q \in \mathbb{R} \tag{31}$$

The threshold value of δ corresponds to the acceleration needed to construct an optimal protocol consisting of two push regions matched at the half control horizon.

Qualitative Properties of the Solution

Equation (28) complemented by (29) and the realizability bound (31) fully specify the solution of the optimization problem we set out to solve. The solution is optimal because it is obtained by composing locally-optimal solutions for a Markovian dynamics. Qualitatively, it states that transitions between equilibrium states are possible at the price of the formation of symmetric boundary layers determined by the occurrence of the switching times. For $\delta \ll 1$, the relative size of the boundary layers is:

$$\frac{t_1}{T} = \frac{T - t_2}{T} \approx \delta$$

In the same limit, the behavior of the current velocity far from the boundaries tends to the optimal value of the refined second law [7]. Namely, for $t \in [t_1, t_f]$, we find:

$$\frac{\dot{K}(q)\, t_1}{T^2} \, \text{sgn}\left(\ell(q) - q\right) \overset{\delta \ll 1}{\approx} \frac{K(q)\, \delta}{T} \, \text{sgn}\left(\ell(q) - q\right) = \frac{\ell(q) - q}{T}$$

More generally, for any $0 \le t_1 \le T/2$, we can couch (28) into the form:

$$\chi_t = q + \left(\ell(q) - q\right) \times \begin{cases} \dfrac{t^2}{2\, t_1\, (T - t_1)} & t \in [0, t_1) \\[2mm] \dfrac{2\, t - t_1}{2\, (T - t_1)} & t \in [t_1, T - t_1] \\[2mm] \left(1 - \dfrac{(T - t)^2}{2\, t_1\, (T - t_1)}\right) & t \in (T - t_1, T] \end{cases} \tag{32}$$

The use of the value of the switching time t_1 to parametrize the bound simplifies the derivation of the Eulerian representation of the current velocity. Namely, in order to find the field $v \colon \mathbb{R} \times [0, T] \mapsto \mathbb{R}$ satisfying:

$$v_t = v(\chi_t, t) \tag{33}$$

We can invert (32) by taking advantage of the fact that all of the arguments of the curly brackets are independent of the position variable q.

We also envisage that the representation (32) may be of use to analyze experimental data when finite measurement resolution may affect the precision with which microscopic forces acting on the system are known.

6. Comparison with Experimental Swift Engineering Protocols

The experiment reported in [4] showed that a micro-sphere immersed in water and trapped in an optical harmonic potential can be driven in finite-time from one equilibrium state to another. The probability distribution of the particle in and out of equilibrium remained Gaussian within the experimental accuracy.

It is therefore expedient to describe more in detail the solution of the optimal control problem in the case when the initial equilibrium distribution in one dimension is normal, i.e., Gaussian with zero mean and variance β^{-1}. We also assume that the final equilibrium state is Gaussian and satisfies (13) with Lagrangian map:

$$\ell(q) = \sigma\, q + h$$

The parameters h and σ respectively describe a change of the mean and of the variance of the distribution. We apply (13) and (32) for any $t \in [0, T]$ to derive the minimum entropy production evolution of the probability density. As a consequence of (22), the running Lagrangian map leaves Gaussian distributions invariant in form with mean value:

$$
\mathrm{E}\,\xi_t = h \times
\begin{cases}
\dfrac{t^2}{2\,t_1\,(T-t_1)} & t \in [0,t_1) \\[2ex]
\dfrac{(2\,t-t_1)}{2\,(T-t_1)} & t \in [t_1, T-t_1] \\[2ex]
\dfrac{2\,t_1\,(T-t_1)-(T-t)^2}{2\,t_1\,(T-t_1)} & t \in (T-t_1, T]
\end{cases}
\tag{34}
$$

and variance:

$$
\mathrm{V}\,\xi_t =
\begin{cases}
\dfrac{\left(2\,t_1\,(T-t_1)+(\sigma-1)\,t^2\right)^2}{4\,\beta\,t_1^2\,(T-t_1)^2} & t \in [0,t_1) \\[3ex]
\dfrac{\left(2\,(T-t_1)+(\sigma-1)(2\,t-t_1)\right)^2}{4\,\beta\,(T-t_1)^2} & t \in [t_1, T-t_1] \\[3ex]
\dfrac{\left(2\,t_1\,(T-t_1)+(\sigma-1)(2\,t_1\,(T-t_1)-(T-t)^2)\right)^2}{4\,\beta\,t_1^2\,(T-t_1)^2} & t \in (T-t_1, T]
\end{cases}
\tag{35}
$$

Finally, we find that the Eulerian representation (33) of the current velocity at $\chi_t = q$ is:

$$
v(q,t) =
\begin{cases}
\dfrac{2\,t\,(h+q\,(\sigma-1))}{2\,t_1\,(T-t_1)+(\sigma-1)\,t^2} & t \in [0,t_1) \\[3ex]
\dfrac{2\left(h+q(\sigma-1)\right)}{2\,(T-t_1)+(\sigma-1)\,(2\,t-t_1)} & t \in [t_1, T-t_1] \\[3ex]
\dfrac{2(T-t)\left(h+q(\sigma-1)\right)}{2\,t_1\,(T-t_1)+(\sigma-1)(2\,t_1\,(T-t_1)-(T-t)^2)} & t \in (T-t_1, T]
\end{cases}
\tag{36}
$$

From (34)–(36), we can derive explicit expressions for all of the thermodynamic quantities governing the energetics of the optimal transition. In particular, we obtain the drift in the Langevin–Smoluchowski dynamics (1) by inverting (8) as in [7]:

$$
(\partial_q U)(q,t) = -\,v(q,t) + \frac{q-\mathrm{E}\,\xi_t}{\mathrm{V}\,\xi_t}
$$

The minimum entropy production is:

$$
\mathrm{E}\int_0^{t_f} \mathrm{d}t\,v^2(\xi_t,t) = \frac{T\,(3\,T-4\,t_1)}{3\,(T-t_1)^2}\,\mathcal{E}_\infty
\tag{37}
$$

with:

$$
\mathcal{E}_\infty = \frac{h^2\,\beta+(\sigma-1)^2}{\beta\,T}
\tag{38}
$$

the value of the minimum entropy production appearing in the refinement of the second law [7].

In Figure 1, we plot the evolution of the running average values of the work done on the system, the heat released and the entropy production during the control horizon. In particular, Figure 1a illustrates the first law of thermodynamics during the control horizon. A transition between Gaussian equilibrium states occurs without any change in the internal energy of the system. The average heat and work must therefore coincide at the end of the control horizon. The theoretical results are consistent with the experimental results of [4].

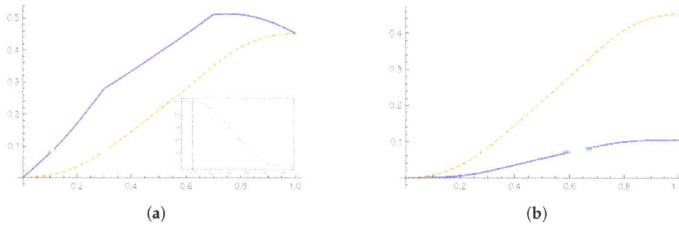

(a)   (b)

Figure 1. First Figure 1a and second law Figure 1b of thermodynamics for the same transition between Gaussian states as in [4]. The initial state is a normal distribution with variance β^{-1}. The final distribution is Gaussian with variance $\beta^{-1}/2$. The condition $K(q) \propto |\ell(q) - q|$ ensures that the probability density remains Gaussian at any time in the control horizon $t \in [0, 1]$. The proportionality factor is chosen such that $t_1 = 0.3$ in (32). The behavior of the variance (inset of Figure 1a) is qualitatively identical to the one observed in [4] (Figure 2). The behavior of the average work and heat also reproduces the one of Figure 3 of [4]. (**a**) Work (continuous curve, blue on-line) and heat release (dashed curve, yellow on-line) during the control horizon. Inset: time evolution of the variance of the process; (**b**) Entropy production (continuous curve, blue on-line) and heat release (dashed curve, yellow on-line) during the control horizon.

7. Optimally-Controlled Nucleation and Landauer Bound

The form of the bound (22) and running Lagrangian map Formula (32) reduce the computational cost of the solution of the optimal entropy production control to the determination of the Lagrangian map (13). In general, the conditions presiding over the qualitative properties of the Lagrangian map have been studied in depth in the context of optimal mass transport [8]. We refer to [11,41] respectively for a self-contained overview from respectively the mathematics an physics slant.

For illustrative purposes, we revisit here the stylized model of nucleation analyzed in [7]. Specifically, we consider the transition between two equilibria in one dimension. The initial state is described by the symmetric double well:

$$p_\iota(q) = Z_\iota^{-1} \exp -\beta \frac{(q^2 - \bar{q}^2)^2}{\sigma^2}$$

In the final state, the probability is concentrated around a single minimum of the potential:

$$p_f(q) = Z_f^{-1} \exp -\beta \frac{(q - \bar{q})^2 \left((q - \bar{q}) + \bar{q}\,(3\,q - \bar{q})\right)}{\sigma^2}$$

In the foregoing expressions, σ is a constant ensuring the consistency of the canonical dimensions.

We used the ensuing elementary algorithm to numerically determine the Lagrangian map. We first computed the median $z(1)$ of the assigned probability distributions and then evaluated first the left and then right branch of the Lagrangian map. For the left branch, we proceeded iteratively in $z(k)$ as follows:

Step 1   We renormalized the distribution restricted to $[-\infty, z(k)]$.
Step 2   We computed the 0.9 quantile $z(k+1) < z(k)$ of the remaining distribution.
Step 3   We solved the ODE:

$$\frac{d\ell}{dq} = \frac{p_\iota(q)}{p_f(\ell(q))}$$

We skipped Step 3 whenever the difference $|z(k) - z(k-1)|$ turned out to be smaller than a given threshold 'resolution'. We plot the results of this computation in Figure 2.

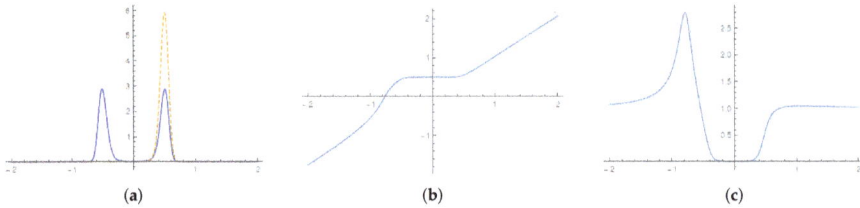

Figure 2. Initial (solid curve, blue on-line) and final (dashed curve, blue on-line) probability distribution of the state of the system for $\beta = 112$ $\sigma = 1$ and $\bar{q} = 1/2$. The evaluation of the Lagrangian map occasions numerical stiffness in the region in between the two minima. (**a**) Boundary conditions for the nucleation problem; (**b**) Lagrangian map; (**c**) Numerical derivative of the Lagrangian map.

Once we know the Lagrangian map, we can numerically evaluate the running Lagrangian map (32) and its spatial derivatives. In Figure 3, we report the evolution of the probability density in the control horizon for two reference values of the switching time.

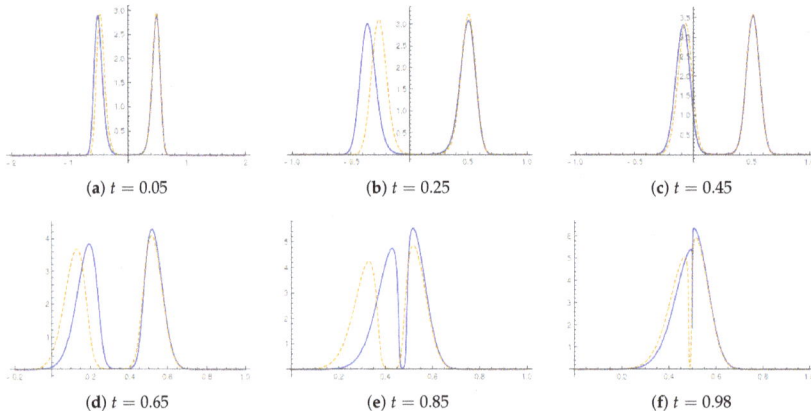

Figure 3. Probability density snapshots at different times within the control horizon. The plots are for $T = 1$ and switching time $t_1 = 10^{-6}$ (dashed interpolation curve, yellow on-line) and $t_1 = 0.3$ (continuous interpolation curve, blue on-line) $\bar{q} = 0.5$, $\sigma = 1$ and $\beta = 112$. We plot the Lagrangian map in the interval $q \in [-2, 2]$.

Figure 4 illustrates the the corresponding evolution of the current velocity.

The qualitative behavior is intuitive. The current velocity starts and ends with a vanishing value; it catches up with the value for $t_1 \downarrow 0$, i.e., when the bound on acceleration tends to infinity, in the bulk of the control horizon. There, the displacement described by the running Lagrangian map occurs at a speed higher than in the $t_1 \downarrow 0$ case. The overall value of the entropy production is always higher than in the $t_1 \downarrow 0$ limit. From (32), we can also write the running values of average heat released by the system. The running average heat is:

$$Q(t) = -\frac{1}{\beta} \int_{\mathbb{R}} d^d q\, p_\iota(q) \ln \frac{d\chi_t(q)}{dq} + \int_0^t ds \int_{\mathbb{R}} d^d q\, p_\iota(q)\, v_t^2(q)$$

and the running average work:

$$W(t) = \int_{\mathbb{R}} dq\, p_{\iota}(q)\, F(\chi_t(q), t) + \int_0^t ds \int_{\mathbb{R}} d^d q\, p_{\iota}(q)\, v_t^2(q)$$

with:

$$F(\chi_t(q), t) = -\int_0^q dy \frac{d\chi_t}{dy}(y)\, v_t(y) - \frac{1}{\beta} \int_{\mathbb{R}} d^d q\, p_{\iota}(q)\, \ln \frac{d\chi_t(q)}{dq} \tag{39}$$

The second summand on the right-hand side of (39) fixes the arbitrary constant in the Helmholtz potential in the same way as in the Gaussian case.

In Figure 5, we plot the running average work, heat and entropy production.

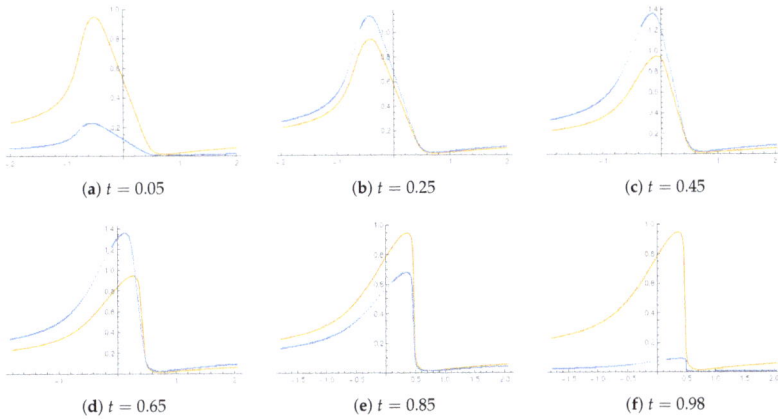

(a) $t = 0.05$ (b) $t = 0.25$ (c) $t = 0.45$

(d) $t = 0.65$ (e) $t = 0.85$ (f) $t = 0.98$

Figure 4. Current velocity snapshots at different times within the control horizon. The plots are for $T = 1$ and switching time $t_1 = 10^{-6}$ (continuous interpolation, yellow on-line) and $t_1 = 0.3$ (points, blue on-line).

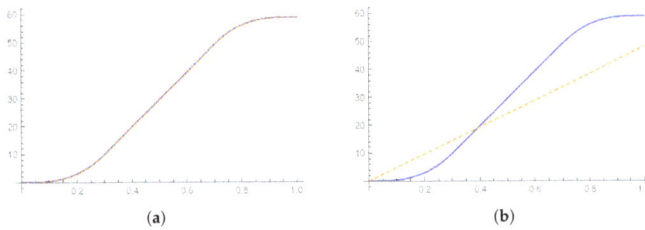

(a)   (b)

Figure 5. First and second law of thermodynamics for the optimally-controlled nucleation transition. All parameters are as in Figure 2. The qualitative picture is the same as in the Gaussian case, Figure 1, with the running average work above the running average heat. The numerical values yield, however, almost overlapping curves. The running average entropy production in Figure 5b is strictly monotonic in the control horizon. The entropy production rate vanishes at the boundary highlighting the reaching of an equilibrium state when the switching time is $t_1 = 0.3$. (a) First law of thermodynamics for the optimally-controlled nucleation. Continuous curve (blue on-line) running average work. Dashed curve (yellow on-line) running average heat; (b) Running average entropy production. The continuous curve (blue on-line) is obtained for switching time at $t_1 = 0.3$, the dashed curve (yellow on-line) for $t_1 = 10^{-6}$.

8. Comparison with the Valley Method Regularization

An alternative formalism to study transitions between equilibrium states in the Langevin–Smoluchowski limit was previously proposed in [25]. As in the present case, Ref. [25] takes advantage of the possibility to map the stochastic optimal control problem into a deterministic one via the current velocity formalism. Physical constraints on admissible controls are, however, enforced by adding to the entropy production rate a penalty term proportional to the squared current acceleration. In terms of the entropy production functional (17), we can couch the regularized functional of [25] into the form:

$$\mathcal{A} = \mathcal{E} + \varepsilon\,\tau^2\,\|\delta_\chi\mathcal{E}\|^2$$

$\delta_\chi\mathcal{E}$ stands for the variation of $\mathcal{E}$ with respect to the running Lagrangian map. The idea behind the approach is the "valley method" advocated by [26] for instanton calculus. The upshot is to approximate field configurations satisfying boundary conditions incompatible with stationary values of classical variational principles by adding extra terms to the action functional. The extra term is proportional to the squared first variation of the classical action. Hence, it vanishes whenever there exists a classical field configurations matching the desired boundary conditions. It otherwise raises the order of the time derivative in the problem, thus permitting one to satisfy extra boundary conditions.

Optimal control problems are well posed if terminal costs are pure functionals of the boundary conditions. The rationale for considering valley method-regularized thermodynamic functionals is to give a non-ambiguous meaning to the optimization of functionals whenever naive formulations of the problem yield boundary conditions or terminal costs as the functional of the controls.

Contrasted with the approach proposed in the present work, [25] has one evident drawback and one edge. The drawback is that the quantities actually minimized are no longer the original thermodynamic functionals. The edge is that the resulting optimal protocol has better analyticity properties. In particular, the running Lagrangian map takes the form:

$$\chi_t = q + \frac{\ell(q) - q}{T - 2\,\tau\,\sqrt{\varepsilon}\,\tanh\frac{T}{2\tau\sqrt{\varepsilon}}}\left(t - \tau\,\sqrt{\varepsilon}\,\frac{\sinh\frac{2t-T}{2\tau\sqrt{\varepsilon}} + \sinh\frac{T}{2\tau\sqrt{\varepsilon}}}{\cosh\frac{T}{2\tau\sqrt{\varepsilon}}}\right) \tag{40}$$

In Figure 6a, we compare the qualitative behavior of the universal part of the running Lagrangian map predicted by the valley method and by the bound (21) on admissible current accelerations. The corresponding values of the running average entropy production are in Figure 6b.

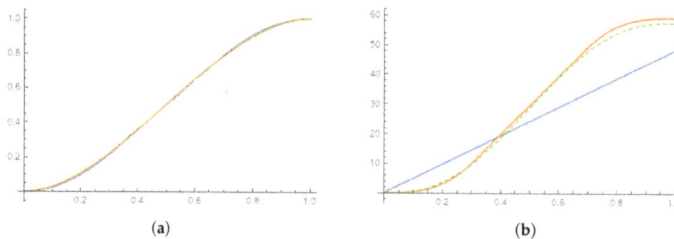

(a)	(b)

Figure 6. Qualitative comparison of universal part of the running Lagrangian maps (32) (continuous curve, blue on line) and (40) (dashed curve, orange on line), Figure 6a. In (40), we choose $\tau = 1$, $\varepsilon = 0.3$. Figure 6b evinces, as to be expected, the qualitatively equivalent behaviors of the entropy production for finite value ($t_1 = 0.3$) of the switching time. The dashed green line is computed from (40). The continuous blue line is the lower bound for the transition as predicted by [7]. (**a**) $\frac{\chi_t - q}{\ell(q) - q}$; (**b**) Running entropy production.

The upshot of the comparison is the weak sensitivity of the optimal protocol to the detail of the optimization once the intensity of the constraint on the admissible control (i.e., the current acceleration) is fixed. We believe that this is an important observation for experimental applications (see, e.g., the discussion in the conclusions of [24]), as the details of how control parameters can be turned on and off in general depend on the detailed laboratory setup and on the restrictions by the available peripherals.

9. Conclusions and Outlooks

We presented a stylized model of engineered equilibration of a micro-system. Owing to explicit integrability modulo numerical reconstruction of the Lagrangian map, we believe that our model may provide a useful benchmark for the devising of efficient experimental setups. Furthermore, extensions of the current model are possible, although at the price of some complications.

The first extension concerns the form of the constraint imposed on admissible protocols. Here, we showed that choosing the current acceleration constraint in the form of (22) greatly simplifies the determination of the switching times. It also guarantees that optimal control with only two switching times exists for all boundary conditions if we allow accelerations to take sufficiently large values. The non-holonomic form of the constraint (21) may turn out to be restrictive for the study of transitions for which admissible controls are specified by given forces. If the current velocity formalism is still applicable to these cases, then the design of optimal control still follows the steps we described here. In particular, uniformly-accelerated Lagrangian displacement at the end of the control horizon correspond to the first terms of the integration of the Newton law in Peano–Picard series. The local form of the acceleration may then occasion some qualitative differences in the form of the running Lagrangian map. Furthermore, the analysis of the realizability conditions of the optimal control may also become more involved.

A further extension is optimal control when constraints on admissible controls are imposed directly on the drift field appearing in the stochastic evolution equation. Constraints of this type are natural when inertial effects become important and the dynamics is governed by the Langevin–Kramers equation in the so-called under-damped approximation. In the Langevin–Kramers framework, finding minimum entropy production thermodynamic transitions requires instead a full-fledged formalism of stochastic optimal control [42]. Nevertheless, it is possible also in that case to proceed in a way analogous to the one of the present paper by applying the stochastic version of the Pontryagin principle [43–45].

We expect that considering these theoretical refinements will be of interest in view of the increasing available experimental resolution for the efficient design of atomic force microscopes [3,46].

Acknowledgments: The authors thank Sergio Ciliberto for useful discussions. The work of KS was mostly performed during his stay at the department of Mathematics and Statistics of the University of Helsinki. P.M.-G. acknowledges support from Academy of Finland via the Centre of Excellence in Analysis and Dynamics Research (Project No. 271983) and the AtMath Collaboration at the University of Helsinki http://wiki.helsinki.fi/display/AtMath/Atmospheric+Mathematics .

Author Contributions: Both authors made substantial contributions to the paper.

Conflicts of Interest: The authors declare no conflict of interest.

Appendix A. Evaluation of Kullback–Leibler Divergences

Let us consider first the drift-less process:

$$\mathrm{d}\xi_t = \sqrt{\frac{2}{\beta}}\,\mathrm{d}\omega_t \tag{A1}$$

with initial data (2). If we denote by P_ω the path-space Wiener measure generated by (A1) in $[t_i, t_f]$, the Girsanov formula yields:

$$\frac{dP}{dP_\omega} = \exp{-\frac{\beta}{2} \int_{t_i}^{t_f} \left(d\xi_t \cdot \partial_{\xi_t} U + dt \frac{\|\partial_{\xi_t} U\|^2}{2} \right)}$$

The Kullback–Leibler divergence is defined as:

$$\mathcal{K}(P||P_\omega) = E \int_{t_i}^{t_f} \ln \frac{dP}{dP_\omega}$$

The expectation value is with respect the measure P generated by (1):

$$\mathcal{K}(P \parallel P_\omega) = -\frac{\beta}{2} E \int_{t_i}^{t_f} \left(d\xi_t \cdot \partial_{\xi_t} U + dt \frac{\|\partial_{\xi_t} U\|^2}{2} \right)$$

$$= -\frac{\beta}{2} E \int_{t_i}^{t_f} \left((d\xi_t + dt \partial_{\xi_t} U) \cdot \partial_{\xi_t} U - dt \frac{\|\partial_{\xi_t} U\|^2}{2} \right)$$

The last expression readily recovers (14) as $d\xi_t + dt\,\partial_{\xi_t} U$ is a Wiener process with respect to P.

To show that the entropy production is proportional to the Kullback–Leibler divergence between the path-space measures of (1) and (16), we observe that:

$$\frac{dP_R}{dP_\omega} = \exp{\frac{\beta}{2} \int_{t_i}^{t_f} \left(d\xi_t \cdot^{1} \partial_{\xi_t} U - dt \frac{\|\partial_{\xi_t} U\|^2}{2} \right)} \tag{A2}$$

The stochastic integral is evaluated in the post-point prescription, as the Radon–Nikodym derivative between backward processes must be a martingale with respect to the filtration of future event (see, e.g., [47] for an elementary discussion). We then avail ourselves of the time reversal invariance of the Wiener process to write:

$$\frac{dP}{dP_R} = \frac{p_i(\xi_{t_i})}{p_f(\xi_{t_f})} \exp{-\beta \int_{t_i}^{t_f} \left(d\xi_t \cdot^{1/2} \partial_{\xi_t} U \right)}$$

$$= \exp{-\beta \int_{t_i}^{t_f} \left(d\xi_t \cdot^{1/2} \partial_{\xi_t} \left(U + \frac{1}{\beta} \ln p \right) + dt\,\partial_t \ln p \right)}$$

Finally, the definition:

$$\mathcal{K}(P \parallel P_R) = E \int_{t_i}^{t_f} \ln \frac{dP}{dP_R}$$

recovers (15) since the probability conservation entails:

$$E\,\partial_t \ln p = 0$$

whilst the properties of the Stratonovich integral [31] yield:

$$E \int_{t_i}^{t_f} d\xi_t \cdot^{1/2} \partial_{\xi_t} \left(U + \frac{1}{\beta} \ln p \right) = -E \int_{t_i}^{t_f} dt\,\|v\|^2$$

We refer to, e.g., [28,38,39,48] for thorough discussions of the significance and applications of the entropy production in stochastic models of non-equilibrium statistical mechanics and to [49,50] for applications to non-equilibrium fluctuating hydrodynamics and granular materials.

Appendix B. Current Velocity and Acceleration in Terms of the Generator of the Stochastic Process

The current velocity is the conditional expectation along the realizations of (1) of the time symmetric conditional increment:

$$v(q,t) = \lim_{\tau \downarrow 0} \frac{E\left(\xi_{t+\tau} - \xi_{t-\tau}\Big|\xi_t = q\right)}{2\tau}$$

A relevant feature of the time symmetry is that the differential can be regarded as the result of the action of a generator including only first order derivatives in space:

$$v(\xi_t, t) = \bar{D}_{\xi_t} \xi_t$$

where:

$$\bar{D}_{\xi_t} := \frac{D_{\xi_t} + D^*_{\xi_t}}{2} \tag{A3}$$

On the right-hand side of (A3), there appear the scalar generator of (1):

$$D_q = \partial_t - (\partial_q U)(q,t) \cdot \partial_q + \frac{1}{\beta}\partial_q^2$$

and the generator of the dual process conjugated by the time-reversal of the probability density in $[t_i, t_f]$ [29,31]:

$$D_q^* = \partial_t - (\partial_q U + \frac{2}{\beta}\partial_q \ln p)(q,t) \cdot \partial_q - \frac{1}{\beta}\partial_q^2$$

The arithmetic averages of these generators readily define a first order differential operator as in the deterministic case. Analogously, we define the current acceleration as:

$$a(q,t) = \lim_{\tau \downarrow 0} \frac{E\left(v(\xi_{t+\tau}, t+\tau) - v(\xi_{t-\tau}, t-\tau)\Big|\xi_t = q\right)}{2\tau}$$

or equivalently:

$$\alpha_t = a(\xi_t, t) = \bar{D}^2_{\xi_t} \xi_t$$

Based on the above definitions, the Fokker–Planck Equation of (1) can be couched into the form:

$$\left(\partial_t + \partial_q \cdot v(q,t)\right) p(q,t) = 0$$

Appendix C. Pontryagin Principle

We recall the statement of Pontryagin's principle for fixed time and fixed boundary conditions [13,51].

Maximum principle: Let the functional:

$$\mathcal{A} = \int_{t_i}^{t_f} dt\, L(\xi_t, \alpha_t, t) \tag{A4}$$

be subject to the dynamical constraint:

$$\dot{\xi}_t = b(\xi_t, \alpha_t, t) \tag{A5}$$

and the endpoint constraints:

$$\xi_{t_i} = q_i \qquad \& \qquad \xi_{t_f} = q_f$$

with the parameter α_t belonging for fixed t to a set $U \subseteq \mathbb{R}^n$, the variable ξ_t taking values in $\mathbb{R}^d$ or in a open subset X of $\mathbb{R}^d$ and the time interval $[t_i, t_f]$ fixed. A necessary condition for a function $\bar{\alpha}_t \colon [t_i, t_f] \mapsto U$ and a corresponding solution $\bar{\xi}_t$ of (A5) to solve the minimization of (A4) is that there exist a function $t\bar{\pi}_t \colon [t_i, t_f] \mapsto \mathbb{R}^d$ and a constant $p_0 \leq 0$, such that:

- $(\bar{\pi}_t, \bar{p}_0) \neq (0, 0) \; \forall t \in [t_i, t_f]$ (non-triviality condition)
- for each fixed t:

$$H_\star(q, p, p_0 t) = \max_{a \in U} \left(p \cdot b(q, a, t) + p_0 L(q, a, t) \right)$$

(maximum condition)
- $(\bar{\xi}_t, \bar{\pi}_t)$ obey the equations:

$$\dot{\bar{\xi}}_t = \partial_{\bar{\pi}_t} H_\star(\bar{\xi}_t, \bar{\pi}_t \cdot \bar{p}_0, t) \qquad \& \qquad \dot{\bar{\pi}}_t = -\partial_{\bar{\xi}_t} H_\star(\bar{\xi}_t, \bar{\pi}_t, \bar{p}_0, t)$$

(Hamilton system condition).

The proof of the maximum principles requires subtle topological considerations culminating with the application of Brouwer's fixed point theorem. The maximum principle has, nevertheless, an intuitive content. Namely, we can reformulate the problem in an extended configuration space by adding the ancillary equation:

$$\dot{\zeta}_t = L(\xi_t, \pi_t, t) \tag{A6a}$$

$$\zeta_{t_i} = 0 \tag{A6b}$$

and looking for the stationary point of the action functional:

$$\tilde{A} = \zeta_{t_f} + \int_{t_i}^{t_f} dt \left(\pi_t \cdot \dot{\xi}_t + \phi_t \dot{\zeta}_t - \left(\pi_t \cdot b(\xi_t, \alpha_t, t) + \phi_t L(\xi_t, \alpha_t, t) \right) \right)$$

Let us make the simplifying assumption that any pair of trajectory and control variables satisfying the boundary has a non-empty open neighborhood where linear variations are well defined. Looking for a stationary point of (A4) entails considering variations of ζ_t under the constraints $\zeta'_{t_i} = \zeta'_{t_f} = 0$. Then, it follows immediately that the stationary value of the Lagrange multiplier ϕ_t must satisfy:

$$\dot{\phi}_t = 0$$

This observation clarifies why the maximum principle is stated for some constant $p_0 \leq 0$, such that $\phi_t = p_0$. In particular, if $p_0 < 0$, we can always rescale it to $p_0 = -1$ and recover the familiar form of the Hamilton equations. Moreover, the maximum principle coincides with the Hamilton form of the stationary action principle if $b = \alpha_t$ and L is quadratic in α_t. If instead, there exist stationary solutions for $p_0 = 0$, they describe abnormal controls.

Abnormal controls do not occur in the optimization problem considered in the main text. In the push regions where the acceleration is non-vanishing abnormal control drive the Lagrange multiplier

θ_t away from zero, thus, they are not compatible with the occurrence of switching times between push and no-action regions. Looking for abnormal control in the no-action region yields the requirement that all Lagrange multipliers vanish against the hypothesis of the maximum principle.

References

1. Blickle, V.; Bechinger, C. Realization of a micrometre-sized stochastic heat-engine. *Nat. Phys.* **2011**, *8*, 143–146.
2. Roßnagel, J.; Abah, O.; Schmidt-Kaler, F.; Singer, K.; Lutz, E. Nanoscale heat engine beyond the carnot limit. *Phys. Rev. Lett.* **2014**, *112*, 030602.
3. Liang, S.; Medich, D.; Czajkowsky, D.M.; Sheng, S.; Yuan, J.Y.; Shao, Z. Thermal noise reduction of mechanical oscillators by actively controlled external dissipative forces. *Ultramicroscopy* **2000**, *84*, 119–125.
4. Martínez, I.A.; Petrosyan, A.; Guéry-Odelin, D.; Trizac, E.; Ciliberto, S. Engineered swift equilibration of a Brownian particle. *Nat. Phys.* **2016**, *12*, 843–846.
5. Trepagnier, E.H.; Jarzynski, C.; Ritort, F.; Crooks, G.E.; Bustamante, C.J.; Liphardt, J. Experimental test of Hatano and Sasa's nonequilibrium steady-state equality. *PNAS* **2004**, *101*, 15038–15041.
6. Jacobs, K. *Stochastic Processes for Physicists: Understanding Noisy Systems*; Cambridge University Press: Cambridge, UK, 2010.
7. Aurell, E.; Gawędzki, K.; Mejía-Monasterio, C.; Mohayaee, R.; Muratore-Ginanneschi, P. Refined Second Law of Thermodynamics for fast random processes. *J. Stat. Phys.* **2012**, *147*, 487–505.
8. Villani, C. *Optimal Transport: Old and New*; Grundlehren der mathematischen Wissenschaften; Springer: Berlin, Germany, 2009.
9. Benamou, J.D.; Brenier, Y. A computational fluid mechanics solution to the Monge-Kantorovich mass transfer problem. *Numer. Math.* **2000**, *84*, 375–393.
10. Brenier, Y.; Frisch, U.; Hénon, M.; Loeper, G.; Matarrese, S.; Mohayaee, R.; Sobolevskiĭ, A. Reconstruction of the early Universe as a convex optimization problem. *Mon. Not. R. Astron. Soc.* **2003**, *346*, 501–524.
11. De Philippis, G.; Figalli, A. The Monge–Ampère equation and its link to optimal transportation. *Bull. Amer. Math. Soc.* **2014**, *51*, 527–580.
12. Alemany, A.; Ribezzi, M.; Ritort, F. Recent progress in fluctuation theorems and free energy recovery. *AIP Conf. Proc.* **2011**, *1332*, 96–110.
13. Liberzon, D. *Calculus of Variations and Optimal Control Theory: A Concise Introduction*; Princeton University Press: Princeton, NJ, USA, 2012.
14. Sekimoto, K. Langevin equation and thermodynamics. *Progr. Theor. Phys. Suppl.* **1998**, *130*, 17–27.
15. Schrödinger, E. Über die umkehrung der naturgesetze. *Sitzungsberichte der Preussischen Akademie der Wissenschaften, Physikalische Mathematische Klasse* **1931**, *8*, 144–153. (In German)
16. Aebi, R. *Schrödinger Diffusion Processes*; Probability and Its Applications; Birkhäuser: Basel, Switzerland, 1996; p. 186.
17. Muratore-Ginanneschi, P. On the use of stochastic differential geometry for non-equilibrium thermodynamics modeling and control. *J. Phys. A* **2013**, *46*, 275002.
18. Arnaudon, M.; Cruzeiro, A.B.; Léonard, C.; Zambrini, J.C. An entropic interpolation problem for incompressible viscid fluids. *arXiv* **2017**, arXiv:1704.02126.
19. Landauer, R. Irreversibility and heat generation in the computing process. *IBM J. Res. Dev.* **1961**, *5*, 183–191.
20. Bennett, C.H. The thermodynamics of computation—A review. *Int. J. Theor. Phys.* **1982**, *21*, 905–940.
21. Dillenschneider, R.; Lutz, E. Memory erasure in small systems. *Phys. Rev. Lett.* **2009**, *102*, 210601.
22. Bérut, A.; Arakelyan, A.; Petrosyan, A.; Ciliberto, S.; Dillenschneider, R.; Lutz, E. Experimental verification of Landauer's principle linking information and thermodynamics. *Nature* **2012**, *483*, 187–189.
23. Koski, J.V.; Maisi, V.F.; Pekola, J.P.; Averin, D.V. Experimental realization of a Szilard engine with a single electron. *Proc. Natl. Acad. Sci. USA* **2014**, *111*, 13786–13789.
24. Jun, Y.; Gavrilov, M.; Bechhoefer, J. High-precision test of Landauer's principle in a feedback trap. *Phys. Rev. Lett.* **2014**, *113*, 190601.
25. Aurell, E.; Mejía-Monasterio, C.; Muratore-Ginanneschi, P. Boundary layers in stochastic thermodynamics. *Phys. Rev. E* **2012**, *85*, 020103(R).
26. Aoyama, H.; Kikuchi, H.; Okouchi, I.; Sato, M.; Wada, S. Valley views: Instantons, large order behaviors, and supersymmetry. *Nucl. Phys. B* **1999**, *553*, 644–710.

27. Zwanzig, R. *Nonequilibrium Statistical Mechanics*; Oxford University Press: New York, NY, USA, 2001; p. 240.
28. Lebowitz, J.L.; Spohn, H. A Gallavotti-Cohen Type Symmetry in the large deviation functional for stochastic dynamics. *J. Stat. Phys.* **1999**, *95*, 333–365.
29. Nelson, E. *Dynamical Theories of Brownian Motion*, 2nd ed.; Princeton University Press: Princeton, NJ, USA, 2001; p. 148.
30. Fényes, I. Eine wahrscheinlichkeitstheoretische Begründung und Interpretation der Quantenmechanik. *Z. Phys.* **1952**, *132*, 81–106. (In German)
31. Nelson, E. *Quantum Fluctuations*; Princeton Series in Physics; Princeton University Press: Princeton, NJ, USA, 1985; p. 146.
32. Qian, H. Mesoscopic nonequilibrium thermodynamics of single macromolecules and dynamic entropy-energy compensation. *Phys. Rev. E* **2001**, *65*, 016102.
33. Shannon, C.E. A mathematical theory of communication. *Bell Syst. Tech. J.* **1948**, *27*, 379, 623–656.
34. Bertini, L.; Sole, A.D.; Gabrielli, D.; Jona-Lasinio, G.; Landim, C. Macroscopic fluctuation theory. *Rev. Mod. Phys.* **2015**, *87*, 593–636.
35. Dai Pra, P. A stochastic control approach to reciprocal diffusion processes. *Appl. Math. Optim.* **1991**, *23*, 313–329.
36. Roelly, S.; Thieullen, M. A characterization of reciprocal processes via an integration by parts formula on the path space. *Probab. Theory Relat. Fields* **2002**, *123*, 97–120.
37. Kullback, S.; Leibler, R. On information and sufficiency. *Ann. Math. Stat.* **1951**, *22*, 79–86.
38. Jiang, D.Q.; Qian, M.; Qian, M.P. *Mathematical Theory of Nonequilibrium Steady States*; Lecture Notes in Mathematics; Springer: Berlin, Germany, 2004; p. 276.
39. Chétrite, R.; Gawędzki, K. Fluctuation relations for diffusion processes. *Commun. Math. Phys.* **2008**, *282*, 469–518.
40. Jordan, R.; Kinderlehrer, D.; Otto, F. The variational formulation of the Fokker–Planck equation. *SIAM J. Math. Anal.* **1998**, *29*, 1–17.
41. Gawędzki, K. Fluctuation relations in stochastic thermodynamics. *arXiv* **2013**, arXiv:1308.1518.
42. Muratore-Ginanneschi, P.; Schwieger, K. How nanomechanical systems can minimize dissipation. *Phys. Rev. E* **2014**, *90*, 060102(R).
43. Bismut, J.M. An introductory approach to duality in optimal stochastic control. *SIAM Rev.* **1978**, *20*, 62–78.
44. Kosmol, P.; Pavon, M. Lagrange approach to the optimal control of diffusions. *Acta Appl. Math.* **1993**, *32*, 101–122.
45. Rogers, L.C.G. Duality in constrained optimal investment and consumption problems: A synthesis. In *Paris-Princeton Lectures on Mathematical Finance*; Bank, P.; Baudoin, F.; Carmona, R.; Föllmer, H.; Rogers, L.C.G.; Touzi, N.; Soner, M., Eds.; Springer: Berlin, Germany, 2003; Vol. 1814, pp. 95–131.
46. Cunuder, A.L.; Martinez, I.; Petrosyan, A.; Guéry-Odelin, D.; Trizac, E.; Ciliberto, S. Fast equilibrium switch of a micro mechanical oscillator. *Appl. Phys. Lett.* **2016**, *109*, 113502.
47. Meyer, P.A. Géométrie différentielle stochastique, II. *Séminaire de Probabilités de Strasbourg* **1982**, *16*, 165–207. (In German)
48. Maes, C.; Redig, F.; Moffaert, A.V. On the definition of entropy production, via examples. *J. Math. Phys.* **2000**, *41*, 1528–1554.
49. Gradenigo, G.; Puglisi, A.; Sarracino, A. Entropy production in non-equilibrium fluctuating hydrodynamics. *J. Phys. Chem.* **2012**, *137*, 014509.
50. Gradenigo, G.; Puglisi, A.; Sarracino, A.; Villamaina, D.; Vulpiani, A. Out-of-equilibrium generalized fluctuation-dissipation relations. In *Nonequilibrium Statistical Physics of Small Systems: Fluctuation Relations and Beyond*; Klages, R., Just, W., Jarzynski, C., Eds.; Wiley: Weinheim, Germany, 2013; Chapter 9.
51. Agrachev, A.A.; Sachkov, Y. *Control Theory from the Geometric Viewpoint*; Encyclopaedia of Mathematical Sciences: Control Theory and Optimization; Springer: Berlin/Heidelberg, Germany, 2004.

entropy

MDPI

Article

On Work and Heat in Time-Dependent Strong Coupling

Erik Aurell [1,2]

[1] KTH—Royal Institute of Technology, AlbaNova University Center, SE-106 91 Stockholm,
    Sweden; eaurell@kth.se
[2] Departments of Computer Science and Applied Physics, Aalto University, FIN-00076 Aalto, Finland

Received: 18 August 2017; Accepted: 28 October 2017; Published: 7 November 2017

Abstract: This paper revisits the classical problem of representing a thermal bath interacting with a system as a large collection of harmonic oscillators initially in thermal equilibrium. As is well known, the system then obeys an equation, which in the bulk and in the suitable limit tends to the Kramers–Langevin equation of physical kinetics. I consider time-dependent system-bath coupling and show that this leads to an additional harmonic force acting on the system. When the coupling is switched on and switched off rapidly, the force has delta-function support at the initial and final time. I further show that the work and heat functionals as recently defined in stochastic thermodynamics at strong coupling contain additional terms depending on the time derivative of the system-bath coupling. I discuss these terms and show that while they can be very large if the system-bath coupling changes quickly, they only give a finite contribution to the work that enters in Jarzynski's equality. I also discuss that these corrections to standard work and heat functionals provide an explanation for non-standard terms in the change of the von Neumann entropy of a quantum bath interacting with a quantum system found in an earlier contribution (Aurell and Eichhorn, 2015).

Keywords: stochastic thermodynamics; strong coupling; Zwanzig model; quantum-classical correspondence for heat

PACS: 03.65.Yz; 05.70.Ln; 05.40.-a

---

1. Introduction

Stochastic thermodynamics has become an established paradigm to describe fluctuating work and heat on mesoscopic scales. Main results in this field are fluctuation relations. The archetypal relation is Jarzynski's Equality (JE) [1]:

$$\left\langle e^{-\beta \delta W} \right\rangle_{eq} = e^{-\beta \Delta F} \tag{1}$$

where the average is over an initial equilibrium state at inverse temperature $\beta = \frac{1}{k_B T}$ and the realization of the process, δW is the work done on the system over some time interval and ΔF is the difference in free energy between the states given by the initial and final values of the parameters. While the initial state in (1) is taken to be in equilibrium with a bath, later states and in particular the final state do not have to be. Another key result is the Integral Fluctuation Theorem (IFT) [2], which states that:

$$\left\langle e^{\Delta \log P - \beta \delta Q} \right\rangle = 1 \tag{2}$$

where the average is over the initial position of the system with respect to the initial state (initial probability density) and over the stochastic trajectory of the system from the initial to the final time. $\Delta \log P$, known as stochastic entropy, is the log-change in probability density from an initial position at the initial time to a final position at the final time, and δQ is the heat given off to the environment during the process. A consequence of (2) is a version of the second law $\Delta S_{sys} + \beta \langle \delta Q \rangle \geq 0$ where $S_{sys} = -\sum_i p_i \log p_i$ is the system entropy and ΔS_{sys} is its change from the initial to the final state. Stochastic thermodynamics is covered by several excellent reviews to which I direct the reader for further information and pointers to the literature [3–7].

A limitation of stochastic thermodynamics in the standard formulation is that it neglects the energy stored in the coupling between the system and the bath. This issue was first discussed from different points of view by Cohen and Mauzerall [8], Jarzynski [9] and later by Esposito and co-workers [10]. The scientific question is whether the theory extends also to this setting and in particular if and when the relevant mesoscopic quantities (heat, work) can be taken as functionals (typically stochastic functionals) of the system only. Recently, this has given rise to an active discussion where on one side, one can find [11–13], and on the other [14]; and an intermediate position taken by [15].

The first goal of the present paper is to circumvent part of the ongoing discussion by taking the system-bath coupling dependent on time so that it vanishes at the beginning and end of a process. Since no energy is then stored in the interactions between the system and the bath, the more involved definitions that can be found in the literature collapse. The work done by the control is simply the sum of the energy changes of the system and the bath between the beginning and the end of the process, while the heat is the change of bath energy. The price to pay are effects arising from the rate-of-change of the system-bath coupling. In particular, there appears from this process a work that is entirely dissipated into heat. When the change of the system-bath coupling is fast, this work is large and can path-wise and on average be (much) larger than the standard Jarzynski work. Due to a cancellation effect, most of it however does not contribute to the average in JE (1).

The second goal is to explain the results on the classical limit of the corresponding quantum problem, which were found in an earlier contribution [16]. The background on the Feynman–Vernon method is well presented in several monographs, e.g., [17], and the extensions to treat heat have now been considered by two other groups; see [18–20]. A time-dependent system-coupling is a further extension of these contributions, but also brings important simplifications. As in the classical case, it allows one to disregard the energy stored in system-bath interaction energy, which is possible, but rather complicated to treat by the Feynman–Vernon-derived method. It will further be shown that the extra term found in [16] was an artifact from assuming that the total density matrix of the system and the bath was initially factorized, while the system and the bath were nevertheless interacting from the beginning of the process. If instead the system-bath coupling goes to zero smoothly at the beginning and the end of the process, the classical limit of heat is recovered fully, and quantum-classical correspondence holds on the level of expected heat.

The paper is organized as follows. In Section 2, I introduce the basic model of a bath of harmonic oscillators with a time-dependent system-bath coupling. Standard material connecting the dynamics of a (large) bath of harmonic oscillators to a stochastic description of the system is given in Appendices A and B. In Section 3, I consider work, internal energy and heat as recently defined in stochastic thermodynamics at strong coupling and relate them to ΔH_B (change of bath energy), and in Section 4, I show how these functionals of the system history are modified when the system-bath coupling depends on time. In Section 5, I discuss the extra work referred to above, and in Section 6, I compare the expectation value of ΔH_B to the earlier computed classical limit of the change of the von Neumann entropy of the bath. In Section 7, I sum up the results. In Appendix C, I discuss for completeness an additional term in the classical limit of the change of the von Neumann entropy that was also found in [16], but which has a different origin than the main focus of this paper.

2. Preliminaries and Definitions of Terms

Throughout this work, I will consider a collection of N harmonic oscillators governed by the Hamiltonian:

$$H_B = \sum_b \frac{p_b^2}{2m_b} + \frac{1}{2}m_b\omega_b^2 q_b^2. \tag{3}$$

The mass and spring constant of oscillator b are m_b and $m_b\omega_b^2$; ω_b is the natural frequency in units of rad/s, and q_b and p_b are respectively the oscillator's coordinate and momentum. The number of oscillators N will be very large, eventually taken to infinity. I will call the collection of oscillators the bath. The bath starts in thermal equilibrium at some inverse temperature β, which will be the bath temperature. I also consider another externally-driven system described by a Hamiltonian:

$$H_S = \frac{P^2}{2M} + V(X, t) \tag{4}$$

where X and P are the position and momentum, and the potential V depends explicitly on time. The system and the bath interact linearly during some finite time interval. The interaction Hamiltonian is thus:

$$H_I(t) = -\sum_b C_b(t)q_b X \tag{5}$$

where the $C_b(t)$ are functions of time. For reasons that will be clear in the following. I add also the Caldeira–Leggett counter-term (a correction to the system potential):

$$H_C(t) = \sum_b \frac{C_b^2(t)X^2}{2m_b\omega_b^2}. \tag{6}$$

The total Hamiltonian of the bath and the system is:

$$H_{TOT} = H_S + H_B + H_I + H_C. \tag{7}$$

The counter-term in (6), the interaction term in (5) and the potential in (3) together form a sum of complete squares, $\sum_b \frac{1}{2}m_b\omega_b^2 \left(q_b - \frac{C_b(t)X}{m_b\omega_b}\right)^2$.

The basic setting in the current paper is that all the $C_b(t)$ together rise rapidly from zero to some steady values C_b after an initial time t_i and decrease rapidly back to zero before some final time t_f. The system and the bath are then disconnected before and at time t_i, as well as at and after time t_f. It is natural to postulate that δQ, the heat exchanged between the system and the bath during the process, is ΔH_B, the change of the energy of the bath in the time interval $[t_i : t_f]$. It is also natural to postulate that the work done during a process is the total energy change of the system and the bath, $\delta W = \Delta H_B + \Delta H_S$. With these definitions, work and heat satisfy the first law $\Delta U = \delta W - \delta Q$ where $\Delta U = \Delta H_S$ is the change of the energy of the system. In Section 3 below, I will show that these simple prescriptions are in agreement with strong-coupling stochastic thermodynamics, for the cases under study.

It is well known that a system coupled to an Ohmic bath obeys in the suitable limit the Fokker–Planck equation with an extra harmonic force when the system-bath coupling depends on time:

$$\dot{X} = \frac{P}{M} \qquad \dot{P} = -V'(X, t) + F_{fric} + F_{if} + \xi(t) \tag{8}$$

where ξ is white noise, and the two deterministic forces from the bath on the system are:

$$F_{fric} = -\eta \dot{X} \tag{9}$$

$$F_{if} = -\frac{1}{2}\frac{\dot{\eta}(t)}{\eta(t)}X \tag{10}$$

For completeness, the derivation of (8)–(10) is summarized in Appendix A and B. A main assumption is that each of the coupling constants $C_b(t)$ is proportional to the square root of a time-dependent friction coefficient $\eta(t)$, which encodes all the time-dependence of the system-bath coupling. For the following discussion, it is convenient to introduce, in analogy with the "Sekimoto force" [3,21],

$$F_S = \xi + F_{fric} + F_{if} \tag{11}$$

It is also convenient to introduce an auxiliary quantity with dimension velocity:

$$\dot{X}_B = -\frac{1}{2}\frac{\dot{\eta}(t)}{\eta(t)}X \tag{12}$$

such that:

$$F_{fric} + F_{if} = -\eta \left(\dot{X} - \dot{X}_B \right). \tag{13}$$

The interpretation of $\dot{X}_B$ is that it is a time- and coordinate-dependent velocity of the system such that the force F_S from the bath on the system vanishes in expectation. The velocity $\dot{X}_B$ is zero when the system-bath coupling is constant in time.

3. Work and Heat of Stochastic Thermodynamics at Strong Coupling

We start by reviewing the derivation of Jarzynski's equality at strong coupling following [9], with adjustments arising from time-dependent system-bath coupling. The system and the bath together constitute a closed system, and the work done on the combined system is the change of the total energy:

$$\Delta H_{TOT} = \Delta H_S + \Delta H_B + \Delta H_I + \Delta H_C. \tag{14}$$

As was first shown in [9], the combined system satisfies a Jarzynski equality:

$$\left\langle e^{-\beta \Delta H_{TOT}} \right\rangle_{eq} = e^{-\beta \Delta F_{TOT}} \tag{15}$$

where the average is over the equilibrium state of the combined system and bath and ΔF_{TOT} is the change of total equilibrium free energy. If F_B is the (constant) free energy of the bath, then $\tilde{F}_s = F_{TOT} - F_B$ is the thermodynamic potential (free energy) of the system appropriate for strong coupling [11–13,22], and the quantity appearing in the exponent of the right-hand side of (15) is $\Delta \tilde{F}_s$.

Since we assume that the system and the bath are uncoupled at the initial and the finite time, ΔF_{TOT} in (1) is however in fact here equal to ΔF, the free energy change of the system only. This is one major simplification and a reason for considering a time-dependent system-bath coupling. For the same reason, we can write:

$$\delta W = \Delta H_{TOT} = \Delta H_S + \Delta H_B \tag{16}$$

the other two terms vanishing on the boundaries. On the other hand, in a mechanical system, the change of the total Hamiltonian (14) is given by the direct time-dependence only, and therefore, we also have:

$$\delta W = \int \partial_t H_S + \partial_t H_I + \partial_t H_C \tag{17}$$

which has two extra terms compared to Jarzynski work when the system-bath coupling depends on time. The bath Hamiltonian H_B does not contribute to (17) as it is independent of time.

From here, the reader may continue directly below to Equation (21). To however connect to the recent discussion of heat in stochastic thermodynamics at strong coupling, we note that it starts from a Hamiltonian of the system at mean force [11,12,14]:

$$\mathcal{H} = H_S + H_C - \beta^{-1} \log \left\langle e^{-\beta H_I} \right\rangle_B \tag{18}$$

where the average is over the Boltzmann distribution of the bath only. From this is derived an internal energy function at mean force defined as:

$$E = \mathcal{H} + \beta \partial_\beta \mathcal{H} \tag{19}$$

A main simplification of the model under consideration here is that for a bath of Hamiltonian oscillators linearly coupled to the system, the average in $\mathcal{H}$ cancels with the counter-term H_C, so that for the case at hand, we have simply:

$$E = \mathcal{H} = H_S. \tag{20}$$

The change in this internal energy can be written as a time integral:

$$\Delta E = \int \partial_t H_S + \dot{X} \partial_X H_S + \dot{P} \partial_P H_S. \tag{21}$$

A definition of a path-wise heat functional that satisfies the first law with (16) and (20) is then:

$$\delta Q = \delta W - \Delta E = \Delta H_B. \tag{22}$$

In a round-about way, we have thus arrived at the same notion of heat as in Section 2. Using (17) and (21), we can write δQ as a time integral as:

$$\delta Q = \int \partial_t H_I + \partial_t H_C - \dot{X} \partial_X H_S - \dot{P} \partial_P H_S \tag{23}$$

4. Work and Heat with a Time-Dependent Ohmic Bath

We start by writing out (21), (17) and (23) explicitly for the model at hand:

$$\Delta E = \int \partial_t H_S + \frac{P}{M} \sum_b C_b \left(q_b - \frac{C_b X}{m_b \omega_b^2} \right) dt \tag{24}$$

$$\delta W = \int \partial_t H_S - X \sum_b \dot{C}_b \left(q_b - \frac{C_b X}{m_b \omega_b^2} \right) dt \tag{25}$$

$$\delta Q = \int \sum_b \left(-\dot{C}_b X - \frac{P}{M} C_b \right) \left(q_b - \frac{C_b X}{m_b \omega_b^2} \right) dt \tag{26}$$

For the change of internal energy, it is seen that the sum over the bath oscillators is the same as that which gives rise to the force F_S in (11) (for details, see Appendixes A and B), and we can thus write:

$$\Delta E = \int (\partial_t H_S) dt + F_S \circ dX \tag{27}$$

The stochastic integral above is in the Stratonovich convention (mid-point prescription in a discretization), as it is the limit of ordinary integrals when the cut-off frequency Ω of the bath is taken to infinity.

Equation (27) is the Jarzynski work ($\int \partial_t H_S dt$) plus a term expressing the work done on the system from the total force F_S that arises from system-bath coupling. It can therefore be read as the sum of the flow of energy into the system from the external system and from the bath, that is the same standard form in stochastic thermodynamics [21]. Equation (27) only involves the time-dependence of the system-bath coupling through the additional force F_{if} entering F_S.

The work functional can be read off by comparing (24) and (25). The sum over oscillators is now not multiplied by $\frac{P}{M}dt = dX$, but by $-X\frac{\dot{C}_b}{C_b}dt$, which is the virtual increment dX_B from (12). The work is therefore:

$$\delta W = \int (\partial_t H_S)dt + F_S \circ dX_B \tag{28}$$

which is the Jarzynski work (the first term) plus the work done by the Sekimoto force through the virtual increment of the system position dX_B. To emphasize the parallelism with (27), the second term has been written as a Stratonovich integral, but it is in fact just an ordinary integral, F_S being a random force acting on momentum variable P and the virtual increment dX_B being proportional to dt. The heat can similarly be written:

$$\delta Q = \int (-F_S) \circ (dX - dX_B) \tag{29}$$

and is the work done by the Sekimoto reaction force $(-F_S)$ through the difference between the actual increment dX and the virtual increment dX_B.

An alternative way to divide up the different contributions to the heat, which will be used in Section 6 below, is:

$$\delta Q = \int -\xi \circ (dX - dX_B) + \left(\frac{d}{dt}\left(\sqrt{\eta}X\right)\right)^2 dt \tag{30}$$

When η is constant in time, this reduces to Sekimoto's heat functional $\int -\xi \circ dX + \eta \dot{X}^2 dt$.

5. The Work Done by the Time-Dependent System-Bath Coupling

It has been seen above that both the heat and the work contain a term $\int F_S \circ dX_B$, which is not reflected in the internal energy change at all. Changing the system-bath coupling can be considered an external control acting on the system, which implies a kind of work. An original feature is that this work is entirely dissipated into heat (change of energy of the bath); nothing remains as internal energy changes. Since $\dot{X}_B$ has been assumed non-zero only near the initial and final time, this work is also approximately a change of state function.

The first part of this work, corresponding to the first of the three terms in F_S, is:

$$\int F_{fric} \circ dX_B = \int \frac{P}{2M}\dot{\eta}X\,dt. \tag{31}$$

Assuming that η changes from zero to a finite value $\bar{\eta}$ over a short time period Δt, the force F_{if} is going to dominate all the other forces in the Kramers–Langevin equation (8). This means that we have:

$$P(t) = P_i - \frac{1}{2}\eta X + \mathcal{O}(\sqrt{\Delta t}) \qquad t \in [t_i, t_i + \Delta t] \tag{32}$$

while at the same time:

$$X(t) = X_i + \mathcal{O}(\Delta t) \tag{33}$$

From this, the contribution to (31), from the beginning of the process, is $\frac{X_i P_i}{2M}\bar{\eta} - \frac{X_i^2}{8M}\bar{\eta}^2$ and analogously just before the final time. The contribution of this part is therefore the change of an auxiliary friction-dependent energy:

$$\int F_{fric} \circ dX_B \;=\; \Delta V_{frict} + \mathcal{O}(\Delta t) \tag{34}$$

$$V_{frict} \;=\; -\frac{\bar{\eta} X P}{2M} + \frac{\bar{\eta}^2 X^2}{8M} \tag{35}$$

The second part of $\int F_S \circ dX_B$ is:

$$\int F_{if} \circ dX_B = \int \frac{1}{4} X^2 \frac{\dot{\eta}^2}{\eta} dt \tag{36}$$

Similarly to above, this can be written as a functional of X_i, P_i and the function η in the interval $[t_i, t_i + \Delta t]$ and the same at the final time. We can therefore write:

$$\int F_{if} \circ dX_B = A_i[X_i, P_i, \eta] + A_f[X_f, P_f, \eta] + \mathcal{O}(\Delta t) \tag{37}$$

with two functionals A_i and A_f. The largest (most divergent) contributions to A_i and A_f are $\frac{1}{4} X^2 \int \frac{\dot{\eta}^2}{\eta} dt$, which diverge as $(\Delta t)^{-1}$. The third part of $\int F_S \circ dX_B$ is:

$$\int \zeta \circ dX_B. \tag{38}$$

This is a random variable that depends on the realization of ζ just after the initial time and just before the final time. If dX_B is considered an externally-given virtual displacement, then (38) is a weighted integral of white noise and then a random variable of mean zero. The same result holds also when dX_B is a combination involving X because X is here a smooth function of time (no Itô correction). The variance of (38) can be computed from the noise-noise correlation and by comparing the increment dX_B with the (similar) boundary force F_{if}, with the result:

$$\left\langle \left(\int \zeta \circ dX_B \right)^2 \right\rangle = 2 k_B T \left(A_i + A_f \right) \tag{39}$$

The second and the third part of $\int F_S \circ dX_B$ are hence potentially both large, and when Δt tends to zero, they can both be larger than standard Jarzynski work by an arbitrary amount.

Let us now consider the contribution of $\int F_S \circ dX_B$ to the Jarzynski equality, which we write:

$$\left\langle e^{-\beta(\int (\partial_t H_S) + F_S \circ dX_B)} \right\rangle_{fact.eq} = e^{-\beta \Delta F} \tag{40}$$

to emphasize that the initial equilibrium distribution is factorized between the system and the bath. For given initial and final coordinates and momenta, the first and second parts contribute simply $e^{-\beta(\Delta V_{fric} + A_i + A_f)}$. The third part (38) on the other hand contributes:

$$\int \frac{1}{N} e^{-\frac{\beta}{2} \frac{x^2}{2(A_i + A_f)}} e^{-\beta x} dx = e^{\beta(A_i + A_f)}. \tag{41}$$

The (potentially divergent) contributions from the second and the third part hence cancel for each given initial and final coordinate and momenta (up to terms as small as Δt), and must therefore cancel in the Jarzynski equality overall.

Combining this result and (34), we have an alternative form of the Jarzynski equality:

$$\left\langle e^{-\beta \int (\partial_t H_S) - \beta \Delta V_{frict}} \right\rangle_{fact.eq} = e^{-\beta \Delta F}. \tag{42}$$

This may be compared to the strong-coupling form of the Jarzynski equality, which holds when η is strictly constant in time:

$$\left\langle e^{-\beta \int (\partial_t H_S)} \right\rangle_{eq} = e^{-\beta \Delta F_{TOT}} \tag{43}$$

In the above, the average is over the system and the bath initially in joint equilibrium with the terms H_I and H_C included, and the free energy change is of the combined system and bath. Path-wise, the contributions to (42) and (43) from the Jarzynski work $\int (\partial_t H_S)$ are the same up to terms $\mathcal{O}(\Delta t)$. The differences between the two averages therefore stem from the additional term ΔV_{frict} and the different distributions over the initial conditions.

6. Comparison to Quantum Heat Exchange

In the present section, I will switch gears and discuss an approach to quantum heat considered in an earlier contribution, there in the guise of a first-order change of the von Neumann entropy of a bath [16]. As in the previous sections, the heat will be related to the change of energy of a bath. In the spirit of Kurchan's approach to quantum fluctuation relations [23,24], it is assumed that the system is measured at the beginning and the end of a process. The outcome of the initial measurement will be state $|i>$ of the system, and the outcome of the final measurement will be state $|f>$; these letters will be used as subscripts to compactify some of the following expressions. The first such notational simplification, introduced by Feynman and Vernon [25], is to write P_{if} for the conditional probability of measuring the system in final state $|f>$ given that it was initially in $|i>$. It is here assumed that between the measurements, the system develops in interaction with the bath, which was initially in equilibrium. Using the shorthand:

$$\int_{if} (\cdots) = \int dX_i dY_i dX_f dY_f \psi_i(X_i) \psi_i^*(Y_i) \psi_f^*(X_f) \psi_f(Y_f) (\cdots)$$

where ψ_i and ψ_f are the wave functions of the states $|i>$ and $|f>$, I will write the transition probability as:

$$P_{if} = \int_{if} \mathcal{D}X\mathcal{D}Y e^{\frac{i}{\hbar}S_S[X] - \frac{i}{\hbar}S_S[Y] + \frac{i}{\hbar}S_i[X,Y] - \frac{1}{\hbar}S_r[X,Y]} \tag{44}$$

where the two path integrals are over respectively the forward path $X(t)$ from X_i to X_f and the backward path $Y(t)$ from Y_i to Y_f. These two path integrals emanate from a representation of the total unitary U and its inverse U^{-1} in the time development of the total density operator of the bath and the system $\rho_f^{TOT} = U\rho_i^{TOT}U^{-1}$ and then integrating out the bath variables. The effects of the bath are thus captured by the two kernels S_i and S_r in (44), which couple the forward and backward paths and which are referred to as the real and the imaginary part of the Feynman–Vernon action. This important theory is lucidly explained in the original 1963 paper [25], by Caldeira and Leggett in [26], as well as in several reviews and monographs; cf. [17,27].

In a slight extension of this well-established theory, we now assume that the Hamiltonians of the system and the bath are described by operators $\hat{H}_S$, $\hat{H}_B$, $\hat{H}_I$ and $\hat{H}_C$, which are translations to the quantum domain of the classical Hamiltonians in Section 2. The expected final energy of the bath, after first measuring the system, is $\mathrm{Tr}\left[\hat{\rho}_B^{post}\hat{H}_B\right]$, where the reduced density matrix of the bath is $\hat{\rho}_B^{post} = \frac{1}{P_{if}} <f|\hat{\rho}_{TOT}|f>$. Using the shorthand:

$$\langle A_B \rangle_{if} = \mathrm{Tr}_B <f|\hat{\rho}_{TOT} A_B \oplus \mathbf{1}_S|f> \tag{45}$$

where A_B is an operator acting on the bath and $\mathbf{1}_S$ is the identity operator on the system, Tr_B is the trace of the bath and the system starts out from pure state $|i>$, the conditional expected quantum heat is:

$$\mathrm{E}\left[\delta Q|if\right] = \frac{1}{P_{if}}\left(\left\langle \hat{H}_B(t_f) \right\rangle_{if} - \left\langle \hat{H}_B(t_i) \right\rangle_{if}\right) \tag{46}$$

Path integral expressions for (46) or closely connected quantities have by now been developed by several groups [16,18,19], notably recently in [20]. I here follow the notation in [16] where the first main result (Equation (15)) is equivalent to:

$$\mathrm{E}\left[\delta Q|if\right] = \frac{1}{P_{if}}\left\langle \mathcal{I}^{(2)} + \mathcal{I}^{(3)} \right\rangle_{if} = \frac{1}{P_{if}} \int_{if} \mathcal{D}X\mathcal{D}Y e^{\frac{i}{\hbar}S_S[X] - \frac{i}{\hbar}S_S[Y] + \frac{i}{\hbar}S_i[X,Y] - \frac{1}{\hbar}S_r[X,Y]}\left(\mathcal{I}^{(2)} + \mathcal{I}^{(3)}\right) \quad (47)$$

where $\mathcal{I}^{(2)}$ and $\mathcal{I}^{(3)}$ are two quadratic functionals. In this section, their classical limit will be discussed and shown to be identical to the expected value of the classical heat discussed above in Section 4. In [16] (Equation (15)), there was also a first functional $\mathcal{I}^{(1)}$, which is related to the choice of subtraction at the initial time and which is for completeness discussed in Appendix C.

Adapting to the setting of the present paper, we from now on take the system-bath coupling coefficients C_b to depend on time as in Appendix A. This leads to:

$$\mathcal{I}^{(2)} = \int_{t_i}^{t_f}\int_{t_i}^{t_f} XY'h^{(2)}ds'ds \quad (48)$$

$$\mathcal{I}^{(3)} = \int_{t_i}^{t_f}\int_{t_i}^{t_f} XY'h^{(3)}ds'ds. \quad (49)$$

where the corresponding two kernels (up to a factor β) are given in [16] Equation 16:

$$h^{(2)} = i\sum_b \frac{C_b(s)C_b(s')}{2m_b}\coth(\frac{\beta\hbar\omega_b}{2})\sin\omega_b(s - s') \quad (50)$$

$$h^{(3)} = \sum_b \frac{C_b(s)C_b(s')}{2m_b}\cos\omega_b(s - s') \quad (51)$$

Caldeira and Leggett in [26] were the first to explicitly consider the classical open system limit of the quantum mechanical transition amplitude P_{if} in (44). That limit is achieved in three steps. The first is taking the bath as large and with the Ohmic spectrum as in Appendix B. When the spectral cut-off Ω is much larger than thermal relaxation time $\hbar\beta$, the kernel S_i in (44) tends to the derivative of the delta function. The second step is to assume a large bath temperature in the sense that $\hbar\beta$ should be much smaller than any characteristic time of the system, which turns the other kernel S_r in (44) into a delta function. The time development of the kernel in (44) is then Markov, i.e., it satisfies a partial differential equation, first derived in [26], and compared to the Lindblad equation in, e.g., [17]. The Wigner function, which is related to the kernel in (44) by a Fourier transform, then also satisfies a partial differential equation and, as found in [26], in fact, a Fokker–Planck equation. The standard interpretation of this result is that the bath "observes" the system and destroys its quantum coherence such that the open system quantum dynamics is fully described (in this limit) by the open system classical dynamics.

We now want to consider the same Caldeira–Leggett limit for the functionals $\mathcal{I}^{(2)}$ and $\mathcal{I}^{(3)}$ in (48) and (49). Similarly to S_i and S_r, the kernels tend to:

$$h^{(2)} \approx -\beta^{-1}\frac{i}{\hbar}2\sqrt{\eta(s)\eta(s')}\frac{d}{d(s - s')}\delta(s - s') \quad (52)$$

$$h^{(3)} \approx -\sqrt{\eta(s)\eta(s')}\frac{d^2}{d(s - s')^2}\delta(s - s') \quad (53)$$

Assuming that both η and $\dot{\eta}$ vanish at the initial and final time, we can integrate by parts freely to find:

$$\left\langle \mathcal{I}^{(2)} + \mathcal{I}^{(3)} \right\rangle_{if} = \left\langle \int \beta^{-1}\frac{i}{\hbar}\left(\left(\frac{d}{dt}(\sqrt{\eta}X)\right)(\sqrt{\eta}Y) - (\sqrt{\eta}X)\left(\frac{d}{dt}(\sqrt{\eta}Y)\right)\right) + \left(\frac{d}{dt}(\sqrt{\eta}X)\right)\left(\frac{d}{dt}(\sqrt{\eta}Y)\right)\right\rangle_{if} \quad (54)$$

In the above, the (quantum) expectation is the same as in (47), and X and Y are the forward and backward quantum paths. If we substitute for X and Y a classical path Q and average over its probability distribution, the term $\left\langle \mathcal{I}^{(3)} \right\rangle_{if}$ in (54) is the same as the expectation value of the corresponding term in (30). The other term can be re-expressed as:

$$\left\langle \mathcal{I}^{(2)} \right\rangle_{if} = \beta^{-1} \frac{i}{\hbar} \left\langle \int \eta \left(\frac{d}{dt}(X+Y) \right)(Y-X) + \frac{1}{2}\dot{\eta}\left(Y^2 - X^2\right) + \frac{1}{2}\frac{d}{dt}\left(\eta X^2\right) - \frac{1}{2}\frac{d}{dt}\left(\eta Y^2\right) \right\rangle_{if} \tag{55}$$

In the above, the last two terms vanish since they are complete integrals, and η has been taken as zero at the initial and final time. A similar separation was used in [16], Section 5, and there expressed as $\frac{2i}{\hbar}S_i^{mid}$ and ΔS_b; in that case (where η was constant), the extra term ΔS_b was however not zero.

To discuss the first two terms in (55), we use the procedure of [16], Sections 6 and 7. The first step is to express the quantum averages through the Wigner transform. For the first term in (55) that leads to the result in [16], we use Equation (34). For the second term in (55), we have that $\frac{1}{2}(Y^2 - X^2) = Q\alpha$ where $Q = \frac{1}{2}(X+Y)$ is the average (eventually classical) path and $\alpha = (Y-X)$ is the quantum deviation; a term $\frac{i\alpha}{\hbar}$ translates to a partial derivative with respect to momentum variable P in the Wigner transform. Collecting both terms, we have hence:

$$\left\langle \mathcal{I}^{(2)} \right\rangle_{if} = \beta^{-1} \int ds\, dQ\, dP\; P(Q, P, Q_i, P_i) \left(-\frac{2\eta}{M}p\partial_p - \frac{\eta}{M} - \dot{\eta}q\partial_p \right) P(Q_f, P_f, Q, P) \tag{56}$$

$P(q', p', q, p)$ in the above is the Wigner function, which in the Caldeira–Leggett limit tends to the transition kernel of the classical stochastic process (8).

The second step is to compare the right-hand side of (55) to the expectation value of the first term in (30) $\left\langle \int -\xi \circ (dQ - dQ_B) \right\rangle_{if}$. The expectation is conditioned on the initial and final states, which are assumed to be given by definite classical coordinates and momenta. It is convenient to re-write this as:

$$\int_{Q_i, P_i}^{Q_f, P_f} \mathcal{D}\text{Prob}(\text{path}) \left[\int \left(-\frac{P}{M} - \frac{\dot{\eta}}{2\eta}Q \right) \circ d\Xi \right] \tag{57}$$

where $\text{Prob}(\text{path})$ is a weight over paths and $d\Xi$ is the increment of the aggregated random force, a Gaussian random variable with mean zero and variance $2k_B T\eta dt$. One can therefore discretize the inner integral in (57) and consider the contribution from the short time interval $[s, s + \Delta s]$:

$$\text{Term from Equation (57)} = \int dQ\, dP\, dQ'\, dP' \left(-\frac{P'+P}{2M} - \frac{\dot{\eta}}{2\eta}Q\Delta s \right) \Delta\Xi\; P(\cdot, Q', P') P_{\Delta s}(Q', P', Q, P) P(Q, P, \cdot). \tag{58}$$

The short-time propagator of the Kramers–Langevin process is:

$$P_{\Delta s}(Q', P', Q, P) = \frac{1}{\mathcal{N}} e^{-\frac{1}{4k_B T\eta\Delta s}\left(P'-P+\left(\partial_Q V + \eta\frac{P+P'}{2M} + \frac{1}{2}\dot{\eta}Q\right)\Delta s \right)^2} \delta\left(Q' - Q - \frac{P'+P}{2M}\Delta s \right) \tag{59}$$

and has the property that:

$$\Delta\Xi P_{\Delta s}(Q', P', Q, P) = -2k_B T\eta\Delta s\partial_{p'} P_{\Delta s}(Q', P', Q, P) + \mathcal{O}\left(\Delta s^2\right) \tag{60}$$

Integration by parts moves over the derivative with respect to P' to respectively $P(Q_f, P_f, Q', P')$ and $\frac{P'+P}{2M}$, which gives the expression in (55). The same analysis through the Wigner function as outlined here can of course also be applied to the last (simpler) term in (54), with the same result as given above up to terms of order $\hbar^2$.

7. Discussion

The study of the interaction of a classical or quantum system with a bath of harmonic oscillators has a long history, and it behooves the author to motivate the need for another paper on the subject.

Consider the assumption of the Feynman–Vernon theory that the system and the bath are originally decoupled and the bath is at equilibrium. If the coupling is not weak, this is questionable because if the system and the bath were in contact precisely at the initial time, they should have been so also slightly before, and then, they could not have been initially fully decoupled. Nevertheless, this assumption is needed to integrate out the bath variables and arrive at the Feynman–Vernon open system development operator of the system only; although the bath oscillators can be integrated out also if the system and the bath are originally in equilibrium together, the analysis is then considerably more complicated [27] and that possibility has not been considered here. It could therefore be argued that the Caldeira–Leggett theory of quantum Brownian motion [26], which is based on Feynman–Vernon, is only valid at weak coupling. Indeed, it has been noted several times that the classical limit (Kramers–Langevin equation) of an Ohmic bath has a delta-function force proportional to the friction acting at the initial time and that the random and deterministic forces from the bath therefore do not obey an Einstein relation. It has also been noted that there is a difference whether one assumes that the decoupled initial conditions pertain exactly at the initial time or only slightly after.

To resolve these (and other) issues, Caldeira and co-workers [28] investigated the possibility that the bath is initially in equilibrium conditional on the position of the system, i.e., with respect to Hamiltonian $\sum_b \frac{p_b^2}{2m_b} + \frac{1}{2}m_b\omega_b^2\left(q_b - \frac{C_b(t)Q}{m_b\omega_b^2}\right)^2$. These initial conditions are mathematically possible and physically allowed, but raise the question of how the system and the bath would find themselves in such a state. If some control would have been exercised on the bath prior to the process, this control must have been aware of the position of the system, and only a very exquisite procedure would have resulted in exactly the assumed initial state.

In this paper, I have aimed to address these issues anew by allowing the system-bath coupling to depend on time. It is then not a problem to have the system and the bath initially decoupled, but the price to pay is the new effects that arise from the time-dependent friction. On the level of classical equations, these effects are just the extra force F_{if} of (10), which when the switching on and off of the bath-system coupling is fast, reduces to the previously known delta-function force. On the level of classical heat and work functionals, the situation is more involved and described by (28) and (29). There appears a part of the work denoted $\int F_S \circ dX_B$, which is entirely dissipated into heat, which has support only at the beginning and the end of the process and which is potentially quite large. Nevertheless, it only makes a finite contribution to the work functional that enters Jarzynski's equality; see (42).

Although the work $\delta W_{if} = \int F_S \circ dX_B$ is (to the author's knowledge) new, it is in retrospect not entirely surprising from a physical point of view. If the bath and the system are initially independent and if they then move so that the bath is in conditional equilibrium with the system, then that is a relaxation process. By analogy with finite-time (macroscopic) thermodynamics, one would expect the heat in such a process to be inversely proportional to the duration of the process. To compare, it is well known that when a system changes state in a finite time interval Δt, the expected heat given off to the environment diverges as $(\Delta t)^{-1}$ [29,30]. The current model is not exactly the same, as the divergent heat arrives to an explicitly-modeled set of oscillators and not to a system coupled to an external and more abstract bath. Nevertheless, in both cases, the leading divergence is inversely proportional to the duration of the process, and the operational meaning is that extra work has to be done on the system to effectuate a transformation in a finite time. The extra work may hence be assimilated to finite-time corrections to Landauer's law [29,30].

On the quantum level of the expected change of the bath energy, the analysis can be carried out using the Feynman–Vernon formalism, as done previously for constant system-bath coupling in [16]. The classical limit of these expressions can then be seen to be the same as the classical heat functional with time-dependent friction. This is hence an example of quantum-classical correspondence [31], on the level of the expected heat. The extra term ΔS_b found in [16] arises from using factorized initial conditions with a finite system-bath interaction present from the very beginning of the process and has therefore here been shown to be a kind of artifact.

Acknowledgments: This paper was motivated by earlier joint work with Ralf Eichhorn, whom I thank for many stimulating discussions on the subject matter. I thank Haitao Quan and Ken Funo for discussions and useful remarks and the Institute of Theoretical Physics at the Chinese Academy of Sciences (Beijing) for the hospitality. This research was supported by the Academy of Finland through its Center of Excellence COIN, Grant No. 251170, and by the Chinese Academy of Sciences (CAS) President's International Fellowship Initiative (PIFI), Grant No. 2016VMA002.

Conflicts of Interest: The author declares no conflict of interest.

Appendix A. Recall of the Zwanzig Theory

All the calculations in this section are straightforward and mostly re-statements of known results [32]. The equations of motion of the bath and the system that follow from (3)–(6) are:

$$\dot{p}_b = -m_b\omega_b^2 q_b + C_b(t)X \tag{A1}$$

$$\dot{q}_b = \frac{p_b}{m_b} \tag{A2}$$

$$\dot{P} = -V'(X,t) + \sum_b C_b(t)\left(q_b - \frac{C_b(t)}{m_b\omega_b^2}X\right) \tag{A3}$$

$$\dot{X} = \frac{P}{M} \tag{A4}$$

The initial position and momentum of the system (X^i, P^i) are assumed known. All randomness in this problem is hence due to lack of knowledge of the initial positions and momenta of the bath. These are $2N$ real numbers, while the time history of the system over the finite time interval, $\{X,P\}_{t_i}^{t_f}$, needs infinitely many variables to be completely specified. One may therefore assume that a complete knowledge of the system history uniquely specifies the initial positions and momenta of the bath; to the equilibrium probability P^{eq} in the bath then corresponds a probability distribution $P\left[\{X,P\}_{t_i}^{t_f}\right]$ on system histories. This will be assumed throughout the following.

The equations of motion for the bath oscillators can be solved as:

$$\begin{pmatrix} q_b(t) \\ p_b(t) \end{pmatrix} = e^{(t-t_i)\begin{pmatrix} 0 & \frac{1}{m_b} \\ -m_b\omega_b^2 & 0 \end{pmatrix}} \begin{pmatrix} q_b(t_i) \\ p_b(t_i) \end{pmatrix} +$$

$$\int_{t_i}^t e^{-(t-s)\begin{pmatrix} 0 & \frac{1}{m_b} \\ -m_b\omega_b^2 & 0 \end{pmatrix}} \begin{pmatrix} 0 \\ C_b Q \end{pmatrix} \tag{A5}$$

Observing that the exponentiated matrix multiplying the initial conditions is:

$$\begin{pmatrix} 1 & 0 \\ 0 & 1 \end{pmatrix} \cos\omega_b(t-t_i) + \begin{pmatrix} 0 & \frac{1}{m_b} \\ -m_b\omega_b^2 & 0 \end{pmatrix} \frac{1}{\omega_b}\sin\omega_b(t-t_i)$$

the solution for the position can then be written:

$$q_b(t) = q_b(t_i)\cos\omega_b(t-t_i) + p_b(t_i)\frac{1}{\omega_b m_b}\sin\omega_b(t-t_i)$$

$$- \int_{t_i}^t [\frac{1}{\omega_b m_b}\sin(\omega_b(t-s))C_b(s)Q(s)]ds \tag{A6}$$

which by one integration by parts can be rewritten as:

$$q_b(t) = \left(q_b(t_i) - \frac{C_b(t_i)X(t_i)}{m_b\omega_b^2} \right) \cos\omega_b(t-t_i) + \frac{C_b(t)X(t)}{m_b\omega_b^2} -$$

$$\frac{1}{\omega_b^2 m_b} \int_{t_i}^{t} [\cos(\omega_b(t-s)) \, (\dot{C}_b(s)X(s) + C_b(s)\dot{X}(s))]ds + p_b(t_i)\frac{1}{\omega_b m_b}\sin\omega_b(t-t_i) \quad \text{(A7)}$$

By the assumption that the system and the bath are initially disconnected, $C_b(t_i)X(t_i)$ vanishes. Inserting $q_b(t)$ into the equations of motion for the system gives:

$$\dot{X} = \frac{P}{M} \qquad \dot{P} = -V'(X,t) + \xi(t)$$

$$- \int_{t_i}^{t} \gamma(t,s)\dot{X}(s) + \chi(t,s)X(s)ds \quad \text{(A8)}$$

using the ancillary quantities:

$$\xi(t) = \sum_b C_b(t)\left[q_b(t_i)\cos\omega_b(t-t_i) + \right.$$

$$\left. p_b(t_i)\frac{1}{\omega_b m_b}\sin\omega_b(t-t_i)\right] \quad \text{(A9)}$$

$$\gamma(t,s) = \sum_b \frac{C_b(t)C_b(s)}{m_b\omega_b^2}\cos\omega_b(t-s) \quad \text{(A10)}$$

$$\chi(t,s) = \sum_b \frac{C_b(t)\dot{C}_b(s)}{m_b\omega_b^2}\cos\omega_b(t-s) \quad \text{(A11)}$$

Equations (8) and (A9)–(A11) are the solutions to the general problem of a system interacting linearly with a bath of harmonic oscillators. The physical meaning of γ is friction, while χ is a reaction force of the bath on the system, a counter-party of the force that the system exerts on each bath oscillator if it is not in equilibrium with the instantaneous position of the system. The term $\xi(t)$ is on the other hand a force of the bath on the system, which is explicitly independent of the system. It is a random force if the initial positions and momenta of the bath are random.

Appendix B. The Time-Dependent Ohmic Bath and Its Forces

In this Appendix, I summarize the additional assumptions needed so that the final Equation (A8) in Appendix A is the classic Kramers–Langevin equation when the system-bath coupling is constant, and I make explicit the additional harmonic force that arises from a variable system-bath interaction and which is given in (8) in the main text.

This extra force is related to the relaxation of the bath to the instantaneous state of the system and has been known for quite some time in the specialized literature [28,33–36]. A convenient starting point is to follow Caldeira and Leggett [26] and assume an Ohmic bath, i.e., that the spectrum of the bath oscillators is continuous up to an upper cut-off Ω and increases quadratically with frequency. The number of oscillators with frequencies in the interval $[\omega, \omega + d\omega]$ is $f(\omega)d\omega$ and:

$$\begin{aligned} f(\omega) &= \frac{2}{\pi}\omega_c^{-3}\omega^2 & \omega < \Omega \\ f(\omega) &= 0 & \omega > \Omega \end{aligned} \quad \text{(A12)}$$

where ω_c is some characteristic frequency less than Ω. The total number of oscillators is hence $\frac{2}{3\pi}\left(\frac{\Omega}{\omega_c}\right)^3$. We also assume a function $\eta(s)$ with dimension (mass/time) such that:

$$C_b(s) = \sqrt{\omega_c^3 m_b \eta(s)} \quad \text{(A13)}$$

which implies, for every bath oscillator,

$$\dot{C}_b(t) = \frac{1}{2}\frac{\dot{\eta}(t)}{\eta(t)}C_b(t). \tag{A14}$$

It is customary to refer to quantities pertaining to a bath oscillator having angular velocity ω by that angular velocity, i.e., m_ω for the mass of the oscillator and C_ω for the coupling coefficient. The two kernels in (A10) and (A11) then follow from:

$$\frac{C_\omega(t)C_\omega(s)f(\omega)}{m_\omega} = \frac{2\sqrt{\eta(t)\eta(s)}}{\pi}\omega^2 \qquad \omega < \Omega \tag{A15}$$

and are:

$$\gamma(t,s) = \frac{2\sqrt{\eta(t)\eta(s)}}{\pi}\int_0^\Omega \cos\omega(t-s)d\omega \tag{A16}$$

$$\chi(t,s) = \frac{\dot{\eta}(t)}{\pi}\sqrt{\frac{\eta(s)}{\eta(t)}}\int_0^\Omega \cos\omega(t-s)d\omega \tag{A17}$$

We now further assume that $C_b(t)$ changes between zero and the full value on a time scale much longer than Ω^{-1} so that the integrals in (A16) and (A17) can be approximated by delta functions. This leads to:

$$\gamma(t,s) \approx 2\eta(t)\delta(t-s) \tag{A18}$$
$$\chi(t,s) \approx \dot{\eta}(t)\delta(t-s) \tag{A19}$$

which means that the two terms inside the integral in (8) are evaluated as:

$$F_{fric} = -\eta(t)\dot{X}(t) \tag{A20}$$
$$F_{if} = -\frac{1}{2}\dot{\eta}(t)X(t)) \tag{A21}$$

as given in Equations (9) and (10) in the main text. The first of the above is a standard friction force, while the second is a time-dependent harmonic force, proportional to the time derivative of the friction coefficient.

Turning now to force term ζ, it varies rapidly in time and depends directly on the initial positions and momenta of the bath oscillators. Its statistical properties follow from averaging over these initial positions and momenta, and it follows from the explicit Formula (A9) that $E[\zeta(t)] = 0$ where the expectation value is taken with respect to $\mathcal{P}_\beta^{eq}$, the equilibrium initial distribution of the bath at inverse temperature β. The second moment for the force is:

$$\begin{aligned} E[\zeta(t)\zeta(s)] &= \int \mathcal{P}_\beta^{eq}(\{q_b(t_i), p_b(t_i)\})\prod_b dq_b(t_i)dq_b(t_i)[\zeta(t)\zeta(s)] \\ &= \sum_b C_b(t)C_b(s)\frac{k_BT}{m_b\omega_b^2}\cos\omega_b(t-s) \approx 2k_BT\eta(t)\delta(t-s) \end{aligned} \tag{A22}$$

The random force ζ therefore satisfies an Einstein relation with only the friction force F_{fric}.

Appendix C. An Extra Term Found in [16]

The purpose of this Appendix is to detail the connection between the path integral representation for heat developed in Section 6 and the one for the first order change in von Neumann entropy studied in the earlier contribution [16]. To translate, we first note that the central quantity computed in the earlier paper (Equation (3) in [16]) was:

$$\overline{\delta S_q} = \mathrm{E}\left[\beta \hat{H}_B(t_f)|if\right] - \beta U(\beta) \tag{A23}$$

where $U(\beta)$ is the equilibrium energy of the bath at inverse temperature β. This differs from (46) by the factor β that makes the quantity dimensionless and by the subtraction at the initial time. We will show that an extra term found [16] is explained by this subtraction.

We then note that the first main result (Equation (15)) of [16] reads:

$$\overline{\delta S_q} = \frac{1}{P_{if}}\left\langle \mathcal{I}^{(1)} + \mathcal{I}^{(2)} + \mathcal{I}^{(3)} \right\rangle_{if} = \frac{1}{P_{if}}\int_{if} \mathcal{D}X\mathcal{D}Y e^{\frac{i}{\hbar}S_S[X] - \frac{i}{\hbar}S_S[Y] + \frac{i}{\hbar}S_i[X,Y] - \frac{1}{\hbar}S_r[X,Y]}\left(\mathcal{I}^{(1)} + \mathcal{I}^{(2)} + \mathcal{I}^{(3)}\right) \tag{A24}$$

where the notation is as in Section 6 with only the slight change that $\mathcal{I}^{(2)}$ and $\mathcal{I}^{(3)}$ include an additional factor β. The first functional is:

$$\mathcal{I}^{(1)} = \int^{t_f}\int^{s}(X - Y)(X' - Y')h^{(1)}ds'ds \tag{A25}$$

with the kernel:

$$h^{(1)} = -\sum_b \frac{C_b^2}{4m_b}\sinh^{-2}\left(\frac{\beta\hbar\omega_b}{2}\right)\cos\omega_b(s - s') \tag{A26}$$

As noted in [16], Equation (A25) equals $\partial_\beta \frac{1}{\hbar} S_r$ where S_r is the real part of the Feynman–Vernon action. Writing the quantum expectation value of $\mathcal{I}^{(1)}$ in the Feynman–Vernon theory therefore means:

$$\frac{1}{P_{if}}\left\langle \mathcal{I}^{(1)} \right\rangle_{if} = \frac{1}{P_{if}}\int_{if}\mathcal{D}X\mathcal{D}Y e^{\frac{i}{\hbar}S_S[X] - \frac{i}{\hbar}S_S[Y] + \frac{i}{\hbar}S_i[X,Y] - \frac{1}{\hbar}S_r[X,Y]}\left(\partial_\beta \frac{1}{\hbar}S_r\right) = -\partial_\beta \log P_{if} \tag{A27}$$

where the notation is as in Section 6 above. The second equality follows because in the exponent, only S_r depends on β.

We want to relate the term in (A27) to the difference between $\mathrm{E}\left[\hat{H}_B(t_i)|if\right]$ and $U(\beta)$, where $U(\beta)$ is the unconditioned expected energy of the bath at inverse temperature β. To do so, we write more formally:

$$P_{if} = \left\langle f|\mathrm{Tr}_B\left[U\left(\rho_B^{eq}(\beta)\oplus|i\rangle\langle i|\right)U^\dagger\right]|f\right\rangle \tag{A28}$$

$$\mathrm{E}\left[\hat{H}_B(t_i)\right] = \left\langle f|\mathrm{Tr}_B\left[U\left((\hat{H}_B\oplus\mathbb{1})\left(\rho_B^{eq}(\beta)\oplus|i\rangle\langle i|\right)\right)U^\dagger\right]|f\right\rangle \tag{A29}$$

$$\mathrm{E}\left[\hat{H}_B(t_f)\right] = \left\langle f|\mathrm{Tr}_B\left[(\hat{H}_B\oplus\mathbb{1})\left(U\left(\rho_B^{eq}(\beta)\oplus|i\rangle\langle i|\right)U^\dagger\right)\right]|f\right\rangle \tag{A30}$$

$$U(\beta) = \mathrm{Tr}_B\left[\hat{H}_B\rho_B^{eq}(\beta)\right] \tag{A31}$$

where U (not to be confused with $U(\beta)$) is the total unitary operation on the system and the bath and $\rho_B^{eq}(\beta)$ is the initial (equilibrium) density matrix of the bath. Using that $\rho_B^{eq}(\beta) = e^{-\beta\hat{H}_B}/Z(\beta)$ and $-\partial_\beta \log Z = U$, we have $\mathrm{E}_{if}\left[\hat{H}_B(t_i)\right] = -\partial_\beta P_{if} + U P_{if}$. Therefore:

$$-\partial_\beta \log P_{if} = \frac{1}{P_{if}}\left\langle \hat{H}_B(t_i)\right\rangle_{if} - U(\beta). \tag{A32}$$

which was to be shown.

As a final remark, the expected energy of the bath energy measured at the initial time, conditioned on the future observation of the final state f of the system, is not the same as the unconditioned equilibrium internal energy of the bath times the transition probability, an (simple) example of quantum retro-diction [37,38].

References

1. Jarzynski, C. Nonequilibrium Equality for Free Energy Differences. *Phys. Rev. Lett.* **1997**, *78*, 2690–2693.
2. Seifert, U. Entropy Production along a Stochastic Trajectory and an Integral Fluctuation Theorem. *Phys. Rev. Lett.* **2005**, *95*, 040602.
3. Sekimoto, K. *Stochastic Energetics*; Springer: Berlin/Heidelberg, Germany, 2010; Volume 799.
4. Sevick, E.; Prabhakar, R.; Williams, S.R.; Searles, D.J. Fluctuation Theorems. *Annu. Rev. Phys. Chem.* **2008**, *59*, 603–633.
5. Esposito, M.; Harbola, U.; Mukamel, S. Nonequilibrium fluctuations, fluctuation theorems, and counting statistics in quantum systems. *Rev. Mod. Phys.* **2009**, *81*, 1665–1702.
6. Jarzynski, C. Equalities and Inequalities: Irreversibility and the Second Law of Thermodynamics at the Nanoscale. *Annu. Rev. Condens. Matter Phys.* **2011**, *2*, 329–351.
7. Seifert, U. Stochastic thermodynamics, fluctuation theorems and molecular machines. *Rep. Prog. Phys.* **2012**, *75*, P126001.
8. Cohen, E.G.D.; Mauzerall, D. A note on the Jarzynski equality. *J. Stat. Mech. Theory Exp.* **2004**, *2004*, P07006.
9. Jarzynski, C. Nonequilibrium work theorem for a system strongly coupled to a thermal environment. *J. Stat. Mech. Theory Exp.* **2004**, *2004*, P09005.
10. Esposito, M.; Ochoa, M.; Galperin, M. Nature of heat in strongly coupled open quantum systems. *Phys. Rev. B* **2015**, *92*, 235440.
11. Seifert, U. First and Second Law of Thermodynamics at Strong Coupling. *Phys. Rev. Lett.* **2016**, *116*, 020601.
12. Jarzynski, C. Stochastic and Macroscopic Thermodynamics of Strongly Coupled Systems. *Phys. Rev. X* **2017**, *7*, 011008.
13. Miller, H.J.D.; Anders, J. Entropy production and time asymmetry in the presence of strong interactions. *Phys. Rev. E* **2017**, *95*, 062123.
14. Talkner, P.; Hänggi, P. Open system trajectories specify fluctuating work but not heat. *Phys. Rev. E* **2016**, *94*, 022143.
15. Strasberg, P.; Esposito, M. Stochastic thermodynamics in the strong coupling regime: An unambiguous approach based on coarse-graining. *Phys. Rev. E* **2017**, *95*, 062101.
16. Aurell, E.; Eichhorn, R. On the von Neumann entropy of a bath linearly coupled to a driven quantum system. *New J. Phys.* **2015**, *17*, 065007.
17. Breuer, H.P.; Petruccione, F. *The Theory of Open Quantum Systems*; Oxford University Press: Oxford, UK, 2002.
18. Carrega, M.; Solinas, P.; Braggio, A.; Sassetti, M.; Weiss, U. Functional integral approach to time-dependent heat exchange in open quantum systems: General method and applications. *New J. Phys.* **2015**, *17*, 045030.
19. Carrega, M.; Solinas, P.; Sassetti, M.; Weiss, U. Energy Exchange in Driven Open Quantum Systems at Strong Coupling. *Phys. Rev. Lett.* **2016**, *116*, 240403.
20. Funo, K.; Quan, H.T. On the thermodynamic implications of path integral formalism of quantum mechanics. *arXiv* **2017**, arXiv:1708.05113.
21. Sekimoto, K. Langevin equation and thermodynamics. *Prog. Theor. Phys. Suppl.* **1998**, *180*, 17–27.
22. Gelin, M.F.; Thoss, M. Thermodynamics of a subensemble of a canonical ensemble. *Phys. Rev. E* **2009**, *79*, 051121.
23. Kurchan, J. A Quantum Fluctuation Theorem. *arXiv* **2000**, arXiv:cond-mat/0007360.
24. Campisi, M.; Hängggi, P.; Talkner, P. TBD. *Rev. Mod. Phys.* **2011**, *83*, 771–791.
25. Feynman, R.P.; Vernon, F. The Theory of a General Quantum System Interacting with a Linear Dissipative System. *Ann. Phys.* **1963**, *24*, 118–173.
26. Caldeira, A.; Leggett, A. Path Integral Approach to Quantum Brownian Motion. *Phys. A Stat. Mech. Appl.* **1983**, *121*, 587–616.
27. Grabert, H.; Schramm, P.; Ingold, G.L. Quantum Brownian Motion: The Functional Intergral Approach. *Phys. Rep.* **1988**, *168*, 115–207.
28. Rosenau da Costa, M.; Caldeira, A.O.; Dutra, S.M.; Westfahl, H. Exact diagonalization of two quantum models for the damped harmonic oscillator. *Phys. Rev. A* **2000**, *61*, 022107.
29. Bérut, A.; Arakelyan, A.; Petrosyan, A.; Ciliberto, S.; Dillenschneider, R.; Lutz, E. Experimental verification of Landauer's principle linking information and thermodynamics. *Nature* **2012**, *483*, 187–189.

30. Aurell, E.; Gawędzki, K.; Mejía-Monasterio, C.; Mohayaee, R.; Muratore-Ginanneschi, P. Refined Second Law of Thermodynamics for fast random processes. *J. Stat. Phys.* **2012**, *147*, 487–505.

31. Jarzynski, C.; Quan, H.T.; Rahav, S. Quantum-Classical Correspondence Principle for Work Distributions. *Phys. Rev. X* **2015**, *5*, 031038.

32. Zwanzig, R. Nonlinear generalized Langevin equations. *J. Stat. Phys.* **1973**, *9*, 215–220.

33. Bez, W. Microscopic preparation and macroscopic motion of a Brownian particle. *Z. Phys. B Condens. Matter* **1980**, *39*, 319–325.

34. Cañizares, J.S.; Sols, F. Translational symmetry and microscopic preparation in oscillator models of quantum dissipation. *Phys. A Stat. Mech. Appl.* **1994**, *212*, 181–193.

35. Hänggi, P. *Generalized Langevin Equations: A Useful Tool for the Perplexed Modeller of Nonequilibrium Fluctuations?* Springer: Berlin/Heidelberg, Germany, 1997; Volume 484, pp. 15–22.

36. Hänggi, P.; Ingold, G.L. Fundamental aspects of quantum Brownian motion. *Chaos* **2005**, *15*, 026105.

37. Gammelmark, S.; Julsgaard, B.; Mølmer, K. Past Quantum States of a Monitored System. *Phys. Rev. Lett.* **2013**, *111*, 160401.

38. Rybarczyk, T.; Peaudecerf, B.; Penasa, M.; Gerlich, S.; Julsgaard, B.; Mølmer, K.; Gleyzes, S.; Brune, M.; Raimond, J.M.; Haroche, S.; et al. Forward-backward analysis of the photon-number evolution in a cavity. *Phys. Rev. A* **2015**, *91*, 062116.

Article

Magnetic Engine for the Single-Particle Landau Problem

Francisco J. Peña [1,*], **Alejandro González** [1], **Alvaro S. Nunez** [2], **Pedro A. Orellana** [1], **René G. Rojas** [3] and **Patricio Vargas** [1,4]

1   Departamento de Física, Universidad Técnica Federico Santa María, Valparaíso 2340000, Chile;
    alephandros@gmail.com (A.G.); pedro.orellana.dinamarca@gmail.com (P.A.O.);
    vargas.patricio@gmail.com (P.V.)
2   Departamento de Física, Facultad de Ciencias Físicas y Matemáticas, Universidad de Chile,
    Santiago 8320000, Chile; alvaro.sebastian.nunez@gmail.com
3   Instituto de Física, Pontificia Universidad Católica de Valparaíso, Valparaíso 2340000, Chile;
    rene.rojas.c@gmail.com
4   Centro para el Desarrollo de la Nanociencia y la Nanotecnología, Santiago 8320000, Chile
*   Correspondence: f.penarecabarren@gmail.com or francisco.penar@usm.cl

Received: 30 September 2017; Accepted: 22 November 2017; Published: 25 November 2017

Abstract: We study the effect of the degeneracy factor in the energy levels of the well-known Landau problem for a magnetic engine. The scheme of the cycle is composed of two adiabatic processes and two isomagnetic processes, driven by a quasi-static modulation of external magnetic field intensity. We derive the analytical expression of the relation between the magnetic field and temperature along the adiabatic process and, in particular, reproduce the expression for the efficiency as a function of the compression ratio.

Keywords: quantum thermodynamics; degeneracy effects; magnetic quantum engine

1. Introduction

Quantum thermodynamics is one of the most interesting topics in physics today. The possibility to create an alternative and efficient nanoscale device, like its macroscopic counterpart, introduces the concept of the quantum engine, which was proposed by Scovil and Schultz-Dubois in the 1950s [1]. The key point here is the quantum nature of the working substance and of course the quantum versions of the laws of thermodynamics [2–18]. The combination of these two simple facts leads to very interesting studies of well-known macroscopic engines of thermodynamics, such as Carnot, Stirling and Otto, among others [2–4].

The classical Otto engine consists of two isochoric processes and two adiabatic processes. If the working substance is a classical ideal gas, the first approximation for efficiency depends on the quotient of the temperatures in the first adiabatic compression [19,20]. This expression is reduced with the specific condition along the adiabatic trajectory for this kind of gas, given by $TV^{\gamma-1} = cnt.$, where $\gamma = C_P/C_V$ obtaining the expression $\eta = 1 - \frac{1}{r^{\gamma-1}}$, where r is defined as a "compression ratio" that is defined as V_1/V_2 (with $V_1 > V_2$) [19]. On the other hand, the quantum Otto engine consists of two quantum adiabatic processes, which keeps invariant the probability occupation for the level of energy, and two quantum isochoric processes, in order to keep constant some parameters in the Hamiltonian. In this context, the quantum harmonic Otto cycle is a hot research topic [21–24], fully addressed by Kosloff and Rezek [25]. In addition, a recent experimental result has been achieved, employing a single ion confined in a linear Paul trap [26]. This research shed lights for a possible realization of a quantum Otto cycle. In the magnetic scenario, it is useful to think that the "isochoric processes" are replaced by "isomagnetic" ones where the constant value of the field in these strokes imply keeping constant value of the cyclotron frequency (or effective frequency depending on the case) due to the proportional relation between both quantities. This kind of approach is developed in the reference [13] for the case of a graphene under strain in the presence of an external magnetic field, exhibiting that the Carnot efficiency is achieved more quickly with the combination of these two effects as opposed to only applying strain to the sample.

The Landau levels of energy in condensed matter physics constitute a very well-known case and a typical academic problem. The thermodynamics is fully addressed in the works of Kumar et al. [27] where one important point is the degeneracy factor present in the partition function, and consequently also in the entropy.

In this way, a thermodynamic cycle, where the magnetic field can be controlled along the adiabatic trajectories will lead to very interesting new results which can be contrasted with the harmonic case. On the other hand, the effects of the degeneracy of energy levels on the efficiency and power of an engine quantum machine have been reported in many works in the past [21,22,28–30]. In this same framework, we highlight the work of Mehta and Ramandeep [31], who worked on a quantum Otto engine in the presence of level degeneracy, finding an enhancement of work and efficiency for two-level particles with a degeneracy in the excited state. Also, Azimi et al. present the study of a quantum Otto engine operating with a working substance of a single phase multiferroic $LiCu_2O_2$ tunable by external electromagnetic fields [22] and is extended by Chotorlishvili et al. [21] under the implementation of shortcuts to adiabaticity, finding a reasonable out power for the proposal machine. Physically it is nowadays possible to confine electrons in 2D. For instance, quantum confinement can be achieved in semiconductor hereojunctions, such as GaAs and AlGaAs. At $T = 300$ K, the band gap of GaAs is 1.43 eV while it is 1.79 eV for AlxGa1-x As ($x = 0.3$). Thus, the electrons in GaAs are confined in a 1-D potential well of length L in the Z-direction. Therefore, electrons are trapped in 2D space, where a magnetic field along Z-axis can be applied [32].

Consequently, the study of electrons under controllable external fields is a topic of growing interest today. This work proposes to study the novel-magnetic cycle presented in the work [13] for the Landau problem and to understand the role of the degeneracy factor along the cycle. In particular, we found an analytical dependence between the magnetic field and temperature along the adiabatic process, and we use these results to calculate the efficiency of this cycle. We compare this efficiency to the one corresponding to the harmonic trap with the same parametrization to study how strong is the influence of this factor on the results.

2. Partition Function for the Single-Particle Landau Problem

We consider the case for an electron with an effective mass m^* and charge e placed in a magnetic field, where the Hamiltonian of this problem working in the symmetric gauge leads to the known expression

$$\hat{H} = \frac{1}{2m^*}\left[\left(p_x - \frac{eBy}{2}\right)^2 + \left(p_y + \frac{exB}{2}\right)\right],$$ (1)

and the corresponding Landau levels display the energy spectrum

$$\mathcal{E}_n = \hbar\omega_B\left(n + \frac{1}{2}\right).$$ (2)

Here, $n = 0, 1, 2, ...$ is the quantum number, and

$$\omega_B = \frac{eB}{m^*},$$ (3)

is the standard definition for the cyclotron frequency [12,13,27]. With the definition of the parameter ω_B, we can define the Landau radius that captures the effect of the intensity of the magnetic field, given by $l_B = \sqrt{\hbar/(m^*\omega_B)}$. The energy spectrum for each level is degenerate with a degeneracy $g(B)$ given by [27]

$$g(B) = \frac{eB}{2\pi\hbar}\mathcal{A}, \tag{4}$$

with $\mathcal{A}$ being the area of the box perpendicular to the magnetic field B. Therefore, with this approach it is straightforward to calculate the partition function to Landau problem, and it turns out to be

$$Z = \frac{m^*\omega_B\mathcal{A}}{4\pi\hbar}\,\text{csch}\left(\frac{\beta\hbar\omega_B}{2}\right), \tag{5}$$

which corresponds to standard partition function for a harmonic oscillator in the canonical ensemble, with a degeneracy for level equal to $g(B)$.

3. Thermodynamics and Magnetic Engine

3.1. The First Law of Thermodynamics: A Microscopic Approach.

The first law of quantum thermodynamics is fully addressed in many works [2–18] and gives us the possibility to explore different quantum cycles and compare them with the classical analogues. We will follow the treatment of [29,33,34], where they conceive a sequence of quasi-static process that drives the subsystem along the sequence of equilibrium states. To derivate this law simply (in a microscopic approach), consider a system describe by a density operator $\hat{\rho}$ and a Hamiltonian with an explicit dependence of some parameter that we will call ξ in a generic form [34]. So, you have a set of eigenvectors of $\hat{H}$ that satisfy the eigenvalue problem

$$\hat{H}|n;\xi\rangle = \mathcal{E}_n|n;\xi\rangle, \tag{6}$$

where n represents a set of indexes that label the spectrum of the Hamiltonian and $|n;\xi\rangle$ constitutes the set of eigenvectors of $\hat{H}$. The variations of work and heat are in general defined as [29]

$$\delta W = Tr\left\{\hat{\rho}d\hat{H}\right\} \tag{7}$$

$$\delta Q = Tr\left\{\hat{H}d\hat{\rho}\right\} \tag{8}$$

On the other hand, when the eigenstates of the system Hamiltonian are used as the basis, and the coupling between the system and the environment is weak, the system can be described by a canonical distribution determined by $\hat{H}$ and the density matrix is diagonal in that representation [29]. Therefore, the ensemble-average energy $E = \langle\hat{H}\rangle$ is reduced to

$$E = \sum_n P_n(\xi)\mathcal{E}_n(\xi), \tag{9}$$

for a given occupation distribution with probabilities $P_n(\xi)$ in the nth eigenstate. So, we can describe the system by the energy levels and states derived from the Hamiltonian [29]. The statistical ensemble just described can be submitted to an arbitrary quasi-static process, involving the modulation of the parameter ξ, and hence the ensemble average energy changes accordingly,

$$\begin{aligned} dE &= \sum_n \left(\mathcal{E}_n(\xi)dP_n(\xi) + P_n(\xi)d\mathcal{E}_n(\xi)\right) \\ &= \delta Q + \delta W. \end{aligned} \tag{10}$$

The last equation corresponds to the microscopic formulation of the first law of thermodynamics [2–18,25–27,31,34,35]. The first term in Equation (10) is associated with the energy exchange, while the second term represents the work done. That is, the work performed corresponds to the change in the eigenenergies $\mathcal{E}_n(\xi)$. This is in agreement with the fact that work can only be carried out through a change in generalized coordinates of the system, which in turn gives rise to a change in the eigenenergies [9,10]. A very important assumption for Equation (10) is very well described in [29], where it is mentioned that: "small changes in $\hat{H}$, which can be considered as a first-order perturbation, only shifts the energy levels of the systems and does not modify its eigenstates".

The usual expression for the entropy is given by the von Neumann form and in the eigenenergy basis can be rewritten as

$$S(\xi) = -k_B \sum_n P_n(\xi) \ln [P_n(\xi)], \tag{11}$$

where the coefficients $P_n(\xi)$ satisfy that $0 \leq P_n(\xi) \leq 1$ with the normalization condition

$$\sum_n P_n(\xi) = 1. \tag{12}$$

3.2. Magnetic Engine

As mentioned above, the result of the efficiency of the conventional Otto cycle can be written in the form that the results only depend on the quotient of the temperatures in the first adiabatic compression. By using the properties of the ideal gas, the efficiency can be rewritten as follows:

$$\eta = 1 - \frac{1}{r^{\gamma-1}}, \tag{13}$$

where $\gamma = C_P/C_V$ is the quotient of the two specific heat (at constant pressure and at constant volume) and r is known as "compression ratio" which is defined as $\frac{V_1}{V_2}$ (with $V_1 > V_2$).

On the other hand, the quantum "conventional" Otto engine is composed of two quantum isochoric processes and two quantum adiabatic processes. For the first mentioned process, the occupation probabilities $P_n(\xi)$ change and thus the entropy S changes, until the working substance finally reaches thermal equilibrium with the heat bath. For the case of the quantum adiabatic process, the population distributions remain unchanged, that is $dP_n(\xi) = 0$. Thus, no transition occurs between levels, and no heat is exchanged during this process. It is important to recall that in a classical adiabatic process the occupation of each level is never invariant (unless the classical thermalization condition is relaxed) [35]. This is a crucial point in the discussion for our magnetic engine, based on the work presented [13]. In general, a quantum adiabatic stroke does not maintain the system in a thermal state [28–30]. A power law potential can guarantee the thermal state condition when the degeneracy does not change, as in the treatment of [36], but not in our case under the study, because the number of states involved in the process should not increase, even if the degeneracy increases. So, if the cycle is strictly quantum, the stage after the quantum adiabatic stroke necessarily is a non-equilibrium state (only in the degenerate case), and cannot be represented for the partition function of Equation (5). Therefore, our approach uses the condition of a classical thermodynamic adiabaticity, where the process is identified in terms of the conservation of the entropy and the isolation of the system from heat exchange with the thermal bath. We recall that we work in a semi-classical scenario where the quantum part is related to the nature of the working substance and the classical part is due to the condition imposed over the adiabatic strokes along the cycle that we propose.

The case of conventional quantum Otto engine has been considered in several works for differents quantum systems [3,9,13,31,36] where the key findings are an expression for the efficiency present in Equation (13) and establishing the value of γ for that case. For the magnetic case, the two isochoric trajectories are replaced by two "isomagnetic" ones, in which the magnetic field intensity along the process remains constant while heat is exchanged between the system and the reservoirs [13].

Let us consider a cycle by devising a sequence of quasi-static trajectories, as depicted in Figure 1. First, the system, while submitted to an external magnetic field B_1, is brought into thermal equilibrium with macroscopic thermostats at temperature T_1. In equilibrium, the probabilities $P_n(\xi)$ take the Boltzmann form and can work with a partition function in the canonical ensemble, $Z(\xi, T)$. So, the Helmholtz free energy can be defined by $F(\xi, T) = -k_B T \ln Z(\xi, B)$ and the entropy given by

Equation (11) can be written as $S(\xi, T) = \frac{E(\xi,T)}{T} + k_B \ln Z(\xi, T)$, where the ensemble average energy is given by

$$E(\xi, T) = k_B T^2 \frac{\partial}{\partial T} \ln Z(\xi, T). \tag{14}$$

Here, the ξ parameter is related to the intensity of the magnetic field, so $\xi \to B$ for the case under the study. Then, the system is submitted to a quantum isoentropic process from $1 \to 2$, increasing the magnitude of the magnetic field from B_1 to B_2. The systems performs work along the isoentropic trajectory according to

$$W_{1\to2} = \int_{B_1}^{B_2} dB \left(\frac{\partial E}{\partial B}\right)_S = E(T_2, B_2) - E(T_1, B_1). \tag{15}$$

For the case of the "isomagnetic" heating process with the intensity of magnetic field equal to B_2 from $2 \to 3$, no work is done, but heat is absorbed. The heat absorbed ($Q_{2\to3}$) is given by the expression

$$Q_{2\to3} = \int_{T_2}^{T_3} dT \left(\frac{\partial E}{\partial T}\right)_{B_2} = E(T_3, B_2) - E(T_2, B_2). \tag{16}$$

In the same way discussed before, the isoentropic trajectory from $3 \to 4$, the system perform work in the form

$$W_{3\to4} = E(T_4, B_1) - E(T_3, B_2). \tag{17}$$

A physical interpretation of the work performed by the engine is obtained by considering the statistical mechanical definition of the ensemble-average magnetization, $M = -\left(\frac{\partial E}{\partial B}\right)_S$. Hence, the works defined in Equations (15) and (17) can also be interpreted as $W = -\int M dB$ [12,13].

Similarly, we obtain the heat released to the low temperature sink in the quantum "isomagnetic" cooling process from $4 \to 1$

$$Q_{4\to1} = \int_{T_4}^{T_1} dT \left(\frac{\partial E}{\partial T}\right)_{B_1} = E(T_1, B_1) - E(T_4, B_1). \tag{18}$$

The efficiency of the engine is then given by the expression

$$\eta = \left|\frac{W_{1\to2} + W_{3\to4}}{Q_H}\right| = 1 - \left|\frac{E(T_1, B_1) - E(T_4, B_1)}{E(T_3, B_2) - E(T_2, B_2)}\right|. \tag{19}$$

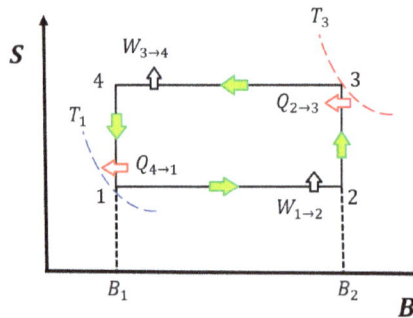

Figure 1. Pictorial description for the novel-magnetic engine represented as an entropy versus a magnetic field diagram.

If we have the analytic function for the entropy, the intermediate temperatures T_2 and T_4 must be determined to reduce the expression for efficiency and can take on two different forms:

- Deducting the relation between the magnetic field and the temperature along an isoentropic trajectory solving the differential equation of first order given by

$$dS(B,T) = \left(\frac{\partial S}{\partial B}\right)_T dB + \left(\frac{\partial S}{\partial T}\right)_B dT = 0, \tag{20}$$

which can be written as

$$\frac{dB}{dT} = -\frac{C_B}{T\left(\frac{\partial S}{\partial B}\right)_T}, \tag{21}$$

where C_B is the specific heat at constant magnetic field.

- The other possibility is connecting the value for the entropy in two isoentropic trajectories in the form

$$\begin{aligned}
S(B_1, T_1) &= S(B_2, T_2) \\
S(B_2, T_3) &= S(B_1, T_4),
\end{aligned} \tag{22}$$

finding the function for the magnetic field in terms of the temperature through numerical calculation. Finally, we parametrize this dependency in the efficiency by defining the ratio

$$r = \frac{l_{B_1}}{l_{B_2}}, \tag{23}$$

which represents the analogy of the compression ratio for the classical case. It is important to remember that the Landau radius is inversely proportional to the magnitude of the magnetic field. Therefore, for a major (minor) magnitude of the field, the Landau radius is smaller (bigger), and the r parameter is well defined.

It is important to highlight the work of Zheng and Poletti [36], where they derived a general form for the efficiency of quantum Otto cycles with power law trapping potentials, corresponding to Equation (11), and showed that γ must be equal to three. We remark that this result requires that two conditions are met. First, it is only valid for the non-degenerate cases, or more specifically, when the degeneracy is independent of the parameter that rules the cycle. The second condition is that the expansion process, when the system goes to $\omega' \to \omega''$, must follow the following relation

$$\kappa \equiv \frac{\mathcal{E}_n(\omega')}{\mathcal{E}_n(\omega'')} = \left(\frac{\omega'}{\omega''}\right)^\alpha, \tag{24}$$

where α depends on the power of the potentials under study. For example, for a conventional harmonic trap, the spectrum of energy is always $\mathcal{E}_n = \hbar\omega\left(n + \frac{1}{2}\right)$ and we quickly obtain the result previously discussed. Moreover, this value of γ is valid for a family of trapping potentials that fulfills the state equation $PV = 2\langle E\rangle$ [36].

The case of the Landau problem is different. The energy spectrum of Equation (2) respects the condition of Equation (24) and has the structure of a harmonic trap; however, the degeneracy factor is a function of the magnetic field and the size of the system. Therefore, the results previously discussed do not hold, because in the first ($1 \to 2$) and third ($3 \to 4$) process, the change in the intensity of magnetic field leads to a change in the degeneracy factor, thus this problem must be analyzed carefully.

Magnetic Engine for the Landau Problem

We show that the representation of the partition function for this case can be taken in the form of Equation (5). The thermodynamic quantities are present in the work of Kumar et al. [27], given by

$$\mathcal{F} = -\frac{1}{\beta} \ln \left[\frac{g(B)}{2} \operatorname{csch} \left(\frac{\beta \hbar \omega_B}{2} \right) \right], \tag{25}$$

$$E = \frac{\hbar \omega_B}{2} \coth \left(\frac{\beta \hbar \omega_B}{2} \right), \tag{26}$$

$$S_L = \frac{\hbar \omega_B}{2T} \coth \left(\frac{\beta \hbar \omega_B}{2} \right) + k_B \ln \left[\frac{g(B)}{2} \operatorname{csch} \left(\frac{\beta \hbar \omega_B}{2} \right) \right], \tag{27}$$

and the specific heat

$$C_B = k_B \beta^2 \left(\frac{\hbar \omega_B}{2} \right)^2 \operatorname{csch}^2 \left(\frac{\beta \hbar \omega_B}{2} \right), \tag{28}$$

where $\beta = \frac{1}{k_B T}$. First, we highlight that Equation (25) imply as natural consequence that the entropy contains the degeneracy terms, due to relation $S = \frac{1}{T}(E - F)$. It is in fact due to the structure of von Neumann entropy, because the probability coefficients contain the information of the degeneracy factor. For example, in thermal equilibrium, this coefficient takes the Boltzmann form, so

$$P_n(\xi) = [Z(\xi, T)]^{-1} g(\xi) e^{-\frac{\varepsilon_n(\xi)}{k_B T}}. \tag{29}$$

An opposite case occurs for the expected value of energy and the specific heat at constant field because these two physical quantities are obtained as the derivative in the temperature of the partition function.

To clarify the importance of the degeneracy, we analyze the following case. Instead of the term $\frac{g(B)}{2}$ appearing in Equations (25) and (27), we put a factor one, corresponding to treat a single oscillator, and we call the entropy for that case just $S(T, B)$. It is easy to show that the dependence of the magnetic field on the temperature for the isoentropic trajectories in the non-degenerate scenario obeys the proportionality $B \propto T$. This trivial relation gives us the possibility to obtain the relations between the temperatures along the cycle given by $\frac{T_1}{T_2} = \frac{T_3}{T_4}$, and the efficiency is reduced to a very well-known expression

$$\eta = 1 - \frac{\omega(B_1)}{\omega(B_2)}, \tag{30}$$

which can be rewritten as

$$\eta = 1 - \frac{1}{\left(\frac{l_{B_1}}{l_{B_2}} \right)^2} \equiv 1 - \frac{1}{r^{3-1}}, \tag{31}$$

and we get the result $\gamma = 3$, as described in the work of Zheng and Poletti [36].

4. Results and Discussion

For the Landau case, it is useful to rewrite the term of the degeneracy factor in the entropy as $\frac{g}{2} = \frac{\Phi(B)}{2\Phi_0}$, where $\Phi(B)$ is the total magnetic flux and Φ_0 is the universal quantum of magnetic flux, given by $h/2e$. Moreover, we define this degeneracy term in the entropy as $\frac{g}{2} = \lambda B$, where $\lambda = \frac{A}{2\Phi_0}$. Thus, the entropy for this case given by Equation (27) depends on three variables, $S_L \equiv S_L(T, B, \lambda)$.

If the dependence of magnetic field and temperature in the adiabatic process for the Landau case is analyzed, we clearly see that the condition for entropy $S_L(T_0, B_0, \lambda) = S_L(T, B, \lambda)$ yields a relation between the magnetic field and temperature which will not depend on λ. This is because the degeneracy term $g(B)$ is associated with a logarithmic term, so the degeneracy effect in the cycle is only reflected in the magnetic field dependence of $g(B)$. To reinforce this idea, we calculate the structure of first order differential equation for the adiabatic processes through Equation (21), which for this case has the form

$$\frac{dB}{dT} = -\frac{C_1^2 \frac{B^2}{T^3} \text{csch}^2\left(C_1 \frac{B}{T}\right)}{\frac{1}{B} - C_1^2 \frac{B}{T^2}\text{csch}^2\left(C_1 \frac{B}{T}\right)}, \tag{32}$$

where C_1 is a constant given by $C_1 = \frac{e\hbar}{2k_B m}$. This previous equation has an analytical solution (see Appendix A for details) given by

$$C_1 \frac{B}{T} \coth\left(C_1 \frac{B}{T}\right) + \ln\left(C_1 B\right) - \ln\left[\sinh\left(C_1 \frac{B}{T}\right)\right] = C_2, \tag{33}$$

where C_2 is an integration constant. Note that the additional term in the differential equation which provides, $g(B)$, is the factor $(1/B)$ in the dominator of Equation (32). If this term does not exist, the differential equation has a simple form $\frac{dB}{dT} = \frac{B}{T}$ and obtains the result previously discussed for the non-degenerate case.

In Figure 2, we see the behavior of the magnetic field versus the temperature along an isoentropic trajectory, showing the linear dependence between the magnetic field and the temperature in the case of $g = 1$ (non-degenerate) and for the case of high degeneracy. In order to see the scale of entropy for $S_L(T, B, \lambda)$ for real values, we select $\lambda \propto 10^8$ T^{-1}, which means an active area of $A \propto 10^{-7}$ m^2, by using the fact that the universal flux quantum has an order of $\Phi_0 \propto 10^{-15}$ Wb. In the left panel of Figure 2, we plot the solution for the case $S(T, B) = S(10, 1)$, and in the right panel we plot the solution for the case $S(T, B, 10^8) = S(10, 1, 10^8)$. The contrast is evident: in the simple scenario an increase in the magnetic field implies an increase in the temperature. However, for the case with degeneracy, the rise in the magnetic field leads to a decrease in the temperature. The explanation of this fact lies in the behavior of the entropy at low temperatures, because of $S(T, B, \lambda)_{T \to 0} \sim k_B \ln(g)$, where g is directly proportional to B. This is discussed in Figure 3 where we show the entropy behavior in these two different scenarios. In the non-degenerate case, when we increase the magnetic field, the function $S(T, B)$ intersects the starting value of the entropy always in a higher value than the initial one, reflected in the left frame of Figure 3. This explains the linearity that we obtain in a plot B vs. T for the left panel of Figure 2. The opposite occurs for the degenerate case, the function $S_L(T, B, 10^8)$, which intersects the starting value of the entropy always in a lower value than the initial one, as we see in the right frame of Figure 3. From this same figure, we can conclude that the entropy function for the degenerate case collapses to approximately the same value for higher temperature for different values of the magnetic field. Therefore, since the magnetic field is the external parameter that makes the engine work, we have a region of temperature and magnetic field where it is valid to carry out this cycle.

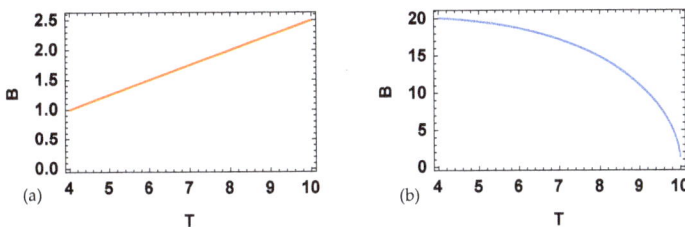

Figure 2. Behavior of the magnetic field versus the temperature for the case without the degeneracy factor (**a**) and the case with the degeneracy factor $\frac{\Phi(B)}{2\Phi_0}$ (**b**). We select the factor $\frac{A}{2\Phi_0} \propto 10^8$ T^{-1} for this example.

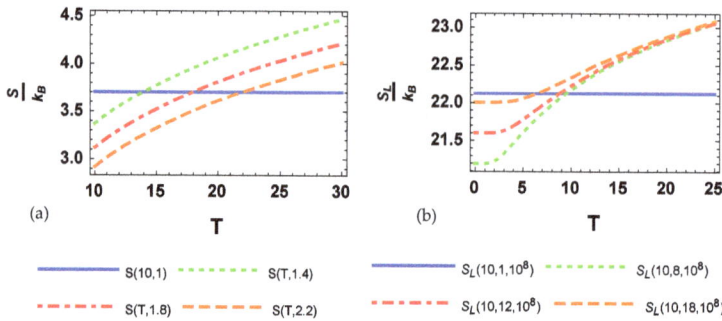

Figure 3. The isoentropic trajectories behavior for the two cases under discussion. In (**a**) we plot the non-degenerate case $S(T,B) = S(10,1)$ and in (**b**) the degenerate case $S_L(T,B,10^8) = S_L(10,1,10^8)$.

As discussed in Appendix A, with an adequate analysis of the asymptotic behaviours of Equation (32), we found a critical temperature, given by

$$T_c = e^{(\mathcal{C}_2 - 1)}, \tag{34}$$

which corresponds to the value of the temperature when the magnetic field goes to zero and a critical value for the magnetic field when it starts to become constant, given by the expression

$$B_c = \frac{e^{\mathcal{C}_2}}{2\mathcal{C}_1}. \tag{35}$$

Therefore, we have the two points for an initial value constant $\mathcal{C}_2$ where it makes sense to carry out the cycle. For the exponential form of Equation (35), the critical value of the constant magnetic field is always a large quantity. For a real example of a starting field and temperature, we can consider the example of Figure 2, where the initial values of the intensity field and temperature are 1 T and a 10 K, respectively. The approximate value for the critical values are $T_c \approx 10.1$ K and $B_c \approx 20.1$ T.

For the starting point previously indicated, we can consider a cycle for the degenerate case like that in Figure 1 operating between the temperatures 4 K and 10 K. However, due to the behavior of temperature along the adiabatic trajectory described in Figure 2, initially we brought the system into thermal equilibrium at $T_1 > T_3$. Thus, for that case, the heat defined by $Q_{4\to1}$ corresponds to the heat absorbed, and for the heat released the correct definition is given by $Q_{2\to3}$, contrary to the non-degenerate case. To reinforce this idea, we display in the right frame of Figure 4 the behavior of heat along the cycle for the degenerate case and non-degenerate case. The convention of the sign (positive for heat absorbed) is satisfied along the entire operation of the engine. The positive work condition, which plays an essential role for a well defined thermal engine, is shown in the left frame of Figure 4 for both cases. From the same figure, we can extract relevant information about the r parameter. For a machine operating between two reservoirs of 4 K and 10 K, we obtain

$$r^{max}_{non-deg} = 1.58 \quad \text{and} \quad r^{max}_{deg} = 4.47, \tag{36}$$

which represents the maximum value that can be taken for the compression ratio along the cycle and corresponds to the point where the Carnot efficiency is obtained. These results are natural only to see Figure 3 for this example. For the non-degenerate case, it is only necessary to increase the field by a factor of 2.5, but for the degenerate case it is necessary to increase the field by a factor of 20 to reach the Carnot efficiency of the problem. However, the two points previously discussed have the average work and power output equal to zero, as we see in the left panel of Figure 4 and the purpose of showing them is only conceptual. From Figure 4 we observe that the two points for the maximum power output in the two different scenarios under discussion are given by

$$r^W_{non-deg} \approx 1.25 \quad \text{and} \quad r^W_{deg} \approx 3.66, \tag{37}$$

with efficiency values $\eta \left(r^W_{non-deg} \right) \approx 0.34$ and $\eta \left(r^W_{deg} \right) \approx 0.26$, as we see in Figure 5. This combination of values is the really important in our problem because it corresponds to the optimal operating region of the proposed thermal machine.

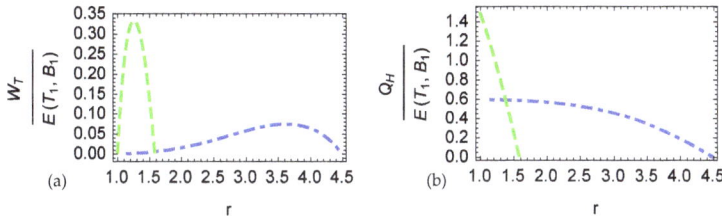

Figure 4. Total work (**a**) and input heat (**b**) versus the r parameter along the cycle for the case with degeneracy (dot dashed line) and without degeneracy (dashed line).

In Figure 5 we compare the three efficiencies, where we see the effect of the degeneracy. One form to understand this behavior corresponds to the approximation that we show in Appendix A for the parametric solution, with the finality to "uncouple" the solution to magnetic field and the temperature for the adiabatic trajectory getting a function in the form

$$B(T) = \frac{k}{2C_1} \left(1 - e^{-k\sqrt{\frac{0.64}{T^2}\left(1-e^{\frac{T}{k}}\right)}} \right), \tag{38}$$

where we define $k = e^{C_2}$. Therefore, for this exponential form for the field as a function of temperature, when we parametrize the efficiency vs. a function of the typical compression ratio (r), we obtain the behavior presented in Figure 5.

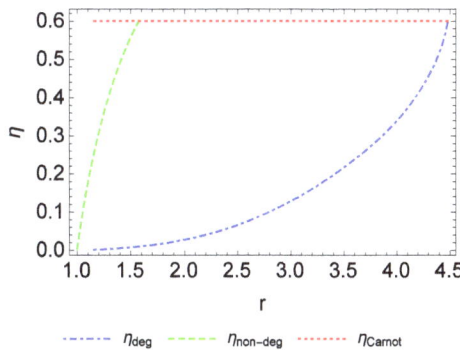

Figure 5. Efficiency for different cases of interest. For this case, the dotted red line corresponds to the value of Carnot cycle for a machine operating between the two temperatures $T_1 = 4$ K and $T_3 = 10$ K.

For the definition of work and its interpretation as $W = - \int M dB$, we study the magnetization along the cycle defined as

$$M = \frac{e\hbar}{2m} \left(\frac{2}{\beta\hbar\omega_B} - \coth\left(\frac{\beta\hbar\omega_B}{2} \right) \right). \tag{39}$$

For the adiabatic trajectory, the temperature and the magnetic field change along the entire process, so we can use a contour plot to see the value of magnetization displayed in Figure 6. Here, we clearly see that the values of magnetization are always negative and the same occurs for the different curves in the "isomagnetic" process, shown in Figure 7, indicating that the response of the system is diamagnetic.

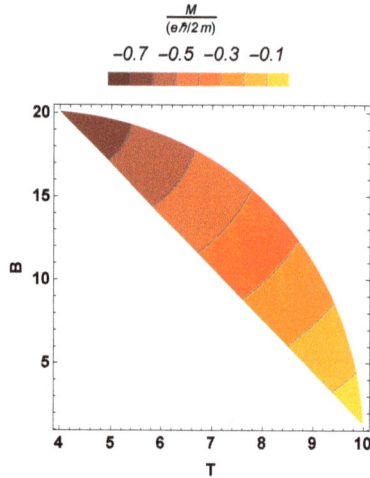

Figure 6. Magnetization as a function of B and T along the adiabatic trajectories.

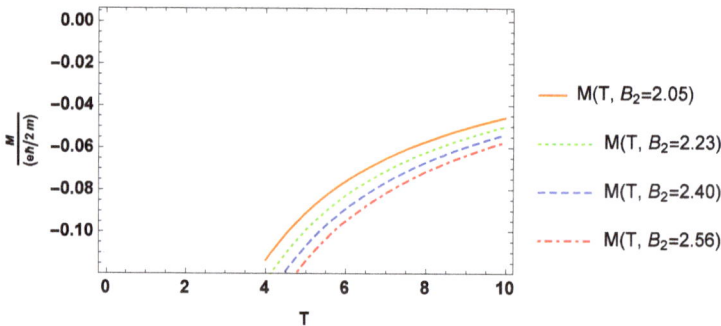

Figure 7. Magnetization along the first iso-magnetic trajectory as a function of T in the range of 4 K to 10 K. We selected the different values for B_2 that we found from numerical calculations.

Our system was studied to prove a concept rather than a practical implementation protocol. However, we believe the readers will find attractive the study of the "full" quantum version of this cycle following the treatment of the works [28–30], where treated the magnetic substance under degenerate conditions using non-equilibrium techniques. Besides, it is promising to treat the quantum version of our formulation optimized following the work of Kosloff and Rezek [25], for the case of frictionless adiabats using the methods of shortcuts to adiabaticity [37–39]. Moreover, this problem can be extended taking in account the edge states of the systems for a more realistic approach. Additionally, it is important to note that the regime of validity of this semi-classical approach deserves further investigation.

5. Conclusions

In this work, we explored the possibility of constructing a single-particle magnetic engine of the Landau problem. In particular, we found an analytical solution for the dependence of the magnetic field and temperature in the adiabatic trajectories. We used this relation to obtain the form of the efficiency showing a radically different behavior of the typical harmonic case and found that a major increase in the external magnetic field to reach the Carnot efficiency is necessary. We remark that the useful work of this engine, related to change in the magnetization along the process, can be used for example in the generation of induction current in other physical systems.

It is important to note that our one-particle approach must be refined to take into account a many electrons scenario, which yields more precise calculations. However, the one electron case is important due to simplicity and the arising of richer physics for comparatives cases. For example, we can work with a one-particle system combining the effects of a cylindrical potential well, which physically represents an accurate model for a semiconductor quantum dot, and an externally imposed magnetic field, where the number of electrons can be controlled without problems; thus, the same analysis presented in this work can be replicated.

Acknowledgments: Francisco J. Peña acknowledges the financial support of FONDECYT-postdoctoral 3170010, as well as G. Alvarado Barrios and F. Albarrán-Arriagada for the important comments to improve the manuscript. Alvaro S. Nunez and Patricio Vargasv acknowledge support from Financiamiento Basal para Centros Científicos y Tecnológicos de Excelencia, under Project No. FB 0807 (Chile), Patricio Vargas acknowledges USM-DGIIP grant number PI-M-17-3 (Chile) and Alvaro S. Nunez acknowledges funding from FONDECYT program under grant 1150072. The authors acknowledge the referees for their constructive comments on our work.

Author Contributions: Francisco J. Peña and Patricio Vargas conceived the idea and formulated the theory. Alejandro González built the computer program and edited figures. René G. Rojas solved the differential equation and their respective asymptotic analysis. Alvaro S. Nunez and Pedro A. Orellana contributed to discussions during the entire work and the corresponding editing of the same. Francisco J. Peña wrote the paper. All authors have read and approved the final manuscript.

Conflicts of Interest: The authors declare no conflict of interest.

Appendix A

We need to solve the differential equation in the form

$$\frac{dB}{dT} = -\frac{C_1^2 \frac{B^2}{T^3} \operatorname{csch}^2\left(C_1 \frac{B}{T}\right)}{\frac{1}{B} - C_1^2 \frac{B}{T^2} \operatorname{csch}^2\left(C_1 \frac{B}{T}\right)}. \tag{A1}$$

We define the parameter $u = C_1\left(\frac{B}{T}\right)$. So, differentiating respect to T, we obtain

$$\frac{du}{dT} = -C_1 \frac{B}{T^2} + \frac{C_1}{T}\frac{dB}{dT}. \tag{A2}$$

Collecting these two last equations, we obtain the first order differential equation in the u parameter in the form

$$\frac{du}{dT} = \frac{u}{T\left(u^2\operatorname{csch}^2(u) - 1\right)}, \tag{A3}$$

which corresponds to a differential equation of separable variables that has a solution given by

$$u\coth(u) + \ln(u) - \ln\left[\sinh(u)\right] + \ln(T) = C_2, \tag{A4}$$

where C_2 is a constant of integration. We can compact this solution if we define the two variables

$$y = \frac{C_1 B}{k} \qquad x = \frac{T}{k}, \tag{A5}$$

where k is given by

$$k = e^{C_2}, \tag{A6}$$

with e as the Euler number. So, the solution takes the parametric form

$$y = e^{-u \coth u} \sinh u \qquad x = \frac{1}{u} e^{-u \coth u} \sinh u. \tag{A7}$$

The asymptotic behaviors of these solutions is very interesting. The expression in the case of $u \ll 1$, which corresponds to high-temperature or small magnetic field limit, takes the form

$$y = \frac{1}{e} \sqrt{6(1 - ex)}, \tag{A8}$$

so, we have a critical value, x_c, when $y \to 0$ given by

$$x_c = \frac{1}{e}. \tag{A9}$$

It gives us a critical temperature T_c when the magnetic field goes to zero, and is strongly dependent on initial values of the problem under study, given by

$$T_c = \frac{k}{e} \equiv e^{(C_2 - 1)}. \tag{A10}$$

In the other case, for $u \gg 1$, which corresponds to low temperature or high magnetic field limit, we obtain

$$y_c = \frac{1}{2}, \tag{A11}$$

and, this represents a critical constant value for the magnetic field, given by

$$B_c = \frac{k}{2C_1} \equiv \frac{e^{C_2}}{2C_1}, \tag{A12}$$

Thus, it is important to keep in mind that, when we consider a variation of the magnetic field as the cause for effective work in the system, the limits discussed before impose physical variable restrictions to operate the quantum machine proposed in the text.

To understand the magnetic field behavior in an explicit form along the adiabatic process, we propose an approximated curve in the form

$$y = \frac{1}{2} \left(1 - e^{\frac{-\sqrt{0.64(1-ex)}}{x}} \right). \tag{A13}$$

The exact parametric solution, the asymptotic behavior for the limiting cases ($u \gg 1$ and $u \ll 1$) and our proposal function are displayed in Figure A1.

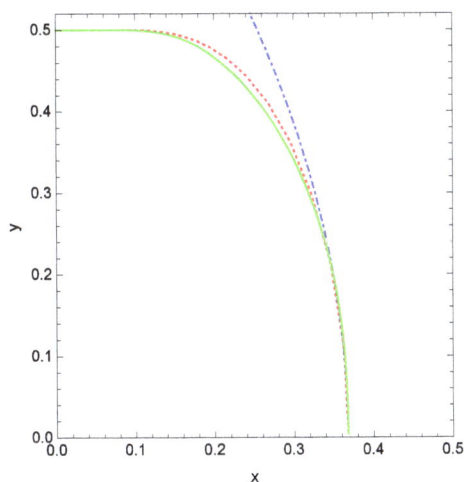

Figure A1. A parametric solution of the differential equation along the adiabatic trajectories for the Landau case. The dotted line represents the exact solution and the dot-dashed line the asymptotic case for $u \ll 1$. We can clearly see the constant value 0.5 for the solution in the case of $u \gg 1$ from the dotted line in the figure. The solid line represents the proposal curve given by Equation (A13) showing a good fit for the problem under study.

References

1. Scovil, H.E.D.; Schulz-DuBois, D.O. Three-Level masers as a heat engines. *Phys. Rev. Lett.* **1959**, *2*, 262–263.
2. Huang, X.L.; Niu, X.Y.; Xiu, X.M.; Yi, X.X. Quantum Stirling heat engine and refrigerator with single and coupled spin systems. *Eur. Phys. J. D* **2014**, *68*, doi:10.1140/epjd/e2013-40536-0.
3. Su, S.H.; Luo, X.Q.; Chen, J.C.; Sun, C.P. Angle-dependent quantum Otto heat engine based on coherent dipole-dipole coupling. *EPL* **2016**, *115*, 30002.
4. Liu, S.; Ou, C. Maximum Power Output of Quantum Heat Engine with Energy Bath. *Entropy* **2016**, *18*, 205.
5. Scully, M.O.; Zubairy, M.S.; Agarwal, G.S.; Walther, H. Extracting work from a single heath bath via vanishing quantum coherence. *Science* **2003**, *299*, 862–864.
6. Scully, M.O.; Zubairy, M.S.; Dorfmann, K.E.; Kim, M.B.; Svidzinsky, A. Quantum heat engine power can be increased by noise-induced coherence. *Proc. Natl. Acad. Sci. USA* **2011**, *108*, 15097–15100.
7. Bender, C.M.; Brody, D.C.; Meister, B.K. Quantum mechanical Carnot engine. *J. Phys. A Math. Gen.* **2000**, *33*, 4427–4436.
8. Bender, C.M.; Brody, D.C.; Meister, B.K. Entropy and temperature of quantum Carnot engine. *Proc. R. Soc. Lond. A* **2002**, *458*, 1519–1526.
9. Wang, J.H.; Wu, Z.Q.; He, J. Quantum Otto engine of a two-level atom with single-mode fields. *Phys. Rev. E* **2012**, *85*, 041148.
10. Huang, X.L.; Xu, H.; Niu, X.Y.; Fu, Y.D. A special entangled quantum heat engine based on the two-qubit Heisenberg XX model. *Phys. Scr.* **2013**, *88*, 065008.
11. Muñoz, E.; Peña, F.J. Quantum heat engine in the relativistic limit: The case of Dirac particle. *Phys. Rev. E* **2012**, *86*, 061108.
12. Muñoz, E.; Peña, F.J. Magnetically driven quantum heat engine. *Phys. Rev. E* **2014**, *89*, 052107.
13. Peña, F.J.; Muñoz, E. Magnetostrain-driven quantum heat engine on a graphene flake. *Phys. Rev. E* **2015**, *91*, 052152.
14. Peña, F.J.; Ferré, M.; Orellana, P.A.; Rojas, R.G.; Vargas, P. Optimization of a relativistic quantum mechanical engine. *Phys. Rev. E* **2016**, *94*, 022109.
15. Wang, J.; He, J.; He, X. Performance analysis of a two-state quantum heat engine working with a single-mode radiation field in a cavity. *Phys. Rev. E* **2011**, *84*, 041127.

16. Abe, S. Maximum-power quantum-mechanical Carnot engine. *Phys. Rev. E* **2011**, *83*, 041117.
17. Wang, J.H.; He, J.Z. Optimization on a three-level heat engine working with two noninteracting fermions in a one-dimensional box trap. *J. App. Phys.* **2012**, *111*, 043505.
18. Wang, R.; Wang, J.; He, J.; Ma, Y. Performance of a multilevel quantum heat engine of an ideal N-particle Fermi system. *Phys. Rev. E* **2012**, *86*, 021133.
19. Callen, H.B. *Thermodynamics and an Introduction to Thermostatistic*; Jhon Wiley & Sons: New York, NY, USA, 1985.
20. Tolman, R.C. *The Principles of Statistical Mechanics*; Oxford University Press: Oxford, UK, 1938.
21. Chotorlishvili, L.; Azimi, M.; Stagraczyński, S.; Toklikishvili, Z.; Schüler, M.; Berakdar, J. Superadiabatic quantum heat engine with a multiferroic working medium. *Phys. Rev. E* **2016**, *94*, 032116.
22. Azimi, M.; Chorotorlisvili, L.; Mishra, S.K.; Vekua, T.; Hübner, W.; Berakdar, J. Quantum Otto heat engine based on a multiferroic chain working substance. *New J. Phys.* **2014**, *16*, 063018.
23. Jaramillo, J.; Beau, M.; del Campo, A. Quantum supremacy of many-particle thermal machines. *New J. Phys.* **2016**, *18*, 075019.
24. Del Campo, A.; Goold, J.; Paternostro, M. More bang for your buck: Super-adiabatic quantum engines. *Sci. Rep.* **2017**, *4*, doi:10.1038/srep06208.
25. Kosloff, R.; Rezek, Y. The Quantum Harmonic Otto Cycle. *Entropy* **2017**, *19*, 136.
26. Roßnagel, J.; Dawkins, T.K.; Tolazzi, N.K.; Abah, O.; Lutz, E.; Kaler-Schmidt, F.; Singer, K. A single-atom heat engine. *Science* **2016**, *352*, 325–329.
27. Kumar, J.; Sreeram, P.A.; Dattagupta, S. Low-temperature thermodynamics in the context of dissipative diamagnetism. *Phys. Rev. E* **2009**, *79*, 021130.
28. Dong, C.D.; Lefkidis, G.; Hübner, W. Quantum Isobaric Process in Ni_2. *J. Supercond. Nov. Magn.* **2013**, *26*, 1589–1594.
29. Dong, C.D.; Lefkidis, G.; Hübner, W. Quantum Magnetic quantum diesel in Ni_2. *Phys. Rev. B* **2013**, *88*, 214421.
30. Hübner, W.; Lefkidis, G.; Dong, C.D.; Chaudhuri, D. Spin-dependent Otto quantum heat engine based on a molecular substance. *Phys. Rev. B* **2014**, *90*, 024401.
31. Mehta, V.; Johal, R.S. Quantum Otto engine with exchange coupling in the presence of level degeneracy. *Phys. Rev. E* **2017**, *96*, 032110.
32. Mani, R.G.; Smet, J.H.; von Klitzing, K.; Narayanamurti, V.; Johnson, W.B.; Umansky, V. Zero-resistance states induced by electromagnetic-wave excitation in GaAs/AlGaAs heterostructures. *Nature* **2002**, *420*, 646–650.
33. Quan, H.T.; Zhang, P.; Sun, C.P. Quantum heat engine with multilevel quantum systems. *Phys. Rev. E* **2005**, *72*, 056110.
34. Muñoz, E.; Peña, F.J.; González, A. Magnetically-Driven Quantum Heat Engines: The Quasi-Static Limit of Their Efficiency. *Entropy* **2016**, *18*, 173.
35. Quan, H.T. Quantum thermodynamic cycles and quantum heat engines (II). *Phys. Rev. E* **2009**, *79*, 041129.
36. Zheng, Y.; Polleti, D. Work and efficiency of quantum Otto cycles in power-law trapping potentials. *Phys. Rev. E* **2014**, *90*, 012145.
37. Cui, Y.Y.; Chem, X.; Muga, J.G. Transient Particle Energies in Shortcuts to Adiabatic Expansions of Harmonic Traps. *J. Phys. Chem. A* **2016**, *120*, 2962–2969.
38. Beau, M.; Jaramillo, J.; del Campo, A. Scaling-up Quantum Heat Engines Efficiently via Shortcuts to Adiabaticity. *Entropy* **2016**, *18*, 168.
39. Deng, J.; Wang, Q.; Liu, Z.; Hänggi, P.; Gong, J. Boosting work characteristics and overall heat-engine performance via shortcuts to adibaticity: Quantum and classical systems. *Phys. Rev. E* **2013**, *88*, 062122.

entropy

MDPI

Article

Information Landscape and Flux, Mutual Information Rate Decomposition and Connections to Entropy Production

Qian Zeng [1] and Jin Wang [1,2,*]

1   State Key Laboratory of Electroanalytical Chemistry, Changchun Institute of Applied Chemistry, Changchun, Jilin 130022, China; qzeng@ciac.ac.cn
2   Department of Chemistry and Physics, State University of New York, Stony Brook, NY 11794, USA
*   Correspondence: jin.wang.1@stonybrook.edu; Tel.: +1-631-632-1185

Received: 29 September 2017; Accepted: 6 December 2017; Published: 11 December 2017

Abstract: We explored the dynamics of two interacting information systems. We show that for the Markovian marginal systems, the driving force for information dynamics is determined by both the information landscape and information flux. While the information landscape can be used to construct the driving force to describe the equilibrium time-reversible information system dynamics, the information flux can be used to describe the nonequilibrium time-irreversible behaviors of the information system dynamics. The information flux explicitly breaks the detailed balance and is a direct measure of the degree of the nonequilibrium or time-irreversibility. We further demonstrate that the mutual information rate between the two subsystems can be decomposed into the equilibrium time-reversible and nonequilibrium time-irreversible parts, respectively. This decomposition of the Mutual Information Rate (MIR) corresponds to the information landscape-flux decomposition explicitly when the two subsystems behave as Markov chains. Finally, we uncover the intimate relationship between the nonequilibrium thermodynamics in terms of the entropy production rates and the time-irreversible part of the mutual information rate. We found that this relationship and MIR decomposition still hold for the more general stationary and ergodic cases. We demonstrate the above features with two examples of the bivariate Markov chains.

Keywords: nonequilibrium thermodynamics; landscape-flux decomposition; mutual information rate; entropy production rate

---

1. Introduction

There is growing interest in studying two interacting information systems in the fields of control theory, information theory, communication theory, nonequilibrium physics and biophysics [1–9]. Significant progresses has been made recently towards the understanding of the information system in terms of information thermodynamics [10–13]. However, the identification of the global driving forces for the information system dynamics is still challenging. Here, we aim to fill this gap by quantifying the driving forces for the information system dynamics. Inspired by the recent development of landscape and flux theory for the continuous nonequilibrium systems [14–16] and the Markov chain decomposition dynamics for the discrete systems [17–23], we show that at least for the underlying marginal Markovian cases, the driving force for information dynamics is determined by both the information landscape and information flux. The information landscape can be used to construct the driving force responsible for the equilibrium time-reversible part of the information dynamics. The information flux explicitly breaks the detailed balance and provides a quantitative measure of the degree of nonequilibrium or time-irreversibility. It is responsible for the time-irreversible part of the information dynamics. The Mutual Information Rate (MIR) [24] represents the correlation between two information subsystems. We uncovered that the MIR between the two subsystems can be decomposed into the time-reversible and time-irreversible parts, respectively. Especially when the two subsystems act as Markov chains, this decomposition can be expressed in terms of information landscape-flux decomposition for Markovian dynamics. An important signature of nonequilibrium is the Entropy Production Rate (EPR) [17,25,26]. We also uncover the intimate relation between the EPRs and the time-irreversible part of the MIR. We demonstrate the above features with two cases of the bivariate Markov chains. Furthermore, we show that the decomposition of the MIR and the relationship between the EPRs and the time-irreversible part of the MIR still hold for more general stationary and ergodic cases.

2. Bivariate Markov Chains

Markov chains have been often assumed for the underlying dynamics of the total system in random environments. When the two subsystems together jointly form a Markov chain in continuous or discrete time, the resulting chain is called the *Bivariate Markov Chain* (BMC, a special case of the multivariate Markov chain with two stochastic variables). The processes of the two subsystems are correspondingly said to be marginal processes or a marginal chain. The BMC was used to model ion channel currents [2]. It was also used to model delays and congestion in a computer network [3]. Recently, different models of BMC appeared in nonequilibrium statistical physics for capturing or implementing Maxwell's demon [4–6], which can be seen as one marginal chain in the BMC playing feedback control to the other marginal chain. Although the BMC has been studied for decades, there are still challenges on quantifying the dynamics of the whole, as well as the two subsystems. This is because neither of them needs to be a Markovian chain in general [7], and the quantifications of the probabilities (densities) for the trajectories of the two subsystems involve the complicated random matrix multiplications [8]. This leads to the problem not exactly being analytically solvable. The corresponding numerical solutions often lack direct mathematical and physical interpretations.

The conventional analysis of the BMC focuses on the mutual information [9] of the two subsystems for quantifying the underlying information correlations. There are three main representations of this. The first one was proposed and emphasized in the works of Sagawa, T. and Ueda, M. [11] and Parrondo, J. M. R., Horowitz, J. M. and Sagawa, T. [10], respectively, for explaining the mechanism of Maxwell's demon in Szilard's engine. In this representation, the mutual information between the demon and controlled system characterizes the observation and the feedback of the demon. This leads to an elegant approach, which includes the increment of the mutual information into a unified fluctuation relation. The second representation was proposed by the work of Horowitz, J. M. and Esposito, M. [12] in an attempt to explain the violation of the second law in a specified BMC, the bipartite model, where the mutual information is divided into two parts corresponding to the two subsystems, respectively, which were said to be the information flows. This representation tries to explain the mechanism of the demon because one can see that the information flows do contribute to the entropy production for both the demon and controlled system. The first two representations are based on the ensembles of the subsystem states. This means that the mutual information is defined only on the time-sliced distributions of the system states, which somehow lack the information of subsystem dynamics: the time-correlations of the observation and feedback of the demon. The last representation was seen in the work of Barato, A. C., Hartich, D. and Seifert, U. [13], where a more general definition of mutual information in information theory was used, which is defined on the trajectories of the two subsystems. More exactly, this is the so-called *Mutual Information Rate* (MIR) [24], which quantifies the correlation between the two subsystem dynamics. However, due to the difficulties from the possible underlying non-Markovian property of the marginal chains, exactly solvable models and comprehensive conclusions are still challenging from this representation.

In this study, we study the discrete-time BMC in both stochastic information dynamics and thermodynamics. To avoid the technical difficulty caused by non-Markovian dynamics, we first assume that the two marginal chains follow the Markovian dynamics. The non-Markovian case will be discussed elsewhere. We explore the time-irreversibility of BMC and marginal processes in the steady state. Then, we decompose the driving force for the underlying dynamics as the information landscape and information flux [14–16], which can be used to describe the time-reversible parts and time-irreversible parts, respectively. We also prove that the non-vanishing flux fully describes the time-irreversibility of BMC and marginal processes.

We focus on the mutual information rate between the two marginal chains. Since the two marginal chains are assumed to be Markov chains here, the mutual information rate is exactly analytically solvable, which can be seen as the averaged conditional correlation between the two subsystem states. Here, the conditional correlations reveal the time correlations between the past states and the future states.

Corresponding to the landscape-flux decomposition in stochastic dynamics, we decompose the MIR into two parts: the time-reversible and time-irreversible parts, respectively. The time-reversible part measures the part of the correlations between the two marginal chains in both forward and backward processes of BMC. The time-irreversible part measures the difference between the correlations in forward and backward processes of BMC, respectively. We can see that a non-vanishing time-irreversible part of the MIR must be driven by a non-vanishing flux in the steady state, and it can be seen as the sufficient condition for a BMC to be time-irreversible.

We also reveal the important fact that the time-irreversible parts of MIR contribute to the nonequilibrium *Entropy Production Rate* (EPR) of the BMC by the simple equality:

EPR of BMC = EPR of 1st marginal chain + EPR of 2nd marginal chain + 2 × time-irreversible part of MIR.

The decomposition of the MIR and the relation between the time-irreversible part of MIR and EPRs can also be found in stationary and ergodic non-Markovian cases, which will be given in the discussions in the Appendix. This may help to develop a general theory of nonequilibrium non-Markovian interacting information systems.

3. Information Landscape and Information Flux for Determining the Information Dynamics, Time-Irreversibility

Consider the case that two interacting information systems form a finite-state, discrete-time, ergodic and irreducible bivariate Markov chain,

$$Z = (X, S) = \{(X(t), S(t)), t \geq 1\}, \tag{1}$$

We assume that the information state space of X is given by $\mathcal{X} = \{1, ..., d\}$ and the information state space of S is given by $\mathcal{S} = \{1, ..., l\}$. The information state space of Z is then given by $\mathcal{Z} = \mathcal{X} \times \mathcal{S}$. The stochastic information dynamics can then be quantitatively described by the time evolution of the probability distribution of information state space Z, characterized by the following master equation (or the information system dynamics) in discrete time,

$$p_z(z; t+1) = \sum_{z'} q_z(z|z') p_z(z'; t), \text{ for } t \geq 1, \text{ and } z \in \mathcal{Z} \tag{2}$$

where $p_z(z; t) = p_z(x, s; t)$ is the probability of observing state z (or joint probability of $X = x$ and $S = s$) at time t; $q_z(z|z') = q_z(x, s|x', s') \geq 0$ are the transition probabilities from $z' = (x', s')$ to $z = (x, s)$, respectively, and have $\sum_z q_z(z|z') = 1$.

We assume that there exists a unique stationary distribution π_z such that $\pi_z(z) = \sum_{z'} q_z(z|z') \pi_z(z')$. Then, given an arbitrary initial probability distribution, the probability distribution goes to π_z exponentially fast in time. If the initial distribution is π_z, we say that Z is in *Steady State* (SS), and our discussion is based on this SS.

The marginal chains of Z, i.e., X and S, do not need to be Markov chains in general. For the simplicity of analysis, we assume that both marginal chains are Markov chains, and the corresponding transition probabilities are given by $q_x(x|x')$ and $q_s(s|s')$ (for $x, x' \in \mathcal{X}$ and $s, s' \in \mathcal{S}$), respectively. Then, we have the following master equations (or the information system dynamics) for X and S, respectively,

$$p_x(x; t+1) = \sum_{x'} q_x(x|x') p_x(x'; t), \tag{3}$$

and,

$$p_s(s; t+1) = \sum_{s'} q_s(s|s') p_s(s'; t), \tag{4}$$

where $p_x(x; t)$ and $p_s(s; t)$ are the probabilities of observing $X = x$ and $S = s$ at time t, respectively.

We consider that both Equations (3) and (4) have unique stationary solutions π_x and π_s, which satisfy $\pi_x(x) = \sum_{x'} q_x(x|x') \pi_x(x')$ and $\pi_s(s) = \sum_{s'} q_s(s|s') \pi_s(s')$ respectively. Furthermore, we assume that when Z is in SS, π_x and π_s are also achieved. The relations between π_x, π_s and π_z read,

$$\begin{cases} \pi_x(x) = \sum_s \pi_z(x, s), \\ \pi_s(s) = \sum_x \pi_z(x, s). \end{cases} \tag{5}$$

In the rest of this paper, we let $X^T = \{X(1), X(2), ..., X(T)\}$, $S^T = \{S(1), S(2), ..., S(T)\}$, and $Z^T = \{Z(1), Z(2), ..., Z(T)\} = (X^T, S^T)$ denote the time sequences of X, S and Z in time T, respectively.

To characterize the time-irreversibility of the Markov chain C in information dynamics in SS, we introduce the concept of probability flux. Here, we let C denote the arbitrary Markov chain in $\{Z, X, S\}$, and let c, π_c, q_c and C^T denote arbitrary state of C, the stationary distribution of C, the transition probabilities of C and a time sequence of C in time T and in SS, respectively.

The averaged number transitions from the state c' to state c, denoted by $N(c' \to c)$, in unit time in SS can be obtained as:

$$N(c' \to c) = \pi_c(c') q_c(c|c').$$

This is also the probability of the time sequence $C^T = \{C(1) = c', C(2) = c\}$, $(T = 2)$. Correspondingly, the averaged number of reverse transitions, denoted by $N(c \to c')$, reads:

$$N(c \to c') = \pi_c(c) q_c(c'|c).$$

This is also the probability of the time-reverse sequence $\widetilde{C}^T = \{C(1) = c, C(2) = c'\}$, $(T = 2)$. The difference between these two transition numbers measures the time-reversibility of the forward sequence C^T in SS,

$$\begin{aligned} J_c(c' \to c) &= N(c' \to c) - N(c \to c') \\ &= P(C^T) - P(\widetilde{C}^T) \\ &= \pi_c(c') q_c(c|c') - \pi_c(c) q_c(c'|c), \text{ for } C = X, S, \text{ or } Z. \end{aligned} \tag{6}$$

Then, $J_c(c' \to c)$ is said to be the probability flux from c' to c in SS. If $J_c(c' \to c) = 0$ for arbitrary c' and c, then C^T $(T = 2)$ is time-reversible; otherwise, when $J_c(c' \to c) \neq 0$, C^T is time-irreversible. Clearly, we have from Equation (6) that:

$$J_c(c' \to c) = -J_c(c \to c'). \tag{7}$$

The transition probability determines the evolution dynamics of the information system. We can decompose the transition probabilities $q_c(c|c')$ into two parts: the time-reversible part D_c and time-irreversible part B_c, which read:

$$q_c(c|c') = D_c(c' \rightarrow c) + B_c(c' \rightarrow c), \text{ with}$$

$$\begin{cases} D_c(c' \rightarrow c) = \frac{1}{2\pi_c(c')}(\pi_c(c')q_c(c|c') + \pi_c(c)q_c(c'|c)), \\ B_c(c' \rightarrow c) = \frac{1}{2\pi_c(c')}J_c(c' \rightarrow c). \end{cases} \qquad (8)$$

From this decomposition, we can see that the information system dynamics is determined by two driving forces. One of the driving forces is determined by the steady state probability distribution. This part of the driving force is time-reversible. The other driving force for the information dynamics is the steady state probability flux, which breaks the detailed balance and quantifies the time-irreversibility. Since the steady state probability distribution measures the weight of the information state, therefore it can be used to quantify the *information landscape*. If we define the potential landscape for the information system as $\phi = -\log \pi$, then the driving force $D_c(c' \rightarrow c) = \frac{1}{2}(q_c(c|c') + \frac{\pi_c(c)}{\pi_c(c')}q_c(c'|c)) = \frac{1}{2}(q_c(c|c') + \exp[-(\phi_c(c) - \phi_c(c'))]q_c(c'|c))$ is expressed in term of the difference of the potential landscape. This is analogous to the landscape-flux decomposition of Langevin dynamics in [15]. Notice that the information landscape is directly related to the steady state probability distribution of the information system. In general, the information landscape is at nonequilibrium since the detailed balance is often broken for general cases. Only when the detailed balance is preserved, the nonequilibrium information landscape is reduced to the equilibrium information landscape. Even though the information landscape is not at equilibrium in general, the driving force $D_c(c' \rightarrow c)$ is time-reversible due to the decomposition construction. The steady state probability flux measures the information flow in the dynamics and therefore can be termed as the *information flux*. In fact, the nonzero information flux explicitly breaks the detailed balance because of the net flow to or from the system. It is therefore a direct measure of the degree of the nonequilibrium or time-irreversibility in terms of the detailed balance breaking.

Note that the decomposition for the discrete Markovian information process can be viewed as the separation of the current corresponding to the $2B_c(c' \rightarrow c)\pi_c(c')$ here and the activity corresponding to the $2D_c(c' \rightarrow c)\pi_c(c')$ in a previous study [19]. The landscape and flux decomposition here for the reduced information dynamics are in a similar spirit as the whole state space decomposition with the information system and the associated environments. When the detailed balance is broken, the information landscape (defined as the negative logarithm of the steady state probability $\phi = -\log \pi$) is not the same as the equilibrium landscape under the detailed balance. There can be uniqueness issue related to the decomposition. To avoid the confusion, we make a physical choice, or in other words, we can fix the gauge so that the information landscape always coincides with the equilibrium landscape when the detailed balance is satisfied. In other words, we want to make sure the Boltzmann law applies at equilibrium with detailed balance. In this way, we can decompose the information landscape and information flux for nonequilibrium information systems without detailed balance. By solving the linear master equation for the steady state, we can quantify the nonequilibrium information landscape, and from that, we can obtain the corresponding steady state probability flux. Some studies discussed various aspects of this issue [18,19,27,28].

By Equations (7) and (8), we have the following relations:

$$\begin{cases} \pi_c(c')D_c(c' \rightarrow c) = \pi_c(c)D_c(c \rightarrow c'), \\ \pi_c(c')B_c(c' \rightarrow c) = -\pi_c(c)B_c(c \rightarrow c'). \end{cases} \qquad (9)$$

As we can see in the next section, D_c and B_c are useful for us to quantify time-reversible and time-irreversible observables of C, respectively.

We give the interpretation that the non-vanishing information flux J_c fully measures the time-irreversibility of the chain C in time T for $T \geq 2$. Let C^T be an arbitrary sequence of C in SS, and without loss of generality, we let $T = 3$. Similar to Equation (6), the measure of the time-irreversibility of C^T can be given by the difference between the probability of $C^T = \{C(1), C(2), C(3)\}$ and that of its time-reversal $\widetilde{C}^T = \{C(3), C(2), C(1)\}$, such as:

$$
\begin{aligned}
&P(C^T) - P(\widetilde{C}^T) \\
&= \pi_c(C(1))q_c(C(2)|C(1))q_c(C(3)|C(2)) - \pi_c(C(3))q_c(C(2)|C(3))q_c(C(1)|C(2)) \\
&= \pi_c(C(1))\,(D_c(C(1) \to C(2)) + B_c(C(1) \to C(2)))\,(D_c(C(2) \to C(3)) + B_c(C(2) \to C(3))) - \\
&\quad \pi_c(C(3))\,(D_c(C(3) \to C(2)) + B_c(C(3) \to C(2)))\,(D_c(C(2) \to C(1)) + B_c(C(2) \to C(1))), \\
&\text{for } C = X, S \text{ or } Z.
\end{aligned}
$$

Then, by the relations given in Equation (9), we have that $P(C^T) - P(\widetilde{C}^T) = 0$ holds for arbitrary C^T if and only if $B_c(C(1) \to C(2)) = B_c(C(2) \to C(3)) = 0$ or equivalently $J_c(C(1) \to C(2)) = J_c(C(2) \to C(3)) = 0$. This conclusion can be made for arbitrary $T > 3$. Thus, non-vanishing J_c can fully describe the time-irreversibility of C for $C = X, S$ or Z.

We show the relations between the fluxes of the whole system J_z and of the subsystem J_x as follows:

$$
\begin{aligned}
J_x(x' \to x) &= \pi_x(x')q_x(x|x') - \pi_x(x)q_x(x'|x) \\
&= P(\{x', x\}) - P(\{x, x'\}) \\
&= \sum_{s,s'} \left(P(\{(x', s'), (x, s)\}) - P(\{(x, s), (x', s')\}) \right) \\
&= \sum_{s,s'} \left(\pi_z(x', s')q_z(x, s|x', s') - \pi_z(x, s)q_z(x', s'|x, s) \right) \\
&= \sum_{s,s'} J_z((x', s') \to (x, s)).
\end{aligned}
\tag{10}
$$

Similarly, we have:

$$
J_s(s' \to s) = \sum_{x,x'} J_z((x', s') \to (x, s)).
\tag{11}
$$

These relations indicate that the subsystem fluxes J_x and J_s can be seen as the coarse-grained levels of total system flux J_z by averaging over the other parts of the system S and X, respectively. We should emphasize that non-vanishing J_z does not mean X or S is time-irreversible and vice versa.

4. Mutual Information Decomposition to Time-Reversible and Time-Irreversible Parts

According to information theory, the two interacting information systems represented by bivariate Markov chain Z can be characterized by the *Mutual Information Rate* (MIR) between the marginal chains X and S in SS. The mutual information rates represent the correlation between two interacting information systems. The MIR is defined on the probabilities of all possible time sequences, $P(Z^T)$, $P(X^T)$ and $P(S^T)$ and is given by [24],

$$
I(X, S) = \lim_{T \to \infty} \frac{1}{T} \sum_{Z^T} P(Z^T) \log \frac{P(Z^T)}{P(X^T)P(S^T)}.
\tag{12}
$$

It measures the correlation between X and S in unit time, or say, the efficient bits of information that X and S exchange with each other in unit time. The MIR must be non-negative, and a vanishing $I(X, S)$ indicates that X and S are independent of each other. More explicitly, the corresponding probabilities of these sequences can be evaluated by using Equations (2)–(4); we have:

$$\begin{cases} P(X^T) = \pi_x(X(1)) \prod_{t=1}^{T-1} q_x(X(t+1)|X(t)), \\ P(S^T) = \pi_s(S(1)) \prod_{t=1}^{T-1} q_s(S(t+1)|S(t)), \\ P(Z^T) = \pi_z(Z(1)) \prod_{t=1}^{T-1} q_z(Z(t+1)|Z(t)). \end{cases}$$

By substituting these probabilities into Equation (12) (see Appendix A), we have the exact expression of MIR as:

$$\begin{aligned} I(X,S) &= \sum_{z,z'} \pi_z(z') q_z(z|z') \log \frac{q_z(z|z')}{q_x(x|x') q_s(s|s')} \\ &= \langle i(z|z') \rangle_{z',z} \geq 0, \text{ for } z = (x,s), \text{ and } z' = (x',s'). \end{aligned} \tag{13}$$

where $i(z|z') = \log \frac{q_z(z|z')}{q_x(x|x') q_s(s|s')}$ is the conditional (Markovian) correlation between the states x and s when the transition $z' = (x',s') \rightarrow z = (x,s)$ occurs. This indicates that when the two marginal processes are both Markovian, the MIR is the average of the conditional (Markovian) correlations. These correlations are measurable when transitions occur, and they can be seen as the observables of Z.

By noting the decomposition of transition probabilities in Equation (8), we have a corresponding decomposition of $I(X,S)$ such as:

$I(X,S) = I_D(X,S) + I_B(X,S)$, with

$$\begin{cases} I_D(X,S) = \sum_{z,z'} \pi_z(z') D_z(z|z') i(z|z') = \frac{1}{2} \sum_{z,z'} (\pi_z(z') q_z(z|z') + \pi_z(z) q_z(z'|z)) i(z|z'), \\ I_B(X,S) = \sum_{z,z'} \pi_z(z') B_z(z|z') i(z|z') = \frac{1}{2} \sum_{z,z'} J_z(z|z') i(z|z') = \frac{1}{4} \sum_{z,z'} J_z(z|z') (i(z|z') - i(z'|z)). \end{cases} \tag{14}$$

This means that the mutual information representing the correlations between the two interacting systems can be decomposed into the time-reversible equilibrium part and the time-irreversible nonequilibrium part. The origin of this is from the fact that the underlying information system dynamics is determined by both the time-reversible information landscape and time-irreversible information flux. These equations are very important to establish the link to the time-irreversibility. We now give further interpretation for $I_D(X,S)$ and $I_B(X,S)$:

Consider a bivariate Markov chain Z in SS wherein X and S are dependent on each other, i.e., $I(X,S) = I_D(X,S) + I_B(X,S) > 0$. By the ergodicity of Z, we have the MIR, which measures the averaged conditional correlation along the time sequences Z^T,

$$\lim_{T \to \infty} \frac{1}{T} \langle i(Z(t+1)|Z(t)) \rangle_{Z^T} = I(X,S), \text{ for } 1 < t < T.$$

Then, $I_B(X,S)$ measures the change of the averaged conditional correlation between X and S when a sequence of Z turns back in time,

$$\lim_{T \to \infty} \frac{1}{T} \langle i(Z(t+1)|Z(t)) - i(Z(t)|Z(t+1)) \rangle_{Z^T} = 2I_B(X,S).$$

A negative $I_B(X,S)$ shows that the correlation between X and S becomes strong in the time-reversal process of Z; A positive $I_B(X,S)$ shows that the correlation becomes weak in the time-reversal process of Z. Both cases show that the Z is time-irreversible since we have a non-vanishing J_z. However, the case of $I_B(X,S) = 0$ is complicated, since it indicates either a vanishing J_z or a non-vanishing J_z. Anyway, we see that a non-vanishing $I_B(X,S)$ is a sufficient condition for Z to be time-irreversible. On the other hand, $I_D(X,S) = I(X,S) - I_B(X,S)$ measures the correlation remaining in the backward process of Z.

The definition of MIR in Equation (12) turns out to be appropriate for even more general stationary and ergodic (Markovian or non-Markovian) processes. Consequentially, the decomposition of MIR is useful to quantify the correlation between two stationary and ergodic processes in a wider sense, i.e., to monitor the changes of the correlation in the forward and the backward processes. As a special case, the analytical expressions in Equation (14) are the reduced results, which are valid for Markovian cases. A brief discussion of the decomposition of MIR of more general processes can be found in Appendix B.

5. Relationship between Mutual Information and Entropy Production

The *Entropy Production Rates* (EPR) or energy dissipation (cost) rate at steady state is a quantitative nonequilibrium measure, which characterizes the time-irreversibility of the underlying processes. The EPR of a stationary and ergodic process C (here $C = Z, X$ or S) can be given by the difference between the averaged surprisal (negative logarithmic probability) of the backward sequences $\widetilde{C}^T$ and that of forward sequences C^T in the long time limit, i.e.,

$$
\begin{aligned}
R_c &= \lim_{T \to \infty} \frac{1}{T} \langle \log P(C^T) - \log P(\widetilde{C}^T) \rangle_{C^T} \\
&= \lim_{T \to \infty} \frac{1}{T} \left\langle \log \frac{P(C^T)}{P(\widetilde{C}^T)} \right\rangle_{C^T} \geq 0,
\end{aligned}
\tag{15}
$$

where R_c is said to be the EPR of C [25]; $-\log P(C^T)$ and $-\log P(\widetilde{C}^T)$ are said to be the surprisal of a forward and a backward sequence of C, respectively. We see that C is time-reversible (i.e., $P(C^T) = P(\widetilde{C}^T)$ for arbitrary C^T for large T) if and only if $R_c = 0$. Additionally, this is due to the form of R_c, which is exactly a Kullback–Leibler divergence. When C is Markovian, then R_c reduces into the following form when Z, X or S is assigned to C, respectively [17,26],

$$
\begin{cases}
R_z = \frac{1}{2} \sum_{z,z'} J_z(z' \to z) \log \frac{q_z(z|z')}{q_z(z'|z)}, \\
R_x = \frac{1}{2} \sum_{x,x'} J_x(x' \to x) \log \frac{q_x(x|x')}{q_x(x'|x)}, \\
R_s = \frac{1}{2} \sum_{s,s'} J_s(s' \to s) \log \frac{q_s(s|s')}{q_s(s'|s)},
\end{cases}
\tag{16}
$$

where total and subsystem entropy productions R_z, R_x and R_s correspond to Z, X and S, respectively. Here, R_z usually contains the detailed interaction information of the system (or subsystems) and environments; R_x and R_s provide the coarse-grained information of time-irreversible observables of X and Z, respectively. Each non-vanishing EPR indicates that the corresponding Markov chain is time-irreversible. Again, we emphasize that a non-vanishing R_z does not mean X or S is time-irreversible and vice versa.

We are interested in the connection between these EPRs and mutual information. We can associate them with $I_B(X,S)$ by noting Equations (10), (11) and (14). We have:

$$
\begin{aligned}
I_B(X,S) &= \frac{1}{4} \sum_{z,z'} J_z(z|z')(i(z|z') - i(z'|z)) \\
&= \frac{1}{4} \sum_{z,z'} J_z(z|z') \log \frac{q_z(z|z')}{q_z(z'|z)} - \frac{1}{4} \sum_{x,x'} J_x(x|x') \log \frac{q_x(x|x')}{q_x(x'|x)} - \frac{1}{4} \sum_{s,s'} J_s(s|s') \log \frac{q_s(s|s')}{q_s(s'|s)} \\
&= \frac{1}{2}(R_z - R_x - R_s).
\end{aligned}
\tag{17}
$$

We note that $I_B(X,S)$ is intimately related to the EPRs. This builds up a bridge between these EPRs and the irreversible part of the mutual information. Moreover, we also have:

$$
\begin{cases}
R_z = R_x + R_s + 2I_B(X,S) \geq 0, \\
R_x + R_s \geq -2I_B(X,S), \\
R_z \geq 2I_B(X,S).
\end{cases}
\tag{18}
$$

This indicates that the time-irreversible MIR contributes to the detailed EPRs. In other words, the differences of the entropy production rate of the whole system and subsystems provide the origin of the time-irreversible part of the mutual information. This reveals the nonequilibrium thermodynamic origin of the irreversible mutual information or correlations. Of course, since the EPR is related to the flux directly as is seen from the above definitions, the origin of the EPR or nonequilibrium thermodynamics is from the non-vanishing information flux for the nonequilibrium dynamics. On the other hand, the irreversible part of the mutual information measures the correlations, and it contributes to the EPRs of the correlated subsystems.

Furthermore, the last expression in Equation (17) (also the expressions in Equation (18)) can be generalized to more general stationary and ergodic processes. A related discussion and demonstration of this can be seen in Appendix B.

6. A Simple Case: The Blind Demon

As a concrete example, we consider a two-state system coupled to two information baths *a* and *b*. The states of the system are denoted by $\mathcal{X} = \{x : x = 0, 1\}$, respectively. Each bath sends an instruction to the system. If the system adopts one of them, it then follows the instruction and makes the change of the state. The instructions generated from one bath are independently and identically distributed (Bernoulli trials). Both the probability distributions of the instructions corresponding to the baths follow Bernoulli distributions and read $\{\epsilon_a(x) : x \in \mathcal{X}, \epsilon_a(x) \geq 0, \sum_x \epsilon_a(x) = 1\}$ for bath *a* and $\{\epsilon_b(x) : x \in \mathcal{X}, \epsilon_b(x) \geq 0, \sum_x \epsilon_b(x) = 1\}$ for bath *b*, respectively. Since the system cannot execute two instructions simultaneously, there exists an information demon that makes choices for the system. The demon is blind to caring about the system, and it makes choices independently and identically distributed. The choices of the demon are denoted by $\mathcal{S} = \{s : s = a, b\}$, respectively. The probability distribution of the demon's choices reads $\{P(s) : s \in \mathcal{S}, P(a) = p, P(b) = 1 - p, p \in [0, 1]\}$. Still, we use $Z = (X, S)$ with $X \in \mathcal{X}$ and $S \in \mathcal{S}$ to denote the BMC of the system and the demon.

Consequentially, the transition probabilities of the system read:

$$q_x(x|x') = p\epsilon_a(x) + (1 - p)\epsilon_b(x).$$

The transition probabilities of the demon read:

$$q_s(s|s') = P(s).$$

Additionally, the transition probabilities of the joint chain read:

$$q_z(x, s|x', s') = P(s)\epsilon_{s'}(x).$$

We have the corresponding steady state distributions or the information landscapes as,

$$\begin{cases} \pi_x(x) = p\epsilon_a(x) + (1 - p)\epsilon_b(x), \\ \pi_s(s) = P(s), \\ \pi_z(x, s) = P(s)\pi_x(x). \end{cases}$$

We obtain the information fluxes as,

$$\begin{cases} J_x(x' \to x) = 0, \text{ for all } x, x' \in \mathcal{X} \\ J_s(s' \to s) = 0, \text{ for all } s, s' \in \mathcal{S} \\ J_z((x', s') \to (x, s)) = P(s)P(s')(\pi_x(x')\epsilon_{s'}(x) - \pi_x(x)\epsilon_s(x')). \end{cases}$$

Here, we use the notations $\epsilon_s(x')$ and $\epsilon_{s'}(x)$ ($s, s' = a$ or b) to denote the probabilities of the instructions x' or x from bath *a* or *b* briefly. We obtain the EPRs as:

$$
\begin{cases}
R_x = 0, \\
R_s = 0, \\
R_z = \sum_x p(1-p)(\epsilon_a(x) - \epsilon_b(x))(\log \epsilon_a(x) - \log \epsilon_b(x)).
\end{cases}
$$

We evaluate the MIR as:

$$
I(X,S) = -\sum_x \pi_x(x) \log \pi_x(x) + p \sum_x \epsilon_a(x) \log \epsilon_a(x) + (1-p) \sum_x \epsilon_b(x) \log \epsilon_b(x).
$$

The time-irreversible part of $I(X,S)$ reads,

$$
I_B(X,S) = \frac{1}{2} R_z.
$$

7. Conclusions

In this work, we identify the driving forces for the information system dynamics. We show that for marginal Markovian information systems, the information dynamics is determined by both the information landscape and information flux. While the information landscape can be used to construct the driving force for describing the time-reversible behavior of the information dynamics, the information flux can be used to describe the time-irreversible behavior of the information dynamics. The information flux explicitly breaks the detailed balance and provides a quantitative measure of the degree of the nonequilibrium or time-irreversibility. We further demonstrate that the mutual information rate, which represents the correlations, can be decomposed into the time-reversible part and the time-irreversible part originated from the landscape and flux decomposition of the information dynamics. Finally, we uncover the intimate relationship between the difference of the entropy productions of the whole system and those of the subsystems and the time-irreversible part of the mutual information. This will help with understanding the non-equilibrium behavior of the interacting information system dynamics in stochastic environments. Furthermore, we verify that our conclusions on the mutual information rate and entropy production rate decomposition can be made more general for the stationary and ergodic processes.

Acknowledgments: This work was supported in part by the National Natural Science Foundation of China (NSFC-91430217) and the National Science Foundation (U.S.) (NSF-PHY-76066).

Author Contributions: Qian Zeng and Jin Wang conceived and designed the experiments; Qian Zeng performed the experiments; Qian Zeng and Jin Wang analyzed the data; Qian Zeng and Jin Wang contributed reagents/materials/analysis tools; Qian Zeng and Jin Wang wrote the paper.

Conflicts of Interest: The authors declare no conflict of interest.

Abbreviations

The following abbreviations are used in this manuscript:

BMC   Bivariate Markov Chain
EPR   Entropy Production Rate
MIR   Mutual Information Rate
SS   Steady State

Appendix A

Here, we derive the exact form of the Mutual Information Rate (MIR, Equation (13)) in the steady state by using the cumulant-generating function.

We write an arbitrary time sequence of Z in time T in the following form:

$$
Z^T = \{Z(1), ..., Z(i), ..., Z(T)\}, \text{ for } T \geq 2,
$$

where $Z(i)$ (for $i \geq 1$) denotes the state at time i. The corresponding probability of Z^T is in the following form:

$$P(Z^T) = \pi_z(Z_1) \left\{ \prod_{i=1}^{T-1} q_z(Z_{i+1}|Z_i) \right\}. \tag{A1}$$

We let the chain $U = (X, S)$ denote a process that X and S follow the same Markov dynamics in Z, but are independent of each other. Then, we have that the transition probabilities of U read:

$$q_u(u|u') = q(x, s|x', s') = q_x(x|x')q_s(s|s'). \tag{A2}$$

Then, the probability of a time sequence of U, U^T, with the same trajectory of Z^T reads:

$$P(U^T) = \pi_u(Z_1) \left\{ \prod_{i=1}^{T-1} q_u(Z_{i+1}|Z_i) \right\}, \tag{A3}$$

with $\pi_u(x, s) = \pi_x(x)\pi_s(s)$ being the stationary probability of U.

For evaluating the exact form of MIR, we introduce the cumulant-generating function of the random variable $\log \frac{P(Z^T)}{P(U^T)}$,

$$K(m, T) = \log \left\langle \exp \left(m \log \frac{P(Z^T)}{P(U^T)} \right) \right\rangle_{Z^T}. \tag{A4}$$

We can see that:

$$\lim_{T \to \infty} \lim_{m \to 0} \frac{1}{T} \frac{\partial K(m, T)}{\partial m}$$
$$= \lim_{T \to \infty} \frac{1}{T} \left\langle \log \frac{P(Z^T)}{P(U^T)} \right\rangle_{Z^T} \tag{A5}$$
$$= I(X, S).$$

Thus, our idea is to evaluate $K(m, T)$ at first. We have:

$$K(m, T) = \log \left\langle \exp \left(m \log \frac{P(Z^T)}{P(U^T)} \right) \right\rangle_{Z^T}$$
$$= \log \left\{ \sum_{Z^T} \frac{(P(Z^T))^{m+1}}{(P(U^T))^m} \right\} \tag{A6}$$
$$= \log \left\{ \sum_{\{Z(0), Z(1), \dots, Z(T)\}} \frac{(\pi_z^{m+1}(Z_0))}{(\pi_u^m(Z_0))} \prod_{i=0}^{T-1} \frac{q_z^{m+1}(Z_{i+1}|Z_i)}{q_u^m(Z_{i+1}|Z_i)} \right\},$$

where we realize that the last equality can be rewritten in the form of matrix multiplication.

We introduce the following matrices and vectors for Equation (A6) such that:

$$\begin{aligned}
\mathbf{Q_z} &= \left\{ (\mathbf{Q_z})_{(z,z')} = q_z(z|z'), \text{ for } z, z' \in \mathcal{Z} \right\}, \\
\mathbf{G}(m) &= \left\{ (\mathbf{G}(m))_{(z,z')} = \frac{q_z^{m+1}(z|z')}{q_u^m(z|z')}, \text{ for } z, z' \in \mathcal{Z} \right\}, \\
\boldsymbol{\pi_z} &= \{ (\boldsymbol{\pi_z})_z = \pi_z(z), \text{ for } z \in \mathcal{Z} \}, \\
\boldsymbol{v}(m) &= \left\{ (\boldsymbol{v}(m))_z = \frac{\pi_z^{m+1}(z)}{\pi_u^m(z)} \right\},
\end{aligned} \tag{A7}$$

where $\mathbf{Q_z}$ is the transition matrix of Z; $\boldsymbol{\pi_z}$ is the stationary distribution of Z. It can be also verified that:

$$Q_z = G(0),$$
$$\pi_z = v(0),$$
$$\pi_z = Q_z\pi_z,$$
$$\mathbf{1}^\dagger Q_z = \mathbf{1}^\dagger,$$

$$\lim_{m\to 0}\frac{dG(m)}{dm} = \left\{\left(\lim_{m\to 0}\frac{dG(m)}{dm}\right)_{(z,z')} = q_z(z|z')\log\frac{q_z(z|z')}{q_u(z|z')}, \text{ for } z, z' \in \mathcal{Z}\right\},$$

$$\lim_{m\to 0}\frac{dv(m)}{dm} = \left\{\left(\lim_{m\to 0}\frac{dv(m)}{dm}\right)_z = \pi_z(z)\log\frac{\pi_z(z)}{\pi_u(z)}, \text{ for } z \in \mathcal{Z}\right\},$$

(A8)

where $\mathbf{1}^\dagger$ is the vector of all ones with appropriate dimension.

Then, $K(m, T)$ can be rewritten in a compact form such that:

$$K(m, T) = \log\left\{\mathbf{1}^\dagger G^{T-1}(m)v(m)\right\}. \tag{A9}$$

Then, we substitute Equation (A9) into Equation (A5) and have:

$$
\begin{aligned}
I(X, S) &= \lim_{T\to\infty}\lim_{m\to 0}\frac{1}{T}\frac{\partial K(m, T)}{\partial m} \\
&= \lim_{T\to\infty}\lim_{m\to 0}\frac{1}{T}\frac{\partial \log\left\{\mathbf{1}^\dagger G^{T-1}(m)v(m)\right\}}{\partial m} \\
&= \lim_{T\to\infty}\lim_{m\to 0}\frac{1}{T}\left\{(T-1)\mathbf{1}^\dagger G^{T-2}(m)\frac{dG(m)}{dm}v(m) + \mathbf{1}^\dagger G^{T-1}(m)\frac{dv(m)}{dm}\right\} \\
&= \lim_{T\to\infty}\frac{1}{T}\left\{(T-1)\mathbf{1}^\dagger G^{T-2}(0)\left(\lim_{m\to 0}\frac{dG(m)}{dm}\right)v(0) + \mathbf{1}^\dagger G^{T-1}(0)\left(\lim_{m\to 0}\frac{dv(m)}{dm}\right)\right\}.
\end{aligned}
$$

(A10)

By noting Equation (A8) and $T \geq 2$, we obtain Equation (13) from Equation (A10) such that:

$$
\begin{aligned}
I(X, S) &= \lim_{T\to\infty}\frac{1}{T}\left\{(T-1)\mathbf{1}^\dagger G^{T-2}(0)\left(\lim_{m\to 0}\frac{dG(m)}{dm}\right)v(0) + \mathbf{1}^\dagger G^{T-1}(0)\left(\lim_{m\to 0}\frac{dv(m)}{dm}\right)\right\} \\
&= \lim_{T\to\infty}\left\{\left(1 - \frac{1}{T}\right)\mathbf{1}^\dagger\left(\lim_{m\to 0}\frac{dG(m)}{dm}\right)\pi_z + \frac{1}{T}\mathbf{1}^\dagger\left(\lim_{m\to 0}\frac{dv(m)}{dm}\right)\right\} \\
&= \mathbf{1}^\dagger\left(\lim_{m\to 0}\frac{dG(m)}{dm}\right)\pi_z \\
&= \sum_{(x,s),(x',s')}\pi_z(x',s')q_z(x,s|x',s')\log\frac{q_z(x,s|x',s')}{q_x(x|x')q_s(s|s')}.
\end{aligned}
$$

(A11)

Appendix B

Appendix B.1 Discussions on the Generality of Mutual Information Rate Decomposition and Connections to Entropy Production in Terms of Equations (14), (17), and (18)

For general cases, indeed, we do not expect that both X and S are Markovian. Even the joint chain Z may be non-Markovian. This means that Equation (2) may fail to depict the dynamics of Z. Then, the landscape-flux decomposition needs to be generalized to this situation. Such decomposition was not developed yet for the non-Markovian cases. This will be discussed in a separate work. However, when Z is a stationary and ergodic process (also assume that both X and S are stationary and ergodic), we show that the MIR can be decomposed into two parts as is shown in Equation (14), and an interesting relation between the MIR and EPRs can still be found in the same form of the last expression in Equation (17).

We are interested in the correlation between the forward sequences of X and S, which can be measured by $\log \frac{P(Z^T)}{P(X^T)P(S^T)}$ ($Z^T = (X^T, S^T)$), then the MIR can be used to quantify the average rate of this correlation in the long time limit as shown in Equation (12). Furthermore, we are interested in the averaged difference between the rate of the correlation of the backward processes and that of the forward processes. This comes to the time-irreversible part of the MIR defined by:

$$I_B(X,S) = \lim_{T \to \infty} \frac{1}{2T} \left\langle \log \frac{P(Z^T)}{P(X^T)P(S^T)} - \log \frac{P(\widetilde{Z}^T)}{P(\widetilde{X}^T)P(\widetilde{S}^T)} \right\rangle_{Z^T}, \tag{A12}$$

where $\log \frac{P(\widetilde{Z}^T)}{P(\widetilde{X}^T)P(\widetilde{S}^T)}$ quantifies the correlation between the backward sequences of X and S. Clearly, the time-irreversible part of MIR depicting the correlation of the forward processes of X and S is enhanced ($I_B(X,S) > 0$) or weakened ($I_B(X,S) < 0$) compared to that of the backward processes. The other important part of the MIR, namely the time-reversible part, shows that the averaged rate of the correlation that remains in both forward and backward processes,

$$I_D(X,S) = \lim_{T \to \infty} \frac{1}{2T} \left\langle \log \frac{P(Z^T)}{P(X^T)P(S^T)} + \log \frac{P(\widetilde{Z}^T)}{P(\widetilde{X}^T)P(\widetilde{S}^T)} \right\rangle_{Z^T}, \tag{A13}$$

Consequentially, the MIR $I(X,S)$ is decomposed into two parts shown as $I(X,S) = I_D(X,S) + I_B(X,S)$. In Markovian cases, each part of the MIR reduces to the form in Equation (14) respectively.

The relation between the time-irreversible part of the MIR and EPRs can be shown as follows,

$$
\begin{aligned}
I_B(X,S) &= \lim_{T \to \infty} \frac{1}{2T} \left\langle \log \frac{P(Z^T)}{P(X^T)P(S^T)} - \log \frac{P(\widetilde{Z}^T)}{P(\widetilde{X}^T)P(\widetilde{S}^T)} \right\rangle_{Z^T} \\
&= \lim_{T \to \infty} \frac{1}{2T} \left\{ \left\langle \log \frac{P(Z^T)}{P(\widetilde{Z}^T)} \right\rangle_{Z^T} - \left\langle \log \frac{P(X^T)}{P(\widetilde{X}^T)} \right\rangle_{X^T} - \left\langle \log \frac{P(S^T)}{P(\widetilde{S}^T)} \right\rangle_{S^T} \right\} \\
&= \frac{1}{2} \left(R_z - R_x - R_s \right),
\end{aligned}
\tag{A14}
$$

which is in the same form as Equation (17). Additionally, due to the non-negativity of the EPRs, the inequalities in (18) still hold for general cases.

Appendix B.2. The Smart Demon

To verify the conclusions in more general cases, we constructed a model of a smart demon, which reflects a more general situation in the nature: the two information subsystems play feedback to each other. A three-state information system is connected to two information baths labeled a and b, respectively. The states of the system are denoted by $\mathcal{X} = \{x : x = 0,1,2\}$, respectively. Each bath sends an instruction to the system. If the system adopts one of them, it then follows the instruction and makes a change of the state. The instructions generated from one arbitrary bath are independent and identically distributed. The probability distributions of the instructions corresponding to the baths read $\{\epsilon_s(x) : \epsilon_s(x) \geq 0, \sum_{x \in \mathcal{X}} \epsilon_s(x) = 1\}$ (for $s = a, b$), respectively. Since the system cannot execute the two incoming instructions simultaneously, there exists an information demon making choices for the system. The choices of the demon are denoted by the labels of the baths $\mathcal{S} = \{s : s = a, b\}$, respectively. The demon observes the state of the system and plays feedback. The (conditional) probability distribution of the demon's choices reads $\{d(s|x',s') : d(s|x',s') \geq 0, \sum_{s \in \mathcal{S}} d(s|x',s') = 1, x' \in \mathcal{X}, s' \in \mathcal{S}\}$. Still, we use X, S and $Z = (X,S)$ to denote the processes of the system, the demon and the corresponding joint chain, a BMC, respectively.

The transition probabilities of the BMC read:

$$q_z(z|z') = q_z(x,s|x',s') = d(s|x',s')\epsilon_s(x),$$

Entropy **2017**, *19*, 678

where $\epsilon_s(x)$ denotes the probability of the instruction x from bath $s = a, b$. We assume that there is a unique stationary distribution of z, π_z such that:

$$\pi_z(z) = \sum_{z'} q_z(z|z')\pi_z(z').$$

The stationary distributions of S and X then read:

$$\begin{cases} \pi_s(s) = \sum_x \pi_z(x, s), \\ \pi_x(x) = \sum_s \pi_z(x, s). \end{cases}$$

The behavior of the demon can be seen as a Markovian process in the steady state. The corresponding transition probabilities of the system read:

$$q_s(s|s') = \frac{1}{\pi_s(s')} \sum_{x'} d(s|x', s')\pi_z(x', s').$$

It can be verified that π_s is the unique stationary distribution of S. However, the dynamics of the system always behaves as a non-Markovian process in general.

To characterize the time-irreversibility of Z, X and S, we use the definition of EPR in Equation (15) and have:

$$\begin{cases} R_z = \frac{1}{2} \sum_{z,z'} J_z(z' \to z) \log \frac{q_z(z|z')}{q_z(z'|z)}, \\ R_s = \frac{1}{2} \sum_{s,s'} J_s(s' \to s) \log \frac{q_s(s|s')}{q_s(s'|s)} = 0, \\ R_x = \lim_{T \to \infty} \frac{1}{T} \sum_{X^T} P(X^T) \log \frac{P(X^T)}{P(\tilde{X}^T)}, \end{cases}$$

where:

$$P(X^T) = \sum_{S^T} P(Z^T = (X^T, S^T)).$$

To quantify the correlation between the system and demon, we use the definition of MIR in Equation (12).

We are also interested in the time-irreversible part of MIR, $I_B(X, S)$, which influences the EPR of the system, R_x. This can be seen from Equation (A14) such that:

$$R_x = R_z - R_s - 2I_B(X, S).$$

We use numerical simulations, which evaluate R_x, $I(X, S)$ and $I_B(X, S)$ directly from the typical sequences of Z (see [7,8]). The corresponding results can be given by:

$$\begin{cases} R_x \approx \frac{1}{T} \log \frac{P(X^T)}{P(\tilde{X}^T)}, \text{ for large } T, \\ I(X, S) \approx \frac{1}{T} \log \frac{P(Z^T)}{P(X^T)P(S^T)}, \text{ for large } T, \\ I_B(X, S) \approx \frac{1}{2T} \log \frac{P(Z^T)}{P(X^T)P(S^T)} - \frac{1}{2T} \log \frac{P(\tilde{Z}^T)}{P(\tilde{X}^T)P(\tilde{S}^T)}, \text{ for large } T, \end{cases}$$

where $Z^T = (X^T, S^T)$ is a typical sequence of Z (hence, X^T and S^T are typical sequences of X and S, respectively). The convergence of this numerical simulation can be observed as T increases. To confirm the result $R_x = R_z - R_s - 2I_B(X, S)$, we use different typical sequences in calculating R_x and $I_B(X, S)$, respectively. R_z and R_s are calculated by using the corresponding analytical results shown above.

For numerical simulations, we randomly choose two groups of the parameters: the probabilities of the instructions of the baths ϵ_a and ϵ_b and the probabilities of the demon's choices d (see Tables A1 and A2).

We evaluate R_x, $I(X, S)$ and $I_B(X, S)$ for both groups. The values of the numerical results are listed in Table A3.

Table A1. Two groups of ϵ_a and ϵ_b.

	$\{\epsilon_a(x=0), \epsilon_a(x=1), \epsilon_a(x=2)\}$	$\{\epsilon_b(x=0), \epsilon_b(x=1), \epsilon_b(x=2)\}$
Group 1	$\{0.2344, 0.2730, 0.4926\}$	$\{0.4217, 0.4094, 0.1689\}$
Group 2	$\{0.1305, 0.3972, 0.4723\}$	$\{0.3358, 0.0010, 0.6633\}$

Table A2. Two groups of d.

	$\{d(s=a|x=0, s=a), d(s=b|x=0, s=a)\}$	$\{d(s=a|x=1, s=b), d(s=b|x=0, s=b)\}$
Group 1	$\{0.3844, 0.6156\}$	$\{0.6811, 0.3189\}$
Group 2	$\{0.1072, 0.8928\}$	$\{0.7473, 0.2527\}$
	$\{d(s=a|x=1, s=a), d(s=b|x=1, s=a)\}$	$\{d(s=a|x=1, s=b), d(s=b|x=1, s=b)\}$
Group 1	$\{0.5195, 0.4805\}$	$\{0.8088, 0.1912\}$
Group 2	$\{0.6595, 0.3405\}$	$\{0.1600, 0.8400\}$
	$\{d(s=a|x=2, s=a), d(s=b|x=2, s=a)\}$	$\{d(s=a|x=2, s=b), d(s=b|x=2, s=b)\}$
Group 1	$\{0.3775, 0.6225\}$	$\{0.3340, 0.6660\}$
Group 2	$\{0.0232, 0.9768\}$	$\{0.0814, 0.9186\}$

Table A3. Numerical results of R_z, R_x, $I(X, S)$ and $I_B(X, S)$.

	R_z	R_x	$I(X,S)$	$I_B(X,S)$
Group 1	0.0645	0.0018	0.0885	0.0313
Group 2	0.5485	0.1291	0.3385	0.2097

References

1. Shannon, C.E. A mathematical theory of communication. *Bell Syst. Tech. J.* **1948**, *27*, 379–423.
2. Ball, F.; Yeo, G.F. Lumpability and Marginalisability for Continuous-Time Markov Chains. *J. Appl. Probab.* **1993**, *30*, 518–528.
3. Wei, W.; Wang, B.; Towsley, D. Continuous-time hidden Markov models for network performance evaluation. *Perform. Eval.* **2002**, *49*, 129–146.
4. Strasberg, P.; Schaller, G.; Brandes, T.; Esposito, M. Thermodynamics of a physical model implementing a Maxwell demon. *Phys. Rev. Lett.* **2013**, *110*, 040601.
5. Koski, J.V.; Kutvonen, A.; Khaymovich, I.M.; Ala-Nissila, T.; Pekola, J.P. On-Chip Maxwell's Demon as an Information-Powered Refrigerator. *Phys. Rev. Lett.* **2015**, *115*, 260602.
6. Mcgrath, T.; Jones, N.S.; Ten Wolde, P.R.; Ouldridge, T.E. Biochemical Machines for the Interconversion of Mutual Information and Work. *Phys. Rev. Lett.* **2017**, *118*, 028101.
7. Mark, B.L.; Ephraim, Y. An EM algorithm for continuous-time bivariate Markov chains. *Comput. Stat. Data Anal.* **2013**, *57*, 504–517.
8. Ephraim, Y.; Mark, B.L. Bivariate Markov Processes and Their Estimation. *Found. Trends Signal Process.* **2012**, *6*, 1–95.
9. Cover, T.M.; Thomas, J.A. *Elements of Information Theory*, 2nd ed.; John Wiley & Sons: Hoboken, NJ, USA, 2006; ISBN 13-978-0-471-24195-9.
10. Parrondo, J.M.R.; Horowitz, J.M.; Sagawa, T. Thermodynamics of information. *Nat. Phys.* **2015**, *11*, 131–139.
11. Sagawa, T.; Ueda, M. Fluctuation theorem with information exchange: Role of correlations in stochastic thermodynamics. *Phys. Rev. Lett.* **2012**, *109*, 180602.
12. Horowitz, J.M.; Esposito, M. Thermodynamics with Continuous Information Flow. *Phys. Rev. X* **2014**, *4*, 031015.
13. Barato, A.C.; Hartich, D.; Seifert, U. Rate of Mutual Information Between Coarse-Grained Non-Markovian Variables. *J. Stat. Phys.* **2013**, *153*, 460–478.

14. Wang, J.; Xu, L.; Wang, E.K. Potential landscape and flux framework of nonequilibrium networks: Robustness, dissipation, and coherence of biochemical oscillations. *Proc. Natl. Acad. Sci. USA* **2008**, *105*, 12271–12276.

15. Wang, J. Landscape and flux theory of non-equilibrium dynamical systems with application to biology. *Adv. Phys.* **2015**, *64*, 1–137.

16. Li, C.H.; Wang, E.K.; Wang, J. Potential flux landscapes determine the global stability of a Lorenz chaotic attractor under intrinsic fluctuations. *J. Chem. Phys.* **2012**, *136*, 194108.

17. Schnakenberg, J. Network theory of microscopic and macroscopic behavior of master equation systems. *Rev. Mod. Phys.* **1976**, *48*, 571–585.

18. Zia, R.K.P.; Schmittmann, B. Probability currents as principal characteristics in the statistical mechanics of non-equilibrium steady states. *J. Stat. Mech.-Theory E* **2007**, *2007*, doi:10.1088/1742-5468/2007/07/p07012.

19. Maes, C.; Netočný, K. Canonical structure of dynamical fluctuations in mesoscopic nonequilibrium steady states. *Europhys. Lett.* **2008**, *82*, doi:10.1209/0295-5075/82/30003.

20. Qian, M.P.; Qian, M. Circulation for recurrent markov chains. *Probab. Theory Relat.* **1982**, *59*, 203–210.

21. Zhang, Z.D.; Wang, J. Curl flux, coherence, and population landscape of molecular systems: Nonequilibrium quantum steady state, energy (charge) transport, and thermodynamics. *J. Chem. Phys.* **2014**, *140*, 245101.

22. Zhang, Z.D.; Wang, J. Landscape, kinetics, paths and statistics of curl flux, coherence, entanglement and energy transfer in non-equilibrium quantum systems. *New J. Phys.* **2015**, *17*, 043053.

23. Luo, X.S.; Xu, L.F.; Han, B.; Wang, J. Funneled potential and flux landscapes dictate the stabilities of both the states and the flow: Fission yeast cell cycle. *PLoS Comput. Biol.* **2017**, *13*, e1005710.

24. Gray, R.; Kieffer, J. Mutual information rate, distortion, and quantization in metric spaces. *IEEE Trans. Inf. Theory* **1980**, *26*, 412–422.

25. Maes, C.; Redig, F.; van Moffaert, A. On the definition of entropy production, via examples. *J. Math. Phys.* **2000**, *41*, 1528–1554.

26. Gaspard, P. Time-reversed dynamical entropy and irreversibility in Markovian random processes. *J. Stat. Phys.* **2004**, *117*, 599–615.

27. Feng, H.D.; Wang, J. Potential and flux decomposition for dynamical systems and non-equilibrium thermodynamics: Curvature, gauge field, and generalized fluctuation-dissipation theorem. *J. Chem. Phys.* **2011**, *135*, 234511.

28. Polettini, M. Nonequilibrium thermodynamics as a gauge theory. *Europhys. Lett.* **2012**, *97*, 30003.

Article

Information Dynamics of a Nonlinear Stochastic Nanopore System

Claire Gilpin [1],*, David Darmon [2], Zuzanna Siwy [1] and Craig Martens [3]

[1] Department of Physics and Astronomy, University of California-Irvine, Irvine, CA 92697-4575, USA; zsiwy@uci.edu
[2] Department of Military and Emergency Medicine, Uniformed Services University, Bethesda, MD 20814, USA; david.m.darmon@gmail.com
[3] Department of Chemistry, University of California-Irvine, Irvine, CA 92697-2025, USA; craig.martens@uci.edu
* Correspondence: gilpinc@uci.edu

Received: 21 February 2018; Accepted: 21 March 2018; Published: 23 March 2018

Abstract: Nanopores have become a subject of interest in the scientific community due to their potential uses in nanometer-scale laboratory and research applications, including infectious disease diagnostics and DNA sequencing. Additionally, they display behavioral similarity to molecular and cellular scale physiological processes. Recent advances in information theory have made it possible to probe the information dynamics of nonlinear stochastic dynamical systems, such as autonomously fluctuating nanopore systems, which has enhanced our understanding of the physical systems they model. We present the results of local (LER) and specific entropy rate (SER) computations from a simulation study of an autonomously fluctuating nanopore system. We learn that both metrics show increases that correspond to fluctuations in the nanopore current, indicating fundamental changes in information generation surrounding these fluctuations.

Keywords: entropy; local entropy rate; specific entropy rate; information dynamics; *k*-nearest neighbor estimation; nanopore

1. Introduction

Over the previous three decades there has been an increasing interest in studying nanopore systems due to their broad applicability to the physical and biological sciences. Advances in nanometer-scale laboratory techniques have made it possible to create and isolate single nanopores, allowing detailed study of their properties. In addition to suggesting applicability as physiological models for molecular and cellular scale processes, and utility for DNA sequencing and rapid drug/microbial detection, their electronic behavior provides opportunities to study the information dynamics of nanoscale nonlinear oscillators [1–17].

Significant advances have been made in the development of tools to study information storage, processing, and transmission. Specifically, our understanding of Shannon entropy measures as metrics for information flow have progressed with notable momentum in the study of conditional (time-dependent) entropies, referred to as entropy rates in the literature. Local and specific entropy rates are among the most recent contributions in this area. These nonlinear measures of information dynamics can reveal interesting properties not accessible by classical linear methods such as autocorrelation and spectral analyses [18,19].

Local and specific entropy rates quantify complementary properties of a dynamical system. If we have a time series representing an observable, the local entropy rate (LER) represents the statistical surprise of seeing an already observed local future, given a specific past [20–23]. It also can be thought of as the rate of information generation at a given time point. The specific entropy rate (SER) represents the statistical uncertainty in an as yet unobserved future, given a specific past [24]. The LER, in isolation, is a retrospective measure, and the SER is a prospective measure, and each yields distinct information about the behavior of the dynamical system.

In [25], the authors use current–voltage data collected from a single, conical nanopore in contact with an ion solution to motivate a mechanistic model for the electronic behavior of the nanopore system. The pore is externally biased, but its fluctuations (openings and closings) are autonomous and result from dynamic formation and dissolution (or passage) of tiny inorganic precipitates [26,27]. They demonstrated that their mathematical model captures important properties of the observable from the nanopore system, including these fluctuations, which appear as rapid transitions between high and low conductance states in the time series. One reason this system is an ideal candidate for analysis via information dynamics is that its fluctuations may be informed by its previous states (transitions may depend on previous states, as they do in a Markov process).

The structure of this paper is as follows. In Section 2, we present the definitions and estimation procedures for local and specific entropy rates. We introduce the model of the nanopore in Section 3. We then investigate the properties of the nanopore system as viewed through the lens of information dynamics in Section 4. Finally, in Section 5, we conclude by considering other potential avenues of study for this and other nanoscale systems.

2. Definition and Estimation of Local and Specific Entropy Rates

2.1. Notation

Before beginning a discussion of the details of local and specific entropy rates, it is valuable to first establish standardized notation that will be used throughout this paper. We will be considering a stochastic process (current), $\{X_t\}_{t \in \mathbb{Z}}$, in an autonomously fluctuating nanopore system. Realizations (or observed values) of this nanopore current will use lower case designation, $x_1, x_2, ..., x_t$. A block of states will be denoted $X_m^n = (X_m, X_{m+1}, ..., X_{n-1}, X_n)$, where $m < n$.

Additionally, conditional densities are relevant to the discussion of local and specific entropy rates. Conditional densities will be expressed in the form $f_{Y|X}(a \mid b)$, which can be read as "the predictive density of Y conditioned on $X = b$". For example, the predictive density of the future X_t, conditioned on a specific p-step past x_{t-p}^{t-1} can be expressed as $f_{X_t|X_{t-p}^{t-1}}(x_t \mid x_{t-p}^{t-1})$. We will assume conditional stationarity [28] of the process and thus denote the predictive density without time dependence as $f(x_t \mid x_{t-p}^{t-1})$. Note that this is a weaker assumption than strong stationarity which requires that joint distributions of any length are invariant under translation in time. Instead, we only assume that the distribution of the future, conditional on a sufficiently long past, is statistically invariant under translation in time.

2.2. Differential Entropy and Total Entropy Rate

In what follows, we assume familiarity with information theory at the level of [29]. Both local and specific entropy rates are born from differential entropy, and it is therefore important to introduce these quantities from the outset. Differential entropy can be thought of as the average uncertainty about the future if the model for a particular stochastic process is known, but the past is not. The definition of differential entropy for a random vector $\mathbf{X}$ is

$$H[\mathbf{X}] = -E[\log f(\mathbf{X})]$$
$$= -\int_{\mathbb{R}^n} f(\mathbf{x}) \log f(\mathbf{x}) \, d\mathbf{x} \tag{1}$$

where $f(\mathbf{x})$ represents the probability density of an n-dimensional distribution. If we know the past, we can condition the entropy, at any particular time point, on the past and obtain the following total entropy rate:

$$H[X_t \mid X_{-\infty}^{t-1}] = -E[\log f(X_t \mid X_{-\infty}^{t-1})]$$
$$= -\int_{\mathbb{R}^\infty} \int_{\mathbb{R}} f(x_t, x_{-\infty}^{t-1}) \log f(x_t \mid x_{-\infty}^{t-1}) \, dx_t dx_{-\infty}^{t-1} \tag{2}$$

where Equation (2) is defined as $\lim_{p \to \infty} H[X_t \mid X_{-p}^{t-1}]$, if the limit exists. This quantity averages over all pasts and all futures to compute the uncertainty, but what if we want to know, at any given time point, our surprise given the *realized* past and future? We can answer this question with the LER.

2.3. Local Entropy Rate

The LER [20–23] is a time-dependent, locally-determined conditional entropy. The LER quantifies the surprise of observing a future x_t given its preceding values. The LER is defined as the expectand (the expression internal to the expectation operator) of the total entropy rate evaluated at a particular past–future pair,

$$H_t^L = H^L(x_t \mid x_{-\infty}^{t-1}) = -\log f(x_t \mid x_{-\infty}^{t-1}). \tag{3}$$

In practice, however, the density f is not known, and Equation (3) is not directly calculable. We must therefore find an estimator for LER through computation of an estimator for the predictive density. Moreover, because we always estimate f using finite data, it is not tractable to take the block of past values infinitely far into the past, and we therefore also require an appropriate model order p.

The first step in computing the estimator for LER is to determine the model order p for the predictive density $f(x_t \mid x_{t-p}^{t-1})$ used in its calculation. Conceptually, p quantifies the number of preceding state values that are necessary to determine the predictive density of X_t. In analyzing stochastic systems, the goal is to determine whether information contained in the past is useful to understanding the immediate future. If we do not look back far enough, we may miss useful information from the past that could help us determine the future. If we look too far back, we may inadvertently include information not relevant to determining the immediate future. As a concrete example, consider the case where the observable is well-modeled by a Markov process of some order p. In that case, after knowing the previous p values of the process, the future is independent of any values further in the past, so those values do not aid in the prediction of the process. However, they do increase the burden of the associated estimation problem, through the curse of dimensionality [30]. That is, geometrically more data is necessary to achieve the same level of precision in the estimate of the density. In the case where the process is not Markovian, a similar argument applies, except with the added consideration for balancing between the contribution of including more of the past and its increasing burden to estimation process. We seek the model order that results in a minimum uncertainty. Using the information theoretic criterion from [31], the model order is chosen to minimize the negative log predictive likelihood (NLPL)

$$\text{NLPL}(p) = -\frac{1}{N-p} \sum_{i=p+1}^{N} \log \hat{f}_{-i}(X_i \mid X_{i-p}^{i-1}) \tag{4}$$

where N is the number of points in the time series, and $\hat{f}_{-i}$ is an estimator of the predictive density estimated holding out the block X_{i-p}^i. We use a kernel-nearest neighbor estimator for f, which performs kernel density estimation over the set of nearest neighbors in the future space [32,33]. The estimator takes the form

$$\hat{f}_{-i}(X_i \mid X_{i-p}^{i-1}) = \frac{1}{J} \sum_{m \in \mathcal{N}_J(X_{i-p}^{i-1})} K_h(X_i - X_m) \tag{5}$$

where K_h is a Gaussian kernel, J is the number of nearest neighbors to X_{i-p}^{i-1}, $\mathcal{N}_J(X_{i-p}^{i-1})$ is the index set of the J-nearest neighbors, and h is a bandwidth for the density estimator over futures. Equation (4) over h, J, and p is optimized using the constrained Nelder–Mead method from the NLopt library, where h is constrained to $(0, \infty)$ and J is constrained to $\{1, ..., J_{max}\}$, where $J_{max} \ll N$ [34].

Once the optimal model order is determined, it is used to compute the estimator for the LER at time t_i:

$$\hat{H}_{t_i}^{L} = -\log\left\{\frac{\hat{f}(X_i, X_{i-p}^{i-1})}{\hat{f}(X_{i-p}^{i-1})}\right\} \tag{6}$$

where the density estimators are computed using a k-th-nearest neighbor method [35]. A free-standing LER estimator (an estimator of the LER that will not be subsequently used to compute an estimator of the SER) must be computed using a value of k that scales with a power of N to ensure consistency in the estimation of the densities [35]. In this work, $k(N) = \lfloor\sqrt{N}\rfloor$ was used. As we will see in the next section, the estimator of LER is also used in the estimator of the SER. In that case, the LER estimator may be more efficiently computed using a smaller k due to the impact of averaging on the bias–variance trade-off. In that case, we use $k = 4$ in this work.

2.4. Specific Entropy Rate

The SER is a measure of the predictive uncertainty in the future conditional on a specific past. Unlike the LER, which is retrospective, the SER is completely predictive because it averages over futures to compute the uncertainty. The SER is an average of LER values conditioned on a specific past. It takes the following form [24,36]:

$$\begin{aligned} H_{t_i}^{S} &= H^{S}[X_i \mid X_{i-p}^{i-1} = x_{i-p}^{i-1}] \\ &= -E[\log f(X_i \mid X_{i-p}^{i-1}) \mid X_{i-p}^{i-1} = x_{i-p}^{i-1}] \\ &= -\int_{\mathbb{R}} f(x_i \mid x_{i-p}^{i-1})\log f(x_i \mid x_{i-p}^{i-1})\,dx_t. \end{aligned} \tag{7}$$

Similarly, for practical purposes, an estimator for the SER is computed via averaging of the estimator for the LER. Once the LER estimator values have been computed, k^* of them are averaged at each time point to obtain the estimator for the SER at time t_i:

$$\hat{H}_{t_i}^{S} = \frac{1}{k^*}\sum_{j\in\mathcal{N}_{k^*}(X_{i-p}^{i-1})} \hat{H}^{L}(X_j \mid X_{j-p}^{j-1}) \tag{8}$$

where again $\mathcal{N}_{k^*}(X_{i-p}^{i-1})$ is the index set of the k^*-nearest neighbors of X_{i-p}^{i-1}. It is important to note that k^* is distinct from k in the previous discussion of LER. k^* is set at $\lfloor\sqrt{N}\rfloor$ in this work to ensure consistency of the estimator [35,37–39].

2.5. Specific Information Dynamics with Python

The code for the determination of model order and the estimation of local, specific, and total entropy rates is available in the Specific Information Dynamics with Python (sidpy) package, hosted on GitHub [40].

3. Model System

We consider the information dynamics of a nanopore current oscillator, empirically observed and modeled in [25–27,41]. A single nanopore current oscillator, which is observed in a single state x as a function of time t, can be modeled as a coupled nonlinear oscillator with bistable potential $U(x, y)$ forced by dynamical noise:

$$dX_t = -\frac{1}{\gamma_x}\frac{\partial U}{\partial x}(X_t, Y_t)\,dt + \sigma_x\,dW_x$$

$$dY_t = \frac{1}{\gamma_y}[k_+\Theta(X_t) - k_-\Theta(-X_t)]dt + \sigma_y dW_y \tag{9}$$

where X_t is the nanopore current, Y_t is an unobserved state variable that controls the opening and closing behavior of the nanopore, γ is a coefficient of friction, Θ is a Heaviside function with amplitude determined by rate constants k_+ and k_-, and W_x and W_y are standard Brownian motions representing other unaccounted for inputs to the system. The double-well potential $U(x, y)$ is taken to be

$$U(x, y) = \frac{1}{4}ax^4 - \frac{1}{2}b(V)x^2 + cxy \tag{10}$$

where $b(V)$ is the voltage-dependent parameter that determines the barrier height:

$$b(V) = b_0\left(\frac{V - V_c}{V_c}\right) \tag{11}$$

with V_c as the critical voltage. The behavior of the potential as it relates to both X and Y is critical to understanding the behavior of the nanopore current and should be examined in more detail. When X takes positive values, Y will be a random walk with drift $\frac{k_+}{\gamma_y}$ plus dynamical noise. As Y drifts to more positive values, the potential tips towards the negative well, making it likely that a transition will occur from a positive to a negative current. This effect can be seen in Figure 1A below. Conversely, when the nanopore current X takes negative values, Y will be a random walk with drift $\frac{-k_-}{\gamma_y}$ plus dynamical noise. Eventually, Y drifts negative to a point of tipping the potential towards the positive well, making it likely that a transition will occur from a negative current to a positive current. This effect can be seen in Figure 1B below. Between fluctuations, Y will pass through $Y = 0$.

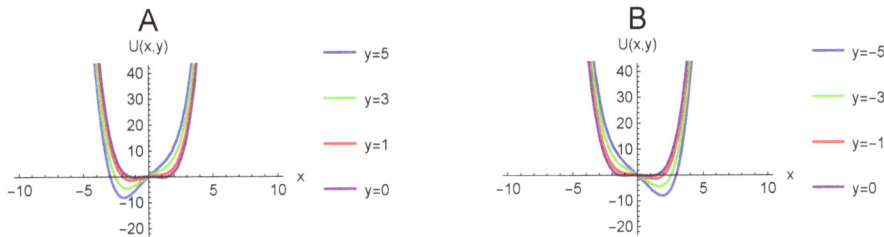

Figure 1. Potential at fixed positive values of y (**A**) and fixed negative values of y (**B**). The $y = 0$ configuration is shown on both plots. As y becomes more positive, it causes the potential to skew towards a transition to negative x. As y becomes more negative, it causes the potential to skew towards a transition to positive x. Physically, positive values of x in this graph correspond to positive current values.

The structure of the potential term as it relates to X and Y acts to ensure that the system undergoes transitions frequently and never becomes stuck indefinitely in one of the wells. Similar behavior of the nanopore current should therefore be expected across realizations of data simulated using this model, despite expected differences in the profiles of individual transitions due to the dynamical noise.

For the simulation, the external bias voltage V is held constant. Values used in this work are $a = b = c = 1$, $\gamma_x = 1$, $\gamma_y = 100$, $k_+ = 1$, $k_- = 5$, and $\sigma_x = \sigma_y = 0.1$ [25]. Realizations from Equation (9) were simulated using the SRI2 stochastic integrator from [42] with a time step of 0.25. The LER and SER were computed for each realization after downsampling to a time step of 0.5, resulting in a total of 40,000 time points.

4. Results

Five separate realizations of the stochastic differential Equation (9) were computed. Figure 2 shows the nanopore current, the LER, and the SER for a group of transitions in the nanopore system (pore open/close events). All three panels are aligned in time. The uppermost panel represents the nanopore current as a function of time. Each orange point is one measurement. Open/close (transition) events can be seen in the rapid switching of the nanopore current from positive to negative values or vice versa. The middle panel is the estimated free-standing LER, computed using Equation (6). We can see that there are peaks in the free-standing LER aligned with the transition events of the nanopore. This indicates that information is generated by these events, and there was some surprise associated with their occurrence. The bottom panel is the SER estimate computed via Equation (8). We see that the SER also increases around the transition events in the nanopore, indicating an increase in uncertainty about future states near the transitions. These results are a subset of results from a single realization, but are representative of the behavior seen in these information measures during transition events across all five realizations, as is expected for this model.

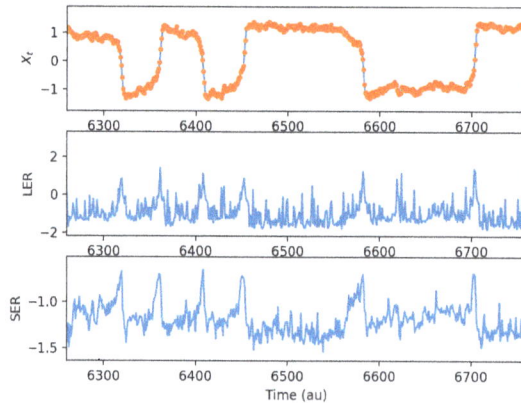

Figure 2. Top: the nanopore current, with each orange dot representing a measurement. **Middle**: the estimate of the local entropy rate (LER) of the nanopore system as a function of time. **Bottom**: the estimate of the specific entropy rate (SER) of the nanopore system as a function of time. This is a representative excerpt from a 40,000 point time series containing on the order of 100 transitions.

The peaks in the LER and SER corresponding to transitions should be considered in the context of the model. It will be easiest to make a transition between positive and negative currents when the slope of the potential is greater (i.e., when the magnitude of y is further from zero). Additionally, when the slope of the potential is large, any small kick from the dynamical noise could lead to a wider array of possible futures (noise amplification). There will accordingly be a greater uncertainty in an unseen future (higher SER) during the transition events. There will also be an elevated LER in these regions because, of the many possible outcomes for current values, when the potential slope is large, individual outcomes may occur only rarely. This will translate to a relatively high surprise.

It should also be noted that there are substantial peaks in the LER that are not always associated with transitions. The LER metric is sensitive to viewing any atypical future. In the relatively flatter (low variation in nanopore current) regions between transitions, any variation above the noise level from the anticipated trajectory may result in high surprise, even though a transition may not occur. This is particularly prominent about 6430 au and 6620 au in Figure 2. If the future is unseen, as in the SER, in these relatively flatter regions there will be low uncertainty about the future. In other words, variation above the noise level is not expected. This is why similar peaks not associated with transitions are rarely seen in the SER.

To further investigate transitions, we consider how the LER and SER vary as a function of the reconstructed state space of the nanopore system. Figure 3 shows a 3D projection of the $p = 4$ reconstructed state space, where each point is shaded by the LER (left) and SER (right). We use the projection (X_{t-2}, X_{t-1}, X_t) for the LER and $(X_{t-3}, X_{t-2}, X_{t-1})$ for the SER. The arrows indicate the direction of the transitions with respect to time. We know that, if the nanopore is in a closed state, it is likely to remain closed and that, if it is in an open state, it is likely to remain open. We thus see relatively low surprise (a low LER) and relatively low uncertainty (a low SER) under those conditions, corresponding to the points in the bottom left and top right of the reconstructed state space. When a transition event is occurring, corresponding to the points along the central "tubes", we are relatively more surprised (a higher LER) and relatively more uncertain about the immediate future (a higher SER). It should also be noted that, if all transitions in this system were identical, the reconstructed state-space trajectory would not show spread about the average path (i.e., the "fuzziness" is due to differences in the profile of the transitions). Although the LER and SER have similar regions of relatively high/low values in their corresponding measures, it is important not to conflate them. The LER measures information generated by seeing the future, and it should not surprise us to see a high degree of symmetry in the LER plot, with maximal information generated for more atypical transitions (on the outsides of the transition "tubes"). The SER, by contrast, measures uncertainty in the future, given a known past, and we might expect high uncertainty in the region of all transitions. It is additionally noteworthy that there is some anti-symmetry between the two transition tubes in the SER plot with respect to the location of the onset of elevation in the SER. This, together with the arrows indicating the direction of the trajectory with time, shows that uncertainty is highest at the beginning of a transition. Uncertainty decreases as the transition proceeds to completion. This is not apparent from examination of the time series.

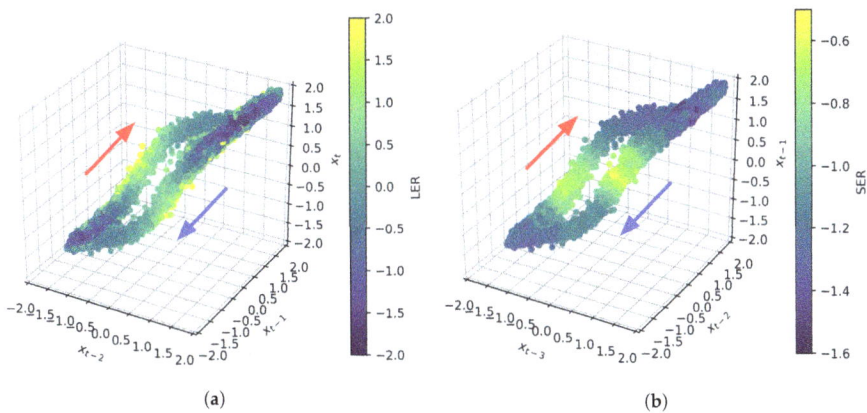

Figure 3. A projection of the reconstructed state space for the nanopore system shaded by the estimates of the LER (**left**) and SER (**right**) associated with the overall state. The plots reveal a clear trajectory in the reconstructed state space, and the arrows indicate the direction along the transitions between open and closed states. Along this trajectory, regions of relatively low surprise (LER) and low uncertainty (SER) occur when the system is in an open/closed state. Conversely, in the central regions, corresponding to transitions, we see increases in both the LER and SER. Anti-symmetry is noted in the onset of increase in SER, which shows that uncertainty is highest at the beginning of a transition and decreases as the transition proceeds to completion. (**a**) LER; (**b**) SER.

It may also be helpful to look directly at the transition region in both the LER and SER schemes, i.e., to look inside the trajectory in the transition regions. To do so, we take a cross section of the plots in Figure 2 at $x_t = 0$ and $x_{t-1} = 0$, respectively, and include points that fall within a tolerance of $\epsilon = \pm 0.05$. This cross section is shown in Figure 4. We can see that the LER is highest in the regions corresponding to less typical transitions (i.e., on the outside of the tubes), as previously mentioned. Additionally, as expected, all transitions in the SER scheme are associated with a similarly elevated SER.

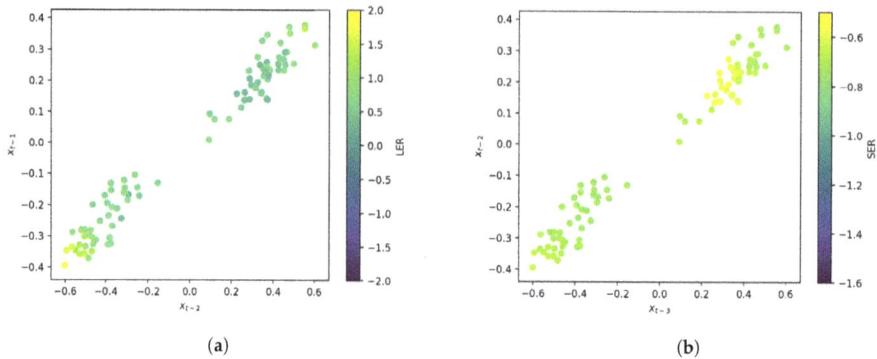

(a)

(b)

Figure 4. A 2D cross section of the reconstructed state space constructed from points within $\epsilon = \pm 0.05$ of the $x_t = 0$ (**left**) and $x_{t-1} = 0$ (**right**) planes for each plot, respectively. These plots show that information is generated most heavily around atypical transition events (the highest LER visible on the periphery of the transition tubes in the LER plot), and there is relatively uniform, high uncertainty for all transitions in the SER plot. (**a**) LER; (**b**) SER.

5. Discussion

At this point it is worth considering what we have learned from calculating both the LER and SER that we could not have otherwise learned from visually examining the time series. It would seem obvious that state transitions in this nanopore system (i.e., pore open or closed) can be observed through such visual examination, but changes in the information dynamics brought about by those transitions cannot be gleaned through such observation. The fact that both the LER and the SER vary with transition behavior suggests that the transitions observed in the time series reflect deeper changes in the information dynamics of the governing system. This is in contrast with systems where transitions in the time series occur in the absence of changes in information status, such as fully deterministic systems or systems where all transitions are alike.

We are able to better understand the complementary value of the LER and SER through visualization of their values in the reconstructed state space. We see that the maximum uncertainty, associated with the maximum SER, occurs when the system is undergoing a transition between the open and closed pore configurations. The uncertainty increased at the start of transitions and falls as they proceed to completion. We also see that surprise, represented by the LER, is concentrated in the transition region, but also has secondary peaks occurring in regions with less typical futures.

Information metrics, such as the LER and SER, provide us with new insights into the underlying dynamics of nanopore systems. This may prove useful in the future to understanding their autonomous fluctuations and the behavior of other systems modeled by similar mathematics. A natural direction for a future study would be to extend these analytical techniques to empirical data from a nanopore system or to empirical data from biological systems, such as gated ion channels. Studying these nanoscale systems using information dynamics is, to the authors' knowledge, a novel activity, and may reveal interesting properties of these systems not previously accessible. With this new exploration comes the possibility that these techniques will help illuminate nanoscale biological processes.

Acknowledgments: The authors acknowledge support from the Uniformed Services University and the Defense Medical Research and Development Program.

Author Contributions: Claire Gilpin and David Darmon conceived and designed the study; Claire Gilpin performed the computations; Claire Gilpin, David Darmon, and Craig Martens analyzed the data; Craig Martens and Zuzanna Siwy contributed the model system; Claire Gilpin wrote the paper. All authors have read and approved the final manuscript.

Conflicts of Interest: The authors declare no conflict of interest.

Abbreviations

The following abbreviations are used in this manuscript:

LER Local entropy rate
SER Specific entropy rate
NLPL Negative log predictive likelihood

References

1. Plett, T.S.; Cai, W.; Le Thai, M.; Vlassiouk, I.V.; Penner, R.M.; Siwy, Z.S. Solid-State Ionic Diodes Demonstrated in Conical Nanopores. *J. Phys. Chem. C* **2017**, *121*, 6170–6176, doi:10.1021/acs.jpcc.7b00258.
2. Kannam, S.K.; Kim, S.C.; Rogers, P.R.; Gunn, N.; Wagner, J.; Harrer, S.; Downton, M.T. Sensing of protein molecules through nanopores: A molecular dynamics study. *Nanotechnology* **2014**, *25*, 155502.
3. Kolmogorov, M.; Kennedy, E.; Dong, Z.; Timp, G.; Pevzner, P.A. Single-molecule protein identification by sub-nanopore sensors. *PLoS Comput. Biol.* **2017**, *13*, e1005356.
4. Howorka, S.; Siwy, Z. Nanopores and Nanochannels: From Gene Sequencing to Genome Mapping. *ACS Nano* **2016**, *10*, 9768–9771, doi:10.1021/acsnano.6b07041.
5. Innes, L.; Gutierrez, D.; Mann, W.; Buchsbaum, S.F.; Siwy, Z.S. Presence of electrolyte promotes wetting and hydrophobic gating in nanopores with residual surface charges. *Analyst* **2015**, *140*, 4804–4812.
6. Schiel, M.; Siwy, Z.S. Diffusion and Trapping of Single Particles in Pores with Combined Pressure and Dynamic Voltage. *J. Phys. Chem. C* **2014**, *118*, 19214–19223, doi:10.1021/jp505823r.
7. Qiu, Y.; Hinkle, P.; Yang, C.; Bakker, H.E.; Schiel, M.; Wang, H.; Melnikov, D.; Gracheva, M.; Toimil-Molares, M.E.; Imhof, A.; et al. Pores with Longitudinal Irregularities Distinguish Objects by Shape. *ACS Nano* **2015**, *9*, 4390–4397, doi:10.1021/acsnano.5b00877.
8. Buchsbaum, S.F.; Nguyen, G.; Howorka, S.; Siwy, Z.S. DNA-Modified Polymer Pores Allow pH- and Voltage-Gated Control of Channel Flux. *J. Am. Chem. Soc.* **2014**, *136*, 9902–9905, doi:10.1021/ja505302q.
9. Buchsbaum, S.F.; Mitchell, N.; Martin, H.; Wiggin, M.; Marziali, A.; Coveney, P.V.; Siwy, Z.; Howorka, S. Disentangling Steric and Electrostatic Factors in Nanoscale Transport Through Confined Space. *Nano Lett.* **2013**, *13*, 3890–3896, doi:10.1021/nl401968r.
10. Howorka, S.; Siwy, Z. Nanopores as protein sensors. *Nat. Biol.* **2012**, *30*, 506–507.
11. Howorka, S.; Siwy, Z. Nanopore analytics: Sensing of single molecules. *Chem. Soc. Rev.* **2009**, *38*, 2360–2384.
12. Mohammad, M.M.; Movileanu, L. Protein sensing with engineered protein nanopores. *Methods Mol. Biol.* **2012**, *870*, 21–37.
13. Zhang, H.; Tian, Y.; Jiang, L. Fundamental studies and practical applications of bio-inspired smart solid-state nanopores and nanochannels. *Nano Today* **2016**, *11*, 61–81.
14. Wanunu, M. Nanopores: A journey towards DNA sequencing. *Phys. Life Rev.* **2012**, *9*, 125–158.
15. Schoch, R.B.; Han, J.; Renaud, P. Transport phenomena in nanofluidics. *Rev. Mod. Phys.* **2008**, *80*, 839.
16. Tian, Y.; Wen, L.; Hou, X.; Hou, G.; Jiang, L. Bioinspired Ion-Transport Properties of Solid-State Single Nanochannels and Their Applications in Sensing. *ChemPhysChem* **2012**, *13*, 2455–2470.
17. Pérez-Mitta, G.; Albesa, A.G.; Trautmann, C.; Toimil-Molares, M.E.; Azzaroni, O. Bioinspired integrated nanosystems based on solid-state nanopores: "Iontronic" transduction of biological, chemical and physical stimuli. *Chem. Sci.* **2017**, *8*, 890–913.

18. Kantz, H.; Schreiber, T. *Nonlinear Time Series Analysis*; Cambridge University Press: Cambridge, UK, 2004; Volume 7.

19. Kaplan, D.; Glass, L. *Understanding Nonlinear Dynamics*; Springer Science & Business Media: New York, NY, USA, 2012.

20. Lizier, J.T.; Prokopenko, M.; Zomaya, A.Y. Local information transfer as a spatiotemporal filter for complex systems. *Phys. Rev. E* **2008**, *77*, 026110.

21. Lizier, J.T. Measuring the dynamics of information processing on a local scale in time and space. In *Directed Information Measures in Neuroscience*; Understanding Complex Systems; Springer: Berlin/Heidelberg, Germany, 2014; pp. 161–193.

22. Lizier, J.T. JIDT: An information-theoretic toolkit for studying the dynamics of complex systems. *arXiv* **2014**, arXiv:1408.3270.

23. Wibral, M.; Lizier, J.T.; Vögler, S.; Priesemann, V.; Galuske, R. Local active information storage as a tool to understand distributed neural information processing. *Front. Neuroinf.* **2014**, *8*, 1.

24. Darmon, D. Specific differential entropy rate estimation for continuous-valued time series. *Entropy* **2016**, *18*, 190.

25. Hyland, B.; Siwy, Z.S.; Martens, C.C. Nanopore Current Oscillations: Nonlinear Dynamics on the Nanoscale. *J. Phys. Chem. Lett.* **2015**, *6*, 1800–1806.

26. Powell, M.R.; Sullivan, M.; Vlassiouk, I.; Constantin, D.; Sudre, O.; Martens, C.C.; Eisenberg, R.S.; Siwy, Z.S. Nanoprecipitation-assisted ion current oscillations. *Nat. Nanotechnol.* **2008**, *3*, 51.

27. Turker Acar, E.; Hinkle, P.; Siwy, Z.S. Concentration Polarization Induced Precipitation and Ionic Current Oscillations with Tunable Frequency. *J. Phys. Chem. C* **2018**, *122*, 3648–3654.

28. Caires, S.; Ferreira, J.A. On the non-parametric prediction of conditionally stationary sequences. *Stat. Inference Stoch. Process.* **2005**, *8*, 151–184.

29. Cover, T.M.; Thomas, J.A. *Elements of Information Theory*; John Wiley & Sons: New York, NY, USA, 2012.

30. Simonoff, J.S. *Smoothing Methods in Statistics*; Springer Science & Business Media: New York, NY, USA, 2012; pp. 124–125.

31. Darmon, D. Information-theoretic model selection for optimal prediction of stochastic dynamical systems from data. *Phys. Rev. E* **2018**, *97*, 032206.

32. Lincheng, Z.; Zhijun, L. Strong consistency of the kernel estimators of conditional density function. *Acta Math. Sin.* **1985**, *1*, 314–318.

33. Izbicki, R.; Lee, A.B. Nonparametric conditional density estimation in a high-dimensional regression setting. *J. Comput. Graph. Stat.* **2016**, *25*, 1297–1316.

34. Nelder, J.A.; Mead, R. A simplex method for function minimization. *Comput. J.* **1965**, *7*, 308–313.

35. Biau, G.; Devroye, L. *Lectures on the Nearest Neighbor Method*; Springer: Berlin, Germany, 2015.

36. Darmon, D.; Rapp, P.E. Specific transfer entropy and other state-dependent transfer entropies for continuous-state input-output systems. *Phys. Rev. E* **2017**, *96*, 022121.

37. Loftsgaarden, D.O.; Quesenberry, C.P. A Nonparametric Estimate of a Multivariate Density Function. *Ann. Math. Stat.* **1965**, *36*, 1049–1051.

38. Wagner, T.J. Strong Consistency of a Nonparametric Estimate of a Density Function. *IEEE Trans. Syst. Man Cybern.* **1973**, *SMC-3*, 289–290.

39. Moore, D.S.; Yackel, J.W. Consistency Properties of Nearest Neighbor Density Function Estimators. *Ann. Stat.* **1977**, *5*, 143–154.

40. Darmon, D. Specific Information Dynamics with Python (sidpy), Version 0.1. 2018. Available online: https://github.com/ddarmon/sidpy (accessed on 28 February 2018).

41. Wen, C.; Zeng, S.; Zhang, Z.; Hjort, K.; Scheicher, R.; Zhang, S.L. On nanopore DNA sequencing by signal and noise analysis of ionic current. *Nanotechnology* **2016**, *27*, 215502.

42. Rößler, A. Runge–Kutta methods for the strong approximation of solutions of stochastic differential equations. *SIAM J. Numer. Anal.* **2010**, *48*, 922–952.

Article

A Chain, a Bath, a Sink, and a Wall

Stefano Iubini [1,2,*], Stefano Lepri [2,3], Roberto Livi [1,2,3], Gian-Luca Oppo [4] and Antonio Politi [5]

[1] Dipartimento di Fisica e Astronomia, Università di Firenze, via G. Sansone 1, I-50019 Sesto Fiorentino, Italy; livi@fi.infn.it

[2] Istituto Nazionale di Fisica Nucleare, Sezione di Firenze, via G. Sansone 1, I-50019 Sesto Fiorentino, Italy; stefano.lepri@isc.cnr.it

[3] Consiglio Nazionale delle Ricerche, Istituto dei Sistemi Complessi, via Madonna del Piano 10, I-50019 Sesto Fiorentino, Italy

[4] SUPA and Department of Physics, University of Strathclyde, Glasgow G4 0NG, UK; g.l.oppo@strath.ac.uk

[5] Institute for Complex Systems and Mathematical Biology & SUPA, University of Aberdeen, Aberdeen AB24 3UE, UK; a.politi@abdn.ac.uk

* Correspondence: stefano.iubini@unifi.it

Received: 22 June 2017; Accepted: 24 August 2017; Published: 25 August 2017

Abstract: We numerically investigate out-of-equilibrium stationary processes emerging in a Discrete Nonlinear Schrödinger chain in contact with a heat reservoir (a bath) at temperature T_L and a pure dissipator (a sink) acting on opposite edges. Long-time molecular-dynamics simulations are performed by evolving the equations of motion within a symplectic integration scheme. Mass and energy are steadily transported through the chain from the heat bath to the sink. We observe two different regimes. For small heat-bath temperatures T_L and chemical-potentials, temperature profiles across the chain display a non-monotonous shape, remain remarkably smooth and even enter the region of negative absolute temperatures. For larger temperatures T_L, the transport of energy is strongly inhibited by the spontaneous emergence of discrete breathers, which act as a thermal wall. A strongly intermittent energy flux is also observed, due to the irregular birth and death of breathers. The corresponding statistics exhibit the typical signature of rare events of processes with large deviations. In particular, the breather lifetime is found to be ruled by a stretched-exponential law.

Keywords: discrete nonlinear schrödinger; discrete breathers; negative temperatures; open systems

PACS: 05.60.-k; 05.70.Ln; 63.20.Pw

1. Introduction

The study of nonequilibrium thermodynamics of systems composed of a relatively small number of particles is motivated by the need for a deeper theoretical understanding of the statistical laws leading to the possibility of manipulating small-scale systems like biomolecules, colloids, or nano-devices. In this framework, statistical fluctuations and size effects play a major role and cannot be ignored as it is customary to do in their macroscopic counterparts.

Arrays of coupled classical oscillators are representative models of such systems and have been studied intensively in this context [1–3]. In particular, the Discrete Nonlinear Schrödinger (DNLS) equation has been widely investigated in various domains of physics as a prototype model for the propagation of nonlinear excitations [4–6]. In fact, it provides an effective description of electronic transport in biomolecules [7] as well as of nonlinear waves propagation in layered photonic or phononic systems [8,9]. More recently, a renewed interest for this multi-purpose equation emerged in the physics of gases of ultra-cold atoms trapped in optical lattices (e.g., see [10] and references therein for a recent survey).

The DNLS dynamics are characterized by two conserved quantities: the energy density h and the mass density a (also termed norm—see the following section for their definitions). Therefore, a generic thermodynamic state of the DNLS equation can be seen as a point in the (a, h) plane. The equilibrium thermodynamics of the DNLS model was studied in a seminal paper by Rasmussen et al. [11] within the grand-canonical formalism. Here, it was shown that an equation of state can be formally derived with the help of transfer–integral techniques. Accordingly, any thermodynamic equilibrium state (a, h) can be equivalently represented in terms of a temperature T and a chemical potential μ which can be numerically determined by implementing suitable microcanonical definitions [12,13]. In Reference [11], it was found that a line of infinite–temperature equilibrium states in the (a, h) plane separates standard thermodynamic states, characterized by a positive T, from those characterized by negative absolute temperatures. The existence of negative temperatures can be traced back to the properties of the two conserved quantities of the system. From the point of view of dynamics, it was realized that negative temperatures can be associated to the presence of intrinsically localized excitations, named discrete breathers (DB) (see e.g., [14–16]). In a series of important papers, Rumpf developed entropy-based arguments to describe the asymptotic states above this infinite-temperature line [17–20]. It has been later found that negative–temperature states can form spontaneously via the dynamics of the DNLS equation. They persist over extremely long time scales, that might grow exponentially with system size as a result of an effective mechanism of ergodicity–breaking [21].

A related question is whether the dynamics of the system influences non-equilibrium properties when the system exchanges energy and/or mass with the environment. In a series of papers [21–23], it has been found that when pure dissipators act at both edges of a DNLS chain (a case sometimes called boundary cooling [24–27]) the typical final state consists of an isolated static breather embedded in an almost empty background. The breather collects a sensible fraction of the initial energy and it is essentially decoupled from the rest of the chain. The spontaneous creation of localized energy spots out of fluctuations has further consequences on the relaxation to equipartition, since the interaction with the remaining part of the chain can become exponentially weak [28,29]. A similar phenomenon occurs after a quench from high to low temperatures in oscillator lattices [30]. Also, boundary driving by external forces may induce non-linear localization [31–33].

When, instead, the chain is put in contact with thermal baths at its edges, non-equilibrium stationary states characterized by a gradient of temperature and chemical potential emerge [13]. Local equilibrium is typically satisfied so that the overall state of the chain can be conveniently represented as a path in the (a, h) plane connecting two points corresponding to the thermodynamic variables imposed by the two reservoirs. The associated transport of mass and energy is typically normal (diffusive) and can be described in terms of the Onsager formalism. However, peculiar features, such as non-monotonous temperature profiles [34] or persistent currents [35], are found as well as a signature of anomalous transport in the low-temperature regime [36,37].

In this paper we consider a set-up where one edge is in contact with a heat reservoir at temperature T_L and chemical potential μ_L, while the other interacts with a pure dissipator, i.e., a mass sink. The original motivation for studying this configuration was to better understand the role of DBs in thermodynamic conditions. At variance with standard set-ups [1,2], this is conceptually closer to a semi-infinite array in contact with a single reservoir. In fact, on the pure-dissipator side, mass can only flow out of the system.

The presence of a pure dissipator forces the corresponding path to terminate close to the point $(a = 0, h = 0)$, which is singular both in T and μ. This is indeed the point where the infinite- and the zero-temperature lines collapse. Therefore, slight deviations may easily lead to crossing the $\beta = 0$ line, where $\beta = 1/T$ is the inverse temperature. This is indeed the typical scenario observed for small T_L, when β smoothly changes sign twice before approaching the dissipator (see Figure 1 for a few sampled paths). The size of the negative-temperature region increases with T_L and appears to be stable over long time scales. Upon further increasing T_L, a different stationary regime is found, characterized by strong fluctuations of mass and energy flux. In practice, the negative-temperature region first extends up to the dissipator edge and then it progressively shrinks in favour of a positive-temperature region (on the other side of the chain). In this regime, the dynamics are controlled by the spontaneous formation (birth) and disappearance (death) of discrete breathers.

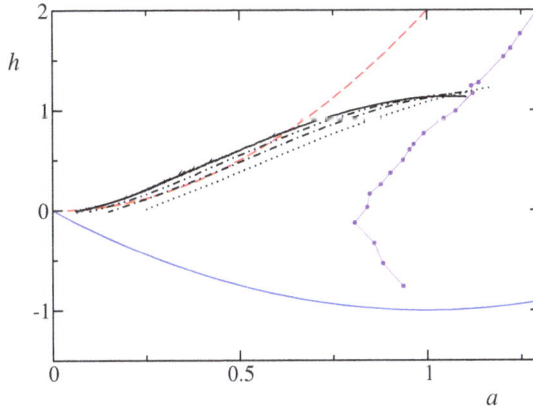

Figure 1. Phase diagram of the DNLS equation in the $(a - h)$ plane of, respectively, energy and mass densities. The positive-temperature region extends between the ground state $\beta = +\infty$ line (solid blue lower curve) and the $\beta = 0$ isothermal (red dashed curve). Purple circles show the $\mu = 0$ line, which has been determined numerically through equilibrium simulations (data are taken from Reference [13]). Black curves refer to nonequilibrium profiles obtained by employing a heat bath with parameters $T_L = 3$ and $\mu_L = 0$ and a pure dissipator located at the left and right edges of the chain, respectively. Dotted, dot-dashed, dot-dot-dashed and solid curves refer to chain sizes $N = 511, 1023, 2047$, and 4095, respectively. Upon increasing N, these above profiles tend to enter the negative temperature region. Simulations are performed by evolving the DNLS chain over 10^7 time units after a transient of 4×10^7 units. For the system size $N = 4095$, a further average over 10 independent trajectories is performed.

In Section 2 we introduce the model and briefly recall the definition of the main observables. Section 3 is devoted to a detailed characterization of the low-temperature phase, while the far-from-equilibrium phase observed for large T_L is discussed in Section 4. This is followed by the analysis of the statistical properties of the birth/death process of large-amplitude DBs, illustrated in Section 5. Finally, in Section 6 we summarize the main results and comment about possible relationships with similar phenomena previously reported in the literature.

2. Model and Methods

We consider a DNLS chain of size N and with open boundary conditions, whose bulk dynamics is ruled by the equation

$$i\dot{z}_n = -2|z_n|^2 z_n - z_{n-1} - z_{n+1} \tag{1}$$

where $(n = 1, \ldots, N)$ and $z_n = (p_n + iq_n)/\sqrt{2}$ are complex variables, with q_n and p_n being standard conjugate canonical variables. The quantity $a_n = |z_n(t)|^2$ can be interpreted as the *number of particles*, or, equivalently, the *mass* in the lattice site n at time t. Upon identifying the set of canonical variables z_n and iz_n^*, Equation (1) can be read as the equation of motion generated by the Hamiltonian functional

$$H = \sum_{n=1}^{N} \left(|z_n|^4 + z_n^* z_{n+1} + z_n z_{n+1}^* \right) \tag{2}$$

through the Hamilton equations $\dot{z}_n = -\partial H/\partial(iz_n^*)$. We are dealing with a dimensionless version of the DNLS equation: the nonlinear coupling constant and the hopping parameters, which are usually indicated explicitly in the Hamiltonian (2), have been set equal to unity. Accordingly, also the time variable t is expressed in arbitrary adimensional units. Without loss of generality, this formulation has the advantage of simplifying numerical simulations.

A relevant property of the DNLS dynamics is the existence of a second conserved quantity beside the total energy H, namely the total mass

$$A = \sum_{n=1}^{N} |z_n|^2 . \tag{3}$$

As a result, an equilibrium state is specified by two parameters, the mass density $a = A/N \geq 0$ and the energy density $h = H/N$. The first reconstruction of the equilibrium phase-diagram (a, h) of the DNLS equation was reported in [11]. It is reproduced in Figure 1 for our choice of parameters, where the lower solid line defines the $(T = 0)$ ground-state line $h = a^2 - 2a$ for different values of the mass density a. The ground-state corresponds to a uniform state with constant amplitude and constant phase-differences $z_n = \sqrt{a}e^{i(\mu t + \pi n)}$, with $\mu = 2(a - 1)$. States below this curve are not physically accessible. The positive-temperature region lies above the ground-state up to the red dashed line $h = 2a^2$, which corresponds to the infinite-temperature $(\beta = 0)$ line. In this limit, the grand-canonical equilibrium distribution becomes proportional to $\exp(\beta\mu A)$, where the finite (negative) product $\beta\mu$ implies a diverging chemical potential. Equilibrium states at infinite temperature are therefore characterized by an exponential distribution of the amplitudes $P(|z_n|^2) = a^{-1}e^{-|z_n|^2/a}$ and random phases. Finally, states above the $\beta = 0$ line belong to the so-called negative-temperature region [11,21].

Finite-temperature equilibrium states do not allow for straightforward analytical treatments. However, one can determine the relation $a(T, \mu)$, $h(T, \mu)$ numerically, by putting the system in interaction with an external reservoir (see below) that imposes T and μ and by measuring the corresponding equilibrium densities. This method was adopted in Reference [13] to identify the $\mu = 0$ line shown in Figure 1. Upon increasing μ, the curve moves to the right in the (a, h) diagram and becomes more and more vertical (data not shown). Isothermal lines $T = c$ can be found analogously: they roughly follow the profile of the $T = 0$ line and, upon increasing c, they span all the positive-temperature region, approaching the infinite-temperature line for $c \to +\infty$ [13].

A non-equilibrium steady state can be represented as a path in the (a, h)-parameter space, where $a(x)$ is the mass density, $h(x)$ the energy density, and $x = n/N$ the rescaled position along the chain. In our set-up, the first site of the chain $(n = 1)$ is in contact with a reservoir at temperature T_L and chemical potential μ_L. This is ensured by implementing the non-conservative Monte-Carlo dynamics described in Reference [13]. In a few words, the reservoir performs random perturbations $\delta z_1 = (\delta p_1 + i\delta q_1)/\sqrt{2}$ of the state variable z_1 that are accepted or rejected according to a grand-canonical Metropolis cost-function $\exp[-T_L(\Delta H - \mu_L\Delta A)]$, where ΔH and ΔA are respectively the variations of energy and mass produced by δz_1. The perturbations δp_1 and δq_1 are independent random variables extracted from a uniform distribution in the interval $[-R, R]$. The opposite site $(n = N)$ interacts with a pure *stochastic dissipator* that sets the variable z_N equal to zero, thus absorbing an amount of mass equal to $|z_N|^2$. Both the heat bath and the dissipator are activated at random times, whose separations are independent and identically distributed variables uniformly distributed within the interval $[t_{min}, t_{max}]$. On average, this corresponds to simulating an interaction process with decay rate $\gamma \sim \bar{t}^{-1}$, where $\bar{t} = (t_{max} + t_{min})/2$. Notice that different prescriptions, such as for example a Poissonian distribution of times with average $\bar{t}$ or a constant pace equal to $\bar{t}$, do not introduce any relevant modification in the dynamical and statistical properties of the model (1). Finally, the Hamiltonian dynamics between successive interactions with the environment has been generated by implementing a symplectic, 4th-order Yoshida algorithm [38]. We have verified that a time step $\Delta t = 2 \times 10^{-2}$ suffices to ensure suitable accuracy.

Throughout the paper we deal with measurements of temperature profiles. Since the DNLS Hamiltonian is not separable, the standard relation between temperature and local kinetic energy does not hold. Due to the existence of the second conserved quantity A, it is necessary to make use of the microcanonical definition provided in [39] and further extended in [12]. Its derivation is founded on the thermodynamic relation

$$\beta = T^{-1} = \left.\frac{\partial S}{\partial E}\right|_{A=M} \tag{4}$$

for a system with total energy $H = E$, total mass $A = M$ and entropy S. The calculation amounts to derive a measure of the hyper-surface at constant energy and mass in the phase space and to compute its variation with respect to E at constant mass. The general expression of T is a thermal average of a non-local function of the dynamical variables z_n and z_n^* and is rather involved; we refer to [13,21] and the related bibliography for theoretical and computational details.

In what follows, we consider a situation where all parameters, other than T_L, are kept fixed. In particular, we have chosen $\mu_L = 0$, $R = 0.8$ and $\bar{t} = 3 \times 10^{-2}$, with t_{max} and t_{min} of order 10^{-2}. We have verified that the results obtained for this choice of the parameter values are general. A more detailed account of the dependence of the results on the thermostat properties will be reported elsewhere.

Finally, we recall the observables that are typically used to characterize a steady-state out of equilibrium: the mass flux

$$j_a = 2\langle \text{Im}(z_n^* z_{n+1}) \rangle , \tag{5}$$

and the energy flux

$$j_h = 2\langle \text{Re}(\dot{z}_n z_{n+1}^*) \rangle , \tag{6}$$

where the angular brackets denote a time average.

3. Low-Temperature Regime: Coupled Transport and Negative Temperatures

In the left panel of Figure 2, we report the average profile of the inverse temperature $\beta(x)$ as a function of the rescaled site position $x = n/N$, for different values of the temperature of the thermostat. A first "anomaly" is already noticeable for relatively small T_L: the profile is non monotonous (see for example the curve for $T_L = 1$). This feature is frequently encountered when a second quantity, besides energy, is transported [13,40,41]. In the present setup, this second thermodynamic observable is the chemical potential $\mu(x)$, set equal to zero at the left edge. Rather than plotting $\mu(x)$, in the right panel of Figure 2, we have preferred to plot the more intuitive mass density $a(x)$. There we see that the profile for $T_L = 1$ deviates substantially from a straight line, indicating that the thermodynamic properties vary significantly along the chain and suggesting that the lattice might not be long enough to ensure an asymptotic behavior.

To clarify this point, we have performed simulations for different values of N. The results for $T_L = 1$ are reported in Figure 3, where we plot the local temperature T as a function of x. All profiles start from $T = 1$, the value imposed by the thermostat and, after an intermediate bump, eventually attain very small values. Since neither the temperature nor the chemical potential are directly imposed by the purely dissipating "thermostat", it is not obvious to predict the asymptotic value of the temperature (and the chemical potential). The data reported in the inset suggest a sort of logarithmic growth with N, but this is not entirely consistent with the results obtained for $T_L = 3$ (see below).

If transport were normal and N were large enough, the various profiles should collapse onto each other, but this is far from the case displayed in Figure 3. The main reason for the lack of convergence is the growth of the temperature bump. This is because, upon further increases of N, the system spontaneously crosses the infinite temperature line and enters the negative-temperature region. For $T_L = 3$, this "transition" has already occurred for $N = 4095$, as it can be seen in Figure 2. The crossings of the infinite temperature points ($\beta = 0$) at the boundaries of the negative temperature region correspond to infinite (negative) values of the chemical potential μ in such a way that the product $\beta\mu$ remains finite at these turning points (data not reported). To our knowledge, this is the first example of negative-temperature states robustly obtained and maintained in nonequilibrium conditions in a chain coupled with a single reservoir at positive temperature. In order to shed light on the thermodynamic limit, we have performed further simulations for different system sizes. In Figure 4, we report the results obtained for $T_L = 3$ and N ranging from 511 to 4095. In Figure 4a, we see that the negative-temperature region is already entered for $N = 1023$. Furthermore, its extension grows with N, suggesting that in the thermodynamic limit it would cover the entire profile but the edges.

Since non-extensive stationary profiles have been previously observed both in a DNLS and a rotor chain (i.e., the XY-model in $d = 1$) at zero temperature and in the presence of chemical potential gradients [42], it is tempting to test to what extent an anomalous scaling (say $n/\sqrt{N}$) can account for the observations. In the inset of Figure 4a, we have rescaled the position along the lattice by $\sqrt{N}$. For relatively small but increasing values of $n/\sqrt{N}$ we do see a convergence towards an asymptotic curve, which smoothly crosses the $\beta = 0$ line. This suggests that close to the left edge, positive temperatures extend over a range of order $\mathcal{O}(\sqrt{N})$, thereby covering a non-extensive fraction of the chain length. The scaling behavior in the rest of the chain is less clear, but it is possibly a standard extensive dynamics characterized by a finite temperature on the right edge. A confirmation of the anomalous scaling in the left part of the chain is obtained by plotting the profiles of h and a again as a function of $n/\sqrt{N}$ (see the panels (b) and(c) in Figure 4, respectively). Further information can be extracted from the scaling behavior of the stationary mass flux j_a. In Figure 5, we report the average value of j_a as a function of the lattice length. There, we see that j_a decreases roughly as $N^{-1/2}$. At a first glance this might be interpreted as a signature of energy super-diffusion, but it is more likely due to the presence of a pure dissipation on the right edge (in analogy to what seen in the XY-model [43]).

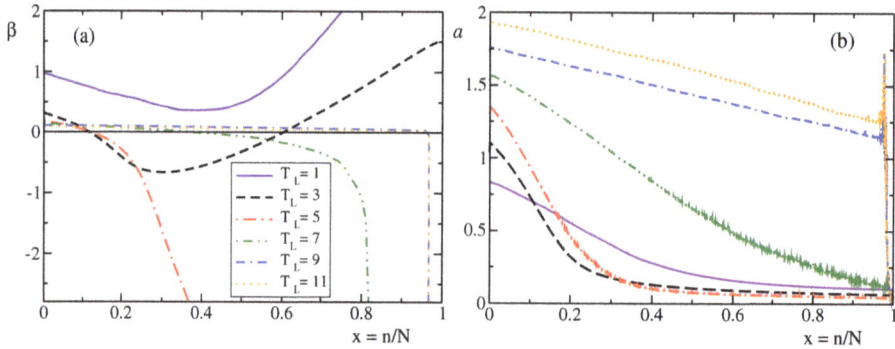

Figure 2. Average profiles of the inverse temperature $\beta(x)$ (panel (**a**)) and mass-density $a(x)$ (panel (**b**)) for a DNLS chain with $N = 4095$ and different temperatures T_L of the reservoir acting at the left edge, where $\mu = 0$. The profile $\beta(x)$ is computed making use of the microcanonical definition of temperature. Simulations are performed evolving the DNLS chain over 10^7 time units after a transient of 4×10^7 units. In order to obtain a reasonable smoothing of these time-averaged profiles we have further averaged each of them over 10 independent trajectories.

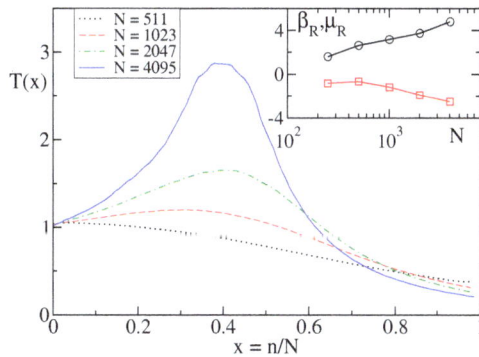

Figure 3. Average profiles of the temperature $T(x)$ for $T_L = 1$ and different system sizes N. The profile $T(x)$ is computed by means of the microcanonical definition of temperature. The inset shows the boundary inverse temperature β_R (black circles) and chemical potential μ_R (red squares) close to the dissipator side as a function of the system size N. The data refers to the microcanonical definitions of temperature and chemical potential computed on the last 10 sites of the chain. Simulations are performed evolving the system over 10^7 time units after a transient of 4×10^7 units. For the system size $N = 4095$ a further average over 10 independent trajectories has been performed.

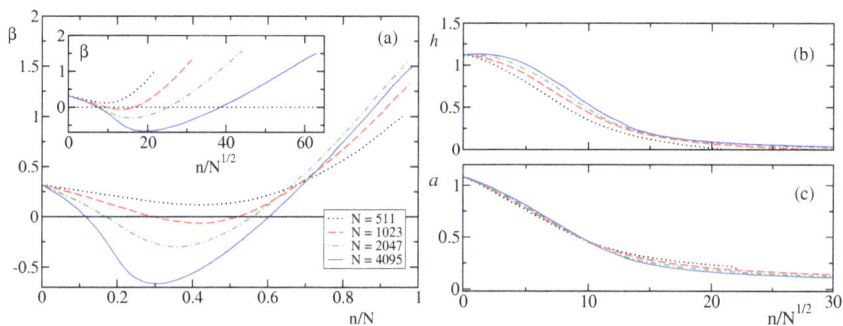

Figure 4. Panel (**a**): average profiles of inverse temperature β for $T_L = 3$ and different system sizes N. The profile β is computed by means of the microcanonical definition of temperature. In the inset, an alternative scaling by $1/\sqrt{N}$ is proposed. For the same setup, panels (**b**,**c**) show the behavior of the energy profile h and the mass profile a, respectively (again scaling the position by $\sqrt{N}$). Simulations are performed evolving the DNLS chain over 10^7 time units after a transient of 4×10^7 units. For the system size $N = 4095$, a further average over ten independent trajectories is performed.

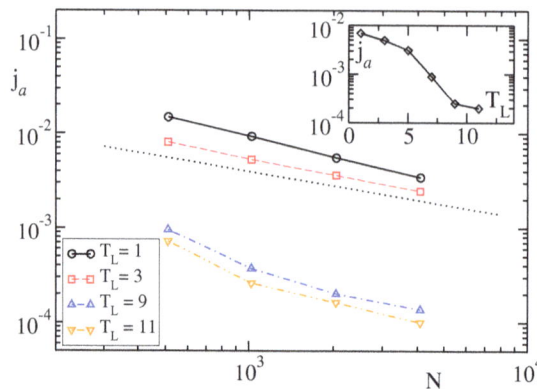

Figure 5. Average mass flux j_a versus system size N for different reservoir temperatures T_L. The black dotted line refers to a power-law decay $j_a \sim N^{-1/2}$. The inset shows the dependence of j_a on the reservoir temperature T_L for the system size $N = 4095$. Simulations are performed evolving the DNLS chain over 10^7 time units after a transient of 4×10^7 units. For the system size $N = 4095$ we have averaged over 10 independent trajectories.

In stationary conditions, mass, and energy fluxes are constant along the chain. This is not necessarily true for the heat flux, as it refers only to the incoherent component of the energy transported across the chain. More precisely, the heat flux is defined as $j_q(x) = j_h - j_a\mu(x)$ [43] Since j_a and j_h are constant, the profile of the heat flux j_q is essentially the same of the μ profile (up to a linear transformation). In Figure 6a, we report the heat flux for $T_L = 1$. It is similar to the temperature profile displayed in Figure 3. It is not a surprise to discover that j_q is larger where the temperature is higher. Figure 6b in the same Figure 6 refers to $T_L = 3$. A very strange shape is found: the flux does not only changes sign twice but exhibits a singular behavior in correspondence of the change of sign, as if a sink and source of heat were present in these two points, where the chemical potential and the local temperature diverge (see, e.g., the red dashed line in Figure 6b representing the β profile). The scenario looks less awkward if the entropy flux $j_s = j_q/T$ is monitored. For $T_L = 1$, the bump disappears and we are in the presence of a more "natural" shape (see Figure 6d) More important is to note that the singularities displayed by j_q for $T_L = 3$ are almost removed since they occur where $T \to \infty$ (we are convinced that the residual peaks are due to a non perfect identification of the singularities). If one removed the singular points, the profile of the entropy flux $j_s = j_q/T$ has a similar shape for the cases $T_L = 1$ and $T_L = 3$. A more detailed analysis of the scenario is, however, necessary in order to provide a solid physical interpretation.

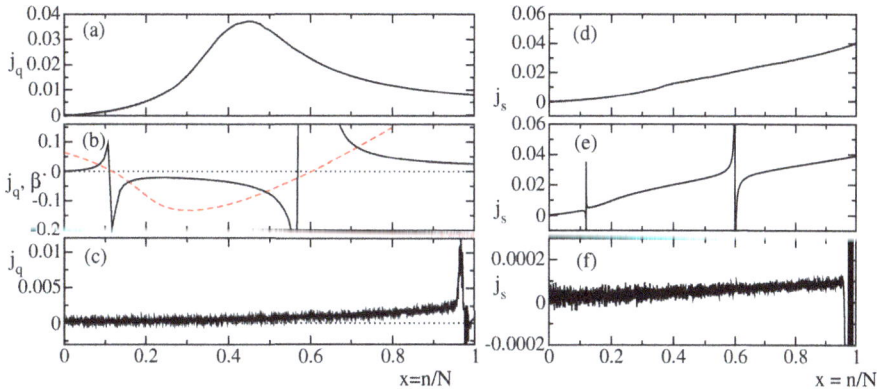

Figure 6. Profiles of heat flux (black lines) for $T_L = 1$ (panel (**a**)), $T_L = 3$ (panel (**b**)) and $T_L = 9$ (panel (**c**)) in a chain with $N = 4095$ lattice sites. For each boundary temperature, we find the following values of mass and energy fluxes: panel (**a**) $j_a = 3.5 \times 10^{-3}$, $j_h = -3.3 \times 10^{-4}$; panel (**b**) $j_a = 2.5 \times 10^{-3}$, $j_h = 3.4 \times 10^{-4}$; panel (**c**) $j_a = 1.2 \times 10^{-4}$, $j_h = 1.5 \times 10^{-4}$. The red dashed line in panel (**b**) refers to the rescaled profile of the inverse temperature $\beta'(x) = \beta(x)/5$ measured along the chain (see Figure 2). Panels (**d–f**) show the profiles of entropy flux for the same temperatures: $T_L = 1$, $T_L = 3$ and $T_L = 9$, respectively. Other simulation details are the same as given in Figure 2.

4. High Temperature Regime: DB Dominated Transport

Let us now turn our attention to the high-temperature case. As shown in Figure 2a, for sufficiently large T_L values, the positive-temperature region close to the dissipator disappears (this is already true for $T_L = 5$) and, at the same time, the positive-temperature region on the left grows. In other words, negative temperatures are eventually restricted to a tiny region close to the dissipator side. This stationary state is induced by the spontaneous formation and the destruction of large DBs close to the dissipator. On average, such a process gives rise to locally steep amplitude profiles that are reminiscent of barriers raised close to the right edge of the chain, see Figure 2b. As it is well known, DBs are localized nonlinear excitations typical of the DNLS chain. Their phenomenology has been widely described in a series of papers where it has been shown that they emerge when energy is dissipated from the boundaries of a DNLS chain [21–23]. In fact, when pure dissipators act at both chain boundaries, the final state turns out to be an isolated DB embedded in an almost empty background. In view of its localized structure and the fast rotation, the DB is essentially uncoupled from the rest of the chain and, a fortiori, from the dissipators. One cannot exclude that a large fluctuation might eventually destroy the DB, but this would be an extremely rare event.

In the setup considered in this paper, DB formation is observed in spite of one of the two dissipators being replaced by a reservoir at finite temperature. DBs are spontaneously produced close to the dissipator edge only for sufficiently high values of T_L. Due to its intrinsic nonlinear character, this phenomenon cannot be described in terms of standard linear-response arguments. In particular, the temperature reported in the various figures cannot be interpreted as the temperature of specific local-equilibrium state: it is at best the average over the many different macrostates visited during the simulation. Actually, it is even possible for some of theses macrostates to deviate substantially from equilibrium. Therefore, we limit ourselves to some phenomenological remarks. The spontaneous formation of small breathers close to the right edge drastically reduces the dissipation and contributed to a further concentration of energy through the merging of the DBs into fewer larger ones and, eventually, to a single DB. Mechanisms of DB-merging have already been encountered under different conditions in the DNLS model [21]. The onset of a DB essentially decouples the left from the right regions of the chain. In particular, it strongly reduces the energy and mass currents fluxes. One can spot DBs simply by looking at the average mass profiles. In Figure 2, the presence of a DB is signaled by the sharp peak close to the right edge for both $T_L = 9$ and $T_L = 11$.

The region between the reservoir and the DB should, in principle, evolve towards an equilibrium state at temperature T_L. However, a close look at the β-profile in Figure 2 reveals the presence of a moderate temperature gradient that is typical of a stationary non-equilibrium state. In fact, DBs are not only born out of fluctuations, but can also collapse due to local energy fluctuations. As shown in Figure 7a, once a DB is formed, it tends to propagate towards the heat reservoir located at the opposite edge. The DB position is tagged by black dots drawn at fixed time intervals over a very long time lapse of $\mathcal{O}(10^6)$, in natural time units of the model. This backward drift comes to a "sudden" end when a suitable energy fluctuation destroys the DB (see Figure 7a). Afterwards, mass and energy start flowing again towards the dissipator, until a new DB is spontaneously formed by a sufficiently large fluctuation close to the dissipator edge (the formation of the DB is signaled by the rightmost black dots in Figure 7a) and the conduction of mass and energy is inhibited again. The DB lifetime is rather stochastic, thus yielding a highly irregular evolution. The statistical properties of such birth/death process are discussed in the following section.

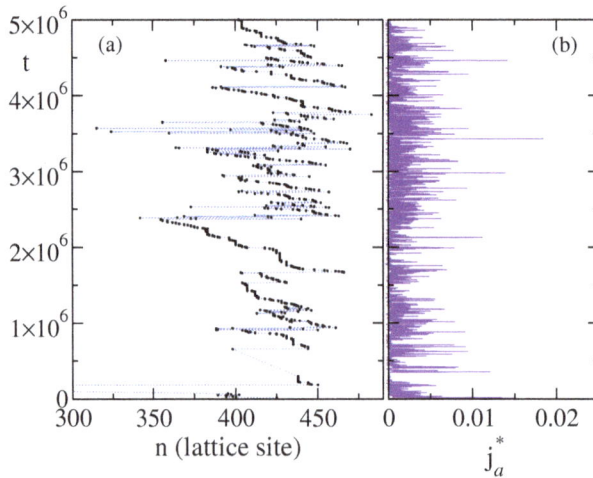

Figure 7. (**a**) Qualitative DB trajectory in a stationary state with $N = 511$ and $T_L = 10$. Each point of the curve corresponds to the position of the maximum average amplitude of the chain in a temporal window of 5×10^3 time units. (**b**) Temporal evolution of the outgoing mass flux j_a^* through the dissipator edge during the same dynamics of panel (**a**). The flux j_a^* is computed every 20 time units as the average amount of mass flowing to the dissipator during such time interval. Higher peaks typically correspond to the breakdown of one or more DBs. Notice that the boundary mass flux can take only positive values because the chain interacts with a pure dissipator.

The statistical process describing the appearance/disappearance of the DB is a complex one. On the one hand, we are in the presence of a stationary regime: the mass and energy currents flowing through the dissipator are found to be constant, when averaged over time intervals much longer than the typical DB lifetime. On the other hand, strong fluctuations in the DB lifetime mean that this regime is not steady but it rather corresponds to a sequence of many different macrostates, some of which are likely to be far from equilibrium. Altogether, in this phase, the presence of long lasting DBs induces a substantial decrease of heat and mass conduction. This is clearly seen in Figure 5, where j_a is plotted for different chain lengths. The two set of data corresponding to $T_L = 9$ and 11 are at least one order of magnitude below those obtained in the low-temperature phase. The sharp crossover that separates the two conduction regimes is neatly highlighted in the inset of Figure 5, where the stationary mass flux is reported as a function of the reservoir temperature for the size $N = 4095$.

The effect of the appearance and disappearance of the DB in the high reservoir temperature regime on the transport of heat and mass along the chain is twofold. During the fast dynamics, it produces bursts in the output fluxes of these quantities as demonstrated in Figure 7b. When the DB is present the boundary flux to the dissipator decreases, while when the DB disappears, avalanches of heat and mass reach the dissipator. In the slow dynamics obtained by averaging over many bursts, the conduction of heat and mass from the heat reservoir to the dissipator is hugely reduced with respect to the low temperature regime. We can conclude that the most important effect of the intermittent DB in the high temperature regime is to act as a thermal wall.

Finally, in Figure 6c,f we plot the heat and entropy profiles observed in the high-temperature phase, respectively. The strong fluctuations in the profiles are a consequence of the large fluctuations in the DB birth/death events and its motion. It is now necessary to average over much longer time scales to obtain sufficiently smooth profiles. It is interesting, however, to observe that the profile of j_s in Figure 6f exhibits an overall shape similar to that observed in the low temperature regimes (see Figure 6d,e). This notwithstanding, there are two main differences with the low temperature behavior. First, close to the right edge of the dissipator we are now in presence of wild fluctuations of j_s, and second the overall scale of the entropy flux profile is heavily reduced.

5. Statistical Analysis

In order to gain information on the high-temperature regime, it is convenient to look at the fluctuations of the boundary mass flux j_a^* and the boundary energy flux j_h^* flowing through the dissipator edge. In Figure 8 we plot the distribution of both j_a^* (panel (a)) and j_h^* (panel (b)) for $T_L = 11$ for different chain lengths. In both cases, power-law tails almost independent of N are clearly visible. This scenario is highly reminiscent of the avalanches occurring in sandpile models. In fact, one such analogy has been previously invoked in the context of DNLS dynamics to characterize the atom leakage from dissipative optical lattices [44].

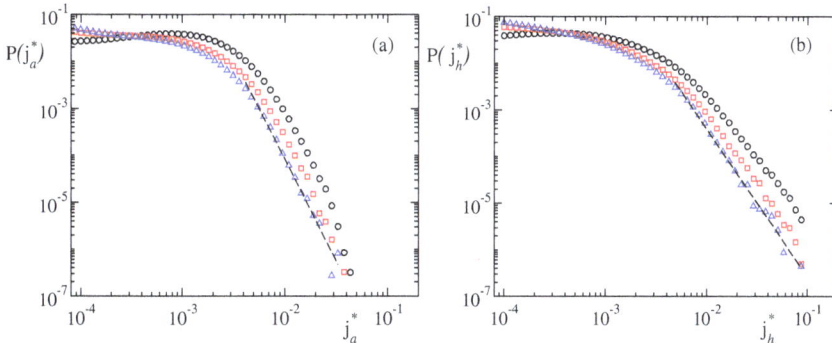

Figure 8. Normalized boundary flux distributions through the dissipator for a stationary state with $T_L = 11$. Panel (**a**) shows the mass flux distribution $P(j_a^*)$, while panel (**b**) refers to the energy flux distribution $P(j_h^*)$. Black circles, red squares, and blue triangles refer to system sizes $N = 511, 1023$, and 2047, respectively. Power-law fits on the largest size $N = 2047$ (see black dashed lines) give $P(j_a^*) \sim j_a^{*-4.32}$ and $P(j_h^*) \sim j_h^{*-3.17}$. Boundary fluxes are sampled by evolving the DNLS chain for a total time t_f after a transient of 4×10^7 temporal units and averaged over time windows of five temporal units. For the sizes $N = 511$ and $N = 1023$ we have considered a single trajectory with $t_f = 10^8$. For the size $N = 2047$ the distributions are extracted from five independent trajectories with $t_f = 2 \times 10^7$.

We processed the time series of the type reported in Figure 7b to determine the duration τ_b of the bursts (avalanches) and τ_l of the "laminar" periods in between consecutive bursts (i.e., the DB life-times). In practice, we have first fixed a flux threshold ($s = 4.25 \times 10^{-3}$) to distinguish between burst and laminar periods. Furthermore, a series of bursts separated by a time shorter than $dt_0 = 10^3$ has been treated as a single burst. This algorithm has been applied to 20 independent realizations of the DNLS dynamics in the high-temperature regime. Each realization has been obtained by simulating a lattice with $N = 511$ sites, $T_L = 10$, and for a total integration time $t = 5 \times 10^6$. In these conditions we have recorded nearly 7000 avalanches.

The probability distribution of the burst duration is plotted in Figure 9a. It follows a a Poissonian distribution, typical of random uncorrelated events. We have also calculated the amount of mass A and energy E associated with each burst, integrating mass, and energy fluxes during each burst. The results are shown as a scatter plot in the inset of Figure 9a. They display a clear (and unsurprising) correlation between these two quantities.

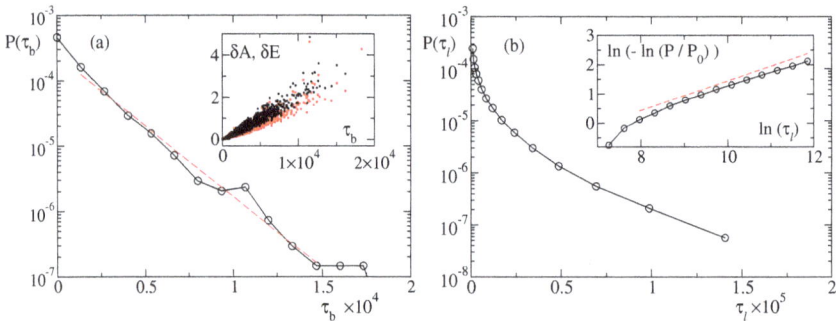

Figure 9. (**a**) Probability distribution of the duration of bursts τ_b. Note the logarithmic scale on the vertical axis. From an exponential fit $P \sim e^{-\gamma \tau_b}$, we find a decay constant $\gamma = 5 \times 10^{-4}$ (red dashed line). The inset shows the relation between the duration τ_b and the amount of mass δA (black dots) and energy δE (red dots) released to the dissipator. (**b**) Probability distribution of DB lifetimes. In the inset we show that this is compatible with a stretched exponential law $P = P_0 e^{-\tau_l^\sigma}$, where $\sigma \simeq 0.5$ (red dashed line) and P_0 is the maximum value of the distribution.

The time interval between two consecutive bursts is characterized by a small mass flux. Typically, during this period the chain develops a stable DB that inhibits the transfer of mass towards the dissipator. Figure 9b shows the probability distribution of the duration of these laminar periods. The distribution displays a stretched-exponential decay with a characteristic constant $\sigma = 0.5$. Such a scenario is consistent with the results obtained in [26] for the Fermi-Pasta-Ulam chain and in [22] for the DNLS lattice. The values of the power σ found in these papers are not far from the one that we have obtained here (see also [45] for similar results on rotor models).

Altogether, there is a clear indication that the statistics of the duration of avalanches and walls is controlled by substantially different mechanisms. It seems that the death of a wall/DB is ruled by rare event statistics [22,26,45], while its birth appears as a standard activation process emerging when an energy barrier is eventually overtaken. In fact, when mass starts flowing through the dissipator edge after the last death of a wall/DB, we have to wait for the spontaneous *activation* of a new wall/DB before the mass flux vanishes again. Conversely, the wall/DB is typically found to persist over much longer time scales and its eventual destruction is determined by a very rare fluctuation, whose amplitude is expected to be sufficiently large to compete with the energy that, in the meanwhile, has been collected by the wall/DB during its motion towards the reservoir.

6. Conclusions

We have investigated the behavior of a discrete nonlinear Schrödinger equation sandwiched between a heat reservoir and a mass/energy dissipator. Two different regimes have been identified upon changing the temperature T_L of the heat reservoir, while keeping fixed the properties of the dissipator. For low T_L and low chemical potential, a smooth β-profile is observed, which extends (in the central part) to negative temperatures without, however, being accompanied by the formation of discrete breathers over the time scales accessible to numerical simulations. In the light of the theoretical achievements by Rumpf [17–20], the negative-temperature regions are incompatible with the assumption of local thermodynamic equilibrium, which instead appears to be satisfied in the positive-temperature part. Therefore, despite the smoothness of the profiles and the stationarity of mass and energy fluxes, such negative-temperature configurations should be better considered as metastable states. Unfortunately, it is not easy to investigate this regime; no single heat bath can impose a fixed negative T in a meaningful way so as to be able to compare with equilibrium states. It is therefore necessary to simulate larger systems in the hope to observe the spontaneous formation of breathers, the only obvious signature of negative temperatures. We plan to undertake such a kind of numerical studies in the near future. It is nevertheless remarkable to see that negative temperatures are steadily sustained for moderately long chain lengths. As a second anomaly, we report the slow decrease of the mass-flux with the chain length: the hallmark of an unconventional type of transport. This feature is, however, not entirely new; a similar scenario has been previously observed in setups with dissipative boundary conditions and no fluctuations [42].

For larger temperatures T_L, we observe an intermittent regime characterized by the alternation of insulating and conducting states, triggered by the appearance/disappearance of discrete breathers. Note that this regime is rather unusual, since it is generated by increasing the amount of energy provided by the heat bath rather than by decreasing the chemical potential, as observed for example in the superfluid/Mott insulator transition in Bose-Einstein condensates in optical lattices. Although, for clarity, we referred to the low and high T_L cases, there is no special difference to be expected in the intermediate regime, except for the typical timescale for DB creation that may be, still, exceedingly large. In fact, for finite N, such a timescale becomes much shorter above some typical T_L.

The intermittent presence of a DB/wall makes the chain to behave as a *rarely leaking pipe*, which releases mass droplets at random times when the DB disappears according to a stretched-exponential distribution. The resulting fluctuations of the fluxes suggest that the regime is stationary but not steady, i.e., locally the chain irregularly oscillates among different macroscopic states characterized, at best, by different values of the thermodynamic variables. A similar scenario is encountered in the XY chain, when both reservoirs are characterized by a purely dissipation accompanied by a deterministic forcing [42]. In such a setup, as discussed in Reference [42], the temperature in the middle of the chain fluctuates over macroscopic scales. Here, however, given the rapidity of changes induced by the DB dynamics, there may be no well-defined values of the thermodynamic observables. For example, during an avalanche it is unlikely that temperature and chemical potentials are well-defined quantities, as there may not even be a local equilibrium. This extremely anomalous behavior is likely to smear out in the thermodynamic limit, since the breather life-time does not probably increase with the system size, however, it is definitely clear that the associated fluctuations strongly affect moderately-long DNLS chains.

Acknowledgments: This research did not receive any funding. Stefano Lepri acknowledges hospitality of the Institut Henri Poincaré-Centre Emile Borel during the trimester Stochastic Dynamics Out of Equilibrium where part of this work was elaborated.

Author Contributions: Stefano Iubini performed the numerical simulations. All Authors contributed to the research work and to writing the paper.

Conflicts of Interest: The authors declare no conflict of interest.

Abbreviations

The following abbreviations are used in this manuscript:

DNLS   Discrete Nonlinear Schrödinger
DB   Discrete Breather

References

1. Lepri, S.; Livi, R.; Politi, A. Thermal conduction in classical low-dimensional lattices. *Phys. Rep.* **2003**, *377*, 1–80.
2. Dhar, A. Heat Transport in low-dimensional systems. *Adv. Phys.* **2008**, *57*, 457–537.
3. Basile, G.; Delfini, L.; Lepri, S.; Livi, R.; Olla, S.; Politi, A. Anomalous transport and relaxation in classical one-dimensional models. *Eur. Phys J. Spec. Top.* **2007**, *151*, 85–93.
4. Eilbeck, J.C.; Lomdahl, P.S.; Scott, A.C. The discrete self-trapping equation. *Physica D* **1985**, *16*, 318–338.
5. Eilbeck, J.C.; Johansson, M. *The Discrete Nonlinear Schrödinger Equation-20 Years on*; World Scientific: Singapore, 2003.
6. Kevrekidis, P.G. *The Discrete Nonlinear Schrödinger Equation*; Springer: Berlin/Heidelberg, Germany, 2009.
7. Scott, A. *Nonlinear Science. Emergence and Dynamics of Coherent Structures*; Oxford University Press: Oxford, UK, 2003.
8. Kosevich, A.M.; Mamalui, M.A. Linear and nonlinear vibrations and waves in optical or acoustic superlattices (photonic or phonon crystals). *J. Exp. Theor. Phys.* **2002**, *95*, 777.
9. Hennig, D.; Tsironis, G. Wave transmission in nonlinear lattices. *Phys. Rep.* **1999**, *307*, 333–432.
10. Franzosi, R.; Livi, R.; Oppo, G.; Politi, A. Discrete breathers in Bose–Einstein condensates. *Nonlinearity* **2011**, *24*, R89.
11. Rasmussen, K.; Cretegny, T.; Kevrekidis, P.G.; Grønbech-Jensen, N. Statistical mechanics of a discrete nonlinear system. *Phys. Rev. Lett.* **2000**, *84*, 3740–3743.
12. Franzosi, R. Microcanonical Entropy and Dynamical Measure of Temperature for Systems with Two First Integrals. *J. Stat. Phys.* **2011**, *143*, 824–830.
13. Iubini, S.; Lepri, S.; Politi, A. Nonequilibrium discrete nonlinear Schrödinger equation. *Phys. Rev. E* **2012**, *86*, 011108.
14. Sievers, A.; Takeno, S. Intrinsic localized modes in anharmonic crystals. *Phys. Rev. Lett.* **1988**, *61*, 970.
15. MacKay, R.; Aubry, S. Proof of existence of breathers for time-reversible or Hamiltonian networks of weakly coupled oscillators. *Nonlinearity* **1994**, *7*, 1623.
16. Flach, S.; Gorbach, A.V. Discrete breathers—Advances in theory and applications. *Phys. Rep.* **2008**, *467*, 1–116.
17. Rumpf, B. Simple statistical explanation for the localization of energy in nonlinear lattices with two conserved quantities. *Phys. Rev. E* **2004**, *69*, 016618.
18. Rumpf, B. Transition behavior of the discrete nonlinear Schrödinger equation. *Phys. Rev. E* **2008**, *77*, 036606.
19. Rumpf, B. Stable and metastable states and the formation and destruction of breathers in the discrete nonlinear Schrödinger equation. *Phys. D Nonlinear Phenom.* **2009**, *238*, 2067–2077.
20. Rumpf, B. Growth and erosion of a discrete breather interacting with Rayleigh-Jeans distributed phonons. *Europhys. Lett.* **2007**, *78*, 26001.
21. Iubini, S.; Franzosi, R.; Livi, R.; Oppo, G.; Politi, A. Discrete breathers and negative-temperature states. *New J. Phys.* **2013**, *15*, 023032.
22. Livi, R.; Franzosi, R.; Oppo, G.L. Self-localization of Bose-Einstein condensates in optical lattices via boundary dissipation. *Phys. Rev. Lett.* **2006**, *97*, 60401.
23. Franzosi, R.; Livi, R.; Oppo, G.L. Probing the dynamics of Bose–Einstein condensates via boundary dissipation. *J. Phys. B At. Mol. Opt. Phys.* **2007**, *40*, 1195.
24. Tsironis, G.; Aubry, S. Slow relaxation phenomena induced by breathers in nonlinear lattices. *Phys. Rev. Lett.* **1996**, *77*, 5225.
25. Piazza, F.; Lepri, S.; Livi, R. Slow energy relaxation and localization in 1D lattices. *J. Phys. A Math. Gen.* **2001**, *34*, 9803.
26. Piazza, F.; Lepri, S.; Livi, R. Cooling nonlinear lattices toward energy localization. *Chaos* **2003**, *13*, 637–645.

27. Reigada, R.; Sarmiento, A.; Lindenberg, K. Breathers and thermal relaxation in Fermi–Pasta–Ulam arrays. *Chaos* **2003**, *13*, 646–656.

28. De Roeck, W.; Huveneers, F. Asymptotic localization of energy in nondisordered oscillator chains. *Commun. Pure Appl. Math.* **2015**, *68*, 1532–1568.

29. Cuneo, N.; Eckmann, J.P. Non-equilibrium steady states for chains of four rotors. *Commun. Math. Phys.* **2016**, *345*, 185–221.

30. Oikonomou, T.; Nergis, A.; Lazarides, N.; Tsironis, G. Stochastic metastability by spontaneous localisation. *Chaos Solitons Fractals* **2014**, *69*, 228–232.

31. Geniet, F.; Leon, J. Energy transmission in the forbidden band gap of a nonlinear chain. *Phys. Rev. Lett.* **2002**, *89*, 134102.

32. Maniadis, P.; Kopidakis, G.; Aubry, S. Energy dissipation threshold and self-induced transparency in systems with discrete breathers. *Phys. D Nonlinear Phenom.* **2006**, *216*, 121–135.

33. Johansson, M.; Kopidakis, G.; Lepri, S.; Aubry, S. Transmission thresholds in time-periodically driven nonlinear disordered systems. *Europhys. Lett.* **2009**, *86*, 10009.

34. Iubini, S.; Lepri, S.; Livi, R.; Politi, A. Off-equilibrium Langevin dynamics of the discrete nonlinear Schroedinger chain. *J. Stat. Mech Theory Exp.* **2013**, *2013*, P08017 .

35. Borlenghi, S.; Iubini, S.; Lepri, S.; Chico, J.; Bergqvist, L.; Delin, A.; Fransson, J. Energy and magnetization transport in nonequilibrium macrospin systems. *Phys. Rev. E* **2015**, *92*, 012116.

36. Kulkarni, M.; Huse, D.A.; Spohn, H. Fluctuating hydrodynamics for a discrete Gross-Pitaevskii equation: Mapping onto the Kardar-Parisi-Zhang universality class. *Phys. Rev. A* **2015**, *92*, 043612.

37. Mendl, C.B.; Spohn, H. Low temperature dynamics of the one-dimensional discrete nonlinear Schroedinger equation. *J. Stat. Mech Theory Exp.* **2015**, *2015*, P08028.

38. Yoshida, H. Construction of higher order symplectic integrators. *Phys. Lett. A* **1990**, *150*, 262–268.

39. Rugh, H.H. Dynamical approach to temperature. *Phys. Rev. Lett.* **1997**, *78*, 772.

40. Iacobucci, A.; Legoll, F.; Olla, S.; Stoltz, G. Negative thermal conductivity of chains of rotors with mechanical forcing. *Phys. Rev. E* **2011**, *84*, 061108.

41. Ke, P.; Zheng, Z.G. Dynamics of rotator chain with dissipative boundary. *Front. Phys.* **2014**, *9*, 511–518.

42. Iubini, S.; Lepri, S.; Livi, R.; Politi, A. Boundary-induced instabilities in coupled oscillators. *Phys. Rev. Lett.* **2014**, *112*, 134101.

43. Iubini, S.; Lepri, S.; Livi, R.; Politi, A. Coupled transport in rotor models. *New J. Phys.* **2016**, *18*, 083023.

44. Ng, G.; Hennig, H.; Fleischmann, R.; Kottos, T.; Geisel, T. Avalanches of Bose–Einstein condensates in leaking optical lattices. *New J. Phys.* **2009**, *11*, 073045.

45. Eleftheriou, M.; Lepri, S.; Livi, R.; Piazza, F. Stretched-exponential relaxation in arrays of coupled rotators. *Phys. D Nonlinear Phenom.* **2005**, *204*, 230–239.

entropy

MDPI

Article

Fourier's Law in a Generalized Piston Model

Lorenzo Caprini [1], **Luca Cerino** [2], **Alessandro Sarracino** [3,*] and **Angelo Vulpiani** [2,4]

[1] GSSI (Gran Sasso Science Institute), Università dell'Aquila, Viale Francesco Crispi, 7, 67100 L'Aquila, Italy; lorenzo.caprini@gssi.it

[2] Dipartimento di Fisica, Università di Roma Sapienza, P.le Aldo Moro 2, 00185 Rome, Italy; cerino.luca@gmail.com (L.C.); Angelo.Vulpiani@roma1.infn.it (A.V.)

[3] ISC-CNR and Dipartimento di Fisica, Università di Roma Sapienza, P.le Aldo Moro 2, 00185 Rome, Italy

[4] Centro Interdisciplinare B. Segre, Accademia dei Lincei, Via della Lungara, 10, 00165 Rome, Italy

* Correspondence: Alessandro.Sarracino@roma1.infn.it

Received: 5 June 2017; Accepted: 6 July 2017; Published: 11 July 2017

Abstract: A simplified, but non trivial, mechanical model—gas of N particles of mass m in a box partitioned by n mobile adiabatic walls of mass M—interacting with two thermal baths at different temperatures, is discussed in the framework of kinetic theory. Following an approach due to Smoluchowski, from an analysis of the collisions particles/walls, we derive the values of the main thermodynamic quantities for the stationary non-equilibrium states. The results are compared with extensive numerical simulations; in the limit of large n, $mN/M \gg 1$ and $m/M \ll 1$, we find a good approximation of Fourier's law.

Keywords: Fourier's law; kinetic theory; non equilibrium statistical mechanics

1. Introduction

Fourier's law, which relates the macroscopic heat flux to the temperature gradient in a solid system, was introduced almost two centuries ago. Nevertheless, its understanding from microscopic basis is still an important open issue of the non-equilibrium statistical mechanics [1]. In particular, among the several theoretical studies on this subject, important results have been derived mainly for $1d$ systems. The prototype model is a chain of masses and non linear springs (Fermi-Pasta-Ulam-like systems), whose ends interacts with thermal baths at different temperatures [1–3]. Other investigated systems are constituted of $1d$ lattices [4,5], chains of cells, with an energy storage device, which exchange energy through tracer particles [6,7], spinning disks [8], systems with local thermalization mechanism [9,10], or a chain of anharmonic oscillators with local energy conserving noise [11]. Despite the apparent simplicity of the problem, and of the considered models, both the analytical approaches and the numerical studies are rather difficult in this context.

The main aim of the present paper is the study of Fourier's law using a mechanical model, which is rather crude (but still non-trivial), allowing for an approach in terms of kinetic theory. More specifically, we consider a generalized piston model, made of a certain number of cells, each containing a non-interacting particle gas. The walls of the cells are mobile, massive objects, that interact with the particles via elastic collisions. The system at its ends interacts with thermal baths at fixed temperatures. It is easy to realize the analogy between such a generalized piston model and the systems of masses and springs: the pistons and the gas compartments play the role of masses and springs, respectively.

Our model is an example of partitioning system (as the adiabatic piston), where previous studies showed that the presence of mobile walls can induce interesting behaviours [12–21]. Basically, in the study of partitioning systems, one can adopt two approaches: in terms of a Boltzmann equation [14] or introducing effective equations (Langevin-like) for suitable observables derived à la Smoluchowski, i.e., from an analysis of the collisions particles/walls. In our study, we will adopt the latter method.

The paper is organized as follows. Section 1 is devoted to the introduction of the model; in Section 2, we show how a thermodynamic approach is not sufficient to determine the values of macroscopic variables in the steady state. In Section 3, we present a coarse-grained Smoluchowski-like description of the system, which provides a good prediction for the main quantities of interest. In Section 4, we compare the theoretical results with numerical simulations and discuss the limit of validity of the proposed approach. In Section 5, some general conclusions on Fourier's Law in mechanical systems are drawn. In Appendix A, we report the details of the analytical computations presented in Section 3.

2. A Generalized Piston Model

We consider a box of length L, partitioned by n mobile adiabatic walls (also called pistons in the following), with mass M, average positions z_j and velocities V_j ($j = 1, 2, \ldots n$). The walls move without friction along the horizontal axis (see Figure 1). The external walls are kept fixed in the positions $z_0 = 0$ and $z_{n+1} = L$. Each of the $n + 1$ compartments separated by the walls contains N non-interacting point-like particles, with mass m, positions x_i and velocities v_i. The particles interact with the pistons via elastic collisions:

$$V' = V + \frac{2m}{m + M} (v_i - V),$$
$$v'_i = v_i - \frac{2M}{m + M} (v_i - V),$$
(1)

where primes denote post-collisional velocities.

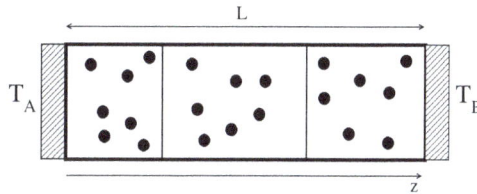

Figure 1. Sketch of the system.

The two external walls in $z = 0$ and $z = L$ act as thermal baths at temperatures T_A and T_B. The interaction of the thermostats with the particles is the following: when a particle collides with the wall, it is reinjected into the system with a new velocity drawn from the probability distribution [22]

$$\rho_{A,B}(v') = \frac{m}{k_B T_{A,B}} |v'| e^{-\frac{mv'^2}{2k_B T_{A,B}}} \Theta(\pm v'),$$
(2)

with $+$ for the case A and $-$ for B, and k_B is the Boltzmann constant, $\Theta(x)$ being the Heaviside step function:

$$\Theta(x) = \begin{cases} 1, & \text{if } x > 0, \\ 0, & \text{if } x \leq 0. \end{cases}$$
(3)

For the following discussion, we define the temperature of the i−th piston as

$$T_i^{(p)} = M \langle V_i^2 \rangle,$$
(4)

and the temperature of the particle gas in the compartment j as an average on the particles between the $(j-1)$-th and the j-th piston

$$T_j = m \frac{1}{N} \sum_{i \in (j-1)N}^{jN} \langle v_i^2 \rangle.$$
(5)

3. Simple Thermodynamic Considerations

We expect, and this is fully in agreement with the numerical computations, that, given a generic initial condition, after a certain transient, the system reaches a stationary state. The positions of the walls fluctuate around their mean values z_j, in a similar way to $T_i^{(p)}$ and T_i. The first non-trivial problem of the non equilibrium statistical mechanics is to determine z_j, $T_i^{(p)}$ and T_i as function of the parameters of the model, i.e., n, L, m, M, T_A and T_B.

We first analyze the simplest case of a single piston, where thermodynamics is sufficient to provide a complete description of the stationary state. Then, we consider the more general case of a multiple piston; in such a situation, thermodynamic relations are not enough to univocally determine the steady state: it is necessary to adopt a statistical mechanics approach. Such an approach will rely on three main assumptions, discussed in more detail below: small mass ratio $m/M \ll 1$, local thermodynamic equilibrium in each compartment, and independence of the collisions particles/pistons.

3.1. Single Piston

If $n = 1$, using the equation for the perfect gas in each compartment, we immediately get the equations

$$
\begin{cases}
pz_1 = Nk_B T_A, \\
p(L - z_1) = Nk_B T_B,
\end{cases}
\tag{6}
$$

where p is the pressure, yielding

$$
z_1 = \frac{T_A}{T_A + T_B} L.
\tag{7}
$$

Therefore, in this case, thermodynamics univocally determines the stationary state of the system.

3.2. Multiple Piston

We now consider the generalized piston model with $n > 1$. An analysis of the case $n = 2$ is enough to understand how the relations obtained from thermodynamics can be not sufficient to fully characterize the non equilibrium steady state. Indeed, we have the following relations:

$$
\begin{cases}
pz_1 = Nk_B T_A, \\
p(z_2 - z_1) = Nk_B T_1, \\
p(L - z_2) = Nk_B T_B,
\end{cases}
\tag{8}
$$

which give the constraints

$$
\begin{aligned}
z_1 &= \frac{T_A}{T_A + T_1 + T_B} L, \\
z_2 &= \frac{T_A + T_1}{T_A + T_1 + T_B} L.
\end{aligned}
\tag{9}
$$

Therefore, since we have three variables (z_1, z_2, T_1) and only two constraints, thermodynamics is not enough to determine the stationary state. The computation can be easily extended to an arbitrary value of n, leading to

$$
\begin{cases}
pz_1 = Nk_B T_A, \\
p(z_2 - z_1) = Nk_B T_1 \\
\dots, \\
p(z_n - z_{n-1}) = Nk_B T_{n-1}, \\
p(L - z_n) = Nk_B T_B,
\end{cases}
\tag{10}
$$

that give the conditions

$$z_1 = \frac{T_A}{T_A + \sum_{k=1}^{n-1} T_k + T_B} L, \quad \cdots \quad z_m = \frac{T_A + \sum_{k=1}^{m-1} T_k}{T_A + \sum_{k=1}^{n-1} T_k + T_B} L, \quad \cdots \quad z_n = \frac{T_A + \sum_{k=1}^{n-1} T_k}{T_A + \sum_{k=1}^{n-1} T_k + T_B} L, \quad (11)$$

with $m = 2, \ldots, n-1$, namely, only n constraints for $2n - 1$ variables.

4. Coarse-Grained Description and Effective Langevin Equations

In order to obtain a statistical description of the system, we now derive effective stochastic equations, governing the dynamics of the relevant variables. Previous theoretical studies based on a Boltzmann equation approach on similar systems were reported in [13–15]. In particular, a generalized piston model was considered in Reference [15]. In those works, theoretical results were not compared to numerical simulations, so that the underlying hypotheses and their range of validity remained unclear.

Here, we present a different analysis. We will assume that the evolution of the observables is described by Langevin equations. The idea, coming back to Smoluchowski, is to integrate out the fast degrees of freedom of the gas particles by computing conditional averages, knowing the macroscopic variables: position z and velocity V of each piston, and temperatures T of the gases. In order to simplify the notation, let us denote by $\langle \cdot \rangle$ this average, meaning the conditional average $\langle \cdot | z, V, T \rangle$. We will compute the average change of a generic observable X in a small time interval Δt, due to the collisions between the particles of the gas and the pistons.

Let us assume that in the stationary state the particles of the gas, in each compartment, have uniform space distribution ρ and a Maxwell–Boltzmann distribution $\phi(v)$ at temperature T:

$$\phi(v) = \sqrt{\frac{m}{2\pi k_B T}} e^{-\frac{m}{2k_B T} v^2}, \qquad \rho(x) = \frac{1}{\Delta z}, \tag{12}$$

where Δz is the length of the box containing the gas. We can obtain the rate of the collisions by considering the following equivalent problem: piston at rest and a particle, which moves with the relative velocity $v - V$. The point particles which collide against the piston in x in the time interval dt are:

$$N\rho(x)(v - V)\Theta(v - V)dt, \qquad N\rho(x)(V - v)\Theta(V - v)dt, \tag{13}$$

respectively, for particles on the left and on the right with respect to the piston. The Heaviside function Θ is necessary to have a collision.

Let us now consider a generic observable X_j, depending in general on the velocities of the gas particles and of the pistons. We want to write down a Langevin equation:

$$\frac{dX_j}{dt} = D_j(\mathbf{X}) + \text{noise}, \tag{14}$$

where both the drift term $D_j(\mathbf{X})$ ($\mathbf{X}$ being the vector of all relevant macroscopic variables in the system) and the noise term are due to collisions with the particles of the gas. We have:

$$D_j(\mathbf{X}) = \lim_{\Delta t \to 0} \frac{\langle \Delta X_j | \mathbf{X} \rangle}{\Delta t} = \lim_{\Delta t \to 0} \frac{1}{\Delta t} \left[\langle \Delta X_j \rangle_{coll}^L + \langle \Delta X_j \rangle_{coll}^R \right], \tag{15}$$

where

$$\langle \Delta X_j \rangle_{coll}^R = N \int_V^\infty dv \int_{z_L - (v-V)\Delta t}^{z_L} dx \, X_j \, \rho(x)\phi(v)\Theta(v - V) = \frac{N}{\Delta z_L} \int_V^\infty dv \, X_j \phi(v)(v - V)\Delta t, \tag{16}$$

$$\langle \Delta X_j \rangle_{coll}^L = N \int_{-\infty}^V dv \int_{z_R}^{z_R - (v-V)\Delta t} dx \, X_j \, \rho(x)\phi(v)\Theta(V - v) = \frac{N}{\Delta z_R} \int_{-\infty}^V dv \, X_j \phi(v)(V - v)\Delta t. \tag{17}$$

$\langle \Delta X_j \rangle_{coll}^R$ is the variation of the observable due to the collisions with the particles that have the pistons on the left, while $\langle \Delta X_j \rangle_{coll}^L$ denotes the variation due to the collisions with the particles which have the piston on the right.

In the stationary state, the macroscopic variables T, V and z do not depend on time. It means that the time derivative of the conditional average of a generic observable X is zero, namely:

$$\lim_{\Delta t \to 0} \frac{1}{\Delta t} \left[\langle \Delta X_j \rangle_{coll}^L + \langle \Delta X_j \rangle_{coll}^R \right] = 0. \tag{18}$$

By using a perturbation development in the small parameter $\epsilon = m/M$, it is possible to derive the relations between the average positions and the temperatures of the pistons and the temperatures of the gas. In order to obtain this result, we need to average on the piston velocity, which is a stochastic variable. Of course, in the steady state, the odd momenta have to be zero:

$$\left\langle V^{2\alpha+1} \right\rangle = 0, \qquad \alpha = 0, 1, 2, \dots \tag{19}$$

The idea is to compute the average force produced by the particles of the gas, which collide against the piston, by computing the average momentum exchanged in the collisions.

The evolution of the velocity of the piston is governed by the following stochastic differential equation:

$$M \frac{d}{dt} V = F_{coll}(\Delta z_L, \Delta z_R, T_L, T_R, V) + K\eta, \tag{20}$$

where η is zero-average white noise, usually K is a constant, and F_{coll} is the average force which acts on the piston. This force is due to the collisions with the gas on the left and on the right, and depends on the average size of the box on the right and on the left of the piston, Δz_L and Δz_R, respectively, and on the temperatures of the gas on the left and on the right, T_L and T_R. Now let us set $\Delta X = V' - V$ in Equation (15). As detailed in the Appendix A, computing the left and right contributions to F_{coll} and expanding in the small mass ratio $\epsilon = \sqrt{m/M}$, we obtain at the lowest orders:

$$O(1): \quad Nk_B \left(\frac{T_L}{\Delta z_L} - \frac{T_R}{\Delta z_R} \right), \tag{21}$$

$$O(\epsilon): \quad -2N\sqrt{2k_B} \sqrt{\frac{M}{\pi}} \left[\frac{T_L^{1/2}}{\Delta z_L} + \frac{T_R^{1/2}}{\Delta z_R} \right] V. \tag{22}$$

In the steady state, when the time derivative vanishes, we have from the order $O(1)$ the following relation between the temperatures of the gas and the average length of the boxes:

$$\Delta z_R = \Delta z_L \frac{T_R}{T_L}. \tag{23}$$

This relation is nothing but the one that can be derived from thermodynamics. From Equation (22), we obtain the obvious result $\langle V \rangle = 0$.

Analogous computations, reported in the Appendix A, can be carried out for the variance of the pistons and the gas temperatures. In particular, for the variable $\Delta X = M(V'^2 - V^2)$, at lowest orders in ϵ, we obtain the relations:

$$O(1): \quad N2k_B \left(\frac{T_L}{\Delta z_L} - \frac{T_R}{\Delta z_R} \right), \tag{24}$$

$$O(\epsilon): \quad \frac{N}{\sqrt{\pi}} \frac{4\sqrt{2}}{M^{1/2}} \left[\left(\frac{(k_B T_L)^{3/2}}{\Delta z_L} + \frac{(k_B T)^{3/2}}{\Delta z_R} \right) - MV^2 \left(\frac{(k_B T_R)^{1/2}}{\Delta z_R} + \frac{(k_B T_L)^{1/2}}{\Delta z_L} \right) \right]. \tag{25}$$

In the steady state, by using the thermodynamic relations, the order $O(1)$ Equation (24) is identically zero. After integrating over all values of the piston velocity and by requiring that the order $O(\epsilon)$ vanishes, we obtain a relation between the temperature of the piston $T^{(p)}$ and those of the gases on the left and on the right:

$$T^{(p)} = M \left\langle V^2 \right\rangle = (T_R T_L)^{1/2}. \tag{26}$$

Finally, a condition on the temperature of the gas as a function of the velocity variance of the pistons on its left and on its right can be obtained by considering the variable $\Delta X = m(v'^2 - v^2)/N$ in Equation (15). This gives (see Appendix A) the relations:

$$O(1): \quad \frac{2k_B T}{\Delta z} \left(V_L - V_R \right), \tag{27}$$

$$O(\epsilon): \quad \frac{\sqrt{2k_B}}{\Delta z} \frac{4}{\sqrt{\pi}} \frac{T^{1/2}}{M^{1/2}} \left[M V_R^2 + M V_L^2 - 2T \right]. \tag{28}$$

By integrating over the velocity of the piston, because $\langle V \rangle$ is zero, the order $O(1)$ is identically zero. By requiring that, in the stationary state, the order $O(\epsilon)$ vanishes, we eventually obtain:

$$0 = \frac{\sqrt{2k_B}}{\Delta z} \frac{4}{\sqrt{\pi}} \frac{T^{1/2}}{M^{1/2}} \left[M \left\langle V_R^2 \right\rangle_0 + M \left\langle V_L^2 \right\rangle_0 - 2T \right] \quad \Longrightarrow \quad T = \frac{M}{2} \left(\left\langle V_R^2 \right\rangle + \left\langle V_L^2 \right\rangle \right). \tag{29}$$

As expected from thermodynamics, one can easily show that the gases in contact with the thermostats reach the temperatures of the thermal baths (see Appendix A for a detailed explanation).

5. Numerical Simulations

In order to check the range of validity of the above approach, we have performed extensive molecular dynamics simulations (using an event driven algorithm) of the system, varying the relevant parameters of the model and comparing the analytical prediction of Section 3 with the actual numerical results. The system is initialized with the pistons placed at equidistant positions, with zero velocity, while the gas particles are randomly distributed in each box, with velocities drawn from a Gaussian distribution at temperature intermediate with respect to those of the thermostats. We have, however, checked that the stationary state reached by the system is independent of the initial conditions. All data presented in the following are measured in the steady state, after the transient relaxation dynamics from the initial state.

In Figure 2, we show the piston temperatures $T_i^{(p)}$, $i = 1, 2, 3$, for a 3-piston system, as measured in numerical simulations (symbols) for a certain choice of the parameters, and compare them with the theoretical predictions of Equation (26) (lines). The approximation is good for the intermediate piston, while it is not very accurate for the side pistons, and is the worst in the case of a large gradient $T_B/T_A \gg 1$. Simulations performed for systems with more pistons give similar results.

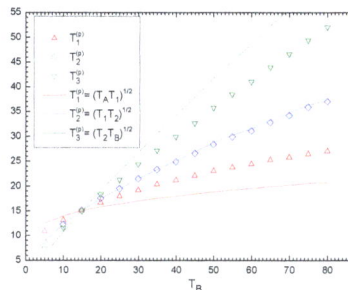

Figure 2. Piston temperatures for a 3-piston system, with parameters $T_A = 15$, $N = M = 100$, $m = 1$, $L = 1$.

In Figure 3, we consider a system with a large number of pistons ($n = 22$), where the temperature profile shows a linear behavior, in agreement with Fourier's law. We report two sets of data, differing in the value of the parameter $R = mN/M$. Notice that the linear behavior extends for a wider range of z in the case $R = 10$. Lines represent linear fits of the data. Let us note that, for large n and $R \gg 1$, $T^{(p)}$ vs. z is in good agreement with a linear behavior in the whole space interval (even close to the thermal baths).

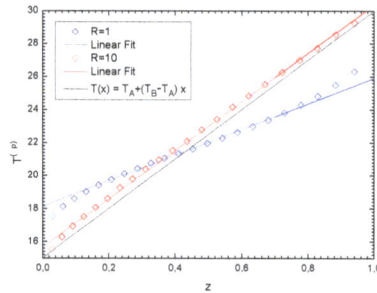

Figure 3. Temperature profiles for a 22-piston system with $R = 1$ and $R = 10$. Other parameters are $T_A = 15, T_B = 30, M = 50, m = 1, L = 1$.

The derivation of the Langevin equation (see Appendix A) is mainly based on the assumptions:

- $m/M \ll 1$;
- a local thermodynamic equilibrium, i.e. in each compartment between the piston $i - 1$ and the piston i, the spatial distribution of the particles is uniform and the velocity probability distribution is a Gaussian function, whose variance is given by the gas temperature;
- the collisions particles/pistons are independent.

The first assumption is easily checked, while the other ones are more difficult. We can expect that a necessary condition for their validity is that N must be large. We have measured the velocity distribution of the gas particles in the numerical simulations, finding that the above hypothesis is verified. Moreover, we expect that the number of recollisions decreases with the ratio m/M, and we have numerically checked that their contribution is negligible (the fraction of recollisions on the total number of collisions is about 0.1%). Figure 4 shows the probability distribution functions of particles colliding from left, $\phi_L(v)$, and from right, $\phi_R(v)$, with a moving wall: the agreement with a Gaussian assumption is rather evident.

In addition, the numerical computations show that the left- and right-moving particles in the same compartment have the following probability distributions:

$$\phi_-(v) = \sqrt{\frac{2m}{\pi k_B T_-}}\, \Theta(v) e^{-\frac{mv^2}{2k_B T_-}}, \tag{30}$$

$$\phi_+(v) = \sqrt{\frac{2m}{\pi k_B T_+}}\, \Theta(-v) e^{-\frac{mv^2}{2k_B T_+}}, \tag{31}$$

where T_- (T_+) denotes the temperature of the particles incident on the piston i ($i - 1$) from the left (right), with $T_- \approx T_+$, resulting in a small but finite heat flow proportional to $(T_- - T_+)$, in agreement with [15]. Let us note that the above probability distributions are different from the distributions $\phi_L(v)$ and $\phi_R(v)$, which are computed for particles actually colliding with the piston; this is the origin of the presence of the factor v appearing in the expressions in the caption of Figure 4.

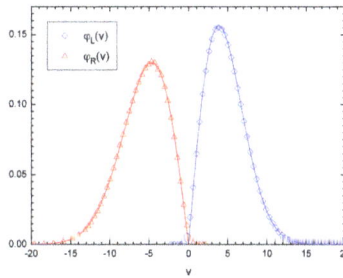

Figure 4. Probability distribution function of the velocities $\phi_L(v)$ ($\phi_R(v)$) of the particles colliding with the first piston from the left (right), for a system with $n = 2$. The curves, red and blue, are respectively: $\phi_L = \frac{m}{k_B T_L} e^{-v^2/k_B T_L} v\,\theta(-v)$ and $\phi_R = \frac{m}{k_B T_R} e^{-v^2/k_B T_R} v\,\theta(v)$, where $T_L = 15.0$ and $T_R = 21.6$ are the temperatures of the gas on the left and on the right with respect to the first piston. Other parameters are: $T_A = 15$, $T_B = 30$, $N = 100$, $M = 100$, $m = 1$, $n = 2$, $L = 1$.

The distribution of v in the compartment between pistons $i - 1$ and i, has the form:

$$\phi(v) \propto A\Theta(v)e^{-\frac{mv^2}{2k_B T_-}} + B\Theta(-v)e^{-\frac{mv^2}{2k_B T_-}},$$

where A and B are constants. However, since

$$T_- = T_+ + O(\epsilon^2),$$

we have a Gaussian distribution for the particle gas in the compartment:

$$\phi_i(v) \propto e^{-\frac{mv^2}{2k_B T_i}} + O(\epsilon^2), \qquad T_i = T_+ + O(\epsilon^2). \tag{32}$$

Let us note that all our results are based on an expansion in powers of ϵ, neglecting $O(\epsilon^2)$. In other words, because the heat rate is very small, this heat flux perturbs the Gaussian form of the probability distribution ϕ in a negligible way.

In order to better understand the dependence of our theoretical results on the parameter R, we have performed numerical simulations of the system for different values of R. From the previous remark, we expect to have an improvement of the agreement between the numerical results and the analytical predictions by increasing R. In Figure 5, we compare the piston temperature in a 4-piston system with theoretical predictions from Equation (26), finding a very good agreement for large values of R. This shows that the total mass of the gas contained in a box has to be larger than the mass of the piston, in order for the kinetic theory presented in the previous section to be accurate.

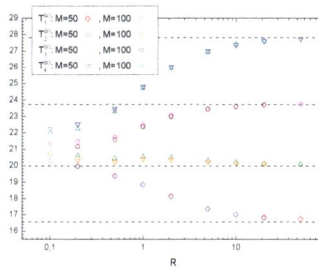

Figure 5. Piston temperatures for a 4-piston system as a function of R. Dashed lines are the theoretical predictions. Other parameters are $T_A = 15$, $T_B = 30$, $m = 1$, $L = 1$.

6. Conclusions

We have studied a generalized piston in contact with two thermal baths at different temperatures. This system represents a simple but interesting case where the emergence of Fourier's law from a microscopic mechanical model can be studied. We have presented a kinetic theory treatment inspired by an approach à la Smoluchowski, and we have investigated the range of validity of these results with molecular dynamics numerical simulations. We have found that, in order for the theory to be accurate, the ratio $R = mN/M$ should be large enough, namely the total mass of the gas in each compartment should be greater than that of the single piston.

We have considered ideal gas in our model, but we do not expect that the introduction of short-range interactions among gas particles, at least in not too dense cases, leads to significant differences in the behavior of the system. An interesting, non-trivial, future line of research in this model is the study of the relaxation to the steady state and the dynamical properties of the system, focusing on correlation and response functions.

Acknowledgments: We thank A. Puglisi for useful discussions.

Author Contributions: All authors contributed equally to this work. All authors have read and approved the final manuscript.

Conflicts of Interest: The authors declare no conflict of interest.

Appendix A

In this appendix, we will derive the relations that determine the stationary state of the multi-component piston model described in the main text.

Appendix A.1. Piston Position

The effective force acting on the piston due to the collisions with gas particles, appearing in Equation (20), is given by two contributions, $F_{coll} = F^L_{coll} + F^R_{coll}$. By taking into account the elastic collisions rule, Equation (1), these terms can be computed as follows:

$$
\begin{aligned}
F^L_{coll} &= \frac{N}{\Delta z_L} \int_{-\infty}^{\infty} dv M(v-V)\Theta(v-V)\phi(v)\,(V'-V) = \frac{N}{\Delta z_L}\frac{2mM}{m+M}\int_V^{\infty} dv\,(v-V)^2 \sqrt{\frac{m}{2\pi k_B T_L}}e^{-\frac{m}{2k_B T_L}v^2} \\
&= \frac{N}{\Delta z_L}\frac{M}{m+M}\left[\mathrm{erfc}\left(\sqrt{\frac{m}{2k_B T_L}}V\right)(mV^2+k_B T_L) - \sqrt{\frac{m}{\pi}}\sqrt{2k_B T_L}Ve^{-\frac{m}{2k_B T_L}V^2}\right],
\end{aligned}
\tag{A1}
$$

where in the first line we have used Equation (12). In the same way, by putting the observable $\Delta X = M(V'-V)$ in the Equation (17), we have:

$$
\begin{aligned}
F^R_{coll} &= \frac{N}{\Delta z_R} \int_{-\infty}^{\infty} dv M(V-v)\Theta(V-v)\phi(v)\,(V'-V) = -\frac{N}{\Delta z_R}\frac{2mM}{m+M}\int_{-\infty}^V dv\,(v-V)^2 \sqrt{\frac{m}{2\pi k_B T_R}}e^{-\frac{m}{2k_B T_R}v^2} \\
&= -\frac{N}{\Delta z_R}\frac{M}{m+M}\left[\left(1+\mathrm{erf}\left(\sqrt{\frac{m}{2k_B T_R}}V\right)\right)(mV^2+k_B T_R) + \sqrt{\frac{m}{\pi}}\sqrt{2k_B T_R}Ve^{-\frac{m}{2k_B T_R}V^2}\right].
\end{aligned}
\tag{A2}
$$

By putting together the previous relations, we obtain:

$$
\begin{aligned}
M\frac{d}{dt}\langle V\rangle_{coll} = F^L_{coll} + F^R_{coll} &= \frac{NM}{m+M}\left[\frac{k_B T_L}{\Delta z_L}\mathrm{erfc}\left(\sqrt{\frac{m}{2k_B T_L}}V\right) - \frac{k_B T_R}{\Delta z_R}\left(\mathrm{erf}\left(\sqrt{\frac{m}{2k_B T_R}}V\right)+1\right)\right] - \\
&\quad -\frac{NM}{m+M}\frac{\sqrt{2m}}{\pi}V\left[\frac{\sqrt{k_B T_L}}{\Delta z_L}e^{-\frac{m}{2k_B T_L}V^2}+\frac{\sqrt{k_B T_R}}{\Delta z_R}e^{-\frac{m}{2k_B T_R}V^2}\right] + \\
&\quad +\frac{NM}{m+M}mV^2\left[\frac{1}{\Delta z_L}\mathrm{erfc}\left(\sqrt{\frac{m}{2k_B T_L}}V\right) - \frac{1}{\Delta z_R}\left(1+\mathrm{erf}\left(\sqrt{\frac{m}{2k_B T_R}}V\right)\right)\right].
\end{aligned}
\tag{A3}
$$

By expanding in the small ratio $\epsilon = \sqrt{m/M} \ll 1$, and assuming that $V/v \sim \epsilon$, we have:

$$O(1): \quad Nk_B \left(\frac{T_L}{\Delta z_L} - \frac{T_R}{\Delta z_R} \right), \tag{A4}$$

$$O(\epsilon): \quad -2N\sqrt{2k_B}\sqrt{\frac{M}{\pi}} \left[\frac{T_L^{1/2}}{\Delta z_L} + \frac{T_R^{1/2}}{\Delta z_R} \right] V. \tag{A5}$$

In the steady state, when the time derivative vanishes, we have from the order $O(1)$ the following relation between the temperatures of the gas and the average lengths of the boxes:

$$\Delta z_R = \Delta z_L \frac{T_R}{T_L}. \tag{A6}$$

Appendix A.2. Piston Fluctuations

Let us now consider the observable $\Delta X = M(V'^2 - V^2)$. From Equation (16), using Equation (1) for the elastic collisions, we obtain:

$$
\begin{aligned}
M\frac{d}{dt}\left\langle V^2 \right\rangle^L_{coll} &= \frac{N}{\Delta z_L} \int_{-\infty}^{\infty} dv\,(v - V)\Theta(v - V)M(V'^2 - V^2)\phi(v) \\
&= \frac{NM}{\Delta z_L} \int_V^{\infty} dv\,(v - V)\left[\frac{(2m)^2}{(m + M)^2}(V - v)^2 + \frac{4m}{m + M}V(V - v) \right]\phi(v).
\end{aligned}
\tag{A7}
$$

Using Equation (12), we have:

$$
\begin{aligned}
M\frac{d}{dt}\left\langle V^2 \right\rangle^L_{coll} &= \frac{MN}{\Delta z_L \sqrt{\pi}} \frac{4m^2}{(m + M)^2}\left(\frac{2k_B T_L}{m} \right)^{3/2} - \\
&\quad - \frac{MN}{\Delta z_L \sqrt{\pi}} \frac{4m}{m + M} \frac{2k_B T_L}{m} \sqrt{\pi}\,\mathrm{erfc}\left(\sqrt{\frac{m}{2k_B T_L}} V \right)\left[\frac{m}{m + M}\frac{3}{4} - \frac{1}{4} \right] V + \\
&\quad + \frac{MN}{\Delta z_L \sqrt{\pi}} \frac{4m}{m + M} \sqrt{\frac{2k_B T_L}{m}}\,e^{-\frac{m}{2k_B T_L}V^2}\left[\frac{m}{m + M}\frac{1}{2} - \frac{1}{2} \right] V^2 - \\
&\quad - \frac{MN}{\Delta z_L \sqrt{\pi}} \frac{4m}{m + M} \frac{\sqrt{\pi}}{2}\,\mathrm{erfc}\left(\sqrt{\frac{m}{2k_B T_L}} V \right)\left[\frac{m}{m + M} - 1 \right] V^3.
\end{aligned}
\tag{A8}
$$

In the same way, from Equation (17), we obtain:

$$
\begin{aligned}
M\frac{d}{dt}\left\langle V^2 \right\rangle^R_{coll} &= \frac{N}{\Delta z_R} \int_{-\infty}^{\infty} dv\,(V - v)\Theta(V - v)M(V'^2 - V^2)\phi(v) \\
&= \frac{NM}{\Delta z_R} \int_{-\infty}^{V} dv\,(V - v)\left[\frac{(2m)^2}{(m + M)^2}(V - v)^2 + \frac{4m}{m + M}V(V - v) \right]\phi(v) \\
&= \frac{MN}{\Delta z_R \sqrt{\pi}} \frac{4m^2}{(m + M)^2}\left(\frac{2k_B T_R}{m} \right)^{3/2} + \\
&\quad + \frac{MN}{\Delta z_L \sqrt{\pi}} \frac{4m}{m + M} \frac{2k_B T_R}{m} \sqrt{\pi}\left(1 + \mathrm{erf}\left(\sqrt{\frac{m}{2k_B T_R}} V \right) \right)\left[\frac{m}{m + M}\frac{3}{4} - \frac{1}{4} \right] V + \\
&\quad + \frac{MN}{\Delta z_R \sqrt{\pi}} \frac{4m}{m + M} \sqrt{\frac{2k_B T_R}{m}}\,e^{-\frac{m}{2k_B T_R}V^2}\left[\frac{m}{m + M}\frac{1}{2} - \frac{1}{2} \right] V^2 - \\
&\quad + \frac{MN}{\Delta z_R \sqrt{\pi}} \frac{4m}{m + M} \frac{\sqrt{\pi}}{2}\left(1 + \mathrm{erfc}\left(\sqrt{\frac{m}{2k_B T_R}} V \right) \right)\left[\frac{m}{m + M} - 1 \right] V^3.
\end{aligned}
\tag{A9}
$$

Putting together the contributions from the relations (A8) and (A9), we have

$$
M\frac{d}{dt}\left\langle V^2\right\rangle_{coll} = \frac{MN}{\sqrt{\pi}}\frac{4m^2}{(m+M)^2}\left(\frac{2k_B}{m}\right)^{3/2}\frac{1}{2}\left[\frac{T_L^{3/2}}{\Delta z_L}e^{-\frac{m}{2k_BT_L}V^2} + \frac{T_R^{3/2}}{\Delta z_R}e^{-\frac{m}{2k_BT_R}V^2}\right] +
$$

$$
+ MN\frac{4m}{m+M}\frac{2k_B}{m}\left[\frac{3}{4}\frac{m}{m+M}-\frac{1}{4}\right]\left\{\frac{T_R}{\Delta z_R}\left(\mathrm{erf}\left(\sqrt{\frac{m}{2k_BT_R}}V\right)+1\right) - \right.
$$

$$
\left. -\frac{T_L}{\Delta z_L}\mathrm{erfc}\left(\sqrt{\frac{m}{2k_BT_L}}V\right)\right\}V + MN\frac{4m}{m+M}\frac{1}{2\sqrt{\pi}}\sqrt{\frac{2k_B}{m}}\left[\frac{m}{m+M}-1\right]
$$

$$
\times\left\{\frac{\sqrt{T_L}}{\Delta z_L}e^{-\frac{m}{2k_BT_L}V^2} + \frac{\sqrt{T_R}}{\Delta z_R}e^{-\frac{m}{2k_BT_R}V^2}\right\}V^2 + MN\frac{4m}{m+M}\frac{1}{2}\left[\frac{m}{m+M}-1\right]
$$

$$
\times\left\{\frac{1}{\Delta z_R}\left(\mathrm{erf}\left(\sqrt{\frac{m}{2k_BT_R}}V\right)+1\right)-\frac{1}{\Delta z_L}\mathrm{erfc}\left(\sqrt{\frac{m}{2k_BT_L}}V\right)\right\}V^3.
$$

(A10)

By using a Taylor expansion in the small parameter $\epsilon = \sqrt{\frac{m}{M}} \ll 1$, we obtain:

$$
O(1): \quad 2N\left(\frac{k_BT_L}{\Delta z_L}-\frac{k_BT_R}{\Delta z_R}\right)V, \tag{A11}
$$

$$
O(\epsilon): \quad \frac{N}{M^{1/2}}\frac{4\sqrt{2}}{\sqrt{\pi}}\left[\left(\frac{(k_BT_L)^{3/2}}{\Delta z_L}+\frac{(k_BT_R)^{3/2}}{\Delta z_R}\right)-MV^2\left(\frac{(k_BT_L)^{1/2}}{\Delta z_L}+\frac{(k_BT_R)^{1/2}}{\Delta z_R}\right)\right]. \tag{A12}
$$

In the steady state, by using the thermodynamic relations, the order $O(1)$ Equation (A11) is identically zero. After integrating over all values of the piston velocity and by requiring that the order $O(\epsilon)$ vanishes, we obtain a relation between the temperature of the right piston and that of the left one:

$$
T^{(p)} \equiv M\left\langle V^2\right\rangle = k_B\left(T_RT_L\right)^{1/2}, \tag{A13}
$$

where we have used also the thermodynamic relation Equation (A6).

Appendix A.3. Temperature of the Gas

Let us now compute the average temperature of the gas. We have to distinguish between two different cases:

- Gas between two moving walls
- Gas between a moving wall and a thermostat

Appendix A.3.1. Gas between Two Moving Walls

Let us consider the observable ΔX equal to the difference of the gas energy before and after the collision: $\Delta X = m(v'^2-v^2)/N$. By putting the observable into the relation (16) and by taking into account the Equation (1), we obtain:

$$
\frac{m}{N}\frac{d}{dt}\left\langle v^2\right\rangle_{coll}^L = \frac{m}{\Delta z}\int_{-\infty}^{\infty}dv\,(v-V_R)\Theta(v-V_R)[v'^2-v^2]\phi(v)
$$

$$
= \frac{m}{\Delta z}\int_{V_R}^{\infty}dv\,(v-V_R)\left[\frac{4M^2}{(m+M)^2}(V_R-v)^2+\frac{4M}{m+M}v(V_R-v)\right]\phi(v).
$$

(A14)

By integrating, we have:

$$
\frac{m}{N}\frac{d}{dt}\left\langle v^2 \right\rangle^L_{coll} = \frac{m}{\Delta z}\frac{1}{\sqrt{\pi}}\frac{4M}{m+M}\left(\frac{2k_BT}{m}\right)^{3/2}e^{-\frac{m}{2k_BT}V_R^2}\left[-\frac{m}{m+M}\right]+
$$
$$
+\frac{m}{\Delta z}\frac{4M}{m+M}\frac{2k_BT}{m}\operatorname{erfc}\left(\sqrt{\frac{m}{2k_BT}}V_R\right)\left[-\frac{M}{m+M}+\frac{2m}{m+M}\right]\frac{V_R}{4}+
$$
$$
+\frac{m}{\Delta z}\frac{1}{\sqrt{\pi}}\frac{4M}{m+M}\sqrt{\frac{2k_BT}{m}}\frac{1}{2}\frac{M}{m+M}e^{-\frac{m}{2k_BT}V_R^2}V_R^2-\frac{m}{\Delta z}\frac{2M^2}{(M+m)^2}\operatorname{erfc}\left(\sqrt{\frac{m}{2k_BT}}V_R\right)V_R^3.
$$

(A15)

In the same way, we obtain:

$$
\frac{m}{N}\frac{d}{dt}\left\langle v^2 \right\rangle^R_{coll} = \frac{m}{\Delta z}\int_{-\infty}^{\infty}(V_L-v)\Theta(V_L-V)[v'^2-v^2]\phi(v)
$$
$$
= \frac{m}{\Delta z}\int_{-\infty}^{V_L}dv\,(V_L-v)\left[\frac{4M^2}{(m+M)^2}(V_L-v)^2+\frac{4M}{m+M}v(V_L-v)\right]\phi(v)
$$
$$
= \frac{m}{\Delta z}\frac{1}{\sqrt{\pi}}\frac{4M}{m+M}\left(\frac{2k_BT}{m}\right)^{3/2}e^{-\frac{m}{2k_BT}V_R^2}\left[-\frac{m}{m+M}\right]-
$$
$$
-\frac{m}{\Delta z}\frac{4M}{m+M}\frac{2k_BT}{m}\left(1+\operatorname{erf}\left(\sqrt{\frac{m}{2k_BT}}V_L\right)\right)\left[-\frac{M}{m+M}+\frac{2m}{m+M}\right]\frac{V_L}{4}+
$$
$$
+\frac{m}{\Delta z}\frac{1}{\sqrt{\pi}}\frac{4M}{m+M}\sqrt{\frac{2k_BT}{m}}\frac{1}{2}\frac{M}{m+M}e^{-\frac{m}{2k_BT}V_L^2}V_L^2+
$$
$$
+2\frac{m}{\Delta z}\frac{M^2}{(M+m)^2}\left(1+\operatorname{erf}\left(\sqrt{\frac{m}{2k_BT}}V_L\right)\right)V_L^3.
$$

(A16)

Putting together Equations (A15) and (A16), one has:

$$
\frac{m}{N}\frac{d}{dt}\left\langle v^2 \right\rangle_{coll} = -\frac{m}{\Delta z}\frac{1}{\sqrt{\pi}}\frac{4M}{m+M}\frac{m}{m+M}\left(\frac{2k_BT}{m}\right)^{3/2}\left[\frac{e^{-\frac{m}{2k_BT}V_L^2}}{2}+\frac{e^{-\frac{m}{2k_BT}V_R^2}}{2}\right]+
$$
$$
+\frac{m}{\Delta z}\frac{4M}{m+M}\frac{2k_BT}{m}\frac{1}{4}\left[\frac{2m}{m+M}-\frac{M}{m+M}\right]\left\{V_R\operatorname{erfc}\left(\sqrt{\frac{m}{2k_BT}}V_R\right)-\right.
$$
$$
\left.-V_L-V_L\operatorname{erf}\left(\sqrt{\frac{m}{2k_BT}}V_L\right)\right\}+\frac{m}{\Delta z}\frac{1}{2}\frac{4M^2}{(m+M)^2}\sqrt{\frac{2k_BT}{m}}\frac{1}{\sqrt{\pi}}\times
$$
$$
\times\left[V_R^2e^{-\frac{m}{2k_BT}V_R^2}+V_L^2e^{-\frac{m}{2k_BT}V_L^2}\right]-\frac{m}{\Delta z}\frac{4M^2}{(m+M)^2}\frac{1}{2}\times
$$
$$
\times\left[V_R^3\operatorname{erfc}\left(\sqrt{\frac{m}{2k_BT}}V_R\right)-V_L^3\left(1+\operatorname{erf}\left(\sqrt{\frac{m}{2k_BT}}V_L\right)\right)\right],
$$

(A17)

and using a Taylor expansion around the small parameter $\epsilon=\sqrt{\frac{m}{M}}\ll 1$ yields:

$$
O(1):\quad \frac{2k_BT}{\Delta z}(V_L-V_R),
$$

(A18)

$$
O(\epsilon):\quad \frac{\sqrt{2k_B}}{\Delta z}\frac{4}{\sqrt{\pi}}\frac{T^{1/2}}{M^{1/2}}\left[MV_R^2+MV_L^2-2T\right].
$$

(A19)

By integrating over the velocity of the piston, because $\langle V \rangle$ is zero, the order $O(1)$ is identically zero. By requiring that in the stationary state the order $O(\epsilon)$ vanishes, we obtain a relation between the temperature of the gas and those of the near pistons:

$$
0=\frac{\sqrt{2k_B}}{\Delta z}\frac{4}{\sqrt{\pi}}\frac{T^{1/2}}{M^{1/2}}\left[M\left\langle V_R^2\right\rangle_0+M\left\langle V_L^2\right\rangle_0-2T\right]\quad\Longrightarrow\quad T=\frac{M}{2}\left(\left\langle V_R^2\right\rangle+\left\langle V_L^2\right\rangle\right).
$$

(A20)

Appendix A.3.2. Gas between a Piston and a Thermostat

Consider the gas that is near a thermostat and piston. The variation of the temperature due to the piston is the same as before and is given by the Equations (A15) or (A16), respectively, for a piston on the right or on left, with respect to the considered gas. On the side of the thermostat, after the collision, the particle takes a velocity according to the distribution given by the Equation (2). For instance, if the thermostat is that one on the left at temperature T_0, we have:

$$
\begin{aligned}
\langle T \rangle_{ther}^L &= \frac{m}{\Delta z} \int_{-\infty}^{\infty} dv \int_{-\infty}^{\infty} dv' \, (v'^2 - v^2) \phi(v) \, |v| \, \Theta(-v) \Phi_{T_0}(v') \\
&\frac{m}{\Delta z} \int_{-\infty}^{\infty} dv \int_{-\infty}^{\infty} dv' \, (v'^2 - v^2) \sqrt{\frac{m}{2\pi k_B T}} e^{-\frac{m}{2k_B T} v^2} \theta(-v) \, |v| \, v' \, \Theta(v') \frac{m}{k_B T_0} e^{-\frac{m}{2k_B T_0} v'^2} \\
&\frac{m}{\Delta z} \frac{m}{k_B T_0} \sqrt{\frac{m}{2\pi k_B T}} \int_{-\infty}^{0} dv \int_{0}^{\infty} dv' v \, v' (v'^2 - v^2) e^{-\frac{m}{2k_B T} v^2} e^{-\frac{m}{2k_B T_0} v'^2}.
\end{aligned}
\tag{A21}
$$

By integrating, we obtain:

$$
\langle T \rangle_{ther}^L = \sqrt{\frac{2}{\pi m}} \frac{k_B^{3/2}}{\Delta z} \sqrt{T} [T_0 - T].
\tag{A22}
$$

In order to compute the variation of the temperature, we have to put together Equations (A22) and (A15):

$$
\begin{aligned}
\frac{d}{dt} \langle T \rangle &= \langle T \rangle_{ther}^L + \frac{m}{N} \frac{d}{dt} \langle v^2 \rangle_{coll}^R = \sqrt{\frac{2}{\pi m}} \frac{k_B^{3/2}}{\Delta z} \sqrt{T} [T_0 - T] + \\
&+ \frac{m}{\Delta z} \frac{1}{\sqrt{\pi}} \frac{4M}{m+M} \left(\frac{m}{2 k_B T} \right)^{3/2} \frac{e^{-\frac{m}{2k_B T} V_R^2}}{2} \left[-\frac{m}{m+M} \right] + \\
&+ \frac{m}{\Delta z} \frac{4M}{m+M} \frac{2 k_B T}{m} \operatorname{erfc}\left(\sqrt{\frac{m}{2 k_B T}} V_R \right) \left[-\frac{M}{m+M} + \frac{2m}{m+M} \right] \frac{V_R}{4} + \\
&+ \frac{m}{\Delta z} \frac{1}{\sqrt{\pi}} \frac{4M}{m+M} \sqrt{\frac{2 k_B T}{m}} \frac{1}{2} \frac{M}{m+M} e^{-\frac{m}{2k_B T} V_R^2} V_R^2 - \\
&- \frac{m}{\Delta z} 2 \frac{M^2}{(M+m)^2} \operatorname{erfc}\left(\sqrt{\frac{m}{2 k_B T}} V_R \right) V_R^3.
\end{aligned}
\tag{A23}
$$

By solving perturbatively around the small parameter $\epsilon = \sqrt{\frac{m}{M}} \ll 1$, we obtain:

$$
O\left(\frac{1}{\epsilon} \right): \quad \sqrt{\frac{2}{\pi}} \frac{k_B^{3/2}}{M^{1/2}} \frac{\sqrt{T}}{\Delta z} (T_0 - T),
\tag{A24}
$$

$$
O(1): \quad -2 \frac{k_B T}{\Delta z} V_R.
\tag{A25}
$$

In the steady state, from relation (A24) at order $O(1/\epsilon)$, we have $T = T_0$. This means that a gas near a thermostat reaches the temperature of the thermal bath: this is the result one can obtain from thermodynamics. We can obtain exactly the same result for a thermostat on the right with respect to the considered gas.

References

1. Lepri, S.; Livi, R.; Politi, A. Thermal conduction in classical low-dimensional lattices. *Phys. Rep.* **2003**, *377*, 1–80.
2. Garrido, P.L.; Hurtado, P.I.; Nadrowski, B. Simple one-dimensional model of heat conduction which obeys Fourier's law. *Phys. Rev. Lett.* **2001**, *86*, 5486.
3. Dhar, A. Heat conduction in the disordered harmonic chain revisited. *Phys. Rev. Lett.* **2001**, *86*, 3554.

4. Lepri, S.; Livi, R.; Politi, A. On the anomalous thermal conductivity of one-dimensional lattices. *EPL* **1998**, *43*, 271–276.

5. Giardinà, C.; Livi, R.; Politi, A.; Vassalli, M. Finite thermal conductivity in 1D lattices. *Phys. Rev. Lett.* **2000**, *84*, 2144.

6. Eckmann, J.P.; Young, L.-S. Temperature profiles in Hamiltonian heat conduction. *EPL* **2004**, *68*, 790–796.

7. Eckmann, J.P.; Young, L.-S. Nonequilibrium energy profiles for a class of 1-D models. *Commun. Math. Phys.* **2005**, *262*, 237–267.

8. Salazar, A.; Larralde, H.; Leyvraz, F. Temperature gradients in equilibrium: Small microcanonical systems in an external field. *Phys. Rev. E* **2014**, *90*, 052127.

9. Gaspard, P.; Gilbert, T. Heat Conduction and Fourier's Law by Consecutive Local Mixing and Thermalization. *Phys. Rev. Lett.* **2008**, *101*, 020601.

10. Gaspard, P.; Gilbert, T. Heat conduction and Fourier's law in a class of many particle dispersing billiards. *New J. Phys.* **2008**, *10*, 103004.

11. Olla, S.; Sasada, M. Macroscopic energy diffusion for a chain of anharmonic oscillators. *Probab. Theory Relat. Fields* **2013**, *157*, 721, doi:10.1007/s00440-012-0469-5.

12. Crosignani, B.; Di Porto, P.; Segev, M. Approach to thermal equilibrium in a system with adiabatic constraints. *Am. J. Phys.* **1996**, *64*, 610–613.

13. Gruber, C.; Piasecki, J. Stationary motion of the adiabatic piston. *Physica A* **1999**, *268*, 412–423.

14. Gruber, C.; Pache, S.; Lesne, A. Two-time-scale relaxation towards thermal equilibrium of the enigmatic piston. *J. Stat. Phys.* **2003**, *112*, 1177–1206.

15. Gruber, C.; Lesne, A. Hamiltonian model of heat conductivity and Fourier law. *Physica A* **2005**, *351*, 358–372.

16. Van den Broeck, C.; Kawai, R.; Meurs, P. Microscopic analysis of a thermal Brownian motor. *Phys. Rev. Lett.* **2004**, *93*, 090601.

17. Cencini, M.; Palatella, L.; Pigolotti, S.; Vulpiani, A. Macroscopic equations for the adiabatic piston. *Phys. Rev. E* **2007**, *76*, 051103.

18. DelRe, E.; Crosignani, B.; Di Porto, P.; Di Sabatino, S. Built-in reduction of statistical fluctuations of partitioning objects. *Phys. Rev. E* **2011**, *84*, 021112.

19. Sarracino, A.; Gnoli, A.; Puglisi, A. Ratchet effect driven by Coulomb friction: the asymmetric Rayleigh piston. *Phys. Rev. E* **2013**, *87*, 040101(R).

20. Cerino, L.; Gradenigo, G.; Sarracino, A.; Villamaina, D.; Vulpiani, A. Fluctuations in partitioning systems with few degrees of freedom. *Phys. Rev. E* **2014**, *89*, 042105.

21. Cerino, L.; Puglisi, A.; Vulpiani, A. Kinetic model for the finite-time thermodynamics of small heat engines. *Phys. Rev. E* **2015**, *91*, 032128.

22. Tehver, R.; Toigo, F.; Koplik, J.; Banavar, J.R. Thermal wall in computer simulations. *Phys. Rev. E* **1998**, *57*, R17–R20.

entropy

MDPI

Article

Equilibration in the Nosé–Hoover Isokinetic Ensemble: Effect of Inter-Particle Interactions

Shamik Gupta [1],* and Stefano Ruffo [2]

[1]   Department of Physics, Ramakrishna Mission Vivekananda University, Belur Math, Howrah 711202, India
[2]   SISSA, INFN and ISC-CNR, Via Bonomea 265, I-34136 Trieste, Italy; ruffo@sissa.it
*   Correspondence: shamik.gupta@rkmvu.ac.in; Tel.: +91-33-2654-9999

Received: 19 September 2017; Accepted: 11 October 2017; Published: 14 October 2017

Abstract: We investigate the stationary and dynamic properties of the celebrated Nosé–Hoover dynamics of many-body interacting Hamiltonian systems, with an emphasis on the effect of inter-particle interactions. To this end, we consider a model system with both short- and long-range interactions. The Nosé–Hoover dynamics aim to generate the canonical equilibrium distribution of a system at a desired temperature by employing a set of time-reversible, deterministic equations of motion. A signature of canonical equilibrium is a single-particle momentum distribution that is Gaussian. We find that the equilibrium properties of the system within the Nosé–Hoover dynamics coincides with that within the canonical ensemble. Moreover, starting from out-of-equilibrium initial conditions, the average kinetic energy of the system relaxes to its target value over a *size-independent* timescale. However, quite surprisingly, our results indicate that under the same conditions and with only long-range interactions present in the system, the momentum distribution relaxes to its Gaussian form in equilibrium over a scale that *diverges with the system size*. On adding short-range interactions, the relaxation is found to occur over a timescale that has a much weaker dependence on system size. This system-size dependence of the timescale vanishes when only short-range interactions are present in the system. An implication of such an ultra-slow relaxation when only long-range interactions are present in the system is that macroscopic observables other than the average kinetic energy when estimated in the Nosé–Hoover dynamics may take an unusually long time to relax to its canonical equilibrium value. Our work underlines the crucial role that interactions play in deciding the equivalence between Nosé–Hoover and canonical equilibrium.

Keywords: Hamiltonian systems; classical statistical mechanics; ensemble equivalence; long-range interacting systems

1. Introduction

Often, one needs in studies in nonlinear dynamics and statistical physics to investigate the dynamical properties of a many-body interacting Hamiltonian system evolving under the condition of a constant temperature. For example, one might be interested in studying the dynamical properties of the system in canonical equilibrium at a certain temperature T, with the temperature being proportional to the average kinetic energy of the system by virtue of the Theorem of Equipartition (In this work, we measure temperatures in units of the Boltzmann constant). To this end, one may devise a dynamics having a temperature T_{target} as a dynamical parameter that is designed to relax an initial configuration of the system to canonical equilibrium at temperature T_{target}, and then make the choice $T_{\text{target}} = T$. A common practice is to employ a Langevin dynamics, i.e., a *noisy, dissipative* dynamics that mimics the interaction of the system with an external heat bath at temperature T_{target} in terms of a deterministic frictional force and an uncorrelated, Gaussian-distributed random force added to the equation of motion [1]. In this approach, one then tunes suitably the strength of the random force such that the

Langevin dynamics relaxes at long times to canonical equilibrium at temperature T_{target}. The presence of dissipation renders the dynamics to be *irreversible in time*. A complementary approach to such a noisy, dissipative dynamics was pioneered by Nosé and Hoover, in which the dynamics is fully *deterministic* and *time-reversible*, while achieving the same objective of relaxing the system to canonical equilibrium at the desired temperature T_{target} [2,3]; for a review, see [4,5]. The time evolution under the condition of relaxation at long times to canonical equilibrium at a given temperature is said to represent isokinetic ensemble dynamics when taking place according to the Nosé–Hoover equation of motion and to represent Langevin/canonical ensemble dynamics when taking place following the Langevin equation of motion.

To illustrate in detail the distinguishing feature of the Nosé–Hoover vis-à-vis Langevin dynamics, consider an interacting N-particle system characterized by the set $\{q_j, \pi_j\}$ of canonical coordinates and conjugated momenta. The particles, which we take for simplicity to have the same mass m, interact with one another via the two-body interaction potential $\Phi(\{q_j\})$. In the following, we consider q_j's and π_j's to be one-dimensional variables for reasons of simplicity. Our analysis, however, extends straightforwardly to higher dimensions. The Hamiltonian of the system is given by

$$\mathcal{H}_{system} = \sum_{j=1}^{N} \frac{\pi_j^2}{2m} + \Phi(\{q_j\}), \tag{1}$$

where the first term on the right-hand side stands for the kinetic energy of the system.

In the approach due to Langevin, the dynamical equations of the system are given by

$$\frac{dq_j}{dt} = \frac{\pi_j}{m}, \quad \frac{d\pi_j}{dt} = -\gamma \frac{\pi_j}{m} - \frac{\partial \Phi(\{q_j\})}{\partial q_j} + \eta_j(t), \tag{2}$$

where t denotes time, $\gamma > 0$ is the dissipation constant, while $\eta_j(t)$ is a Gaussian, white noise satisfying

$$\overline{\eta_j(t)} = 0, \quad \overline{\eta_j(t)\eta_k(t')} = 2D\delta_{jk}\delta(t - t'). \tag{3}$$

Here, the overbars denote averaging over noise realizations, while $D > 0$ characterizes the strength of the noise. The dynamics (2) are evidently not time-reversal invariant. Choosing $D = \gamma T_{target}$ ensures that the dynamics (2) relaxes at long times to the canonical distribution at T_{target} given by [1]

$$P(\{q_j, \pi_j\}) \propto \exp(-\mathcal{H}_{system}/T_{target}), \tag{4}$$

in which the kinetic energy density of the system fluctuates around the average value $T_{target}/2$.

In the approach due to Nosé and Hoover, a degree of freedom s augmenting the set $\{q_j, \pi_j\}$ is introduced, which is taken to characterize an external heat reservoir that interacts with the system through the momenta π_j's. The Hamiltonian of the combined system is given by

$$\mathcal{H} = \sum_{j=1}^{N} \frac{\pi_j^2}{2ms^2} + \Phi(\{q_j\}) + \frac{p_s^2}{2Q} + (N+1)T_{target} \ln s, \tag{5}$$

where Q is the mass and p_s is the conjugated momentum of the additional degree of freedom. The dynamics of the system is given by the following Hamilton equations of motion:

$$\frac{dq_j}{dt} = \frac{\pi_j}{ms^2}, \quad \frac{d\pi_j}{dt} = -\frac{\partial \Phi(\{q_j\})}{\partial q_j},$$

$$\frac{ds}{dt} = \frac{p_s}{Q}, \quad \frac{dp_s}{dt} = \sum_{j=1}^{N} \frac{\pi_j^2}{ms^3} - (N+1)\frac{T_{target}}{s}. \tag{6}$$

It may be easily checked that unlike dynamics (2), dynamics (6) is invariant under time reversal. In terms of new variables

$$p_j \equiv \frac{\pi_j}{s}, \; \zeta \equiv \frac{p_s}{Q}, \tag{7}$$

and rescaled time

$$\tilde{t} \equiv \frac{t}{s}, \tag{8}$$

one obtains from the Hamilton Equations (6) the following dynamics:

$$\frac{dq_j}{d\tilde{t}} = \frac{p_j}{m}, \tag{9}$$

$$\frac{dp_j}{d\tilde{t}} = -\frac{\partial \Phi(\{q_j\})}{\partial q_j} - \zeta p_j, \tag{10}$$

$$\frac{ds}{d\tilde{t}} = \zeta s, \tag{11}$$

$$\frac{d\zeta}{d\tilde{t}} = \frac{1}{Q} \left(\sum_{j=1}^{N} \frac{p_j^2}{m} - (N+1)T_{\text{target}} \right) = \frac{1}{\tau^2} \left(\frac{K(P)}{K_0} - 1 \right), \tag{12}$$

where $K(P) \equiv \sum_{j=1}^{N} p_j^2/(2m)$ is the kinetic energy, while we have defined

$$K_0 \equiv (N+1)\frac{T_{\text{target}}}{2}, \; \tau^2 \equiv \frac{Q}{2K_0}. \tag{13}$$

From Equations (9)–(12), we observe that a complete description of the time evolution of the system is given in terms of Equations (9), (10), and (12), without any reference to Equation (11) for s, so that, as far as the description of the system is concerned, the variable s is an irrelevant one that may be ignored. We note in passing that a different, but closely related, Hamiltonian giving directly the Nosé-Hoover equations of motion but without any time scaling, as in Equation (8), is discussed in [6]. We will from now on drop the tilde over time in order not to overload the notation. Let us note that, in terms of the variables p_j's, the Hamiltonian (5) takes the form

$$\mathcal{H} = \sum_{j=1}^{N} \frac{p_j^2}{2m} + \Phi(\{q_j\}) + \frac{Q\zeta^2}{2} + (N+1)T \ln s. \tag{14}$$

From Equation (12), we find that, in the stationary state ($d\zeta/dt = 0$), the kinetic energy of the system equals $(N+1)T_{\text{target}}/2$ (the extra factor of unity takes care of the presence of the additional degree of freedom s). For large $N \gg 1$, we then have the desired result: an ensemble of initial conditions under the evolution given by Equations (9), (10), and (12) evolves at long times to a stationary state in which the average kinetic energy density has the value $T_{\text{target}}/2$. The quantity τ in Equation (12) denotes a relaxation timescale over which the kinetic energy relaxes to its target value. Beyond the average kinetic energy, it has been demonstrated by invoking the phase space continuity equation that the distribution

$$f \propto \exp \left[- \left(\sum_{j=1}^{N} \frac{p_j^2}{2m} + \Phi(\{q_j\}) + Q\zeta^2/2 \right) / T_{\text{target}} \right] \tag{15}$$

is a stationary state of the Nosé–Hoover dynamics [3]. It then follows that the corresponding stationary distribution for the system variables $\{q_j, p_j\}$ is the canonical equilibrium distribution:

$$P(\{q_j, p_j\}) \propto \exp \left[- \left(\sum_{j=1}^{N} \frac{p_j^2}{2m} + \Phi(\{q_j\}) \right) / T_{\text{target}} \right], \tag{16}$$

normalized as $\int \left(\prod_{j=1}^{N} \mathrm{d}q_j \mathrm{d}p_j \right) P(\{q_j, p_j\}) = 1$. Thus, the dynamics (9)–(12) that includes the additional dynamical variable s nevertheless preseves the canonical equilibrium distribution of the system. A general formalism for constructing modified Hamiltonian dynamical systems that preserve a canonical equilibrium distribution on adding a time evolution equation for a single additional thermostat variable is discussed in [7].

Equation (16) implies that the single-particle momentum distribution $P(p)$, defined such that $P(p)\mathrm{d}p$ gives the probability that a randomly chosen particle has its momentum between p and $p + \mathrm{d}p$, is a Gaussian distribution with mean zero and width equal to T_{target}:

$$P(p) = \frac{1}{\sqrt{2\pi m T_{\text{target}}}} \exp\left(-\frac{p^2}{2m T_{\text{target}}} \right). \tag{17}$$

Consequently, the moments $\langle p^n \rangle \equiv \int_{-\infty}^{\infty} \mathrm{d}p \, p^n P(p)$, with $n = 1, 2, 3, \ldots$, satisfy $\langle p^4 \rangle / \langle p^2 \rangle^2 = 3$.

In the above backdrop, the principal objective of this work is to answer the question: what is the effect of inter-particle interactions on the relaxation properties of the Nosé–Hoover dynamics? More specifically, considering a system embedded in a d-dimensional space, we ask: do systems with long-range interactions, in which the inter-particle interaction decays slower than $1/r^d$, behave in a similar way to short-range systems that have the inter-particle interaction decaying faster than $1/r^d$? How does the timescale over which the phase space distribution relaxes to its canonical equilibrium form behave in the two cases, and, in particular, is there a system-size dependence in the timescale for long-range systems with respect to short-range ones? Studying these issues is particularly relevant and timely in the wake of recent surge in interest across physics in long-range interacting (LRI) systems.

LRI systems may display a notably distinct thermodynamic behavior with respect to short-range ones [8–12]. These systems are characterized by a two-body interaction potential $V(r)$ that decays asymptotically with inter-particle separation r as $V(r) \sim r^{-\alpha}$, with $0 \leq \alpha \leq d$ in d spatial dimensions. The limit $\alpha \to 0$ corresponds to the case of mean-field interaction. Examples of LRI systems are self-gravitating systems, plasmas, fluid dynamical systems, and some spin systems. One of the striking dynamical features resulting from long-range interactions is the occurrence of non-equilibrium quasi-stationary states (QSSs) during relaxation of LRI systems towards equilibrium. These states have lifetimes that diverge with the number of particles constituting the system, so that, in the thermodynamic limit, the system remains trapped in QSSs and does not attain equilibrium. Only for a finite number of particles do the QSSs eventually evolve towards equilibrium. Even in equilibrium, LRI systems may exhibit features such as ensemble inequivalence and a negative heat capacity in the microcanonical ensemble that are unusual for short-range systems.

In this work, we address our aforementioned queries within the ambit of a model system comprising classical XY-spins occupying the sites of a one-dimensional periodic lattice and interacting via a long-range (specifically, a mean–field interaction in which every spin interacts with every other and a short-range (specifically, a nearest-neighbor interaction in which every spin interacts with its left and right neighbors) interaction. With an aim to study the equilibrium properties as well as relaxation towards equilibrium, we simulate the Nosé–Hoover dynamics of the model by integrating the corresponding equations of motion in time. A signature of canonical equilibrium is a single-particle momentum distribution that is Gaussian (see Equation (17)). We find that the equilibrium properties of our model system evolving under the Nosé–Hoover dynamics coincide with those within the canonical ensemble. As regards relaxation towards canonical equilibrium, we observe that starting from out-of-equilibrium initial conditions, the average kinetic energy of the system relaxes to its target canonical-equilibrium value over a *size-independent* timescale. However, quite surprisingly, our results indicate that under the same conditions and with only long-range interactions present in the system, the momentum distribution relaxes to its Gaussian form in equilibrium over a scale that *diverges* with the system size. On adding short-range interactions, the relaxation is found to occur over a timescale that has a much weaker dependence on system size. This system-size dependence vanishes when only short-range interactions are present in the system. An implication of such an ultra-slow relaxation when only long-range interactions are present in the system is that macroscopic observables other than the average kinetic energy when estimated in the Nosé–Hoover dynamics may take an unusually long time to relax to its canonical equilibrium value. Our work underlines the crucial role that interactions play in deciding the equivalence between Nosé–Hoover and canonical equilibrium.

The paper is organized as follows. In Section 2, we describe the model of study. In Section 3, we obtain the so-called caloric curve of the model within the canonical ensemble, which we eventually invoke in later parts of the paper to decide on the equivalence of the equilibrium properties of the Nosé–Hoover dynamics and canonical equilibrium. In Section 4, we present results from simulations of the Nosé–Hoover dynamics of the model, and discuss the implications and relevance of the results. The paper ends with conclusions in Section 5.

2. Model of Study

Our system of study comprises a one-dimensional periodic lattice of N sites. Each site of the lattice is occupied by a unit-inertia rotor characterized by its angular coordinate $\theta_j \in [0, 2\pi)$ and the corresponding conjugated momentum p_j, with $j = 1, 2, \ldots, N$. One may also think of the rotors as representing classical XY-spins. Note that both of the θ_j's and the p_j's are one-dimensional variables. There exist both a long-range (specifically, a global or a mean–field) coupling and a short-range (specifically, nearest-neighbor) coupling between the rotors. Thus, a rotor on site j interacts with strength $J/(2N)$ with rotors on all the other sites and with strength K with the rotor occupying the $(j-1)$-th and the $(j+1)$-th site. The Hamiltonian of the system is given by [13,14]

$$\mathcal{H} = \sum_{j=1}^{N} \frac{p_j^2}{2} + \frac{J}{2N} \sum_{j,k=1}^{N} \left[1 - \cos(\theta_j - \theta_k) \right] + K \sum_{j=1}^{N} \left[1 - \cos(\theta_{j+1} - \theta_j) \right] ; \quad \theta_{N+1} \equiv \theta_1, p_{N+1} \equiv p_1. \quad (18)$$

Note that, for $K = 0$, the Hamiltonian (18) reduces to that of the widely-studied Hamiltonian mean–field (HMF) model [15], which is regarded as a paradigmatic model to study statics and dynamics of LRI systems [10]. On the other hand, for $J = 0$, the model (18) reduces to a short-ranged XY model in one dimension.

In the following, we take both the mean–field coupling J and the short-range coupling K to be positive, thereby modeling ferromagnetic global and nearest-neighbor couplings. Consequently, both the long-range and the short-range coupling between the rotors favor an ordered state in which all the rotor angles are equal, thereby minimizing the potential energy contribution to the total energy. Such a tendency is, however, opposed by the kinetic energy contribution whose average in equilibrium may be characterized by a temperature by invoking the Theorem of Equipartition. Noting that, for a given N, the total potential energy is bounded from above while the total kinetic energy is not, one expects the system to show in equilibrium an ordered/magnetized phase at low energies/temperatures and a disordered/unmagnetized phase at high energies/temperatures. This scenario holds even with $K = 0$.

The amount of order in the system is characterized by the XY magnetization

$$\mathbf{m} \equiv \frac{1}{N} \left(\sum_{j=1}^{N} \cos \theta_j, \sum_{j=1}^{N} \sin \theta_j \right), \quad (19)$$

which is a vector whose length m has the thermodynamic value in equilibrium denoted by m^{eq} that is nonzero in the ordered phase and zero in the disordered phase. For $K = 0$, the corresponding HMF model is known to display a second-order phase transition between a high-temperature unmagnetized phase and a low-temperature magnetized phase at the critical temperature $T_c = J/2$, with the corresponding critical energy density being $u_c = 3J/4$ [10]. On the other hand, invoking the Landau's argument for the absence of any phase transition at a finite temperature in a one-dimensional model with only short-range interactions, one may conclude for $J = 0$ that the corresponding short-ranged XY model does not display any phase transitions, though it has been shown to have interesting dynamical effects [16]. For general $J \neq 0, K \neq 0$, when both long-range and short-range interactions are present, the model displays a second-order phase transition between an ordered and a disordered phase [13,14]. Note that all the mentioned phase transitions are continuous. Although ensemble equivalence is not guaranteed for LRI systems, it has been argued that inequivalence arises when one has a first-order phase transition in the canonical ensemble, and not when one has a second-order transition [17]. Consequently, we may regard the phase diagram of model (18) to be equivalent within microcanonical and canonical ensembles. For an explicit demonstration of ensemble equivalence for the model (18), one may refer to [14].

In the following section, we will obtain the caloric curve of model (18) that relates the equilibrium internal energy with the equilibrium temperature of the system.

3. The Caloric Curve within the Canonical Ensemble

As mentioned in the preceding section, model (18) is known to have equivalent microcanonical and canonical ensemble descriptions in equilibrium. Consequently, in obtaining the caloric curve of the model, which will be invoked to decide the equivalence between the equilibrium properties of the Nosé–Hoover dynamics and canonical equilibrium, it will suffice to restrict our analysis to the canonical ensemble description of the model.

The Langevin/canonical ensemble dynamics (2) for the model (18) comprises the set of time-evolution equations

$$\frac{d\theta_j}{dt} = p_j,$$

$$\frac{dp_j}{dt} = -\gamma p_j + \frac{J}{N} \sum_{k=1}^{N} \sin(\theta_k - \theta_j) + K \left[\sin(\theta_{j+1} - \theta_j) + \sin(\theta_{j-1} - \theta_j) \right] + \eta_j(t),$$

(20)

with the properties of the noise $\eta_j(t)$ given by Equation (3) with $D = \gamma T$. Within the microcanonical ensemble description of the system, the time evolution of the variables $\{\theta_j, p_j\}$ is given by Hamilton equations obtained from Equation (20) by setting γ to zero. The Nosé–Hoover dynamics of the variables $\{\theta_j, p_j\}$ is obtained from Equations (9) and (10) as

$$\frac{d\theta_j}{dt} = p_j,$$

$$\frac{dp_j}{dt} = \frac{J}{N} \sum_{k=1}^{N} \sin(\theta_k - \theta_j) + K \left[\sin(\theta_{j+1} - \theta_j) + \sin(\theta_{j-1} - \theta_j) \right] - \zeta p_j,$$

(21)

where the time evolution of the variable ζ is given by Equation (12).

In order to derive the desired caloric curve of model (18) within the canonical ensemble, we start with the canonical partition function of the system at temperature T given by $Z_N \equiv \int \left(\prod_{j=1}^{N} d\theta_j dp_j \right) \exp[-\beta \mathcal{H}(\{\theta_j, p_j\})]$, with $\beta \equiv 1/T$. Using Equation (18), we get

$$Z_N = \left(\frac{2\pi}{\beta} \right)^{N/2} e^{-\beta JN/2 - \beta KN} \int \left(\prod_{j=1}^{N} d\theta_j \right) \exp \left[\frac{\beta J}{2N} \left\{ \left(\sum_{j=1}^{N} \cos\theta_j \right)^2 + \left(\sum_{j=1}^{N} \sin\theta_j \right)^2 \right\} + \beta K \sum_{j=1}^{N} \cos(\theta_{j+1} - \theta_j) \right]. \quad (22)$$

Using the Hubbard–Stranovich transformation $\exp(ax^2)$ = $1/(\sqrt{4\pi a}) \int_{-\infty}^{\infty} dz \, \exp\left(-\frac{z^2}{4a} + zx\right)$, $a > 0$ in Equation (22), we obtain

$$
Z_N = \left(\frac{2\pi}{\beta}\right)^{N/2} e^{-\beta J N/2 - \beta K N} \frac{N\beta J}{2\pi} \int_{-\infty}^{\infty} dz_1 \int_{-\infty}^{\infty} dz_2 \int \left(\prod_{j=1}^{N} d\theta_j\right) \exp\left[-\frac{N\beta J}{2}(z_1^2 + z_2^2)\right.
$$
$$
\left. + \beta J z_1 \sum_{j=1}^{N} \cos\theta_j + \beta J z_2 \sum_{j=1}^{N} \sin\theta_j + \beta K \sum_{j=1}^{N} \cos(\theta_{j+1} - \theta_j)\right]. \tag{23}
$$

Writing $z_1 = z\cos\phi, z_2 = z\sin\phi$, with real $z = (z_1^2 + z_2^2)^{1/2} > 0$ and $\phi \in [0, 2\pi)$ given by $\phi = \tan^{-1}(z_2/z_1)$, we get

$$
Z_N = \left(\frac{2\pi}{\beta}\right)^{N/2} e^{-\beta J N/2 - \beta K N} \frac{N\beta J}{2\pi} \int_0^{2\pi} d\phi \int_0^{\infty} dz \, z \int \left(\prod_{j=1}^{N} d\theta_j\right) \exp\left[-\frac{N\beta J}{2} z^2\right.
$$
$$
\left. + \beta J z \sum_{j=1}^{N} \cos(\theta_j - \phi) + \beta K \sum_{j=1}^{N} \cos(\theta_{j+1} - \theta_j)\right]. \tag{24}
$$

In view of the invariance of the Hamiltonian (18) under rotation by an equal amount of all the θ_j's, we get [18]

$$
Z_N = \left(\frac{2\pi}{\beta}\right)^{N/2} e^{-\beta J N/2 - \beta K N} N\beta J \int_0^{\infty} dz \, z \int \left(\prod_{j=1}^{N} d\theta_j\right) \exp\left[-\frac{N\beta J}{2} z^2 + \beta J z \sum_{j=1}^{N} \cos\theta_j + \beta K \sum_{j=1}^{N} \cos(\theta_{j+1} - \theta_j)\right]. \tag{25}
$$

In order to proceed further, we consider separately the cases $K = 0$ and $K \neq 0$ in the following.

3.1. K = 0

For $K = 0$, Equation (25) yields

$$
Z_N = \left(\frac{2\pi}{\beta}\right)^{N/2} N\beta J \int_0^{\infty} dz \, z \exp\left[-N\left\{\frac{\beta J}{2}(1 + z^2) - \ln\left(\int_0^{2\pi} d\theta \, \exp(\beta J z \cos\theta)\right)\right\}\right]. \tag{26}
$$

In the thermodynamic limit, Z_N may be approximated by invoking the saddle-point method to perform the integration in z on the right-hand side; one gets

$$
Z_N = \left(\frac{2\pi}{\beta}\right)^{N/2} N\beta J z_s \exp\left[-N\left\{\frac{\beta J}{2}(1 + z_s^2) - \ln\left(\int_0^{2\pi} d\theta \, \exp(\beta J z_s \cos\theta)\right)\right\}\right], \tag{27}
$$

where the saddle-point value z_s solves the equation

$$
z_s = \frac{I_1(\beta J z_s)}{I_0(\beta J z_s)}, \tag{28}
$$

with $I_n(x) = (1/(2\pi)) \int_0^{2\pi} d\theta \, \exp(x\cos\theta)\cos(n\theta)$ being the modified Bessel function of first kind and of order n. It may be shown by following the arguments given in [18] that z_s is nothing but the stationary magnetization m^{eq}. Equation (28) has a trivial solution $m^{\mathrm{eq}} = 0$ valid at all temperatures, while a non-zero solution exists for $\beta \geq \beta_c = 2/J$ [10]. In fact, the system shows a continuous transition, from a magnetized phase ($m^{\mathrm{eq}} \neq 0$) at low temperatures to an unmagnetized phase ($m^{\mathrm{eq}} = 0$) at high temperatures at the critical temperature $T_c = J/2$ [10].

In the thermodynamic limit, the internal energy density of the system $u = -\lim_{N\to\infty}(1/N)\mathrm{d}\ln Z_N/\mathrm{d}\beta$ is obtained by using Equations (27) and (28) as

$$
u = \frac{1}{2\beta} + \frac{J}{2}\left(1 - (m^{\mathrm{eq}})^2\right); \, m^{\mathrm{eq}} = \frac{I_1(\beta J m^{\mathrm{eq}})}{I_0(\beta J m^{\mathrm{eq}})}, \tag{29}
$$

yielding the critical energy density

$$u_c = \frac{3J}{4}.\tag{30}$$

Equation (29) gives the caloric curve of the model (18) at canonical equilibrium for $J \neq 0, K = 0$.

3.2. $K \neq 0$

For $K \neq 0$, Equation (25) gives

$$Z_N = \left(\frac{2\pi}{\beta}\right)^{N/2} N\beta J \int_0^\infty dz\, z \exp\left[-\frac{N\beta J}{2}(1+z^2) - \beta K N\right] \mathcal{Z}_N,\tag{31}$$

$$\mathcal{Z}_N \equiv \int \left(\prod_{j=1}^N d\theta_j\right) \exp\left[\beta J z \sum_{j=1}^N \cos\theta_j + \beta K \sum_{j=1}^N \cos(\theta_{j+1} - \theta_j)\right],\tag{32}$$

where we may identify the factor $\mathcal{Z}_N$ with the canonical partition function of a $1d$ periodic chain of N interacting angle-only rotors, where a rotor on each site interacts with strength K with the rotor on the left nearest-neighbor and the right nearest-neighbor site, and also with an external field of strength Jz along the x direction.

One may evaluate $\mathcal{Z}_N$ by rewriting it in terms of a transfer operator $\mathcal{T}(\theta, \theta')$ as

$$\mathcal{Z}_N = \int \left(\prod_{j=1}^N d\theta_j\right) \mathcal{T}(\theta_1, \theta_2)\mathcal{T}(\theta_2, \theta_3)\dots\mathcal{T}(\theta_N, \theta_1),\tag{33}$$

$$\mathcal{T}(\theta_j, \theta_{j+1}) \equiv \exp\left[\beta J z\left\{\frac{\cos\theta_j + \cos\theta_{j+1}}{2}\right\} + \beta K \cos(\theta_{j+1} - \theta_j)\right].\tag{34}$$

Let $\{\lambda_m\}$ denote the set of eigenvalues of the transfer operator $\mathcal{T}(\theta, \theta')$. In other words, denoting the eigenfunctions of $\mathcal{T}(\theta, \theta')$ as $f_m(\theta)$, we have $\int d\theta'\, \mathcal{T}(\theta, \theta') f_m(\theta') = \lambda_m f_m(\theta)$. In terms of $\{\lambda_m\}$, we obtain

$$\mathcal{Z}_N = \sum_m [\lambda_m(\beta J z, \beta K)]^N.\tag{35}$$

For large N, the sum in Equation (35) is dominated by the largest eigenvalue $\lambda_{\max} = \lambda_{\max}(\beta J z, \beta K)$, yielding

$$\mathcal{Z}_N = \lambda_{\max}^N.\tag{36}$$

Substituting Equation (36) in Equation (31), and approximating the integral on the right-hand side of the latter by the saddle-point method, one gets

$$Z_N = \left(\frac{2\pi}{\beta}\right)^{N/2} N\beta J z_s \exp\left[-N\left\{\frac{\beta J}{2}(1+z_s^2) + \beta K - \ln\lambda_{\max}(\beta J z_s, \beta K)\right\}\right],\tag{37}$$

where z_s solves the saddle-point equation $z_s \equiv \sup_z \tilde{\phi}(\beta, z)$, with $\tilde{\phi}(\beta, z)$ being the free-energy function:

$$-\tilde{\phi}(\beta, z) \equiv -\frac{1}{2}\ln\beta - \frac{\beta J}{2}(1+z^2) - \beta K + \ln\lambda_{\max}(\beta J z, \beta K).\tag{38}$$

The saddle-point equation may thus be written as

$$z_s = \frac{\partial \ln\lambda_{\max}(\beta J z, \beta K)}{\partial(\beta J z)}\bigg|_{z=z_s}.\tag{39}$$

Equation (37) gives the dimensionless free energy per rotor, $\phi(\beta) \equiv -\lim_{N\to\infty}(\ln Z_N)/N$, as $-\phi(\beta) = \sup_z\left[-\tilde{\phi}(\beta, z)\right]$, where we have suppressed the dependence of $\phi(\beta)$ on K. We thus have

$$-\phi(\beta) \equiv -\frac{1}{2}\ln\beta - \frac{\beta J}{2}(1 + z_s^2) - \beta K + \ln\lambda_{\max}(\beta J z_s, \beta K). \tag{40}$$

Note that the free energy at a given temperature has a definite value given by Equation (40), and is obtained by substituting the saddle-point solution z_s into the expression for the free-energy function $\tilde{\phi}(\beta, z)$.

In the thermodynamic limit, the internal energy density of the system $u = -\lim_{N\to\infty}(1/N)\mathrm{d}\ln Z_N/\mathrm{d}\beta$ is obtained as

$$u = \frac{1}{2\beta} + \frac{J}{2}(1 + z_s^2) + \beta J z_s\frac{\mathrm{d}z_s}{\mathrm{d}\beta} + K - \frac{\mathrm{d}\ln\lambda_{\max}(\beta J z, \beta K)}{\mathrm{d}\beta}\bigg|_{z=z_s}. \tag{41}$$

Using Equation (39), and the fact that, as for $K = 0$, the quantity z_s is nothing but the stationary magnetization m^{eq}, we get

$$u = \frac{1}{2\beta} + \frac{J}{2}\left(1 - (m^{\mathrm{eq}})^2\right) + \beta J m^{\mathrm{eq}}\frac{\mathrm{d}m^{\mathrm{eq}}}{\mathrm{d}\beta} + K - K\frac{\partial\ln\lambda_{\max}(\beta J m^{\mathrm{eq}}, \beta K)}{\partial(\beta K)}, \tag{42}$$

with m^{eq} satisfying

$$m^{\mathrm{eq}} = \frac{\partial\ln\lambda_{\max}(\beta J z, \beta K)}{\partial(\beta J z)}\bigg|_{z=m^{\mathrm{eq}}}. \tag{43}$$

To proceed, we need to find $\lambda_{\max}(\beta J z, \beta K)$. We consider separately the cases $J = 0$ and $J \neq 0$.

3.2.1. $J = 0$

In this case, it may be easily checked that the eigenvalues of $\mathcal{T}$ are given by $2\pi I_m(\beta K)$ with the corresponding eigenvector given by plane waves $\exp(iq\theta)/\sqrt{2\pi}$ [14]. Using $I_0(x) > I_1(x) > I_2(x)\ldots$, we conclude that $\lambda_{\max}(0, \beta K) = I_0(\beta K)$. Equation (43) then yields $m^{\mathrm{eq}} = 0$, while Equation (42) gives

$$u = \frac{1}{2\beta} + K\left(1 - \frac{I_1(\beta K)}{I_0(\beta K)}\right), \tag{44}$$

where we have used the result $\mathrm{d}I_0(x)/\mathrm{d}x = I_1(x)$. Equation (44) is the desired caloric curve of the model (18) within the canonical ensemble for $J = 0, K \neq 0$.

3.2.2. $J \neq 0$

In this case, not knowing the analytic forms of the eigenvalues of $\mathcal{T}$, we resort to a numerical scheme to estimate the largest eigenvalue $\lambda_{\max}(\beta J z, \beta K)$. To this end, we discretize the angles over the interval $[0, 2\pi)$ as $\theta_j^{(a_j)} = a_j\Delta\theta$, with $a_j = 1, 2, \ldots, P$ and $\Delta\theta = 2\pi/P$ for any large positive integer P (we choose $P = 30$). The operator $\mathcal{T}(\theta, \theta')$ then takes the form of a matrix of size $P \times P$, whose largest eigenvalue may be estimated numerically by employing the so-called power method [19] (A FORTRAN90 library that implements the power method and is distributed under the GNU Lesser General Public License (GPL) is available at [20]). Noting that $\mathcal{T}(\theta, \theta')$ is a finite-dimensional real square matrix with positive entries, the application of the Perron–Frobenius theorem implies the existence of its largest eigenvalue that is real and non-degenerate. At given values of T, K, J, z, once $\lambda_{\max}(\beta J z, \beta K)$ has been estimated numerically, we compute the free-energy function $\tilde{\phi}(\beta, z)$ as a function of z by using Equation (38). We then find numerically the value of z at which the computed free-energy function attains its minimum value. As discussed above, this minimizer is the equilibrium magnetization of the system at the given values of T, K, J. In order to obtain the caloric curve, one has to estimate numerically the derivative $\partial\ln\lambda_{\max}(\beta J m^{\mathrm{eq}}, \beta K)/\partial(\beta K)$, and then use Equation (42).

4. Results and Discussion

In this section, we discuss the results on equilibrium as well as relaxation properties of model (18) obtained by performing numerical integration of the Nosé–Hoover equations of motion (21). The numerical integration involved using a fourth-order Runge–Kutta method with timestep $dt = 0.01$.

4.1. Results in Equilibrium

Here, we discuss the Nosé–Hoover equilibrium properties for model (18). The initial condition corresponds to the θ_j's independently and uniformly distributed in $[0, 2\pi)$ and the p_j's independently sampled from a Gaussian distribution with zero mean and width equal to 0.5. The initial value of the parameter ζ is 2.0, while we have taken $\tau = 0.01$. In Figure 1, we consider the case when only long-range interactions are present in system ($J = 1.0, K = 0.0$). Figure 1a shows for $T_{\text{target}} = 2.5$ that the average kinetic energy relaxes at long times to the value $T_{\text{target}}/2$, as desired. Figure 1b shows for the same value of T_{target} that the average internal energy has the same value in the stationary state as the one in canonical equilibrium given by Equation (29); Figure 1c shows the single-particle momentum distribution $P(p)$ in the stationary state. We observe that $P(p)$ has the correct canonical-equilibrium form of a Gaussian distribution, which further corroborates the property of the Nosé–Hoover dynamics that the canonical distribution (16) is a stationary state of the dynamics. Figure 1d shows for a range of values of the temperature $T = T_{\text{target}}$ that the caloric curve obtained within the Nosé–Hoover dynamics in equilibrium coincides with that within the canonical ensemble given by Equation (29). Figure 1a–c refer to the system size $N = 128$, while Figure 1d refers to two system sizes, namely, $N = 128$ and $N = 512$. The aforementioned observed properties of the Nosé–Hoover dynamics have been checked to hold for (i) the case when only short-range interactions are present in the system (see Figure 2 that corresponds to $J = 0.0, K = 1.0$), in which case the caloric curve within the canonical ensemble is given by Equation (44), and (ii) when both long- and short-range interactions are present in the system (data not shown; see, however, Figure 3c).

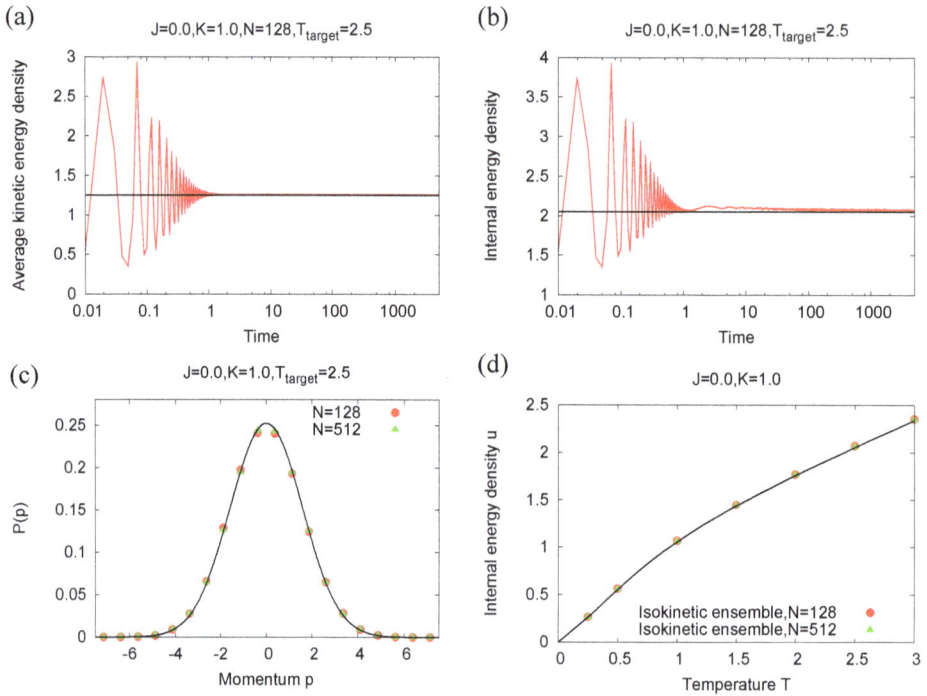

Figure 1. Comparison of Nosé–Hoover and canonical equilibrium results for model (18) with $J = 0.0, K = 1.0$ (that is, with only short-range interactions). (**a**) variation of the average kinetic energy density with time. The black line denotes the value $T_{target}/2$; (**b**) variation of the internal energy density with time. The black line denotes the average internal energy density within the canonical ensemble given by Equation (44); (**c**) stationary single-particle momentum distribution obtained from momentum values measured at time $t = 5000$. The black line denotes a Gaussian distribution with zero mean and width equal to T_{target}; (**d**) caloric curve for two system sizes, $N = 128$ and $N = 512$. The black line shows the caloric curve within the canonical ensemble given by Equation (44). The data for the Nosé–Hoover dynamics are generated by integrating the equations of motion (21) using a fourth-order Runge–Kutta method with timestep equal to 0.01. The initial condition corresponds to the θ_j's independently and uniformly distributed in $[0, 2\pi)$ and the p_j's independently sampled from a Gaussian distribution with zero mean and width equal to 0.5. The initial value of the parameter ζ is 2, while we have taken $\tau = 0.01$.

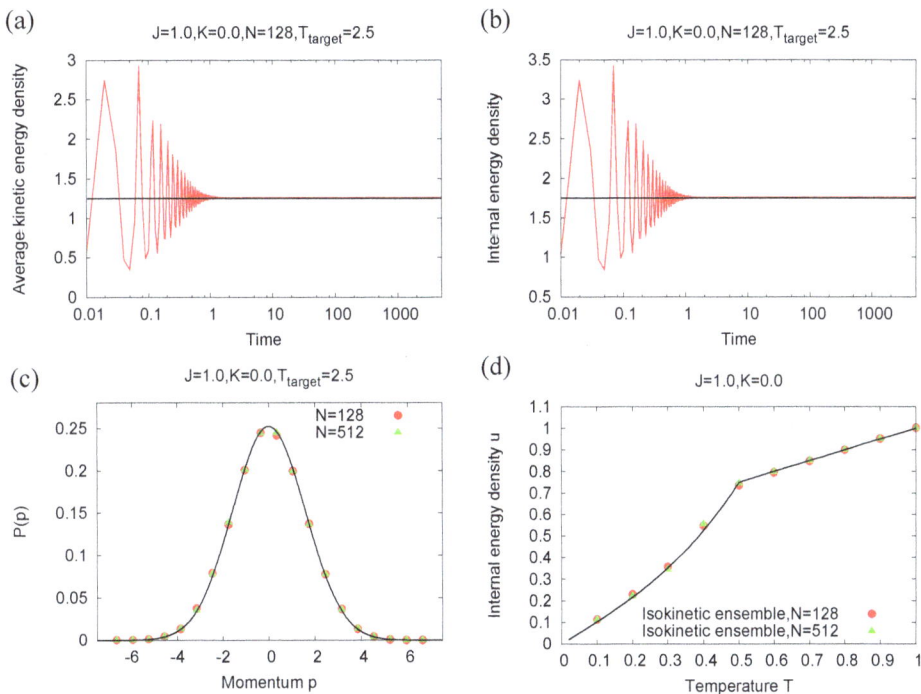

Figure 2. Comparison of Nosé–Hoover and canonical equilibrium results for the model (18) with $J = 1.0, K = 0.0$ (that is, with only long-range interactions); (**a**) variation of the average kinetic energy density with time. The black line denotes the value $T_{target}/2$; (**b**) variation of the internal energy density with time. The black line denotes the average internal energy density within the canonical ensemble given by Equation (29); (**c**) stationary single-particle momentum distribution obtained from momentum values measured at time $t = 5000$. The black line denotes a Gaussian distribution with zero mean and width equal to T_{target}; (**d**) caloric curve for two system sizes, $N = 128$ and $N = 512$. The black line shows the caloric curve within the canonical ensemble given by Equation (29). The data for the Nosé–Hoover dynamics are generated by integrating the equations of motion (21) using a fourth-order Runge–Kutta method with timestep equal to 0.01. The initial condition corresponds to the θ_j's independently and uniformly distributed in $[0, 2\pi)$ and the p_j's independently sampled from a Gaussian distribution with zero mean and width equal to 0.5. The initial value of the parameter ζ is 2, while we have taken $\tau = 0.01$.

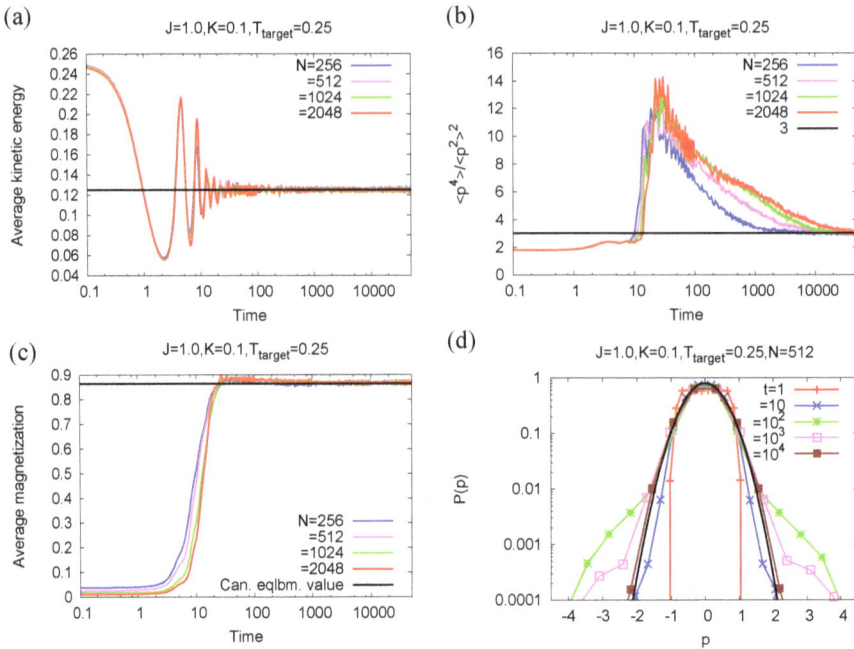

Figure 3. Relaxation properties of the Nosé–Hoover dynamics for model (18) with $J = 1.0, K = 0.1$. (**a**) variation of the average kinetic energy density with time, for four different system sizes. The black line denotes the value $T_{target}/2$; (**b**) variation of the ratio $\langle p^4 \rangle / \langle p^2 \rangle^2$ with time, for four different system sizes. The black line denotes the value 3 corresponding to a Gaussian distribution; (**c**) variation of the magnetization with time, again for four different system sizes. The black line denotes the canonical equilibriu m value obtained by the method described in Section 3.2.2; (**d**) single-particle momentum distribution as a function of time, for system size $N = 512$. The black line denotes a Gaussian distribution with zero mean and width equal to T_{target}, Equation (17). The data for the Nosé–Hoover dynamics are generated by integrating the equations of motion (21) using a fourth-order Runge–Kutta method with timestep equal to 0.01. The initial condition corresponds to the θ_j's independently and uniformly distributed in $[0, 2\pi)$ and the p_j's independently and uniformly distributed in $[-\sqrt{1.5}, \sqrt{1.5}]$. The initial value of the parameter ζ is 2, while we have taken $\tau = 1.0$.

4.2. Results out of Equilibrium

Here, we discuss the relaxation properties of the Nosé–Hoover dynamics for model (18). The initial condition corresponds to the so-called water-bag distribution that has both θ and p uniformly distributed over given intervals [10]. We consider θ_j's to be independently and uniformly distributed in $[0, 2\pi)$ and the p_j's to be independently and uniformly distributed in $[-\sqrt{1.5}, \sqrt{1.5}]$. The initial value of the parameter ζ is 2.0, while we have taken $\tau = 1.0$.

Let us start with a discussion of the results in Figure 4 that corresponds to the case when only long-range interactions are present in the system (18). In Figure 4a, we see that, for four different system sizes, the average kinetic energy density relaxes at long times to the target value $T_{target}/2$ over a timescale that *does not depend on the system size*. A Gaussian distribution for the momentum, expected in canonical equilibrium, is characterized by a value 3 of the ratio $\langle p^4 \rangle / \langle p^2 \rangle^2$ (see Equation (17)). We see in Figure 4b that, in contrast to Figure 4a, this ratio, however, relaxes to the canonical equilibrium value over a time that *depends on the system size*, and which grows with increase of N. Figure 4c shows that the long-time magnetization value reached by the Nosé–Hoover dynamics coincides with the canonical equilibrium value for all system sizes. On the basis of these results, we conclude that, with only long-range interactions in system (18), only the second moment of the momentum distribution relaxes to its canonical equilibrium value over a size-independent timescale, while higher moments (and consequently, the whole of the momentum distribution) relax to their canonical equilibrium values over a time that grows with the system size. The latter fact is demonstrated in Figure 4d that shows for $N = 512$ the time evolution of the single-particle momentum distribution.

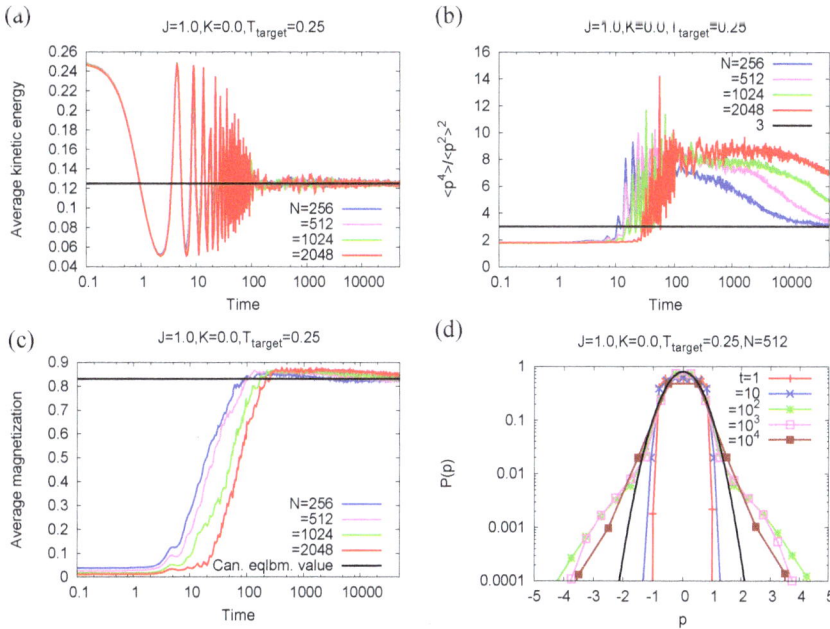

Figure 4. Relaxation properties of the Nosé–Hoover dynamics for the model (18) with $J = 1.0, K = 0.0$ (that is, with only long-range interactions). (**a**) variation of the average kinetic energy density with time, for four different system sizes. The black line denotes the value $T_{target}/2$; (**b**) variation of the ratio $\langle p^4 \rangle / \langle p^2 \rangle^2$ with time, for four different system sizes. The black line denotes the value 3 corresponding to a Gaussian distribution; (**c**) variation of the magnetization with time, again for four different system sizes. The black line denotes the canonical equilibrium value given by Equation (28); (**d**) single-particle momentum distribution as a function of time, for system size $N = 512$. The black line denotes a Gaussian distribution with zero mean and width equal to T_{target}, Equation (17). The data for the Nosé–Hoover dynamics are generated by integrating the equations of motion (21) using a fourth-order Runge–Kutta method with timestep equal to 0.01. The initial condition corresponds to the θ_j's independently and uniformly distributed in $[0, 2\pi)$ and the p_j's independently and uniformly distributed in $[-\sqrt{1.5}, \sqrt{1.5}]$. The initial value of the parameter ζ is 2, while we have taken $\tau = 1.0$.

The feature of a size-independent timescale for the relaxation of the average kinetic energy density to its canonical equilibrium value, observed in the case of purely long-range interactions in model (18), also holds on adding short-range interactions to the model and when the latter are the only interactions present in the system (see Figures 3a and 5a). Moreover, in all cases, the long-time value of the magnetization matches with its canonical equilibrium value (see Figures 3c and 5c). The most significant difference in the relaxation properties that is observed on adding short-range interactions may be inferred by comparing Figures 3b and 4b: the very strong size-dependence observed in the relaxation of the ratio $\langle p^4 \rangle / \langle p^2 \rangle^2$ to its canonical equilibrium value gets substantially weakened on adding short-range interactions with coupling strength as low as $K = 0.1$ compared to the value of the long-range coupling constant $J = 1.0$. Similar inference may be drawn from a comparison of Figures 3d and 4d. This observation has an immediate and an important implication: additional short-range interactions speed up the relaxation of the momentum distribution towards canonical equilibrium. The aforementioned system-size dependence vanishes on turning off long-range interactions, so that the only remnant interactions in the system are the short-range ones (see Figure 5b,d).

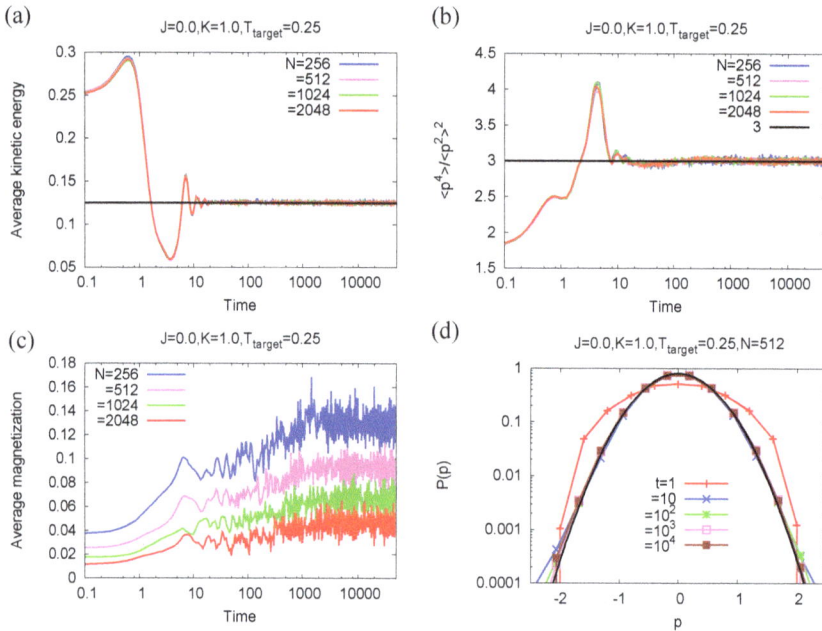

Figure 5. Relaxation properties of the Nosé–Hoover dynamics for the model (18) with $J = 0.0, K = 1.0$ (that is, with only short-range interactions). (**a**) variation of the average kinetic energy density with time, for four different system sizes. The black line denotes the value $T_{\text{target}}/2$; (**b**) variation of the ratio $\langle p^4 \rangle / \langle p^2 \rangle^2$ with time, for four different system sizes. The black line denotes the value 3 corresponding to a Gaussian distribution; (**c**) variation of the magnetization with time, again for four different system sizes. The equilibrium magnetization goes to zero with increase of N as $m^{\text{eq}} \sim 1/\sqrt{N}$; (**d**) single-particle momentum distribution as a function of time, for system size $N = 512$. The black line denotes a Gaussian distribution with zero mean and width equal to T_{target}, Equation (17). The data for the Nosé–Hoover dynamics are generated by integrating the equations of motion (21) using a fourth-order Runge–Kutta method with timestep equal to 0.01. The initial condition corresponds to the θ_j's independently and uniformly distributed in $[0, 2\pi)$ and the p_j's independently and uniformly distributed in $[-\sqrt{1.5}, \sqrt{1.5}]$. The initial value of the parameter ζ is 2, while we have taken $\tau = 1.0$.

5. Conclusions

In this paper, we investigated the relaxation properties of the Nosé–Hoover dynamics of many-body interacting Hamiltonian systems, with an emphasis on the effect of inter-particle interactions. The dynamics aim to generate the canonical equilibrium distribution of a system at the desired temperature by employing time-reversible, deterministic dynamics. To pursue our study, we considered a representative model comprising N classical XY-spins occupying the sites of a one-dimensional periodic lattice. The spins interact with one another via both a long-range interaction, modelled as a mean–field interaction in which every spin interacts with every other, and a short-range one, modelled as a nearest-neighbor interaction in which every spin interacts with its left and right neighboring spins. We studied the Nosé–Hoover dynamics of the model through N-body integration of the corresponding equations of motion. Canonical equilibrium is characterized by a momentum distribution that is Gaussian. We found that the equilibrium properties of our model system evolving according to Nosé–Hoover dynamics are in excellent agreement with exact analytic results for the equilibrium properties derived within the canonical ensemble. Moreover, while starting from out-of-equilibrium initial conditions, the average kinetic energy of the system relaxes to its target value over a *size-independent* timescale. However, quite unexpectedly, we found that under the same conditions and with only long-range interactions present in the system, the momentum distribution relaxes to its Gaussian form in equilibrium over a scale that *grows* with N. The N-dependence gets weaker on adding short-range interactions, and vanishes when the latter are the only inter-particle interactions present in the system.

Viewed from the perspective of LRI systems, the slow relaxation observed within the Nosé–Hoover dynamics allows for drawing an analogy with a similar slow relaxation observed within the microcanonical dynamics of isolated LRI systems, a phenomenon that leads to the occurrence of nonequilibrium quasistationary states (QSSs) that have lifetimes diverging with the system size [10,21]. Within a kinetic theory approach, the QSSs are understood as stable, stationary solutions of the so-called Vlasov equation that governs the time evolution of the single-particle phase space distribution. The Vlasov equation is obtained as the first equation of the Bogoliubov–Born–Green–Yvon–Kirkwood (BBGKY) hierarchy by neglecting the correlation between particle trajectories, with corrections that decrease with an increase of N. For large but finite N, the eventual relaxation of QSSs towards equilibrium is understood as arising due to these finite-N corrections, the so-called collisional terms, to the Vlasov equation. In models in which the momentum variables are one-dimensional, it has been shown by analyzing the behavior of the dominant collisional term that Vlasov-stable phase-space distributions that are homogeneous in the coordinates evolve on times much larger than N, thereby leading for the distributions to characterize QSSs that have lifetimes diverging with N [8,11,22]. In light of the foregoing discussions, it is evidently pertinent and of immediate interest to invoke a kinetic theory approach and investigate in the context of the Nosé–Hoover dynamics of long-range systems whether additional short-range interactions play the role of collisional dynamics that speed up the relaxation of the system towards canonical equilibrium. Work in this direction is in progress and will be reported elsewhere.

The agreement reported in this paper in the value of the average kinetic energy computed in canonical equilibrium and within the Nosé–Hoover dynamics is reminiscent of a similar agreement in the large-system limit between ensemble and time averages predicted by Khinchin for the so-called sum-functions, that is, functions such as the kinetic energy that are sums of single-particle contributions [23]. The result was obtained for rarefied gases, which was later observed to also hold for systems with short-range interactions [24,25]. Our work hints at the validity of such a result even for long-range systems, as is evident from the agreement in the value of the average kinetic energy computed within the Nosé–Hoover dynamics and in canonical equilibrium (see Figure 4a). This point warrants a more detailed investigation that will be left for future studies.

Acknowledgments: This research did not receive any funding. The authors thank W. G. Hoover for fruitful email exchanges. Thanks are due to Leticia Cugliandolo for pointing out that a relaxation phenomenon similar to what is reported in this paper, namely, a fast relaxation of the kinetic energy and a much slower one of other observables, has also been observed in spin-glass systems.

Author Contributions: Shamik Gupta performed the analytical computations and numerical simulations and wrote the paper with inputs from Stefano Ruffo. Both of the authors contributed to the research work. All authors have read and approved the final manuscript.

Conflicts of Interest: The authors declare no conflict of interest.

Entropy **2017**, *19*, 544

References

1. Zwanzig, R. *Nonequilibrium Statistical Mechanics*; Oxford University Press: Oxford, UK, 2001.
2. Nosé, S. A unified formulation of the constant temperature molecular-dynamics methods. *J. Chem. Phys.* **1984**, *81*, 511–519.
3. Hoover, W.G. Canonical dynamics: Equilibrium phase-space distributions. *Phys. Rev. A* **1985**, *31*, 1695–1697.
4. Morriss, G.P.; Dettmann, C.P. Thermostats: Analysis and application. *Chaos* **1998**, *8*, 321–336.
5. Klages, R. *Microscopic Chaos, Fractals and Transport in Nonequilibrium Statistical Mechanics, monograph, Advanced Series in Nonlinear Dynamics Vol. 24*; World Scientific, Singapore, 2007.
6. Dettmann, C.P.; Morriss, G.P. Hamiltonian reformulation and pairing of Lyapunov exponents for Nosé-Hoover dynamics. *Phys. Rev. E* **1997**, *55*, 3693–3696.
7. Ramshaw, J.D. General formalism for singly thermostated Hamiltonian dynamics. *Phys. Rev. E* **2015**, *92*, 052138–052143.
8. Campa, A.; Dauxois, T.; Ruffo, S. Statistical mechanics and dynamics of solvable models with long-range interactions. *Phys. Rep.* **2009**, *480*, 57–159.
9. Bouchet, F.; Gupta, S.; Mukamel, D. Thermodynamics and dynamics of systems with long-range interactions. *Physica A* **2010**, *389*, 4389–4405.
10. Campa, A.; Dauxois, T.; Fanelli, D.; Ruffo, S. *Physics of Long-Range Interacting Systems*; Oxford University Press: Oxford, UK, 2014.
11. Levin, Y.; Pakter, R.; Rizzato, F.B.; Teles, T.N.; Benetti, F.P.C. Nonequilibrium statistical mechanics of systems with long-range interactions. *Phys. Rep.* **2014**, *535*, 1–60.
12. Gupta, S.; Ruffo, S. The world of long-range interactions: A bird's eye view. *Int. J. Mod. Phys. A* **2017**, *32*, 1741018.
13. Campa, A.; Giansanti, A.; Mukamel, D.; Ruffo, S. Dynamics and thermodynamics of rotators interacting with both long- and short-range couplings. *Physica A* **2006**, *365*, 120–127.
14. Dauxois, T.; de Buyl, P.; Lori, L.; Ruffo, S. Models with short- and long-range interactions: The phase diagram and the reentrant phase. *J. Stat. Mech. Theory Exp.* **2010**, *2010*, P06015.
15. Antoni, M.; Ruffo, S. Clustering and relaxation in Hamiltonian long-range dynamics. *Phys. Rev. E* **1995**, *52*, 2361–2373.
16. Escande, D.; Kantz, H.; Livi, R.; Ruffo, S. Self-consistent check of the validity of Gibbs calculus using dynamical variables. *J. Stat. Phys.* **1994**, *76*, 605–626.
17. Bouchet, F.; Barré, J. Classification of phase transitions and ensemble inequivalence, in systems with long range interactions. *J. Stat. Phys.* **2005**, *118*, 1073–1105.
18. Gupta, S. Spontaneous collective synchronization in the Kuramoto model with additional non-local interactions. *J. Phys. A: Math. Theor.* **2017**, *50*, 424001.
19. Larson, R. *Elementary Linear Algebra*, 8th ed.; Cengage Learning: Boston, MA, USA, 2017.
20. POWER_METHOD: The Power Method for Eigenvalues and Eigenvectors. Available online: http://people.sc.fsu.edu/~jburkardt/f_src/power_method/power_method.html (accessed on 14 October 2017).
21. Yamaguchi, Y.Y.; Barré, J.; Bouchet, F.; Dauxois, T.; Ruffo, S. Stability criteria of the Vlasov equation and quasi-stationary states of the HMF model. *Physica A* **2004**, *337*, 36–66.
22. Bouchet, F.; Dauxois, T. Prediction of anomalous diffusion and algebraic relaxations for long-range interacting systems, using classical statistical mechanics. *Phys. Rev. E* **2005**, *72*, 045103.
23. Khinchin, A.I. *Mathematical Foundations of Statistical Mechanics*; Dover: New York, NY, USA, 1949.
24. Mazur, P.; van der Linden, J. Asymptotic form of the structure function for real systems. *J. Math. Phys.* **1963**, *4*, 271–277.
25. Livi, R.; Pettini, M.; Ruffo, S.; Vulpiani, A. Chaotic behavior in nonlinear Hamiltonian systems and equilibrium statistical mechanics. *J. Stat. Phys.* **1987**, *48*, 539–559.

entropy

MDPI

Article

Lyapunov Spectra of Coulombic and Gravitational Periodic Systems

Pankaj Kumar and Bruce N. Miller *

Department of Physics and Astronomy, Texas Christian University, Fort Worth, TX 76129, USA;
pankaj.kumar@tcu.edu
* Correspondence: b.miller@tcu.edu; Tel.: +1-817-257-7123

Academic Editors: Andrea Puglisi, Alessandro Sarracino and Angelo Vulpiani
Received: 2 April 2017; Accepted: 15 May 2017; Published: 20 May 2017

Abstract: An open question in nonlinear dynamics is the relation between the Kolmogorov entropy and the largest Lyapunov exponent of a given orbit. Both have been shown to have diagnostic capability for phase transitions in thermodynamic systems. For systems with long-range interactions, the choice of boundary plays a critical role and appropriate boundary conditions must be invoked. In this work, we compute Lyapunov spectra for Coulombic and gravitational versions of the one-dimensional systems of parallel sheets with periodic boundary conditions. Exact expressions for time evolution of the tangent-space vectors are derived and are utilized toward computing Lypaunov characteristic exponents using an event-driven algorithm. The results indicate that the energy dependence of the largest Lyapunov exponent emulates that of Kolmogorov entropy for each system for a given system size. Our approach forms an effective and approximation-free instrument for studying the dynamical properties exhibited by the Coulombic and gravitational systems and finds applications in investigating indications of thermodynamic transitions in small as well as large versions of the spatially periodic systems. When a phase transition exists, we find that the largest Lyapunov exponent serves as a precursor of the transition that becomes more pronounced as the system size increases.

Keywords: Kolmogorov–Sinai entropy, Lyapunov exponents; periodic boundary conditions; chaotic dynamics; *N*-body simulation; stochastic thermodynamics

PACS: 52.27.Aj; 05.10.-a; 05.45.Pq; 05.45.Ac; 05.70.Fh

1. Introduction

Our understanding of the collective behavior of macroscopic systems is encompassed in the laws of thermodynamics. Phase transitions demonstrate that the state space of macroscopic systems is segmented into regions of strongly differing thermodynamic properties. While phase transitions are rigorously defined in the thermodynamic limit, it is remarkable that their manifestations can sometimes be observed in very small systems (see [1] and references therein). In addition to investigating the occurrence of equilibrium phase transitions, small systems have also been employed to study the breakdown of the second law [2], as well as various equilibrium and nonequilibrium phenomena [3–5]. Besides the usual changes in equilibrium properties—such as pressure, energy, and density—at phase transitions typically associated with macroscopic systems, the search for dynamical indicators of phase transitions has also born fruitful results. In particular, in ergodic systems, it has been demonstrated that the largest Lyapunov characteristic exponent (LCE) plays a pivotal role in announcing the existence of a phase transition. While the theory of phase transitions in systems with short-range interactions is well understood, systems with long-range interactions do not satisfy the usual convergence properties [6].

Consequently, surprising behavior, such as ensemble inequivalence, can occur. In contrast with short-range systems, the nature of boundary conditions for long-range systems becomes crucial since the interaction range can be of the same order as that of the system size. In studying possible indicators of phase transitions in small systems, it is useful to know the exact behavior of the system in the thermodynamic limit. One-dimensional systems have proven especially useful in this regard (see [7] for a review of relevant work).

One dimensional systems are of great interest to physicists in terms of their intrinsic properties and as a starting point in the analysis of their more-complicated higher-dimensional counterparts (see [8–12] and references therein). Not only can one-dimensional systems map the behaviors of experimental results [13], one-dimensional interactions have, in fact, been recently emulated in the laboratory [14]. In systems with long-range forces, establishing the exact form of the potential energy is difficult when periodic boundary conditions are involved. In two or three dimensions, Ewald sums are required and the potential cannot be deduced in a closed form. However, in one dimension, it is possible to express the inter-particle potential energy analytically [8]. In the analysis of large systems considered in plasma and gravitational physics, periodic boundary conditions are preferred [15–18] and have been utilized in the study of one-dimensional Coulombic and gravitational systems [12,19–21]. Such studies often rely on numerical simulations to validate the predictions made by theory. Moreover, in cases where theoretical relations have not been mathematically formulated, numerical simulations serve as a powerful approach in characterizing the dynamical behaviors and thermodynamic properties [12].

Simulation studies usually employ the molecular-dynamics (MD) approach for studying dynamical systems that undergo phase-space mixing to exhibit ergodic-like behavior. Phase-space mixing is a necessary condition for equilibrium statistical mechanics to apply and is often characterized by the existence of positive Lyapunov characteristic exponents (LCEs) [22]. LCEs represent the average rates of exponential divergence of nearby trajectories from a reference trajectory in different directions of the phase space and quantify the degree of chaos in a dynamical system [23–27]. In addition, LCEs have also been reported to serve as indicators of phase transitions [28–33].

Numerical calculation of LCEs may be realized by studying the geometry of the phase-space trajectories [29,30] for smooth systems. In general, however, if the time evolution of each particles' position and velocity can be followed for a given system, the largest LCE may be calculated by finding the rate of divergence between a reference trajectory and a nearby test trajectory obtained by pertubing the former [26]. This numerical approach was extended to the case of systems with periodic boundary conditions by Kumar and Miller and was applied to find the largest LCE for a spatially-periodic one-dimensional Coulombic system [12].

While the largest LCE is a good indicator of the degree of maximum dynamical instability in a system, the mixing speed, that is, the rate at which a given phase-space volume element diffuses across the allowed regions of the phase space is indicated by the Kolmogorov–Sinai (KS) entropy. For ergodic-like Hamiltonian systems, KS entropy is obtained as the sum of the positive LCEs [23]. In simulation, a full spectrum of LCEs may be obtained by finding the time-averaged exponential rates of growth of perturbation vectors applied to the phase-space flows in the tangent space [25,34–36]. An exact numerical method of calculating the full Lyapunov spectrum was proposed for the case of one-dimensional gravitation gas [25].

Even though a full spectrum of Lyapunov exponents is highly desirable, the calculation becomes computationally challenging for systems with a large number of degrees of freedom and it is usually impractical to aim for a full spectrum through N-body simulations [25]. An improved version of the approach was presented in [37] for the case of free-boundary conditions (also see [38]). In this paper, we extend the numerical approach presented in [25] to compute the complete Lyapunov spectra of spatially-periodic one-dimensional Coulombic and gravitational systems and show that the energy dependence of the largest LCE emulates that of the sum of all the positive LCEs for both versions of the system.

The paper is organized as follows: in Section 2, we describe the Coulombic and gravitational versions of the model and discuss the potential interactions as well as their implications on the phase-space characteristics. In Section 3, we recall some theoretical results for LCEs and the general approach for their numerical computation. Section 4 presents derivations for the time evolution of the tangent vectors and the results of the *N*-body simulations. Finally, in Section 5, we discuss the results and provide concluding remarks.

2. Model

We consider two versions of a spatially periodic lineal system on an *x*-axis with the primitive cell extending in $[-L, L]$ and which contains N infinite sheets, each with a surface mass density m. The positions of the sheets are given by $x_1, \ldots, x_N$ with respect to the center of the cell and their corresponding velocities by $v_1, \ldots, v_N$. In one version, the sheets are uncharged and are only interacting gravitationally. The other version is essentially a one-dimensional Coulombic system in which the sheets are charged with a surface charge density q and are immersed in a uniformly distributed negative background such that the net charge is zero. For the case of the charged version of the system, we neglect the gravitational effects and take into account only the Coulomb interactions. If we denote momenta as $p_i \equiv m v_i$, the Hamiltonian of the system may be expressed as

$$\mathcal{H} = \frac{1}{2m} \sum_{i=1}^{N} p_i^2 + \kappa \sum_{i<j}^{N} \left(\frac{(x_j - x_i)^2}{2L} - |x_j - x_i| \right), \tag{1}$$

where $\kappa = 2\pi k q^2$ for the case of the Coulombic system [12] (with each sheet, henceforth referred to as a *particle* or a *body*, having a surface charge density q in addition to the surface mass density m) and $\kappa = -2\pi G m^2$ for the gravitational system [21].

3. Lyapunov Characteristic Exponents

3.1. Theoretical Overview

We provide here a brief overview of dynamical system theory that will be helpful in developing the formulations in subsequent sections. While more general and comprehensive discussions are provided in [25,37,39], we restrict this overview to a smooth Hamiltonian system. We will see later how the concept may be extended to flows that take place on non-differentiable manifolds. Although the mathematical definitions used in the following discussions are rather general, we have adopted—and, for the sake of completion and easy comparison with relevant work, reproduced—their versions as presented in [25,37] unless noted otherwise.

Let the phase-space flow, $\phi^t(z)$, be a one-parameter group of measure-preserving diffeomorphisms $M \to M$, where M is an *n*-dimensional compact differentiable manifold and $z \in M$. For a Hamiltonian system with a phase-space dimensionality of $2N$, $n = 2N - 1$. If $T_z M$ is the tangent space to M at z, then we can define $\mathcal{D}\phi_z^t(w)$ as a linearized flow in the tangent space ($T_z M \to T_{\phi_z^t} M$), where w is a vector in the tangent space. For a non-zero w, there are n independent eigenvectors $e_1, \ldots, e_n$ with $\chi_1, \ldots, \chi_n$ as the corresponding eigenvalues such that $|\chi_1| \geq |\chi_2| \geq \cdots \geq |\chi_n|$. For a periodic orbit with period t_0, if we define $\lambda_i \equiv t_0^{-1} \ln |\chi_i|$, then

$$\frac{\left\| \mathcal{D}\phi_z^{kt_0}(e_i) \right\|}{\|e_i\|} = e^{\lambda_i k t_0}, \tag{2}$$

where $\| \ \|$ represents the Euclidean norm on $T_z M$ and k is a positive integer [25]. For a tangent-space vector w with a non-zero component along e_1, the divergence for large t will be dominated by $e^{\lambda_1 t}$, and, therefore,

$$\lim_{t \to \infty} \frac{1}{t} \ln \frac{\left\| \mathcal{D}\phi_z^t(w) \right\|}{\|w\|} = \lambda_1. \tag{3}$$

λ_1 is usually called the largest Lyapunov characteristic exponent (LCE) of the orbit represented by the flow $\phi^t(z)$ and is a measure of the overall stability of the orbit; if $\lambda_1 \geq 0$, then the nearby trajectories diverge exponentially. Note that even though we have used a periodic orbit to define the largest LCE, it may be shown that the limit on the left-hand side of Equation (3) exists and is finite for any given dynamical system and the result applies rather generally under very weak smoothness conditions [40].

LCE defined in Equation (3) may be thought of as the mean exponential growth rate of a one-dimensional "volume" (length of a vector w) in the tangent space. Therefore, λ_1 is often referred to as an LCE of order 1. Similarly, λ_p, that is, LCE of order p (where $1 \leq p \leq n$, $p \in \mathbb{Z}_+$), may be related to the mean exponential rate of growth of a p-dimensional hyperparallelepiped formed by the evolution of p linearly independent tangent-space vectors $w_1, \ldots, w_p$. We first find the rate of volume divergence as

$$\lambda^{\mathcal{P}} = \lim_{t \to \infty} \frac{1}{t} \ln \frac{\text{Vol}^{\mathcal{P}} \left[\mathcal{D}\phi_z^t(w_1), \ldots, \mathcal{D}\phi_z^t(w_p) \right]}{\text{Vol}^{\mathcal{P}} \left[w_1, \ldots, w_p \right]}, \tag{4}$$

where $\text{Vol}^{\mathcal{P}}$ represents the volume spanned by a set of p tangent-space vectors. Finally, following [40], λ_p is found as

$$\lambda_p = \begin{cases} \lambda^{\mathcal{P}}, & p = 1, \\ \lambda^{\mathcal{P}} - \lambda^{\mathcal{P}-1}, & 1 < p \leq n. \end{cases} \tag{5}$$

3.2. Numerical Approach

We start with a randomly chosen set of n orthonormal tangent vectors $\{\hat{w}_1^0, \ldots, \hat{w}_n^0\}$. Clearly, for each $p \leq n$, $\text{Vol}^{\mathcal{P}} \left[\hat{w}_1^0, \ldots, \hat{w}_p^0 \right] = 1$. After a fixed time interval τ, the evolved tangent vectors—which we denote by $\{w_1^1, \ldots, w_n^1\}$, where $w_i^1 = \mathcal{D}\phi_z^{t=\tau}(\hat{w}_i^0)$—are, in general, no longer mutually orthogonal. This is because the component of each $\hat{w}_i^0$ along the direction of maximum divergence e_1 (that is, $\hat{w}_i^0 \cdot e_1$) will witness a disproportionately larger growth in its value as compared to the remaining components. In order to avoid numerical errors arising from one component getting increasingly large in comparison to the others, a new orthonormal set of tangent vectors $\{\hat{w}_1^1, \ldots, \hat{w}_n^1\}$ is defined after time τ through Gram–Schmidt reduction on the set of evolved $\{w_1^0, \ldots, w_n^0\}$. This new set of ornothormal tangent vectors are then used for the following iteration, and the process is recursively repeated until $\lambda^{\mathcal{P}}$ has converged [25,39].

Numerical calculation of $\lambda^{\mathcal{P}}$ involves finding the corresponding p-volume for each iteration. If at the end of the j-th iteration, $w_i^j = \mathcal{D}\phi_z^{t=j\tau}(\hat{w}_i^{j-1})$ represents the evolved versions of the orthornormal tangent vectors $\hat{w}_i^{j-1}$, the p-volume may be found as the norm of the the exterior product involving the corresponding p vectors, that is,

$$\text{Vol}^{\mathcal{P}}[w_1^j, \ldots, w_p^j] = \left\| w_1^j \wedge w_2^j \wedge \cdots \wedge w_p^j \right\|. \tag{6}$$

Finally, the average exponential growth rate of the p-volume is found as

$$\lambda^{\mathcal{P}} = \lim_{l \to \infty} \frac{1}{l\tau} \sum_{j=1}^{l} \left(\ln \text{Vol}^{\mathcal{P}}[w_1^j, \ldots, w_p^j] \right), \tag{7}$$

where l is the total number of iterations. A complete set of LCEs $\{\lambda_1, \ldots, \lambda_n\}$, also known as the Lyapunov spectrum, may then be obtained for the trajectory by utilizing Equation (5) for all permissible values of p. Finally, an upper limit on the KS entropy h_{KS} for Hamiltonian systems may be obtained as the sum of positive LCEs [23,25]:

$$h_{KS} \le \lambda_S = \sum_{p=1}^{p_{max}^+} \lambda_p, \tag{8}$$

where p_{max}^+ is the largest value of p for which λ_p is positive and where the equality $h_{KS} = \lambda_S$ holds for ergodic-like systems. The sum of the positive LCEs λ_S is often termed as the density of KS entropy.

In the following section, we discuss how we employ the theory and the numerical approach presented thus far to obtain Lyapunov spectra for the Coulombic and gravitational systems discussed in Section 2. While most of the presented theory applies to the two systems in its original form, non-smoothness arising from the absolute valued linear terms in the potential demand additional consideration. As we shall see, following the time evolution of the tangent-space vectors involves treating the motion as a flow in between two consecutive events of interparticle crossings and as a mapping at each event of such crossings. Consequently, it is indispensable to have the ability to find the exact time corresponding to each crossing.

4. *N*-Body Simulation

4.1. Equations of Motion

Positions and velocities of the particles are obtained using event-driven algorithms based on the approaches proposed in [12] for the Coulombic system and in [21] for the gravitational system. The algorithms employ analytic expressions for the time dependencies of the relative separations $Z_j(t)$ and relative velocities $W_j(t)$ between two consecutive particles in the primitive cell, where $Z_j = (x_{j+1} - x_j)$ and $W_j = (v_{j+1} - v_j)$, with x_j and v_j representing, respectively, the position and velocity of the j-th particle, whereas x_{j+1} and v_{j+1} representing those of the $(j+1)$-th particle. Combining the results of [12,21], we find that

$$\frac{d}{dt}W_j(t) = -\frac{\kappa}{m}\left\{\frac{N}{L}Z_j(t) - 2\right\}. \tag{9}$$

Crossing times may be found by solving $Z_j(t) = 0$ for t. The corresponding positions x_j and velocities v_j are obtained using a matrix-inversion subroutine as described in [12].

4.2. Time Evolution of Tangent-Space Vectors

In order to follow the time evolution of the tangent vectors, we adopt an approach based on the "exact" numerical method proposed in [25]. The method invokes that, for a one-dimensional Hamiltonian system with N particles, one does not have to restrict to the $(2N - 1)$-dimensional manifold Γ_E. One may alternatively choose to represent the flow ϕ^t in the entire $2N$-dimensional phase space (say, Ω) whereby the tangent space $T_z\Gamma_E$ becomes a subspace of $T_z\Omega$.

Let $z(x,v)$ be a point in the phase space Ω, where $x = (x_1, \ldots, x_N)$ and $v = (v_1, \ldots, v_N)$. The equations of motion representing the system are given by

$$\dot{x}_j = v_j, \tag{10}$$

and

$$\dot{v}_j = -\frac{1}{m}\frac{\partial}{\partial x_j}V(x), \tag{11}$$

with

$$V(x) = \kappa \sum_{i<j}^{N}\left(\frac{(x_j - x_i)^2}{2L} - |x_j - x_i|\right). \tag{12}$$

Similarly, if we have a vector $w(\xi, \eta)$ in the tangent space $T_z\Omega$, then the variational equations [25] governing the evolution of w are given by

$$
\begin{pmatrix} \dot{\xi} \\ \dot{\eta} \end{pmatrix} = \begin{pmatrix} 0 & I_N \\ A(x) & 0 \end{pmatrix} \begin{pmatrix} \xi \\ \eta \end{pmatrix},
\tag{13}
$$

where I_N is the $N \times N$ identity matrix and $A(x)$ is an $N \times N$ matrix whose elements are given by

$$
A_{ij}(x) = -\frac{1}{m} \frac{\partial^2}{\partial x_i \partial x_j} V(x).
\tag{14}
$$

For the potential expressed in Equation (12), one finds that

$$
A_{ij}(x) = \begin{cases} -\frac{\kappa}{m} \left[-\frac{1}{L} - 2\delta(x_i - x_j) \right], & i \neq j, \\[2ex] -\frac{\kappa}{m} \left[\frac{N-1}{L} - \sum_{i \neq k=1}^{N} 2\delta(x_i - x_k) \right], & i = j. \end{cases}
\tag{15}
$$

Using Equations (13) and (15), we may deduce that

$$
\dot{\xi}_j = \eta_j,
\tag{16}
$$

and

$$
\dot{\eta}_j = -\frac{\kappa}{m} \left[\frac{N\xi_j - \Xi_S}{L} - 2 \sum_{i \neq j=1}^{N} (\xi_j - \xi_i)\delta(x_j - x_i) \right],
\tag{17}
$$

where

$$
\Xi_S = \sum_{j=1}^{N} \xi_j.
\tag{18}
$$

Equations (16) and (17) imply that

$$
\frac{d^2\xi_j}{dt^2} = -\frac{\kappa}{m} \left[\frac{N\xi_j - \Xi_S}{L} - 2 \sum_{i \neq j=1}^{N} (\xi_j - \xi_i)\delta(x_j - x_i) \right].
\tag{19}
$$

Time evolution of Ξ_S between the events of interparticle crossings may be deduced by adding Equation (19) for all values of j as

$$
\sum_{j=1}^{N} \frac{d^2\xi_j}{dt^2} = 0,
\tag{20}
$$

which implies that

$$
\frac{d^2\Xi_S}{dt^2} = 0.
\tag{21}
$$

Solution to Equation (21) yields

$$
\Xi_S(t) = H_S(0)t + \Xi_S(0),
\tag{22}
$$

where

$$
H_S = \sum_{j=1}^{N} \eta_j.
\tag{23}
$$

Hence, in between the events of crossings, the evolution of tangent vectors takes the form

$$\frac{d^2\xi_j}{dt^2} = -\frac{\kappa}{m}\left[\frac{N\xi_j - H_S(0)t + \Xi_S(0)}{L}\right]. \tag{24}$$

With the two values of κ, one for the Coulombic system and the other for the gravitational system, one may solve Equation (24) to find the exact dependencies of ξ_j and η_j on time.

4.2.1. Coulombic System

Utilizing the initial conditions for the Coulombic system with $\kappa = 2\pi k q^2$, we obtain the solutions to Equation (24) for ξ_j and η_j in between crossings as

$$\xi_j(t) = \frac{H_S(0)}{N}t + \frac{\Xi_S(0)}{N} + \frac{1}{\omega}\left\{\eta_j(0) - \frac{H_S(0)}{N}\right\}\sin\omega t + \left\{\xi_j(0) - \frac{\Xi_S(0)}{N}\right\}\cos\omega t, \tag{25}$$

and,

$$\eta_j(t) = \frac{H_S(0)}{N} + \left\{\eta_j(0) - \frac{H_S(0)}{N}\right\}\cos\omega t - \omega\left\{\xi_j(0) - \frac{\Xi_S(0)}{N}\right\}\sin\omega t, \tag{26}$$

where $\omega \equiv \sqrt{\frac{\kappa N}{mL}} = \sqrt{\frac{2\pi k q^2 N}{mL}}$.

If the r-th and s-th particles undergo a crossing at time $t = t_c$, and t^- and t^+, respectively, denote the instants just before and after $t = t_c$, then

$$\eta_r(t^+) = \eta_r(t^-) + \frac{4\pi k q^2}{m}\frac{(\xi_r(t^-) - \xi_s(t^-))}{|v_r(t^-) - v_s(t^-)|}, \tag{27}$$

$$\eta_s(t^+) = \eta_s(t^-) - \frac{4\pi k q^2}{m}\frac{(\xi_r(t^-) - \xi_s(t^-))}{|v_r(t^-) - v_s(t^-)|}. \tag{28}$$

4.2.2. Gravitational System

For the gravitational system, $\kappa = -2\pi G m^2$. Similar to the Coulombic case, we utilize the initial conditions to find solutions to Equation (24) for ξ_i and η_j as functions of time for the gravitational case in between the events of crossings:

$$\xi_j(t) = \frac{1}{2\Lambda}\{A_g e^{\Lambda t} + B_g e^{-\Lambda t}\} + \frac{H_S(0)}{N}t + \frac{\Xi_S(0)}{N}, \tag{29}$$

$$\eta_j(t) = \frac{1}{2}\{A_g e^{\Lambda t} - B_g e^{-\Lambda t}\} + \frac{H_S(0)}{N}, \tag{30}$$

where,

$$A_g = \Lambda\xi_j(0) + \eta_j(0) - \frac{H_S(0)}{N} - \Lambda\frac{\Xi_S(0)}{N}, \tag{31}$$

$$B_g = \Lambda\xi_j(0) - \eta_j(0) + \frac{H_S(0)}{N} - \Lambda\frac{\Xi_S(0)}{N}, \tag{32}$$

and $\Lambda \equiv \sqrt{\frac{kN}{mL}} = \sqrt{\frac{2\pi G m N}{L}}$. For the r-th and s-th particle involved in a crossing, we get

$$\eta_r(t^+) = \eta_r(t^-) - 4\pi G m\frac{(\xi_r(t^-) - \xi_s(t^-))}{|v_r(t^-) - v_s(t^-)|}, \tag{33}$$

$$\eta_s(t^+) = \eta_s(t^-) + 4\pi G m\frac{(\xi_r(t^-) - \xi_s(t^-))}{|v_r(t^-) - v_s(t^-)|}, \tag{34}$$

where we have used the same definitions of t_c, t^-, and t^+ as we did in the expressions' Coulombic counterparts.

4.3. p-Volume and Lyapunov Spectrum

We perform numerical computations using algorithms that are driven by tracking the events of interparticle crossings. Since the time derivatives of the velocities of the particles involved in a crossing undergo abrupt changes, tracking of crossing becomes indispensable. In other words, even if we were to sample the positions and velocities at fixed intervals of time, we would still need to track every crossing [25]. Therefore, instead of choosing fixed time intervals for Gram–Schmidt orthonormalization of the tangent-space vectors, we choose to do it after each crossing. It should be noted that the duration of each iteration, fixed or variable, is irrelevant as long it remains short enough for the various components of the tangent vectors to remain comparable with a given precision offered by the computing platform utilized.

After an iteration, say, the $(j-1)$-th iteration, ending at time $t = t_c^{j-1}$, each of the $2N$ orthonormal tangent vectors $\hat{w}_1^{j-1}, \ldots, \hat{w}_N^{j-1}$ is allowed to evolve for the duration δt_c^j of the j-th iteration. δt_c^j may be thought of as the time elapsed between the instants right after the $(j-1)$-th crossing and right before the j-th crossing. Then, the evolved vectors may be given by $w_i^j = \mathcal{D}\phi_z^{t_c^j}(\hat{w}_i^{j-1})$, where

$$t_c^j = t_c^{j-1} + \delta t_c^j = \delta t_c^1 + \delta t_c^2 + \cdots + \delta t_c^j. \tag{35}$$

For the j-th iteration, p-volume is found as follows: we define a $p \times p$ symmetric matrix $\mathcal{G}_p^j$ whose elements $\mathcal{G}_{p\,\mu\nu}^j$ are inner products between w_μ^j and w_ν^j, where $1 \leq \mu, \nu \leq p$ (with $\mu, \nu \in \mathbb{Z}_+$). That is,

$$\mathcal{G}_p^j = \begin{pmatrix} \langle w_1^j, w_1^j \rangle & \langle w_1^j, w_2^j \rangle & \cdots & \langle w_1^j, w_p^j \rangle \\ \langle w_2^j, w_1^j \rangle & \langle w_2^j, w_2^j \rangle & \cdots & \langle w_2^j, w_p^j \rangle \\ \vdots & \vdots & \ddots & \vdots \\ \langle w_p^j, w_1^j \rangle & \langle w_p^j, w_2^j \rangle & \cdots & \langle w_p^j, w_p^j \rangle \end{pmatrix}. \tag{36}$$

The matrix $\mathcal{G}_p^j$, also referred to as *Gram matrix*, encapsulates the necessary geometric information about the subspace spanned by the set of vectors $\{w_1^j, \ldots, w_p^j\}$ such as the lengths of the vectors and the angles between them. The absolute value of the determinant of $\mathcal{G}_p^j$, known as the *Gramian*, is essentially the square of the norm of the exterior product [41]. Using Equation (6), the p-volume is found as

$$\text{Vol}^{\mathcal{P}}[w_1^j, \ldots, w_p^j] = \sqrt{\left|\det\left(\mathcal{G}_p^j\right)\right|}. \tag{37}$$

With the ability to find each p-volume for a given iteration, the final value of the corresponding $\lambda^{\mathcal{P}}$ are found as

$$\lambda^{\mathcal{P}} = \lim_{l \to \infty} \frac{1}{t_c^l} \sum_{j=1}^l \ln\left(\text{Vol}^{\mathcal{P}}[w_1^j, \ldots, w_p^j]\right), \tag{38}$$

where t_c^l, as defined in Equation (35), is the total time elapsed for l crossings to occur. Finally, LCEs λ_p and Kolmogorov-entropy density λ_S are obtained using Equations (5) and (8), respectively.

4.4. Results

In our simulation, the initial conditions are chosen as follows: for a given number of particles N and per-particle energy $\mathcal{H}$, if $\mathcal{H}$ is lower than the maximum allowed value of the potential energy V_{max}, then the particle positions are chosen randomly such that the potential energy is slightly smaller than the target value of $\mathcal{H}$. For $\mathcal{H}$ greater than V_{max}, the particle positions are randomly selected such that the potential energy is close to V_{max}. Velocities are chosen randomly from a Gaussian distribution and are scaled such that the sum of the potential and kinetic energies exactly equals the target value of $\mathcal{H}$ [12].

Simulations are performed by rescaling the system parameters using dimensionless units such that the number density $N/2L = 1$ and the characteristic frequencies ω, Λ equal unity [12,21]. Energies $\mathcal{H}_c$ and $\mathcal{H}_g$ (respectively for Coulombic and gravitational systems) are measured with respect to the minimum values of the corresponding potential energies allowed for each N.

Before running the Lyapunov algorithm, we allowed the system to evolve for a relaxation period of 500 time units (in terms of $1/\omega$ or $1/\Lambda$). In the calculation of LCEs, the total number of iterations l for each $\lambda^{\mathcal{P}}$ in Equation (38) is decided by an adaptive algorithm that runs a minimum preliminary number of iterations assigned beforehand and then continues running until the value of $\lambda^{\mathcal{P}}$ has converged to within a pre-specified tolerance for the standard deviation. The tolerance value that we specified was 0.1 percent of the mean from the newest $500,000$ iterations, with a minimum of 1 million iterations. In each run, we found that the values converged to within our specified tolerance after the preliminary run of 1 million iterations.

We computed Lyapunov spectra for the Coulombic and gravitational systems with a varying number of particles N with $5 \leq N \leq 20$. Figure 1 shows examples of the dependence of the various LCEs on the per-particle energy for the Coulombic and gravitational systems with $N = 11$. The figure also shows the sum of all LCEs for each case. As expected, the sums were found to be close to zero. It should be noted that versions of the systems with $N < 5$ may exhibit largely segmented phase-space distributions with a coexistence of chaotic and stable regions for a given energy value ($N = 2$ being purely trivial with completely integrable phase space) [11,12,25,42]. Hence, there is a considerable probability that a randomly chosen initial condition ends up in one of the stable regions whose trajectories exhibit no stochasticity and which results in values of LCEs close to zero in simulation.

The energy dependencies of the largest LCE and the Kolmogorov-entropy density with $N = 5, 8, 11, 15,$ and 20 have been shown for the Coulombic and gravitational systems in Figures 2 and 3, respectively. It can be seen in each case that the behavior of λ_S versus $\mathcal{H}/N$ resembles, in general, a scaled version of λ_1 versus $\mathcal{H}/N$. To elucidate this, we have presented comparative plots of the normalized versions, $\hat{\lambda}_1$ and $\hat{\lambda}_S$, of λ_1 and λ_S against per-particle energy for $N = 11$ in Figure 4, where we have divided λ_1 and λ_S by their respective maximum values to get the normalized values.

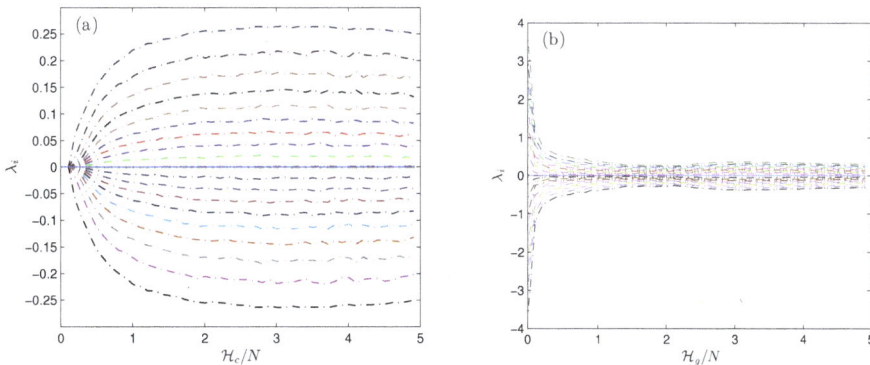

Figure 1. Full spectra of Lyapunov characteristic exponents (LCEs) plotted against per-particle energy for (**a**) Coulombic system, and (**b**) gravitational system, with $N = 11$. The topmost curve shows λ_1, the second to top curve shows λ_2, and so on all the way to the curve on the very bottom representing λ_{22} in (**a**,**b**). The central solid (blue) line in each plot shows the sum of LCEs. $\mathcal{H}_c$ and $\mathcal{H}_g$ are expressed in units of $\frac{2L}{N}|\kappa|$. λ_i are expressed in units of (**a**) ω, and (**b**) Λ.

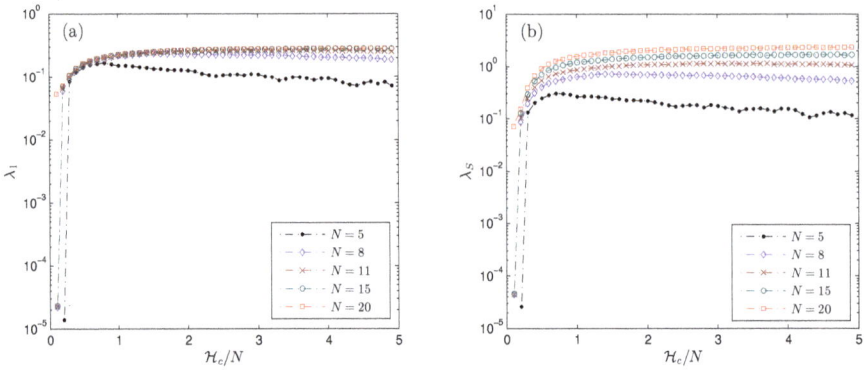

Figure 2. Energy dependence of (**a**) the largest LCE, and (**b**) Kolmogorov-entropy density for Coulombic system with different degrees of freedom. λ_1 and λ_S are expressed in units of ω, whereas $\mathcal{H}_c$ is expressed in units of $\frac{2L}{N}|\kappa|$.

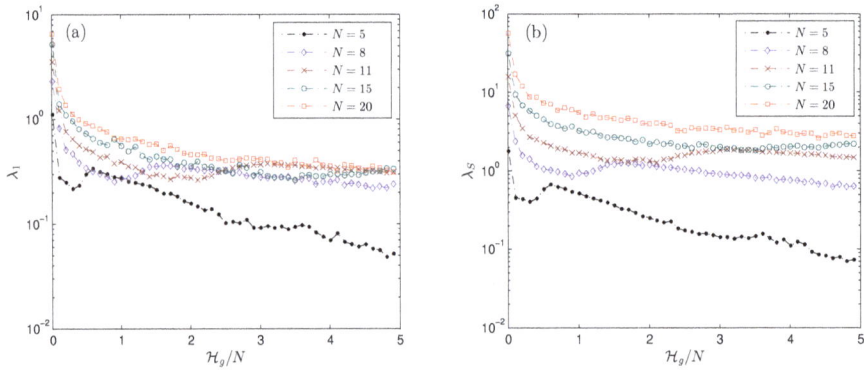

Figure 3. Energy dependence of (**a**) the largest LCE, and (**b**) Kolmogorov-entropy density for gravitational system with different degrees of freedom. λ_1 and λ_S are expressed in units of Λ, whereas $\mathcal{H}_g$ is expressed in units of $\frac{2L}{N}|\kappa|$.

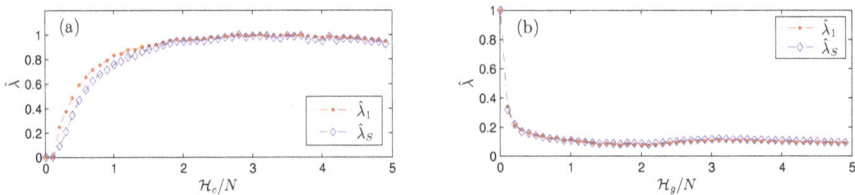

Figure 4. Energy dependence of the normalized values of the largest LCE and Kolmogorov-entropy density for (**a**) Coulombic system, and (**b**) gravitational system for $N = 11$. $\mathcal{H}_c$ and $\mathcal{H}_g$ are expressed in units of $\frac{2L}{N}|\kappa|$, whereas $\hat{\lambda}$ is dimensionless.

5. Discussion

Our study provides insight into the chaotic dynamics of the two versions of the periodic system under conditions of varying energy and degrees of freedom. The results for the Coulombic system are consistent with those provided in [12]. λ_1 for the Coulombic system stays zero as long as the energy is low enough for the particles to not undergo crossings. As the energy is increased from a low value, λ_1 sees an initial increase, reaches a maximum and then decreases as the energy is progressively raised. However, with an increase in the number of particles, the right edge of the "hill" near the maximum opens up to form a "plateau" and flattens out asymptotically as shown in Figure 2a. Interestingly, our results also suggest that all the other LCEs are, more or less, scaled (and inverted, for the case of negative LCEs) versions of the largest LCE, as we can see in Figure 1a for $N = 11$.

Unlike the Coulombic version, the gravitational system shows a maximum degree of chaos at low energies. Our results suggest that the largest LCE starts off with a high value at low energies and decreases to a minimum as the energy is increased for a given N. With further rise in energy, λ_1 increases, reaches a maximum and then decreases asymptotically. As the number of particles is increased, the local trough near the local minimum and the hill near the local maximum start opening up to the right with the asymptotic edge rising up to a form a plateau as shown in Figure 3b. Moreover, it can also be seen in Figure 3b that the local minimum and maximum themselves shift to the right as well with an increase in the number of particles. It should be noted that, in the thermodynamic limit, that is, for versions of the system with sufficiently large N, the LCEs may exhibit discontinuities in their values or their slopes near the troughs and crests when plotted against temperature. Such an observation may indicate toward the existence of phase transitions [31,32].

While the energy dependencies of λ_1 differs dramatically for each version of the system, the other LCEs for the gravitational system are also, in general, scaled versions of λ_1, similar to the Coulombic system, as exemplified in Figure 1b for $N = 11$. Moreover, in both versions of the spatially-periodic system, the energy dependence of LCEs tends to approach a common limiting behavior as N is increased. This is in contrast to the free-boundary gravitating case in which the values of LCEs were shown to increase linearly with an increase in the number of particles [25].

From a dynamical perspective, the results are consistent with the theoretical predictions for Hamiltonian systems [25] as outlined below:

(a) The sum of LCEs was found to converge to zero for all energies and N.

(b) For the ordered set $\{\lambda_i\}$ ($\lambda_1 \geq \lambda_2 \geq \ldots \geq \lambda_{2N}$), the results show that

$$\lambda_i \sim -\lambda_{2N-i+1}, \quad i = 1, 2, \ldots, 2N. \tag{39}$$

(c) In addition, our results show that

$$\lambda_{N-1} \sim \lambda_N \sim \lambda_{N+1} \sim \lambda_{N+2} \sim 0. \tag{40}$$

Result (c) is nothing but a consequence of the conservation of momentum [25].

To summarize, we have provided an exact method to compute the full spectra of LCEs for the spatially periodic versions of the one-dimensional Coulombic and gravitational systems. Analytic expressions for the time evolution of the tangent vectors were derived and used in numerically computing LCEs using an efficient event-driven algorithm. While the resulting values of the largest LCE for the Coulombic system agree with those reported previously [12], our exact approach offers striking advantages over the method used in [12] in that it allows one to calculate a full spectrum of LCEs rather than just the largest LCE. Second, the results of the exact method do not depend on the size of the perturbation. In finding the largest LCE using finite perturbations to a reference orbit as discussed in [12], one has to first make sure that the value chosen for initial perturbation is small enough. This becomes challenging for systems in which particles tend to stay clumped together. An example of such behavior is seen in the gravitational system at low energies. Finally, the exact approach circumvents the need for defining a test trajectory altogether, which, as discussed in [12], poses difficulties in expressing phase-space separations for systems with periodic boundary conditions. Nevertheless, the method discussed in [12] still remains powerful, and perhaps the only resort, in dealing with spatially-periodic systems for which analytic evolution of tangent-space vectors may not be obtained.

In our exposition, we started with the general mathematical definitions for the computation of Lyapunov exponents arising in nonlinear dynamics that have been known for quite some time. We subsequently developed those general analytic expressions to account for the particular changes brought upon by the introduction of periodic boundary conditions. As it became evident early in the treatment, the resulting mathematical relations, although somewhat reminiscent of those arrived at in [25] in their differential form, are considerably more complex in their final form.

The results of our study further indicate that the energy dependence of the largest LCE captures the general behavior of the dependence of the Kolmogorov-entropy density on energy for both Coulombic and gravitational systems. This result is particularly significant because of the numerical difficulties encountered while calculating the full Lyapunov spectra of large systems. For the two versions of the spatially-periodic system, our study suggests that one may gain insights into the full spectrum of LCEs by simply looking at the largest LCE, thereby allowing one to avoid the computational complications faced when calculating a full spectrum. Note that this is in contrast with the recent study of Lyapunov spectra in the HMF model [43]. A possible explanation lies in the fact that, unlike the HMF model, the systems considered in the present work are truly one-dimensional and rigorously obey Poisson's equation [21].

It should be noted that, for a given number of particles, the energy dependence of the LCEs roughly followed the same behavior for any randomly selected initial conditions with only slight deviations. As the number of particles was increased, the deviations became smaller, leading the behaviors to converge to a single universal one, suggesting that the systems approach ergodicity. Moreover, the convergence times of the LCNs for different randomly-chosen initial conditions also showed uniformity for a larger number of particles, thereby pointing toward a consistent relaxation to equilibrium with increasing degrees of freedom. However, the exact dependence of relaxation time on the number of particles requires further investigation and we plan to pursue it in our future work.

Finally, it is worth emphasizing that if a phase transition occurs in either of the two spatially-periodic systems, the *temperature dependence* of the largest LCE is expected to show a transitioning behavior in the thermodynamic limit (large-N limit) [31,32,44]. Although of only technical interest for short-range interactions, boundary conditions have a strong influence on the behavior of matter for long-range interactions. In a related work [7], we showed, using both mean field theory as well as the simulation approach presented in this paper, the occurrence of a phase transition in a spatially-periodic, one-dimensional gravitational system. It should be noted that no phase transition occurs in the free-boundary version of the same system. For the periodic case studied here with finite N, it turns out that the local peak observed in the largest LCE is the precursor of the phase transition. When plotted versus temperature, this precursor transforms into a cusp characterized by a discontinuity of slope for the gravitational system in the thermodynamics limit [7]. Similar behavior was reported for a self-gravitating ring system [33]. However, as conjectured by Kunz [19], no such transitioning behavior is indicated for the Coulombic system. As we mentioned earlier, the ability to emulate a many body, one-dimensional, gravitational system in the laboratory is a recent development [14]. As experimental techniques progress, it may become possible to control the type of boundary conditions employed in the laboratory. This offers the promise of experimentally observing the phase transition that we predicted in our recent work [7].

Acknowledgments: The authors thank Harald Posch of the University of Vienna and Igor Prokhorenkov of Texas Christian University for valuable insights and helpful discussions.

Author Contributions: Pankaj Kumar and Bruce Miller made contributions to the conception of the research project and formulation of its scope; Pankaj Kumar carried out mathematical derivations, algorithm development, numerical simulations, data acquisition, and data analysis; Pankaj Kumar drafted the paper; Bruce Miller revised the paper critically for relevant background content and added important scientific remarks concerning the behaviors exhibited by the results.

Conflicts of Interest: The authors declare no conflict of interest.

Abbreviations

The following abbreviations are used in this manuscript:

LCE   Lyapunov characterstic exponent
KS   Kolmogorov–Sinai
MD   Molecular dynamics
HMF   Hamiltonian Mean Field

References

1. Klinko, P.; Miller, B.N. Dynamical study of a first order gravitational phase transition. *Phys. Lett. A* **2004**, *333*, 187–192.
2. Wang, G.; Sevick, E.M.; Mittag, E.; Searles, D.J.; Evans, D.J. Experimental demonstration of violations of the second law of thermodynamics for small systems and short time scales. *Phys. Rev. Lett.* **2002**, *89*, 050601.
3. Bustamante, C.; Liphardt, J.; Ritort, F. The nonequilibrium thermodynamics of small systems. *Phys. Today* **2005**, *58*, 43–48.
4. Ritort, F. The nonequilibrium thermodynamics of small systems. *C. R. Phys.* **2007**, *8*, 528–539.
5. Schnell, S.K.; Vlugt, T.J.; Simon, J.M.; Bedeaux, D.; Kjelstrup, S. Thermodynamics of a small system in a μT reservoir. *Chem. Phys. Lett.* **2011**, *504*, 199–201.
6. Frisch, H.L.; Lebowitz, J.L. *The Equilibrium Theory of Classical Fluids: A Lecture Note and Reprint Volume*; WA Benjamin: Los Angeles, CA, USA, 1964.
7. Kumar, P.; Miller, B.N.; Pirjol, D. Thermodynamics of a one-dimensional self-gravitating gas with periodic boundary conditions. *Phys. Rev. E* **2017**, *95*, 022116.
8. Rybicki, G.B. Exact statistical mechanics of a one-dimensional self-gravitating system. *Astrophys. Space Sci.* **1971**, *14*, 56–72.
9. Yawn, K.R.; Miller, B.N. Equipartition and Mass Segregation in a One-Dimensional Self-Gravitating System. *Phys. Rev. Lett.* **1997**, *79*, 3561–3564.
10. Miller, B.N.; Youngkins, P. Phase Transition in a Model Gravitating System. *Phys. Rev. Lett.* **1998**, *81*, 4794–4797.
11. Lauritzen, A.; Gustainis, P.; Mann, R.B. The 4-body problem in a (1+1)-dimensional self-gravitating system. *J. Math. Phys.* **2013**, *54*, 072703.
12. Kumar, P.; Miller, B.N. Chaotic dynamics of one-dimensional systems with periodic boundary conditions. *Phys. Rev. E* **2014**, *90*, 062918.
13. Milner, V.; Hanssen, J.L.; Campbell, W.C.; Raizen, M.G. Optical Billiards for Atoms. *Phys. Rev. Lett.* **2001**, *86*, 1514–1517.
14. Chalony, M.; Barré, J.; Marcos, B.; Olivetti, A.; Wilkowski, D. Long-range one-dimensional gravitational-like interaction in a neutral atomic cold gas. *Phys. Rev. A* **2013**, *87*, 013401.
15. Springiel, V.; Frenk, C.S.; White, S.D.M. The large-scale structure of the universe. *Nature* **2006**, *440*, 1137–1144.
16. Bertschinger, E. Simulations of structure formation in the universe. *Annu. Rev. Astron. Astrophys.* **1998**, *36*, 599–654.
17. Hockney, R.W.; Eastwood, J.W. *Computer Simulation Using Particles*; CPC Press: Boca Raton, FL, USA, 1988.
18. Hernquist, L.; Bouchet, F.R.; Suto, Y. Application of the Ewald method to cosmological N-body simulations. *Astrophys. J. Suppl.* **1991**, *75*, 231–240.
19. Kunz, H. The one-dimensional classical electron gas. *Ann. Phys.* **1974**, *85*, 303–335.
20. Schotte, K.D.; Truong, T.T. Phase transition of a one-dimensional Coulomb system. *Phys. Rev. A* **1980**, *22*, 2183–2188.
21. Miller, B.N.; Rouet, J.L. Ewald sums for one dimension. *Phys. Rev. E* **2010**, *82*, 066203.
22. Krylov, N.; Migdal, A.; Sinai, Y.G.; Zeeman, Y.L. *Works on the Foundations of Statistical Physics by Nikolai Sergeevich Krylov*; Princeton Series in Physics; Princeton University Press: Princeton, NJ, USA, 1979.
23. Pesin, J.B. Characteristic Liapunov indices and ergodic properties of smooth dynamical systems with invariant measure. *Sov. Math. Dokl.* **1976**, *17*, 196.

24. Pesin, Y.B. Characteristic Lyapunov exponents and smooth ergodic theory. *Russ. Math. Surv.* **1977**, *32*, 55–114.

25. Benettin, G.; Froeschle, C.; Scheidecker, J.P. Kolmogorov entropy of a dynamical system with an increasing number of degrees of freedom. *Phys. Rev. A* **1979**, *19*, 2454–2460.

26. Ott, E. *Chaos in Dynamical Systems*; Cambridge University Press: Cambridge, UK, 2002; pp. 137–145.

27. Sprott, J. *Chaos and Time-Series Analysis*; Oxford University Press: Oxford, UK, 2003; pp. 116–117.

28. Butera, P.; Caravati, G. Phase transitions and Lyapunov characteristic exponents. *Phys. Rev. A* **1987**, *36*, 962–964.

29. Caiani, L.; Casetti, L.; Clementi, C.; Pettini, M. Geometry of Dynamics, Lyapunov Exponents, and Phase Transitions. *Phys. Rev. Lett.* **1997**, *79*, 4361–4364.

30. Casetti, L.; Pettini, M.; Cohen, E.G.D. Geometric approach to Hamiltonian dynamics and statistical mechanics. *Phys. Rep.* **2000**, *337*, 237–341.

31. Dellago, C.; Posch, H.A. Lyapunov instability, local curvature, and the fluid-solid phase transition in two-dimensional particle systems. *Phys. A Stat. Mech. Appl.* **1996**, *230*, 364–387.

32. Barre, J.; Dauxois, T. Lyapunov exponents as a dynamical indicator of a phase transition. *Europhys. Lett.* **2001**, *55*, 2.

33. Monechi, B.; Casetti, L. Geometry of the energy landscape of the self-gravitating ring. *Phys. Rev. E* **2012**, *86*, 041136.

34. Dellago, C.; Posch, H. Kolmogorov-Sinai entropy and Lyapunov spectra of a hard-sphere gas. *Phys. A Stat. Mech. Appl.* **1997**, *240*, 68–83.

35. Milanović, L.; Posch, H.A.; Thirring, W. Statistical mechanics and computer simulation of systems with attractive positive power-law potentials. *Phys. Rev. E* **1998**, *57*, 2763–2775.

36. Tsuchiya, T.; Gouda, N. Relaxation and Lyapunov time scales in a one-dimensional gravitating sheet system. *Phys. Rev. E* **2000**, *61*, 948–951.

37. Benettin, G.; Galgani, L.; Giorgilli, A.; Strelcyn, J.M. Lyapunov Characteristic Exponents for smooth dynamical systems and for hamiltonian systems; a method for computing all of them. Part 1: Theory. *Meccanica* **1980**, *15*, 9–20.

38. Shimada, I.; Nagashima, T. A numerical approach to ergodic problem of dissipative dynamical systems. *Prog. Theor. Phys.* **1979**, *61*, 1605–1616.

39. Sandri, M. Numerical calculation of Lyapunov exponents. *Math. J.* **1996**, *6*, 78–84.

40. Oseledec, V.I. A multiplicative ergodic theorem. Lyapunov characteristic numbers for dynamical systems. *Trans. Mosc. Math. Soc.* **1968**, *19*, 197–231.

41. Shilov, G.E.; Silverman, R.A. *An Introduction to the Theory of Linear Spaces*; Dover: Mineola, NY, USA, 2012.

42. Kumar, P.; Miller, B.N. Dynamics of Coulombic and gravitational periodic systems. *Phys. Rev. E* **2016**, *93*, 040202.

43. Ginelli, F.; Takeuchi, K.A.; Chaté, H.; Politi, A.; Torcini, A. Chaos in the Hamiltonian mean-field model. *Phys. Rev. E* **2011**, *84*, 066211.

44. Bonasera, A.; Latora, V.; Rapisarda, A. Universal Behavior of Lyapunov Exponents in Unstable Systems. *Phys. Rev. Lett.* **1995**, *75*, 3434–3437.

MDPI

Article

Kovacs-Like Memory Effect in Athermal Systems: Linear Response Analysis

Carlos A. Plata and Antonio Prados *

Física Teórica, Universidad de Sevilla, Apartado de Correos 1065, E-41080 Sevilla, Spain; cplata1@us.es
* Correspondence: prados@us.es; Tel.: +34-954-559-514

Received: 19 September 2017; Accepted: 11 October 2017; Published: 13 October 2017

Abstract: We analyze the emergence of Kovacs-like memory effects in athermal systems within the linear response regime. This is done by starting from both the master equation for the probability distribution and the equations for the physically-relevant moments. The general results are applied to a general class of models with conserved momentum and non-conserved energy. Our theoretical predictions, obtained within the first Sonine approximation, show an excellent agreement with the numerical results. Furthermore, we prove that the observed non-monotonic relaxation is consistent with the monotonic decay of the non-equilibrium entropy.

Keywords: linear response; Kovacs effect; athermal system; master equation; *H*-functional

1. Introduction

The equilibrium state of physical systems is characterized by the value of a few macroscopic variables, for example pressure, volume and temperature in fluids. This characterization of the equilibrium state is complete, in the sense that different samples sharing the same values of the macroscopic variables respond identically to an external perturbation. On the contrary, a system in a nonequilibrium state, even if it is stationary, is not completely characterized by the value of the macroscopic variables: the response to an external perturbation depends also on additional variables or, equivalently, on its entire thermal history. Therefore, it is often said that the response depends on the way the system has been aged.

A pioneering work in the field of memory effects in nonequilibrium systems was carried out by Kovacs [1]. The Kovacs experiment showed that pressure, volume and temperature did not completely characterize the state of a sample of polyvinyl acetate that had been aged for a long time at a certain temperature T_1. The pressure was fixed during the whole experiment, and the time evolution of the volume was recorded. At a certain time t_w, the temperature was suddenly changed to T, for which the equilibrium value of the volume equaled its instantaneous value at t_w. Counterintuitively (thinking in equilibrium terms), the volume did not remain constant. Instead, it displayed a hump, passing through a maximum before tending back to its equilibrium (and initial) value.

We look into the Kovacs experiment in a more detailed way in Figure 1. In recent studies of the effect in glassy systems, the relevant physical variable is the energy instead of the volume [2–8]. The system is equilibrated at a "high" temperature T_0, and at $t = 0$, the temperature is suddenly quenched to a lower temperature T, after which the relaxation function $\phi(t)$ of the energy E is recorded. Specifically, $\phi(t) = \langle E(t) \rangle - \langle E \rangle_e$, where $\langle E \rangle_e$ is the average equilibrium energy at temperature T. Then, a similar procedure is followed, equilibrating the system again at T_0, but at $t = 0$, the temperature is changed to an even lower value T_1, $T_1 < T < T_0$. The system relaxes isothermally at T_1 for a certain time t_w, such that $\langle E \rangle(t = t_w)$ equals $\langle E \rangle_e$. At this time t_w, the temperature is increased to T, but the energy does not remain constant: it displays the behavior that is qualitatively shown by $K(t)$. At first, $K(t)$ increases from zero until a maximum is attained for $t = t_k$, and only afterwards, it goes back to zero. Similarly to the relaxation function, we have defined $K(t) = \langle E(t) \rangle - \langle E \rangle_e$, for $t \geq t_w$. Note that $K(t) \leq \phi(t)$ for all times, with the equality being only asymptotically approached in the long time limit.

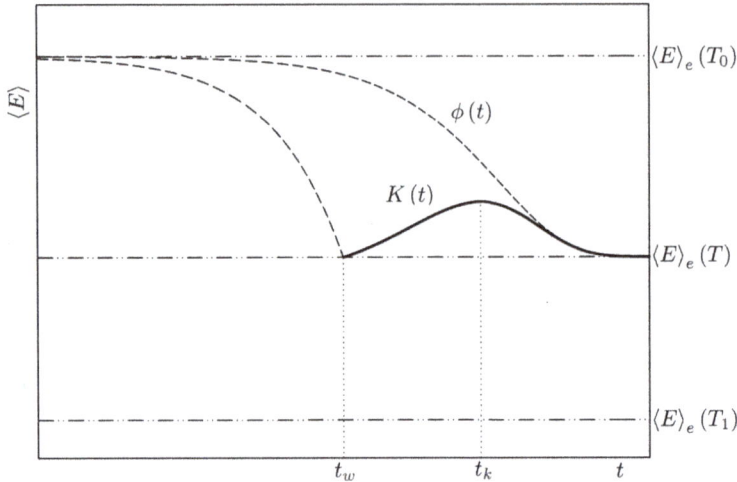

Figure 1. Scheme of the Kovacs experiment described in the text. The dashed curve on the right, labeled by $\phi(t)$, represents the direct relaxation from T_0 to T. The dashed curve on the left stands for the part of the relaxation from T_0 to T_1 that is interrupted at $t = t_w$ by the second temperature jump, changing abruptly the temperature from T_1 to T. After this second jump, the system follows the non-monotonic response $K(t)$ (solid line), which reaches a maximum at $t = t_k$ and, afterwards, approaches $\phi(t)$ for very long times.

For molecular (thermal) systems, the equilibrium distribution has the canonical form, and it has been shown that in linear response theory: [6]

$$K(t) = \frac{T_0 - T_1}{T_0 - T} \phi(t) - \frac{T - T_1}{T_0 - T} \phi(t - t_w), \tag{1}$$

where the final temperature T and the waiting time t_w are related by:

$$\frac{T - T_1}{T_0 - T_1} = \frac{\phi(t_w)}{\phi(0)}. \tag{2}$$

In linear response, the relaxation function $\phi(t)$ decays monotonically in time because it is proportional to the equilibrium time correlation function (fluctuation-dissipation theorem) $\langle E(0)E(t) \rangle_e - \langle E \rangle_e^2 = \sum_i c_i \exp(\lambda_i t)$, with $c_i > 0$ and $\lambda_i < 0$ for all i [9].

The linear response results above make it possible to understand the crux of the observed Kovacs hump in experiments [6]: (i) the inequality $0 \leq K(t) \leq \phi(t)$, which assures that the hump always has a positive sign (from now on, "normal" behavior), (ii) the existence of only one maximum in the hump and (iii) the increase of the maximum height and the shift of its position to smaller times as t_w is decreased. Nevertheless, it must be noted that the experiments, both real [1] and numerical [2–5,7], are mostly done out of the linear response regime: thus, it seems that the validity of these results extends beyond expectations. In fact, it has been checked in simple models that the linear approximation still gives a fair description of the hump for not-so-small temperature jumps [8].

Very recently, the investigation of Kovacs-like effects in athermal systems has been started. A granular fluid provides a prototypical example of an athermal system, which is intrinsically out-of-equilibrium [10]. A physical mechanism that inputs energy into the system and balances on average the energy loss in collisions, for instance the so-called stochastic thermostat [11], must be considered to reach a nonequilibrium steady state (NESS). Moreover, in general, fluctuations are far more important in granular systems than in molecular systems because of their smallness. The number of particles N, ranging from 10^2 to 10^4, is large enough to make it possible to apply the methods of statistical mechanics, but definitely much smaller than Avogadro's number.

The simplest case is that of uniformly-driven granular gases considered in [12,13]. The value of the kinetic energy (granular temperature T_g) at the NESS is controlled by the driving intensity ξ of the stochastic thermostat. Therefore, a Kovacs-like protocol can be implemented in a completely analogous way to the one described above, with the changes $T \to \xi$, $E \to T_g$. One of the main differences found is the emergence of "anomalous" Kovacs behavior for large enough dissipation, when $K(t)$ becomes negative and displays a minimum instead of a maximum. It must be stressed that these results have been obtained in the nonlinear regime, that is for driving jumps $\xi_0 - \xi$, $\xi_0 - \xi_1$ that are not small.

More recently, Kovacs-like behavior has been observed in other, more complex, athermal systems. This is the case of disordered mechanical systems [14] and also of active matter [15]. In the latter, a "giant" Kovacs hump has been reported, in the sense that the numerically observed maximum is much larger than the one predicted by the extrapolation of the linear approximation expression (1) to the considered protocol. Moreover, an alternative derivation of (1) has been provided in the supplemental material of [15]. This derivation holds for athermal systems, since it does not make use of either the explicit form of the probability distribution or the relationship between response functions and time correlations at the steady state, but is restricted to discrete-time dynamics at the macroscopic (average) level of description.

The objectives of our paper are two-fold. Firstly, we put forward a rigorous and general derivation of the linear response expression for the Kovacs hump for systems with a realistic continuous time dynamics. This is done at both the mesoscopic and macroscopic levels of description, starting from the master equation for the probability distribution and from the hierarchy of equations for the moments, respectively. Our proof is also valid for athermal systems, since no hypothesis is needed with regard to the form of either the stationary probability distribution or the fluctuation-dissipation relation. Secondly, we apply our results to a simple class of dissipative models that mimic the shear component of a granular fluid, recently introduced [16–18]. Therein, we obtain explicit expressions for the Kovacs hump within the first Sonine approximation and compare the theoretical predictions with numerical results. We also discuss the compatibility of the non-monotonic behavior displayed by the granular temperature in the Kovacs experiment and the monotonic approach to the NESS of the nonequilibrium entropy or H-functional [19–21].

2. Linear Theory for Kovacs-Like Memory Effects

2.1. General Markovian Dynamics

Here, we consider a general system, whose state is completely characterized by a vector x with M components, $x = \{x_1, x_2, \ldots, x_M\}$. For example, in a one-dimensional Ising chain of N spins, $x_i = \sigma_i = \pm 1$, and $M = N$; for a gas comprising N particles with positions r_i and velocities v_i, $M = 6N$, and $x = \{r_1, v_1, \ldots, r_N, v_N\}$. As is customary and for the sake of simplicity, from now on, we use a notation suitable for systems in which the states can be labeled with a discrete index α, $1 \le \alpha \le \Omega$. For example, this is the case of the Ising system above, where $\Omega = 2^N$. The generalization for a continuous index is straightforward, by changing sums into integrals and the Kronecker delta by the Dirac delta [9].

At the mesoscopic level of description, we assume that x is a Markov process, and its dynamics is governed by a master equation for the probabilities $P(x_\alpha, t)$,

$$\partial_t P(x_\alpha, t) = \sum_{\tilde{\alpha}} \left[W(x_\alpha | x_{\tilde{\alpha}}; \xi) P(x_{\tilde{\alpha}}, t) - W(x_{\tilde{\alpha}} | x_\alpha; \xi) P(x_\alpha, t) \right], \tag{3}$$

where $W(x_\alpha | x_{\tilde{\alpha}}; \xi)$ are the transition rates from state $x_{\tilde{\alpha}}$ to state x_α, and our notation explicitly marks their dependence on some control parameter ξ. Equation (3) can be formally written as:

$$\partial_t |\mathcal{P}(t)\rangle = \mathbb{W}(\xi) |\mathcal{P}(t)\rangle, \tag{4}$$

where $|\mathcal{P}(t)\rangle$ is a vector (column matrix) whose components are the probabilities $P(x_\alpha, t)$ and $\mathbb{W}(\xi)$ is the linear operator (square matrix) that generates the dynamical evolution of $|\mathcal{P}(t)\rangle$,

$$\mathbb{W}(x_\alpha | x_{\tilde{\alpha}}; \xi) = W(x_\alpha | x_{\tilde{\alpha}}; \xi) - \delta_{\alpha, \tilde{\alpha}} \sum_{\alpha'} W(x_{\alpha'} | x_\alpha; \xi). \tag{5}$$

Let us assume that the Markovian dynamics is ergodic (or irreducible [9]), that is all the states are dynamically connected through a chain of transitions with non-zero probability. Therefore, there is a unique steady solution of the master equation $|\mathcal{P}_s(\xi)\rangle$, which verifies:

$$\mathbb{W}(\xi) |\mathcal{P}_s(\xi)\rangle = 0, \tag{6}$$

and depends on the parameter ξ controlling the system dynamics. Note that ergodicity does not imply detailed balance, so we can have non-zero currents in the steady state. In general, this means that the system approaches an NESS in the long time limit, not an equilibrium state.

Now, we consider the system evolving from a certain initial state at time t_0, characterized by the distribution $|\mathcal{P}(t_0)\rangle$. The formal solution of the master equation can be written as:

$$|\mathcal{P}(t) - \mathcal{P}_s(\xi)\rangle = e^{(t-t_0)\mathbb{W}(\xi)} |\mathcal{P}(t_0) - \mathcal{P}_s(\xi)\rangle. \tag{7}$$

This is the starting point for our derivation of the expression for the Kovacs effect in the linear response approximation, which is carried out in the next section.

The time evolution of any physical property Y is obtained right away. Let us denote the value of Y for a given configuration x by $Y(x)$: its expected or average value is given by :

$$\langle Y(t)\rangle = \sum_\alpha Y(x_\alpha) P(x_\alpha, t) = \langle \mathcal{Y} | \mathcal{P}(t)\rangle, \tag{8}$$

where $|\mathcal{Y}\rangle$ is a ket whose components are $Y(x_\alpha)$, and $\langle \mathcal{Y}|$ its corresponding bra (row matrix with the same components). Note that we are assuming that Y is a real quantity for all the configurations. By substituting (7) into (8), it is obtained that:

$$\Delta Y(t; \xi) \equiv \langle Y(t)\rangle - \langle Y\rangle_s(\xi) = \langle \mathcal{Y} | e^{(t-t_0)\mathbb{W}(\xi)} | \mathcal{P}(t_0) - \mathcal{P}_s(\xi)\rangle, \tag{9}$$

in which $\langle Y\rangle_s(\xi)$ is the average value at the steady state of the system corresponding to ξ.

Now, we investigate the relaxation of the system from the steady state for $\xi_0 = \xi + \Delta\xi$ to the steady state for ξ. We do so in linear response, that is $\Delta\xi$ is considered to be small, and we neglect all terms beyond those linear in $\Delta\xi$. Thus, at $t = 0$ we have that the probability distribution is:

$$|\mathcal{P}(t = 0)\rangle = |P_s(\xi + \Delta\xi)\rangle = |P_s(\xi)\rangle + \Delta\xi \left| \frac{dP_s(\xi)}{d\xi} \right\rangle + O(\Delta\xi)^2. \tag{10}$$

Substitution of (10) into (9) yields the formal expression for the relaxation of Y in linear response,

$$\Delta Y(t; \xi) = \Delta\xi \left\langle \mathcal{Y} \left| e^{(t-t_0)\mathbb{W}(\xi)} \right| \frac{dP_s(\xi)}{d\xi} \right\rangle. \tag{11}$$

In order to have an order of unity function, one may define a normalized relaxation function:

$$\phi_Y(t; \xi) \equiv \lim_{\Delta \xi \to 0} \frac{\Delta Y(t; \xi)}{\Delta \xi} = \left\langle \mathcal{Y} \middle| e^{(t-t_0)\mathbb{W}(\xi)} \middle| \frac{dP_s(\xi)}{d\xi} \right\rangle. \tag{12}$$

Sometimes, the relaxation function is further normalized by considering $\phi_Y(t)/\phi_Y(t = t_0)$ (see for instance [6,22]), but such a multiplying factor is clearly not physically relevant and will not be introduced here.

2.2. The Kovacs Protocol: Linear Response Analysis from the Master Equation

Here, it is a Kovacs-like protocol that we introduce, by considering that the parameter ξ controlling the dynamics is changed in the following stepwise manner:

$$\xi(t) = \begin{cases} \xi_0, & -\infty < t < 0, \\ \xi_1, & 0 < t < t_w, \\ \xi, & t > t_w. \end{cases} \tag{13}$$

Therefore, since ξ_0 is kept for an infinite time, at $t = 0$, the system is prepared in the corresponding steady state, $\mathcal{P}(t = 0) = \mathcal{P}_s(\xi_0)$. Our idea is to consider that the jumps $\xi_1 - \xi_0$ and $\xi - \xi_1$ are small, in the sense that all expressions can be linearized in the magnitude of these jumps. Note that this protocol is completely analogous to the Kovacs protocol described in the Introduction (see Figure 1), but with ξ playing the role of the temperature.

We start by analyzing the relaxation in the first time window, $0 < t < t_w$. Therein, we apply (7) with the substitutions $t_0 \to 0$ and $\xi \to \xi_1$, that is,

$$|\mathcal{P}(t) - \mathcal{P}_s(\xi_1)\rangle = e^{t\mathbb{W}(\xi_1)} |\mathcal{P}_s(\xi_0) - \mathcal{P}_s(\xi_1)\rangle, \quad 0 \leq t \leq t_w. \tag{14}$$

The final distribution function, at $t = t_w$, is the initial condition for the next stage, $t > t_w$, in which the system relaxes towards the steady state corresponding to ξ. Making use again of (7) with $t_0 \to t_w$,

$$\begin{aligned} |\mathcal{P}(t) - \mathcal{P}_s(\xi)\rangle &= e^{(t-t_w)\mathbb{W}(\xi)} |\mathcal{P}(t_w) - \mathcal{P}_s(\xi)\rangle \\ &= e^{(t-t_w)\mathbb{W}(\xi)} \left[e^{t_w \mathbb{W}(\xi_1)} |\mathcal{P}_s(\xi_0) - \mathcal{P}_s(\xi_1)\rangle + |\mathcal{P}_s(\xi_1) - \mathcal{P}_s(\xi)\rangle \right], \quad t \geq t_w. \end{aligned} \tag{15}$$

It must be stressed that the above expressions, Equations (14) and (15), are exact, and no approximation has been made.

The linear response approximation is introduced now: we assume that both jumps $\xi_0 - \xi_1$ and $\xi - \xi_1$ are small, so we can expand both $|\mathcal{P}_s(\xi_0) - \mathcal{P}_s(\xi_1)\rangle$ and $|\mathcal{P}_s(\xi_1) - \mathcal{P}_s(\xi)\rangle$, similarly to what was done in Equation (10). Namely,

$$|\mathcal{P}_s(\xi_0) - \mathcal{P}_s(\xi_1)\rangle = (\xi_0 - \xi_1) \left| \frac{dP_s(\xi)}{d\xi} \right\rangle + O(\xi_0 - \xi_1)^2, \tag{16a}$$

$$|\mathcal{P}_s(\xi_1) - \mathcal{P}_s(\xi)\rangle = (\xi_1 - \xi) \left| \frac{dP_s(\xi)}{d\xi} \right\rangle + O(\xi_1 - \xi)^2. \tag{16b}$$

Note that in both (16a) and (16b), the derivatives are evaluated at ξ; the difference introduced by evaluating them at either ξ_1 or ξ_0 is second order in the deviations. Then,

$$|\mathcal{P}(t) - \mathcal{P}_s(\xi)\rangle = (\xi_0 - \xi_1) e^{(t-t_w)\mathbb{W}(\xi)} e^{t_w \mathbb{W}(\xi_1)} \left| \frac{dP_s(\xi)}{d\xi} \right\rangle + (\xi_1 - \xi) e^{(t-t_w)\mathbb{W}(\xi)} \left| \frac{dP_s(\xi)}{d\xi} \right\rangle. \tag{17}$$

This equation can be further simplified: since the two terms on its rhs are first order in the jumps, we can substitute $\mathbb{W}(\xi_1)$ with $\mathbb{W}(\xi)$, with the result:

$$|\mathcal{P}(t) - \mathcal{P}_s(\xi)\rangle = (\xi_0 - \xi_1) e^{t\mathbb{W}(\xi)} \left| \frac{dP_s(\xi)}{d\xi} \right\rangle - (\xi - \xi_1) e^{(t-t_w)\mathbb{W}(\xi)} \left| \frac{dP_s(\xi)}{d\xi} \right\rangle. \tag{18}$$

This is the superposition of two responses: the first term on the rhs gives the relaxation from ξ_0 to ξ_1, which starts at $t = 0$, whereas the second term stands for the relaxation from ξ_1 to ξ, which starts at $t = t_w$. We have chosen to write $-(\xi - \xi_1)$ in the second term because $\xi > \xi_1$ in the Kovacs protocol.

The same structure in Equation (18) is transferred to the average value. Taking into account Equation (9):

$$\Delta Y(t) = (\xi_0 - \xi_1) \left\langle \mathcal{Y} \left| e^{t \mathbb{W}(\xi)} \right| \frac{d\mathcal{P}_s(\xi)}{d\xi} \right\rangle - (\xi - \xi_1) \left\langle \mathcal{Y} \left| e^{(t-t_w)\mathbb{W}(\xi)} \right| \frac{d\mathcal{P}_s(\xi)}{d\xi} \right\rangle, \quad t \geq t_w, \quad (19)$$

in which we recognize the relaxation function in linear response, defined in Equation (12). We can also normalize the response in this experiment, by defining a function $K(t)$ as follows,

$$K_Y(t) \equiv \lim_{\xi_0 \to \xi} \frac{\Delta Y(t)}{\xi_0 - \xi} = \frac{\xi_0 - \xi_1}{\xi_0 - \xi} \phi_Y(t) - \frac{\xi - \xi_1}{\xi_0 - \xi} \phi_Y(t - t_w). \quad (20)$$

It is understood that, as $\xi_0 - \xi \to 0$, both prefactors $\frac{\xi_0 - \xi_1}{\xi_0 - \xi}$ and $\frac{\xi - \xi_1}{\xi_0 - \xi}$ are kept of the order of unity.

A few comments on Equation (20) are in order. Hitherto, no restriction has been imposed on the state of the system at $t = t_w$; therefore, (20) is valid for arbitrary (ξ_0, ξ_1, ξ), provided that the jumps are small enough and the ratio of the jumps is of the order of unity. The function $K(t)$ corresponds to a Kovacs-like experiment when ξ is chosen as a function of t_w in such a way that $\langle Y(t_w) \rangle = \langle Y \rangle_s(\xi)$ or $K_Y(t_w) = 0$, that is,

$$\frac{\xi - \xi_1}{\xi_0 - \xi_1} = \frac{\phi_Y(t_w)}{\phi_Y(0)}. \quad (21)$$

Alternatively, one may consider that (21) defines t_w as a function of ξ.

The complete analogy between Equations (20) and (21) and Equations (1) and (2) is apparent. Nevertheless, we have made use neither of the explicit form of the steady state distribution (in general non-canonical), nor of the relation between response functions and time correlation functions (fluctuation-dissipation relation), which were necessary in [6] to demonstrate Equation (1). Therefore, the proof developed here is more general, being valid for any steady state, equilibrium or nonequilibrium, and thus, it specifically holds in athermal systems. Furthermore, it must be noted that it can be easily extended to the Fokker–Planck, or the equivalent Langevin, equation.

2.3. Linear Response from the Equations for the Moments

In this section, we do not start from the equation for the probability distribution, but from the equations for the relevant physical properties of the considered system. For example, one may think of the hydrodynamic equations for a fluid or the law of mass action equations for chemical reactions. Of course, these equations can be derived in a certain "macroscopic" approximation [9], which typically involves neglecting fluctuations, from the equation for the probability distribution by taking moments, but this is not our approach here. Anyhow, we borrow this term to call our starting point "equations for the moments".

We denote the relevant moments by z_i, $i = 1, \ldots, J$, where J is the number of relevant moments. The equations for the moments have the general form:

$$\frac{d}{dt} z_i = f_i(z_1, \ldots, z_J; \xi), \quad (22)$$

where f_i are continuous, in general nonlinear, functions of the moments. This is a key difference between moment equations and the master (or Fokker–Planck) equation, since the latter is always linear in the probability distribution. Therefore, unlike the master equation, Equation (22) cannot be formally solved for arbitrary initial conditions. This notwithstanding, in the linear response approximation, we show here that a procedure similar to the one carried out in the previous section leads to the same expression for the Kovacs hump.

We assume that there is only one steady solution of Equation (22) that is globally stable, at which the corresponding values of the moments are $z_i^s(\xi)$. Linearization of the dynamical system around the steady state gives:

$$\frac{d}{dt}|\Delta z(t)\rangle = \mathbb{M}(\xi)|\Delta z(t)\rangle, \qquad |\Delta z(t)\rangle \equiv |z(t) - z_s(\xi)\rangle. \tag{23}$$

We are using a notation completely similar to that in the previous section: $|z\rangle$ is a vector, represented by a column matrix with components z_i, and $\mathbb{M}(\xi)$ is a linear operator, represented by a square matrix with elements:

$$M_{ij}(\xi) = \partial_{z_j} f_i \Big|_{|z\rangle = |z_s(\xi)\rangle}. \tag{24}$$

Note that the dimensions of these matrices are much smaller than those for the master equation, since J is of the order of unity and does not diverge in the thermodynamic limit. In general, $M_{ij} \neq M_{ji}$, and the operator $\mathbb{M}$ is not Hermitian. However, we do not need $\mathbb{M}$ to be Hermitian to solve the linearized system in a formal way, as shown below.

Analogously to what was done for the master equation, the formal solution of Equation (23) is:

$$|\Delta z(t)\rangle = e^{(t-t_0)\mathbb{M}(\xi)}|\Delta z(t_0)\rangle. \tag{25}$$

In particular, if the initial condition is chosen to correspond to the steady state for $\xi_0 = \xi + \Delta\xi$, one has:

$$|\Delta z(t)\rangle = \Delta\xi\, e^{(t-t_0)\mathbb{M}(\xi)}\left|\frac{dz_s(\xi)}{d\xi}\right\rangle. \tag{26}$$

The response for any of the relevant moments can be extracted by projecting the above result onto the "natural" basis $|u_i\rangle$, whose components are $u_{ij} = \delta_{ij}$. Then, the normalized linear response function for z_i can be defined by:

$$\phi_{z_i}(t) = \lim_{\Delta\xi \to 0} \frac{\langle u_i|\Delta z(t)\rangle}{\Delta\xi} = \left\langle u_i\left|e^{(t-t_0)\mathbb{M}(\xi)}\right|\frac{dz_s(\xi)}{d\xi}\right\rangle. \tag{27}$$

Note the utter formal analogy of Expression (27) with Equation (12), which was obtained from the master equation. The proof of the expression for the Kovacs hump follows exactly the same line of reasoning, and the result is exactly that in Equations (20) and (21); thus, it is not repeated here.

3. A Lattice Model with Conserved Momentum and Non-Conserved Energy

3.1. Definition of the Model: Kinetic Description

Here, we briefly put forward a class of models for the shear component of the velocity in a granular gas, focusing on the features that are needed for the discussion of memory effects. A more complete description of the model, including its physical motivation, can be found in [16–18,23].

The system is defined on a 1d lattice: there is a particle with velocity v_l at each site l. The system configuration is thus given by $v \equiv \{v_1, ..., v_N\}$. The dynamics proceeds through inelastic nearest-neighbor binary collisions: each pair $(l, l+1)$ collides inelastically with a characteristic rate proportional to $|v_l - v_{l+1}|^\beta$, with $\beta \geq 0$. For $\beta = 0$, nearest-neighbor particles collide independently of their relative velocity (the so-called Maxwell-molecule model [24]), whereas for $\beta = 1$ and $\beta = 2$, we have a collision rate analogous to that of hard spheres and very hard spheres, respectively [10,25]. The pre-collisional velocities are transformed into the post-collisional ones by the operator $\hat{b}_l$,

$$\hat{b}_l v_l \;=\; v_l - \frac{1+\alpha}{2}\left(v_l - v_{l+1}\right), \tag{28a}$$

$$\hat{b}_l v_{l+1} \;=\; v_{l+1} + \frac{1+\alpha}{2}\left(v_l - v_{l+1}\right), \tag{28b}$$

where $0 < \alpha \le 1$ is the normal restitution coefficient. Momentum is always conserved in collisions, $(\hat{b}_l - 1)(v_l + v_{l+1}) = 0$, but energy is not; in fact, $(\hat{b}_l - 1)(v_l^2 + v_{l+1}^2) = (\alpha^2 - 1)(v_l - v_{l+1})^2/2 \le 0$, with equality holding only in the elastic limit $\alpha = 1$.

In order to have a steady state, a mechanism that injects energy into the system, and thus compensates on average the energy loss in collisions, must be introduced. For the sake of simplicity, we consider here that the system is "heated" by a white noise force, that is the so-called stochastic thermostat [11,26]. The velocity change introduced by the stochastic forcing is:

$$\partial_\tau v_i(\tau)\big|_{\text{noise}} = \xi_i(\tau) - \frac{1}{N}\sum_{j=1}^{N}\xi_j(\tau), \tag{29}$$

and $\xi_i(\tau)$ are Gaussian white noises of amplitude χ:

$$\langle \xi_i(\tau)\rangle_{\text{noise}} = 0, \quad \langle \xi_i(\tau)\xi_j(\tau')\rangle_{\text{noise}} = \chi \delta_{ij}\delta(\tau - \tau'), \tag{30}$$

for $i, j = 1, \dots, N$. This version of the stochastic thermostat conserves total momentum, which is needed to assure the existence of a well-defined steady state [27,28].

We do not write here the master-Fokker–Planck equation for the N-particle $P_N(v, \tau)$ probability density of finding the system in state v at time τ. Instead, we directly write the "kinetic" equation for the one-particle distribution function, namely $P_1(v; l, \tau) = \int dv P_N(v, \tau)\delta(v_l - v)$. Therefrom, all the one-site velocity moments, which are the relevant physical quantities for our present purposes, can be derived, $\langle v_l^n(\tau)\rangle \equiv \int_{-\infty}^{+\infty} dv\, v^n P_1(v; l, \tau)$. As usual in kinetic theory, a closed equation for P_1 can be written only after introducing the molecular chaos assumption: spatial correlations at different sites are of the order of N^{-1} and then negligible.

We analyze here homogeneous states; if the initial state is homogeneous, the above dynamics preserves homogeneity. Moreover, we employ the usual notation in kinetic theory, $f(v, \tau) \equiv P_1(v, l; \tau)$ and consider the quasi-elastic limit, that is $\epsilon \equiv 1 - \alpha^2 \ll 1$. This allows us to write a simpler expression for the inelastic collision term. Proceeding along the same lines as in [17,21], it is obtained that [29]:

$$\partial_\tau f(v, \tau) = \frac{\omega\epsilon}{2}\partial_v \int_{-\infty}^{+\infty} dv'(v - v')|v - v'|^\beta f(v, \tau)f(v', \tau) + \frac{\chi}{2}\partial_v^2 f(v, \tau). \tag{31}$$

We can define a dimensionless time scale, $t = \omega\epsilon\tau$, over which:

$$\partial_t f(v, t) = \frac{1}{2}\partial_v \int_{-\infty}^{+\infty} dv'(v - v')|v - v'|^\beta f(v, t)f(v', t) + \frac{\tilde{\xi}}{2}\partial_v^2 f(v, t), \tag{32}$$

where $\tilde{\xi}$ is the rescaled strength of the noise, $\tilde{\xi} = \frac{\chi}{\omega\epsilon}$. Note that this kinetic equation is, like Boltzmann's or Enskog's, nonlinear in $f(v, t)$.

The main physical magnitude is the granular temperature T, which we define as:

$$T \equiv \langle v^2 \rangle = \int_{-\infty}^{+\infty} dv\, v^2 f(v, t). \tag{33}$$

We used the notation T_g for the granular temperature in the Introduction, to differentiate it from the usual thermodynamic temperature T. Since the latter plays no role in our system, we employ the usual notation T for the granular temperature hereafter. It is also customary to define the thermal velocity $v_0 = \sqrt{2T}$ to make v dimensionless, with the change:

$$v = v_0 c, \qquad f(v,t)dv = \varphi(c,t)dc \Leftrightarrow \varphi(c,t) = v_0 f(v,t). \tag{34}$$

Taking moments in (32) and making the change of variables above, one gets:

$$\frac{d}{dt}T = -\zeta\, T^{1+\frac{\beta}{2}} + \xi, \qquad \zeta = 2^{\frac{\beta}{2}} \int_{-\infty}^{+\infty} dc \int_{-\infty}^{+\infty} dc' |c - c'|^{2+\beta} \varphi(c,t)\varphi(c',t) \tag{35}$$

The first term on the rhs stems from collisions and cools the system, in the sense that it always makes the granular temperature decrease. The second term stems from the stochastic thermostat and heats the system, and thus, in the long time limit, an NESS is attained in which both terms counterbalance each other.

3.2. First Sonine Approximation

The equation for the granular temperature is not closed in general, and then an expansion in Sonine (or Laguerre) polynomials is typically introduced,

$$\varphi(c,t) = \frac{e^{-c^2}}{\sqrt{\pi}} \left[1 + \sum_{k=2}^{\infty} a_k(t) L_k^{\left(-\frac{1}{2}\right)}(c^2) \right], \tag{36}$$

where $L_k^{(m)}(x)$ are the associated Laguerre polynomials [30]. In kinetic theory, $m = \frac{d}{2} - 1$, with d being the spatial dimension, and often, the notation $S_k(x) \equiv L_k^{\left(\frac{d}{2}-1\right)}$ is used. Here, we mainly use the so-called first Sonine approximation, in which (i) only the term with $k = 2$ is retained and (ii) nonlinear terms in a_2 are neglected. The coefficient a_2 is the excess kurtosis, $\langle c^4 \rangle = 3(1 + a_2)/4$.

Although the linearization in a_2 is quite standard in kinetic theory, we derive firstly the evolution equations considering just Step (i) of the first Sonine approximation, that is we truncate the expansion for the scaled distribution (36) after the $k = 2$ term. Henceforth, we call this approximation the nonlinear first Sonine approximation. Afterwards, in the numerical results, we will discuss how both approximations, nonlinear and standard, give almost indistinguishable results.

In the nonlinear first Sonine approximation, the evolution equation for the temperature is readily obtained [29]:

$$\frac{d}{dt}T = -\zeta_0\, T^{1+\frac{\beta}{2}} \left[1 + \frac{\beta(2+\beta)}{16} a_2 + \frac{\beta(2+\beta)(2-\beta)(4-\beta)}{1024} a_2^2 \right] + \xi, \tag{37a}$$

where $\zeta_0 = \pi^{-1/2}\, 2^{1+\beta} \Gamma\left(\frac{3+\beta}{2}\right)$. Unless $\beta = 0$ (Maxwell molecules), the equation for the temperature is not closed. Then, we write down the equation for a_2: again, after a lengthy, but straightforward calculation, we derive:

$$T\frac{d}{dt}a_2 = -2\xi a_2 - \frac{\zeta_0}{3}\beta\, T^{1+\frac{\beta}{2}}$$
$$\times \left[1 + \frac{56 + \beta(6+\beta)}{16} a_2 - \frac{(2+\beta)[384 + (2-\beta)\beta(4+\beta)]}{1024} a_2^2 - \frac{3(4-\beta)(2-\beta)(2+\beta)}{512} a_2^3 \right]. \tag{37b}$$

The evolution equations in the standard first Sonine approximation are easily reached just neglecting nonlinear terms in a_2 in Equations (37), that is:

$$\frac{d}{dt}T = -\zeta_0\, T^{1+\frac{\beta}{2}} \left[1 + \frac{\beta(2+\beta)}{16} a_2 \right] + \xi, \tag{38a}$$

$$T\frac{d}{dt}a_2 = -\frac{\zeta_0}{3}\beta\, T^{1+\frac{\beta}{2}}\left[1 + \frac{56+\beta(6+\beta)}{16}a_2\right] - 2\xi a_2. \tag{38b}$$

For $\xi \neq 0$, the steady solution of these equations is:

$$T_s = \left(\frac{\xi}{\zeta_0\left[1 + \frac{\beta(2+\beta)}{16}a_2^s\right]}\right)^{\frac{2}{2+\beta}}, \qquad a_2^s = -\frac{16\beta}{96+56\beta+6\beta^2+\beta^3}. \tag{39}$$

Note that (i) $0 \leq |a_2^s| \leq 0.133$ for $0 \leq \beta \leq 2$, which makes it reasonable to use the first Sonine approximation, and (ii) a_2^s is independent of the driving intensity ξ. This will be useful in the linear response analysis, to be developed below, because a sudden change in the driving only changes the stationary value of the temperature, but not that of the excess kurtosis. If $\xi = 0$, the system evolves towards the homogeneous cooling state, in which the excess kurtosis tends to the value:

$$a_2^{\mathrm{HCS}} = -\frac{16}{56+\beta(6+\beta)}, \tag{40}$$

as predicted by Equation (38b) and the temperature decays following Haff's law, $dT/dt \propto -T^{1+\frac{\beta}{2}}$.

From now on, we use reduced temperature and time,

$$\theta = \frac{T}{T_s}, \qquad s = \zeta_0 T_s^{\beta/2} t. \tag{41}$$

The steady temperature T_s plays the role of a natural energy (or granular temperature) unit. In reduced variables, the evolution equations are:

$$\frac{d}{ds}\theta = 1 - \theta^{1+\frac{\beta}{2}} + \frac{\beta(2+\beta)}{16}\left(a_2^s - a_2\theta^{1+\frac{\beta}{2}}\right), \tag{42a}$$

$$\theta\frac{d}{ds}a_2 = \kappa_1\left(a_2 - a_2^{\mathrm{HCS}}\right)\left(1 - \theta^{1+\frac{\beta}{2}}\right) - \kappa_2\left(a_2 - a_2^s\right), \tag{42b}$$

where we have introduced two parameters of the order of unity,

$$\kappa_1 = -\frac{\beta}{3a_2^{\mathrm{HCS}}}, \qquad \kappa_2 = -\frac{\beta}{3a_2^s}, \tag{43}$$

$0 \leq \kappa_1 \leq 3$ and $2 \leq \kappa_2 \leq 5$ for $0 \leq \beta \leq 2$.

The evolution equations in the first Sonine approximation, (38) or (42), are the particularization of the equations for the moments (22) to our model: $J = 2$, and $z_1 = T$ (or θ), $z_2 = a_2$. Consistently, they are nonlinear, although here, due to the simplifications introduced in the first Sonine approximation, only nonlinear in θ. When the system is close to the NESS, Equation (42) can be linearized around it by writing $\theta = 1 + \Delta\theta$, $a_2 = a_2^s + \Delta a_2$,

$$\frac{d}{ds}\begin{pmatrix}\Delta\theta\\\Delta a_2\end{pmatrix} = M\cdot\begin{pmatrix}\Delta\theta\\\Delta a_2\end{pmatrix}, \qquad M = \begin{pmatrix}-\frac{2(2+\beta)(12+\beta)}{48+4\beta+\beta^2} & -\frac{\beta(2+\beta)}{16}\\-\kappa_1\left(1+\frac{\beta}{2}\right)\left(a_2^s - a_2^{\mathrm{HCS}}\right) & -\kappa_2\end{pmatrix}. \tag{44}$$

Of course, the general solution of this linear system for arbitrary initial conditions $\Delta\theta(0)$ and $\Delta a_2(0)$ can be immediately written, but we omit it here.

3.3. Kovacs Hump in Linear Response

Now, we look into the Kovacs hump in the linear response approximation. Following the discussion leading to Equation (20), first we have to calculate the relaxation function ϕ_T for the granular temperature. The system is at the steady state corresponding to a driving ξ_0 for $t < 0$; at $t = 0$, the driving is instantaneously changed to ξ, and only the linear terms in $\Delta\xi = \xi - \xi_0$ are retained. We choose the normalization of $\phi_T(s)$ in such a way that $\phi_T(0) = 1$, that is:

$$\phi_T(s) \equiv \lim_{\Delta T(0) \to 0} \frac{\Delta T(s)}{\Delta T(0)} = \lim_{\Delta\theta(0) \to 0} \frac{\Delta\theta(s)}{\Delta\theta(0)}. \tag{45}$$

Since T_s changes with ξ, but a_2 does not, we have to solve Equation (44) for $\Delta a_2(0) = 0$ and arbitrary (small enough) $\Delta\theta(0)$. The solution is:

$$\phi_T(s) = c_+ e^{\lambda_+ s} + c_- e^{\lambda_- s}, \tag{46a}$$

$$c_+ = \frac{M_{11} - \lambda_-}{\lambda_+ - \lambda_-}, \qquad c_- = \frac{\lambda_+ - M_{11}}{\lambda_+ - \lambda_-}, \tag{46b}$$

where M_{ij} is the (i, j) element of the matrix M and $\lambda_\pm$ its eigenvalues,

$$\lambda_\pm = \frac{\operatorname{Tr} M \pm \sqrt{(\operatorname{Tr} M)^2 - 4 \det M}}{2} = \frac{\operatorname{Tr} M \pm \sqrt{(M_{11} - M_{22})^2 + 4 M_{12} M_{21}}}{2}. \tag{47}$$

Both eigenvalues $\lambda_\pm$ are negative, since $\operatorname{Tr} M < 0$ and $\det M > 0$ for all $\beta > 0$. Therefore, $|\lambda_+| < |\lambda_-|$, and it is λ_+ that dominates the relaxation of the granular temperature for long times. Moreover, $c_\pm > 0$, and thus, the linear relaxation function $\phi_T(s)$ is always positive and decays monotonically to zero.

Next, we consider a Kovacs-like experiment: the system was at the NESS corresponding to a driving ξ_0, with granular temperature $T_{s,0}$ for $t < 0$; the driving is suddenly changed to ξ_1 at $t = 0$ so that the system starts to relax towards a new steady temperature $T_{s,1}$ for $0 \leq t \leq t_w$, and this relaxation is interrupted at $t = t_w$, because the driving is again suddenly changed to the value ξ such that the stationary granular temperature T_s equals its instantaneous value at t_w. The time evolution of the granular temperature for $t \geq t_w$ is given by the particularization of Equations (20) and (21) to our situation, that is,

$$K_T(s) = \frac{\xi_0 - \xi_1}{\xi_0 - \xi} \phi_T(s) - \frac{\xi - \xi_1}{\xi_0 - \xi} \phi_T(s - s_w), \qquad \frac{\xi - \xi_1}{\xi_0 - \xi_1} = \phi_T(s_w), \tag{48}$$

where we have made use of the normalization $\phi_T(0) = 1$. In the linear response approximation, the jumps in the driving values can be substituted by the corresponding jumps in the stationary values of the granular temperature.

The structure of the linear relaxation function $\phi_T(s)$, as a linear combination of decreasing exponentials $\exp(\lambda_\pm t)$, $\lambda_\pm < 0$, with positive weights $c_\pm$, assures that the Kovacs behavior is normal: (i) $K_T(s)$ is always positive and bounded from above by $\phi_T(s)$ and (ii) there is only one maximum at a certain time $s_k > s_w$ [6]. The anomalous behavior found in the uniformly heated hard-sphere granular for large enough inelasticity is thus not present here. This is consistent with the quasi-elastic limit we have introduced to simplify the collision operator.

3.4. Nonlinear Kovacs Hump

Here, we consider the Kovacs hump for arbitrary large driving jumps. In our model, we can make use of the smallness of a_2, which is assumed in the first Sonine approximation, in order to introduce a perturbative expansion of Equations (42) in powers of a_2^s. The procedure is completely analogous to that performed in [12,13] for a dilute gas of inelastic hard spheres, and thus, we omit the details here. We start by writing $a_2 = a_2^s A_2$, with A_2 of the order of unity, and:

$$\theta(s) = \theta_0(s) + a_2^s \theta_1(s) + \dots, \qquad A_2(s) = A_{20}(s) + a_2^s A_{21}(s) + \dots. \tag{49}$$

These expansions are inserted into Equations (42), which have to be solved with initial conditions $\theta(s_w) = 1$, $A_2(s_w) = A_2^{\text{ini}}$. To the lowest order, $\theta_0(s) = 1$, whereas $A_{20}(s)$ decays exponentially to one,

$$A_{20}(s) - 1 \sim \left(A_2^{\text{ini}} - 1 \right) e^{-\kappa_2(s - s_w)}. \tag{50}$$

In order to describe the Kovacs hump, we compute $\theta_1(s)$ that verifies the differential equation:

$$\frac{d\theta_1}{ds} = -\left(1 + \frac{\beta}{2} \right) \theta_1 + \frac{\beta(2 + \beta)}{16} \left(A_{20} - 1 \right), \tag{51}$$

which can be immediately integrated to give:

$$\theta(s) - 1 \sim \left(a_2^{\text{ini}} - a_2^{\text{s}} \right) \frac{\beta(2 + \beta)}{8(2 + \beta - 2\kappa_2)} \left[e^{-\kappa_2(s - s_w)} - e^{-\left(1 + \frac{\beta}{2} \right)(s - s_w)} \right], \qquad s \geq s_w, \tag{52}$$

The structure of this result is completely analogous to those in [12,13], and thus, the conclusions can also be drawn in a similar way. In particular, we want to highlight that (i) the factor that controls the size of the hump is proportional to $a_2^{\text{ini}} - a_2^{\text{s}}$ and (ii) the shape of the hump is codified in the factor between brackets that only depends on β. Note that $(a_2^{\text{ini}} - a_2^{\text{s}}) > 0$ for the cooling protocols $(\xi_1 < \xi < \xi_0)$ considered here, and thus, no anomalous Kovacs hump is expected in the nonlinear regime either.

4. Numerical Results

Here, we compare the theoretical approach above to numerical results of our model. Specifically, we focus on the case $\beta = 1$ that gives a collision rate similar to that of hard-spheres. All simulations have been carried out with a restitution coefficient $\alpha = 0.999$, which corresponds to the quasi-elastic limit in which our simplified kinetic description holds. Furthermore, we set $\omega = 1$ without loss of generality.

4.1. Validation of the Evolution Equations and Linear Relaxation

First of all, we check the validity of the first Sonine approximation, as given by Equations (38), to describe the time evolution of our system. In order to do so, we compare several relaxation curves between the NESS corresponding to two different noise strengths. In particular, we always depart from the stationary state corresponding to $\chi_0 = 1$ and afterwards let the system evolve with $\chi = \{0.2, 0.6, 0.8, 1\}$ for $t > 0$. In Figure 2, we compare the Monte Carlo simulation (see Appendix A for details) of the system with the numerical solution of the evolution equations in the first Sonine approximation (38). In addition, we also have plotted the analytical solution of the linear response system, Equation (44). The agreement is complete between simulation and theory, and it can be observed how the linear response result becomes more accurate as the temperature jump decreases.

In order not to clutter the plot in Figure 2, we have not shown the results for the nonlinear first Sonine approximation, Equation (37). The relative error between their numerical solution and that of the standard first Sonine approximation (38) is at most of order 0.1%, for all the cases we have considered. Henceforth, we always use the latter, which is the usual approach in kinetic theory.

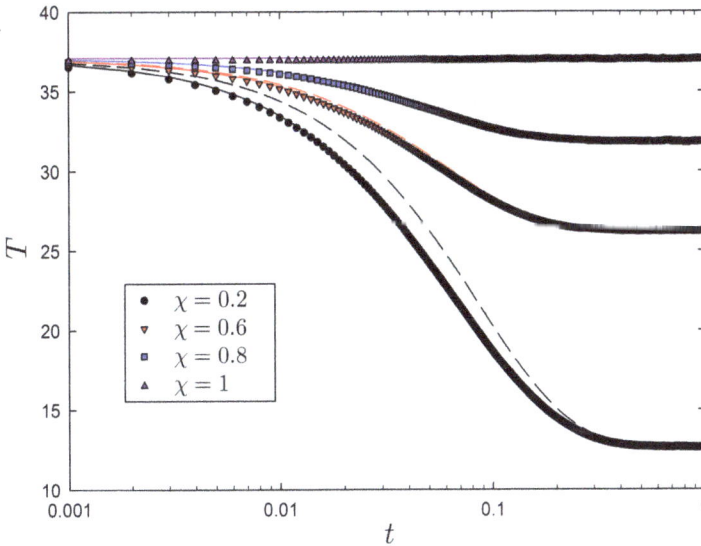

Figure 2. (Color online) Direct relaxation of the granular temperature T for different final noise amplitudes. All curves start from the stationary state corresponding to $\chi_0 = 1$. We compare Monte Carlo simulation results for a system of $N = 100$ sites (symbols) with the numerical solution of the first Sonine approximation, Equation (38) (solid lines), and the analytic solution of the linear response system, Equation (44) (dashed lines).

4.2. Kovacs Hump

Since the numerical integration of the first Sonine approximation perfectly agrees with Monte Carlo simulations, we compare the analytical results for the Kovacs hump with the former. Specifically, we work in reduced variables, and therefore, we integrate numerically Equations (42).

4.2.1. Linear Response

It is convenient to rewrite the expression for the Kovacs hump in an alternative form to compare our theory to numerical results. We take advantage of the simple structure of the relaxation function in the first Sonine approximation, which is the sum of two exponentials, to introduce the factorization [6]:

$$K_T(s) = K_0(s_w)K_1(s - s_w), \tag{53a}$$

where:

$$K_0(s_w) = c_+ c_- \frac{e^{\lambda_+ s_w} - e^{\lambda_- s_w}}{1 - \phi_T(s_w)}, \qquad K_1(s - s_w) = e^{\lambda_+(s-s_w)} - e^{\lambda_-(s-s_w)}. \tag{53b}$$

Firstly, this factorization property shows that the position s_k of the maximum relative to the waiting time s_w, that is, $s_k - s_w$, is controlled by the function K_1. Thus, $s_k - s_w$ does not depend on the waiting time, but only on the two eigenvalues $\lambda_\pm$. Namely,

$$s_k - s_w = \frac{1}{\lambda_+ - \lambda_-} \ln\left(\frac{\lambda_-}{\lambda_+}\right) \underset{\beta=1}{\simeq} 0.442. \tag{54}$$

Secondly, the height of the maximum $K_{\max}$ does depend on the waiting time s_w due to the factor $K_0(s_w)$. Specifically, it can be shown that $K_{\max}$ is a monotonically decreasing function of the waiting time s_w that vanishes in the limit as $s_w \to \infty$.

In order to check the above results, we have fixed the initial and final drivings in the Kovacs protocol χ_0 and χ and changed the intermediate driving value χ_1. We do so to simplify the comparison, because the time scale s involves the steady value of the temperature; see Equation (41). Note that the smaller χ_1 is, the shorter the waiting time becomes. Therefore, one expects to get a Kovacs hump whose maximum remains at $s - s_w \simeq 0.44$, but raises as χ_1 decreases. This is shown in Figure 3, where the numerical solution of the first Sonine approximation equations (42) and the analytical result (53) are compared. Their agreement is almost perfect for the two lowest curves, corresponding to $\chi_1 = 0.99$ and $\chi_1 = 0.95$, as expected, but is still remarkably good for the two topmost ones, corresponding to the not-so-small jumps for $\chi_1 = 0.8$ and $\chi_1 = 0.5$.

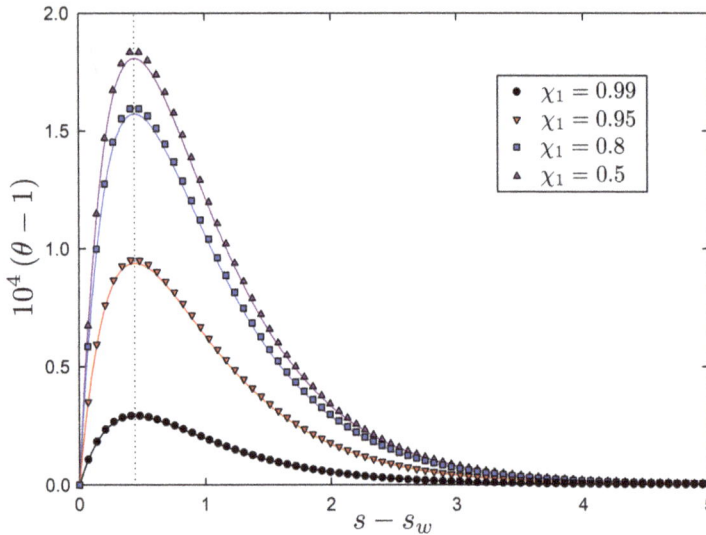

Figure 3. (Color online) Kovacs hump in linear response. We have fixed the initial and final drivings, $\chi_0 = 1.05$ and $\chi = 1$, and considered four values for the intermediate driving $\chi_1 = \{0.5, 0.8, 0.95, 0.99\}$. The linear response theory result (53) (solid line) perfectly agrees with the numerical solution of the first Sonine approximation (symbols), Equation (42). Furthermore, the theoretical prediction for the maximum (54), which again agrees with the numerical results, is plotted (dotted line).

4.2.2. Nonlinear Regime

Furthermore, we explore the Kovacs effect out of the linear regime. Figure 4 is similar to Figure 3, but for larger temperature (or driving) jumps. We have also fixed the initial and final values of the driving, $\chi_0 = 10$ and $\chi = 1$. The intermediate values of the driving are the same as in the linear case except for the largest one, $\chi_1 = 0.99$, which we have omitted for the sake of clarity (its hump is too small in the scale of the figure). Now, the linear response theory results just provide the qualitative behavior of the hump, correctly predicting the position of the maximum, but not its height. On the one hand, and consistent with the numerical results in an active matter model [15], the Kovacs hump out of the linear response regime is larger than the prediction of linear response theory. On the other hand, the position of the maximum remains basically unchanged, and its height still increases as χ_1 decreases.

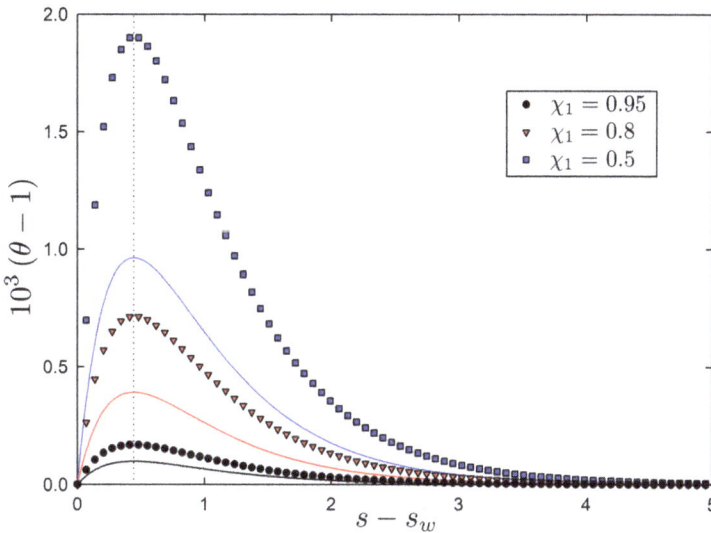

Figure 4. (Color online) Kovacs hump out of the linear regime. The initial driving is much higher than that in Figure 3, $\chi_0 = 10$, whereas the final and intermediate values of the driving are again $\chi = 1$ and $\chi_1 = \{0.5, 0.8, 0.95\}$. The linear response theoretical expression (48) (solid line) remains quite below the numerical solutions of the first Sonine approximation (42) (symbols). The theoretical expression for the maximum in linear response (54) (dotted line) still gives a good description thereof; see also Figure 5.

We also compare our analytical expansion in a_2^s with the numerical solutions of Equations (42) for large jumps. Specifically, in order to make things as simple as possible, we choose $\chi_1 = 0$. If the waiting time is long enough, the system reaches the homogeneous cooling state and $a_2(s_w) = a_2^{HCS}$, which is then the initial condition for the final stage of the Kovacs protocol. Moreover, we can compute the location of the maximum of the hump from Equation (52), obtaining:

$$s_k - s_w = \frac{2}{\kappa_2 - 2 - \beta} \ln\left(\frac{2\kappa_2}{2 + \beta}\right) \underset{\beta=1}{\simeq} 0.437. \tag{55}$$

This result is numerically indistinguishable from that of linear response, as given by Equation (54), since the relative error is around 1%.

In Figure 5, we put forward a comparison between the Kovacs hump obtained from the numerical solution of the first Sonine approximation equations and our theoretical expression for the nonlinear regime, Equation (52). Fixing $\xi_1 = 0$ and $\xi = 1$, as ξ_0 increases (as the waiting time is increased), the hump approaches Equation (52) with $a_2^{ini} = a_2^{HCS}$. Moreover, our theory perfectly reproduces all the numerical curves when we substitute the actual values of a_2^{ini} into Equation (52).

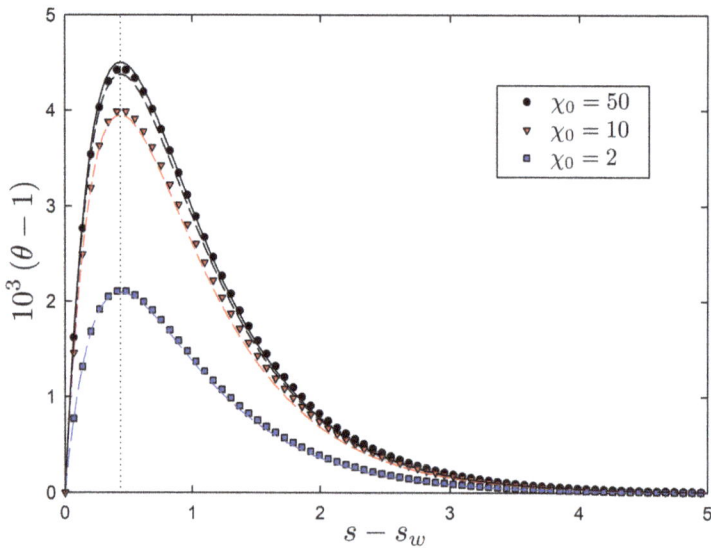

Figure 5. (Color online) Kovacs hump in the nonlinear regime and prediction of the perturbative expansion in a_2^s. We have considered the following values of the drivings: $\chi_0 = \{2, 10, 50\}$, $\chi_1 = 0$ and $\chi = 1$. Symbols stand for the numerical solutions of the first Sonine approximation (42), whereas lines correspond to the theoretical expression (52). For the solid line, $a_2^{ini} = a_2^{HCS}$, while for the dashed lines, we have used the value of a_2^{ini} in the numerical solution. A perfect agreement is observed. Finally, we have plotted the theoretical expression for the maximum position in nonlinear response (dotted line), Equation (55), which also shows an excellent agreement with numerics.

4.3. Monotonicity of an H-Functional

The non-monotonicity in the relaxation of the granular temperature that is brought about by the Kovacs protocol is not automatically transferred to other relevant physical magnitudes. Specifically, here, we deal with the H-functional:

$$H(t) = \int_{-\infty}^{+\infty} dv f(v, t) \log \left[\frac{f(v, t)}{f_s(v)} \right],$$ (56)

There is strong numerical evidence about $H(t)$ being a Lyapunov functional for granular fluids, thus allowing it to be considered as an out-of-equilibrium entropy relative to that of the NESS, in this context [19,20]. Moreover, it has been analytically proven that $H(t)$ is a Lyapunov functional in our system for the Maxwell collision rule $\beta = 0$ [21].

We have computed $H(t)$ numerically from Equation (56) within the first Sonine approximation (that is, we have substituted both $f(v, t)$ and $f_s(v)$ by their expressions in the first Sonine approximation and calculated the integral numerically) for the Kovacs protocols considered in Figures 3 and 4. Once more, we have taken $\beta = 1$, for which the analytical proof in [21] does not hold. The results are shown in Figure 6 and make it clear that $H(t)$ still monotonically decreases for the Kovacs-like protocols, in both the linear (left panel) and nonlinear (right panel) regimes. At the time of the maximum in the hump, $s - s_w \simeq 0.44$, no special signature is observed in the entropy.

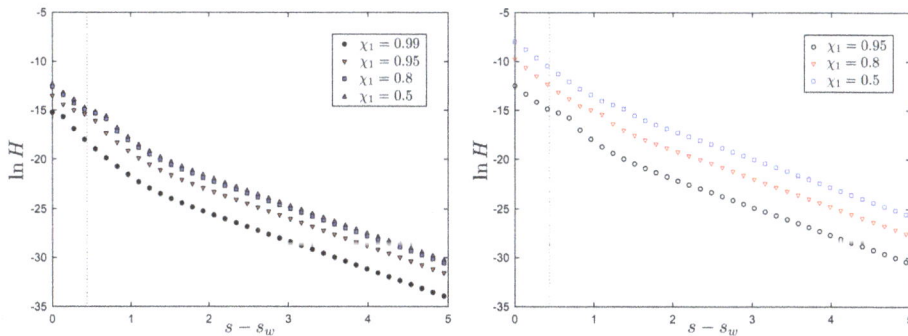

Figure 6. (Color online) Time evolution of the H-functional. The relaxation of H is shown to be monotonic even for Kovacs-like experiments. The left and right panels correspond to the protocols in Figures 3 (filled symbols) and 4 (open symbols), that is to the linear and nonlinear regimes. The vertical dotted line marks the theoretical position of the maximum in the corresponding regime.

5. Discussion

One of the main results in our paper is the extension of the linear response expression for the Kovacs hump in thermal systems, as given by Equations (1) and (2), to the realm of athermal systems, Equations (20) and (21). This extension is (i) not trivial, since athermal systems tend in the long time limit to an NESS, not to an equilibrium state, and (ii) very general, since it can be done starting from the evolution equations at either the mesoscopic or the macroscopic level of description. Therefore, it means that a Kovacs-like effect is to be expected for a very wide class of physical properties Y in a very wide class of systems, basically as long as the relaxation function $\phi_Y(t)$ to the NESS is monotonic.

This theoretical result has been checked in a class of systems that mimic the dynamics of the shear component of the velocity of a granular fluid. In the linear response regime, we have found a good agreement between the theoretical prediction and the numerical results. Furthermore, we have investigated how the linear response result extends to the nonlinear regime. In this region, the linear response theory results remain useful at a qualitative level, but the actual values for the hump lie well above the linear prediction. Interestingly, this kind of giant Kovacs hump has been recently reported in active systems [15]. In addition, the nonlinear Kovacs hump can be theoretically explained by an expansion in the excess kurtosis.

Our work also opens new interesting perspectives for future research. Remarkably, the Kovacs-like memory effects analyzed here involve only jump protocols for the control parameter, which is time-independent both in the waiting time window $0 < t < t_w$ (value ξ_1) and in the return to stationarity region $t > t_w$ (value ξ). Very recently, the response of the system to continuous perturbations [31,32] has been studied, which raises the interesting question of building up a theory, similar to ours, in which the control is continuously varied up to t_w in a first stage and then abruptly changed to the value that corresponds to the stationary value of the relaxing physical observable. To the best of our knowledge, a thorough study of this kind of response is lacking. Moreover, the analysis of the nonlinear response in light of generalized fluctuation-dissipation relations [33] is worthy of consideration.

On another note, our analysis of the Kovacs effect in the model, being restricted to the quasi-elastic limit, has not found the anomalous behavior shown by a gas of inelastic hard spheres for large enough inelasticity in the nonlinear regime [12,13]. The possibility of such a behavior in linear response, either in the model or in the granular gas, deserves further investigation. In addition, our work clearly shows the compatibility of the non-monotonic decay of the granular temperature (or the corresponding relevant physical variable) and the monotonic decay of the nonequilibrium entropy or

H-functional [19–21,34]. In this respect, to elucidate if the hump leaves some signature in the decay of the nonequilibrium entropy is compelling.

Acknowledgments: We acknowledge the support of the Spanish Ministerio de Economía y Competitividad through Grant FIS2014-53808-P. Carlos A. Plata also acknowledges the support from the FPUFellowship Programme of the Spanish Ministerio de Educación, Cultura y Deporte through Grant FPU14/00241.

Author Contributions: All authors contributed equally to this work. All authors have read and approved the final manuscript.

Conflicts of Interest: The authors declare no conflict of interest. The founding sponsors had no role in the design of the study; in the collection, analyses or interpretation of data; in the writing of the manuscript; nor in the decision to publish the results.

Appendix A. Simulation Algorithm

We have made use of a residence time algorithm that gives the numerical integration of a master equation in the limit of infinite trajectories [35,36]. The basic numerical recipe is as follows:

1. At time τ, a random "free time" $\tau_f > 0$ is extracted with an exponential probability density $\Omega(v) \exp\left[-\Omega(v)\tau_f\right]$, where $\Omega(v) = \sum_l \omega |v_l - v_{l+1}|^\beta$ depends on the state of the system v;
2. Time is advanced by such a free time $\tau \to \tau + \tau_f$;
3. A pair $(l, l+1)$ is chosen to collide with probability $\omega |v_l - v_{l+1}|^\beta / \Omega(v)$;
4. All particles are heated by the stochastic thermostat, by adding independent Gaussian random numbers of zero mean and variance $\chi \tau_f$ to their velocities;
5. In order to conserve momentum, the mean value of the random numbers generated in the previous step is subtracted from the velocities of all particles;
6. The process is repeated from Step 1.

References

1. Kovacs, A.J.; Aklonis, J.J.; Hutchinson, J.M.; Ramos, A.R. Isobaric volume and enthalpy recovery of glasses. II. A transparent multiparameter theory. *J. Polym. Sci.* **1979**, *17*, 1097–1162.
2. Bertin, E.M.; Bouchaud, J.P.; Drouffe, J.M.; Godreche, C. The Kovacs effect in model glasses. *J. Phys. A* **2003**, *36*, 10701.
3. Buhot, A. Kovacs effect and fluctuation–dissipation relations in 1D kinetically constrained models. *J. Phys. A* **2003**, *36*, 12367.
4. Mossa, S.; Sciortino, F. Crossover (or Kovacs) effect in an aging molecular liquid. *Phys. Rev. Lett.* **2004**, *92*, 045504.
5. Aquino, G.; Leuzzi, L.; Nieuwenhuizen, T.M. Kovacs effect in a model for a fragile glass. *Phys. Rev. B* **2006**, *73*, 094205.
6. Prados, A.; Brey, J.J. The Kovacs effect: A master equation analysis. *J. Stat. Mech.* **2010**, *2010*, 368–369.
7. Diezemann, G.; Heuer, A. Memory effects in the relaxation of the Gaussian trap model. *Phys. Rev. E* **2011**, *83*, 031505.
8. Ruiz-García, M.; Prados, A. Kovacs effect in the one-dimensional Ising model: A linear response analysis. *Phys. Rev. E* **2014**, *89*, 012140.
9. Van Kampen, N.G. *Stochastic Processes in Physics and Chemistry*; North-Holland: Amsterdam, The Netherlands, 1992.
10. Pöschel, T.; Luding, S. *Granular Gases Vol. 564 of Lecture Notes in Physics*; Springer: Berlin/Heidelberg, Germany, 2001.
11. Van Noije, T.P.C.; Ernst, M.H. Velocity distributions in homogeneous granular fluids: The free and the heated case. *Granul. Matter* **1998**, *1*, 57–64.
12. Prados, A.; Trizac, E. Kovacs-Like Memory Effect in Driven Granular Gases. *Phys. Rev. Lett.* **2014**, *112*, 198001.
13. Trizac, E.; Prados, A. Memory effect in uniformly heated granular gases. *Phys. Rev. E* **2014**, *90*, 012204.
14. Lahini, Y.; Gottesman, O.; Amir, A.; Rubinstein, S.M. Nonmonotonic Aging and Memory Retention in Disordered Mechanical Systems. *Phys. Rev. Lett.* **2017**, *118*, 085501.

15. Kürsten, R.; Sushkov, V.; Ihle, T. Giant Kovacs-Like Memory Effect for Active Particles. *Phys. Rev. Lett.* **2017**, in press.

16. Lasanta, A.; Manacorda, A.; Prados, A.; Puglisi, A. Fluctuating hydrodynamics and mesoscopic effects of spatial correlations in dissipative systems with conserved momentum. *New J. Phys.* **2015**, *17*, 083039.

17. Manacorda, A.; Plata, C.A.; Lasanta, A.; Puglisi, A.; Prados, A. Lattice Models for Granular-Like Velocity Fields: Hydrodynamic Description. *J. Stat. Phys.* **2016**, *164*, 810–841.

18. Plata, C.A.; Manacorda, A.; Lasanta, A.; Puglisi, A.; Prados, A. Lattice models for granular-like velocity fields: Finite-size effects. *J. Stat. Mech.* **2016**, *2016*, 093203.

19. Marconi, U.M.B.; Puglisi, A.; Vulpiani, A. About an H-theorem for systems with non-conservative interactions. *J. Stat. Mech.* **2013**, *2013*, 114–124.

20. García de Soria, M.I.; Maynar, P.; Mischler, S.; Mouhot, C.; Rey, T.; Trizac, E. Towards an H-theorem for granular gases. *J. Stat. Mech.* **2015**, *2014*, P11009.

21. Plata, C.A.; Prados, A. Global stability and *H*-theorem in lattice models with nonconservative interactions. *Phys. Rev. E* **2017**, *95*, 052121.

22. Brey, J.J.; Prados, A. Stretched exponential decay at intermediate times in the one-dimentional Ising model at low temperatures. *Physica A* **1993**, *197*, 569–582.

23. Prasad, V.V.; Sabhapandit, S.; Dhar, A.; Narayan, O. Driven inelastic Maxwell gas in one dimension. *Phys. Rev. E* **2017**, *95*, 022115.

24. Ben-Naim, E.; Krapivsky, P.L. The inelastic Maxwell model. In *Granular Gas Dynamics*; Pöschel, T., Brilliantov, N., Eds.; Lecture Notes in Physics; Springer: Berlin/Heidelberg, Germany, 2003; Volume 624, pp. 65–94.

25. Ernst, M.H.; Trizac, E.; Barrat, A. The rich behavior of the Boltzmann equation for dissipative gases. *EPL* **2006**, *76*, 56–62.

26. Montanero, J.M.; Santos, A. Computer simulation of uniformly heated granular fluids. *Granul. Matter* **2000**, *2*, 53–64.

27. Maynar, P.; García de Soria, M.I.; Trizac, E. Fluctuating hydrodynamics for driven granular gases. *Eur. Phys. J. Spec. Top.* **2009**, *179*, 123–139.

28. Prasad, V.V.; Sabhapandit, S.; Dhar, A. High-energy tail of the velocity distribution of driven inelastic Maxwell gases. *EPL* **2013**, *104*, 54003.

29. Plata, C.A.; Prados, A. Kinetic description of a class of systems with non-conservative interactions. **2018**, unpublished.

30. Abramowitz, M.; Stegun, I.A. *Handbook of Mathematical Functions*; Dover: New York, NY, USA, 1972.

31. Falasco, G.; Baiesi, M. Temperature response in nonequilibrium stochastic systems. *EPL* **2016**, *113*, 20005.

32. Falasco, G.; Baiesi, M. Nonequilibrium temperature response for stochastic overdamped systems. *New J. Phys.* **2016**, *18*, 043039.

33. Lippiello, E.; Corberi, F.; Sarracino, A.; Zannetti, M. Nonlinear response and fluctuation-dissipation relations. *Phys. Rev. E* **2008**, *78*, 041120.

34. Brey, J.J.; Prados, A. Normal solutions for master equations with time-dependent transition rates: Application to heating processes. *Phys. Rev. E* **1993**, *47*, 1541.

35. Bortz, A.B.; Kalos, M.H.; Lebowitz, J.L. A new algorithm for Monte Carlo simulation of Ising spin systems. *J. Comput. Phys.* **1975**, *17*, 10–18.

36. Prados, A.; Brey, J.J.; Sánchez-Rey, B. A dynamical Monte Carlo algorithm for master equations with time-dependent transition rates. *J. Stat. Phys.* **1997**, *89*, 709–734.

entropy

MDPI

Article

Hydrodynamics of a Granular Gas in a Heterogeneous Environment

Francisco Vega Reyes [1,*] and Antonio Lasanta [2]

[1] Departamento de Física, Universidad de Extremadura, 06071 Badajoz, Spain
[2] G. Millán Institute, Fluid Dynamics, Nanoscience and Industrial Mathematics, Department of Materials Science and Engineering and Chemical Engineering, Universidad Carlos III de Madrid, 28911 Leganés, Spain; alasanta@ing.uc3m.es
*   Correspondence: fvega@unex.es

Received: 21 August 2017; Accepted: 9 October 2017; Published: 11 October 2017

Abstract: We analyze the transport properties of a low density ensemble of identical macroscopic particles immersed in an active fluid. The particles are modeled as inelastic hard spheres (granular gas). The non-homogeneous active fluid is modeled by means of a non-uniform stochastic thermostat. The theoretical results are validated with a numerical solution of the corresponding the kinetic equation (direct simulation Monte Carlo method). We show a steady flow in the system that is accurately described by Navier-Stokes (NS) hydrodynamics, even for high inelasticity. Surprisingly, we find that the deviations from NS hydrodynamics for this flow are stronger as the inelasticity decreases. The active fluid action is modeled here with a non-uniform fluctuating volume force. This is a relevant result given that hydrodynamics of particles in complex environments, such as biological crowded environments, is still a question under intense debate.

Keywords: granular gas; heterogeneous media; hydrodynamics; active matter non-equilibrium

1. Introduction

Biological Systems, and more specifically, systems with active particles, are in general out of equilibrium, which implies that for these systems the results of equilibrium statistical mechanics do not in general apply. The term *active matter* refers to a system with an ensemble of particles that are able to transform energy—internal or environmental—into movement [1]. A correct description of transport phenomena in this type of systems can be achieved *via* a specific kinetic theory, that takes into account the peculiar energy processing of these particles, see for example [2–5] for the well known Vicsek-like models [6].

In recent years, the study of active particles in crowded environments has become the subject of intense study [7]. One of the motivations is the large number of promising applications derived from the use of active particles as micro-machines [7]. In addition, the behaviour of inert particles inside complex active environments remains an open question. For instance, there are experimental studies on inert Brownian particles trapped by a harmonic potential in a fluid composed by bacteria [8]. Also, Sándor et al. [9] study active Brownian particles on a travelling-wave substrate. Other works study particles which can be trapped by holes [10,11]. More realistic active systems such as biological tissues, share important analogies with granular matter, namely glass transition, cage effect, fluidization, solidification and the appearance of giant density fluctuations [12–15]. There, active complex non-homogeneous forces appear. Also the non-Gaussian behaviour of diffusing non-active particles in heterogeneous media has been studied recently for a spatially varying diffusion coefficient [16].

On the other hand, the transport properties of granular matter are of interest at a fundamental level and also for applications, since granular matter is present in a variety of industry and technology sectors [17–20]. Granular transport theories usually make a connection with traditional continuum mechanics theories, due to fundamental similarities observed with classical fluids [17,19]. Granular particles loose a fraction of their kinetic energy after collision and for this reason we need to continuously excite the system, in order to sustain the dynamics. This excitation usually comes from the system boundaries, which results in inhomogeneous flows that may eventually become steady [17,19,21].

We propose in this work a study that tries to model a system of macroscopic inert particles that are immersed in an active fluid with inhomogeneous temperature. For this, we model the system as a granular fluid [22] subject to a nonuniform stochastic volume force (in the form of a white noise [23]). The inelastic cooling term, inherent in a granular fluid [19], and a stochastic force allow us in this work to model the energy sink and source terms that are characteristic to active particle systems [1,7]. More specifically, we focus on a very low density granular system (i.e., a *granular gas*) surrounded by a low density (inhomogeneous) fluid that can be considered as a coarse grained version of an active matter system. Since the system has very low density, particle velocity correlations can be neglected and thus the the inelastic Boltzmann equation applies [19,24]). Moreover, systems of solid particles suspended in a low density interstitial fluid usually fall into the grain-grain collision dominated regime [25], as opposed to the interstitial fluid viscosity-dominated regime, that applies for most cases of interstitial liquids [26].

In previous theoretical works, granular gas fluidization has been studied only in the simplistic case of a homogeneous interstitial fluid [27–32]. And yet this theoretical homogeneous state is not quite a realistic situation [33]. Neat examples of inhomogeneous granular gas suspensions are the large sand plumes that can be observed in the Earth's atmosphere for weeks [34]. A homogeneous energy source is obviously not a realistic situation either in active fluids, at a biological level [1]. For this reason we focus in this work on the more elaborate case of a non-uniform interstitial fluid.

However, the intent of this work is not experimental validation but to extract a more clear picture of the relevant physics of this kind of systems. As a result, we show there are granular gas flows immersed in an inhomogeneous active fluid that can be accurately described with Navier-Stokes (NS) hydrodynamics, even at high inelasticities. Furthermore, we surprisingly found cases where NS hydrodynamics is a better approximation for the granular gas than for a perfectly elastic gas subjected to the same boundary forcing conditions. In particular, we will show that, once the granular gas is fluidized, the intensity of the spatial gradients in response to external excitations (those coming from the boundaries) can actually be weaker for granular gases than for elastic gases.

In a previous experimental work [35], local balance between the total energy input (fluctuating force plus boundary heating) and the energy sink was found. We use now this kind of balance condition. As we will show below, this local balance results, in the specific set-up considered in the present work, in a steady flow with uniform heat flux throughout the system. This balance condition is in fact analogous to the one occurring in well known non-Newtonian granular flows like the uniform shear flow [36,37].

The structure of the paper is as follows: in Section 2 we present a description of the system and the corresponding steady state equations. In Section 3 we describe the flows where there is a balance between inelastic cooling coming from particle collisions and volume energy input (and with no shear). Section 4 is devoted to a brief discussion on the steady flows resulting from adding a weak shear Section 5 presents a discussion on the transport coefficients and rheology properties of the studied flows and the paper conclusions are drawn in Section 6. The simulation method and transport coefficient equations are discussed in the Appendices A and B respectively.

2. System and Steady Base State Equations

2.1. Description of the System

Our system consists of a set of identical inert smooth hard spheres (or disks) immersed in an active fluid. The system is limited by two infinite parallel hard walls, located at planes $y = \pm L/2$ respectively. The walls act like two distinct kinetic energy sources, characterized with temperatures $T_\pm$. They may have relative movement (wall velocities $U_\pm$ respectively), eventually inducing the particles to continuously flow along the channel between them.

Collisions between the inert particles do not preserve energy (i.e., particle collisions are *inelastic*). The particles have a diameter (radius) that we denote as σ and their mass is m. For inelastic smooth hard spheres/disks, inelasticity may be characterized accurately, in a range of experimental situations, by a constant coefficient of normal restitution α [38,39]. This coefficient ranges from 1 for purely elastic collisions to 0 for purely inelastic collisions. In addition, we model the interstitial active fluid injection of thermal energy onto the granular gas particles as a stochastic force $\mathbf{F}^{st}$. The equation of motion for a particle i can be written

$$m\dot{\mathbf{v}}_i = \mathbf{F}_i^{st}(\mathbf{r}, t) + \mathbf{F}_i^{coll}, \tag{1}$$

where $\mathbf{F}_i^{coll}$ is the force due to inelastic collisions, $\mathbf{F}_i^{st}(\mathbf{r}, t)$ is the force exerted by the heterogeneous medium and m the mass of particles (set to 1 for simplicity).

We can model this interaction by means of a fluctuating volume force [40], that we denote as $\mathbf{F}^{st}(\mathbf{r})$ and fulfills the conditions [32,35,40,41]

$$\begin{aligned} \langle \mathbf{F}_i^{st}(\mathbf{r}, t) \rangle &= \mathbf{0}, \\ \langle \mathbf{F}_i^{st}(\mathbf{r}, t) \mathbf{F}_j^{st}(\mathbf{r}, t') \rangle &= \mathbf{1} m^2 \xi(\mathbf{r})^2 \delta_{ij} \delta(t - t'), \end{aligned} \tag{2}$$

being $\xi^2(\mathbf{r})$ the fluctuating force intensity [32,41] (that has dimensions of velocity squared times time), $\mathbf{1}$ the unit matrix in d dimensional space, δ_{ij} is the Kronecker delta function and $\langle A \rangle$ indicates average of magnitude A over a sufficiently long time interval (long compared with the characteristic microscopic time scale). It has been shown in the case of the large mass limit and homogeous energy injection, that Equation (1) for a particle can be written as a Langevin equation corresponding to a granular brownian particle [42].

On the other hand, collisions between two grains follow the inelastic smooth hard sphere collisional model

$$\mathbf{v}_1' = \mathbf{v}_1 - \frac{1 + \alpha}{2} (\sigma \cdot \mathbf{g}) \sigma \tag{3}$$

$$\mathbf{v}_2' = \mathbf{v}_2 - \frac{1 + \alpha}{2} (\sigma \cdot \mathbf{g}) \sigma \tag{4}$$

with $\sigma = (\mathbf{r}_1 - \mathbf{r}_2)/|\mathbf{r}_1 - \mathbf{r}_2|$ and $\mathbf{g} = \mathbf{v}_1 - \mathbf{v}_2$. The primes denote pre-collisional velocities.

Notice that in (2) we have $\xi = \xi(\mathbf{r})$ (a space-dependent noise intensity). The noisy term appears in the inelastic Boltzmann equation as a Fokker-Planck-like operator [43,44]

$$\left(\frac{\partial}{\partial t} + \mathbf{v} \cdot \nabla - \frac{1}{2} \xi(\mathbf{r})^2 \frac{\partial^2}{\partial v^2} \right) f(\mathbf{r}, \mathbf{v}; t) = J[\mathbf{v} | f, f], \tag{5}$$

where J is the collisional integral for inelastic hard spheres and whose expression may be found elsewhere [19,45], $\mathbf{v}$ is particle velocity, and v its modulus.

Taking into account the system geometries, we consider only $\xi(\mathbf{r}) = \xi(y)$, that is the simplest space-dependence our geometry configuration allows for an inhomogeneous active fluid. The steady base states may be deduced from the reduction of the kinetic Equation (5)

$$v_y \frac{\partial f}{\partial y} - \frac{1}{2} \xi(y)^2 \frac{\partial^2 f}{\partial v^2} = J[f, f]. \tag{6}$$

We have numerically solved the kinetic Equation (6) by means of the Direct Simulation Monte Carlo method [18,46], applying in each case the corresponding boundary conditions and heterogeneous fluctuating force properties. For more details on the simulation algorithm, see the corresponding section in the Appendix A.

2.2. Steady Base State Equations

Multiplying by velocity momenta the kinetic Equation (6) and performing velocity integrals, we may easily obtain from (6) the mass, momentum and energy balance equations [43,47]

$$\frac{\partial P_{yy}}{\partial y} = 0, \quad 0 = -\frac{1}{mn}\frac{\partial P_{xy}}{\partial y} \tag{7}$$

$$T\zeta(\alpha) - m\xi^2(y) = -\frac{2}{dn}\left(P_{xy}\frac{\partial u_x}{\partial y} + \frac{\partial q_y}{\partial y}\right), \tag{8}$$

where **u** is the flow velocity field (with u_y its y-component), n is the particle density, and P and q are momentum flux tensor (also called stress tensor) and heat flux vector respectively, and $\zeta(\alpha)$ is the cooling rate, a magnitude that emerges from the energy balance due to inelasticity in the collisions [18,19]. Equations (7) and (8) contain no approximations; i.e., they are exact for this type of geometry, even if the steady state were not hydrodynamic. Notice also that the stress tensor element P_{yy} is homogeneous and is not a function of time. This condition breaks if the flow is not laminar, but we will only consider laminar flows in this work.

We take now into account the characteristics of NS hydrodynamics in (7) and (8). In (7) we use that all diagonal elements of the stress tensor are equal and we also use that the horizontal heat flux q_x is null (they both raise from fluxes terms that are of second order in the gradients [48]) and thus, from the first equation in (7), we obtain that $P_{yy} = p = $ constant. We also use the fluxes expressions at NS order [45]: $P_{xy} = \eta(\alpha)\partial u_x/\partial y$, $q_y = -\lambda \partial T/\partial y - \mu \partial n/\partial y$, where η is the viscosity and λ, μ are the heat flux transport coefficients. With this, the steady state forms of (7) and (8) are

$$p = \text{constant}, \quad P_{xy} = -\eta(\alpha)\frac{\partial u_x}{\partial y} = \text{constant} \tag{9}$$

$$T\zeta(\alpha) - m\xi^2(y) =$$
$$\frac{2}{dn}\left(\eta(\alpha)\left(\frac{\partial u_x}{\partial y}\right)^2 + \kappa(\alpha)\frac{\partial T}{\partial y} + \mu(\alpha)\frac{\partial n}{\partial y}\right). \tag{10}$$

Next, we take into account that the NS transport coefficients and cooling rate scale in the following forms with temperature [45]: $\eta(\alpha) = \eta^*(\alpha)\eta_0' T^{1/2}$, $\kappa(\alpha) = \kappa^*(\alpha)\kappa_0' T^{1/2}$, $\mu = \mu^*(\alpha)\kappa_0' T^{3/2}/n$, $\zeta = \zeta^*(\alpha)p/(\eta_0' T^{1/2})$. Here, the coefficients η_0', κ_0' are related to the elastic gas [49] NS viscosity $\eta_0 = \eta_0' T^{1/2}$ and thermal conductivity $\kappa_0 = \kappa_0' T^{1/2}$, and thus

$$\eta_0' = \frac{d+2}{8}\Gamma(d/2)\pi^{-(d-1)/2}m^{1/2}\sigma^{-(d-1)}/8, \tag{11}$$

$$\kappa_0' = \frac{d(d+2)^2}{16(d-1)}\Gamma(d/2)\pi^{-(d-1)/2}m^{-1/2}\sigma^{-(d-1)}. \tag{12}$$

where Γ is the gamma function [50] and $d = 2$ for disks and $d = 3$ for spheres. Thus, the steady state balance Equations (9) and (10) yield

$$-\eta(\alpha)^* \eta_0' \sqrt{\frac{T}{T_r}} \frac{\partial u_x}{\partial y} = \text{constant},$$ (13)

$$\sqrt{\frac{T}{T_r}} \frac{\partial}{\partial y} \left(\sqrt{\frac{T}{T_r}} \frac{\partial T}{\partial y} \right) =$$
$$\frac{d}{2} \frac{\zeta_0^* \, p^2}{\beta_0^* \, T_r} - \frac{d}{2} \frac{m\tilde{\zeta}^2(y) n T^{1/2}}{\beta_0^* T_r} - \frac{\eta_0^*}{\beta_0^*} \frac{T}{T_r} \left(\frac{\partial u_x}{\partial y} \right)^2,$$ (14)

where $\beta^*(\alpha) \equiv \kappa^*(\alpha) - \mu^*(\alpha)$, and we call it *effective thermal conductivity* and T_r is the reference unit of temperature, that we choose as $T_r = T_-$ (temperature at the lower, and colder, wall). Here, $\zeta^*(\alpha)$ is the dimensionless cooling rate [18,51] and the expressions of the reduced transport coefficients viscosity $\eta^*(\alpha)$, thermal conductivity $\kappa^*(\alpha)$, and $\mu^*(\alpha)$ that we used are the ones obtained by Garzó and Montanero [52] for a granular gas heated by a white noise. Their expressions are displayed in the transport coefficients section in the Appendix B.

We use a scaled space variable l that allows us to find an analytical solution for the flow velocity (13) and the temperature (14) profiles. This variable is defined by the relation $\sqrt{T/T_r} \partial/\partial y = \partial/\partial l$. When expressed in terms of this scaled variable, Equations (9) and (10) read

$$\frac{\partial \hat{u}_y}{\partial \hat{l}} = a,$$ (15)

$$\frac{\partial^2 \hat{T}}{\partial \hat{l}^2} = -\gamma(\alpha, l),$$ (16)

where a and $-\gamma$ are dimensionless magnitudes in the system and we call them *local shear rate* and *temperature curvature*, respectively. In (15), (16) and henceforth we use dimensionless variables indicated with a hat, where we have chosen as set of units: T_r for temperature (already defined), m for mass, p for pressure (the steady state hydrostatic pressure, that as we said is a global constant for the system). As a unit for length we use

$$\lambda_r = \sqrt{2}(2 + d)\Gamma(d/2)\pi^{-(d-1)/2}(\sqrt{2}n_r\sigma^{(d-1)})^{-1},$$ (17)

that is the reference mean free path [49] and for time we use

$$\nu_r^{-1} = \lambda_r(\sqrt{m/T_r})$$ (18)

(inverse of reference collision frequency). These two definitions imply that the unit for velocity is $\sqrt{T_r/m}$. We indicate the spatial coordinate y in dimensionless form as $\hat{y}$, and analogously for all dimensionless magnitudes.

The reader may infer the physical meaning of a and γ with the aid of (7) and (8): a (reduced shear rate) measures how rapidly varies the flow velocity field, in units of the mean free path; $|-\gamma|^{1/2}$ measures the degree of imbalance between energy volume sinks and sources per unit time as we explain above in Equation (18), this unit being the inverse of the collision frequency). The differential Equation (16) has been observed for the granular temperature in similar experimental configurations [33].

With the presence of the stochastic volume force in the energy balance Equation (8), it is straightforward to deduce from (14)–(16) and the definitions above that $-\gamma(\alpha, l)$ has the form

$$-\gamma(\alpha) \equiv \frac{2(d-1)}{d(d+2)\beta^*(\alpha)} \times \tag{19}$$
$$\left(\frac{d}{2}(\zeta^*(\alpha) - \hat{\xi}^2(\hat{l})\hat{T}(\hat{l})^{-1/2}) - \eta^*(\alpha)a^2 \right),$$

that, as it can be noticed, depends in general on position through $\hat{l}$.

3. Steady Base States with Energy Balance and no Shear

We now want to look up for steady states in a granular gas fluidized by an heterogeneous source that may be accurately described by hydrodynamics at NS order. We will denote them as 'reference states'. As in rarefied gases, a close to unity Knudsen number (Kn) signals hydrodynamics failure. Moreover, hydrodynamics at first order in the gradients (NS equations) are usually only valid if Kn takes low values [19,46]. The Knudsen number Kn is defined as a representative microscopic time or length scale over a macroscopic time or length scale. However, the choice of reference scales is not unique and picking an appropriate Knudsen number for each specific flow problem is a subtle question [53].

It was shown in a previous work [47] that Couette-Fourier flows (i.e., those occurring in our system, under the appropriate circumstances) are characterized by three representative and independent microscopic vs. macroscopic relative scales (or Knudsen numbers) that happen to be constant throughout the system: a, $|\gamma|^{1/2}$ (look at Equations (15) and (16)) and $\Delta T/L$, with $\Delta T \equiv T_+ - T_-$, where $\Delta T/L$ measures the size of the temperature gradient imposed by the boundaries, this gradient being referred to mean free path units. These numbers arise from the specific form of the relevant differential equations for steady states, and the boundary geometry. Since the geometry of the differential Equations (15) and (16) and boundary conditions are the same in this work, we may safely use the same set of reference Knudsen numbers (see previous work [47] for more details on this issue). Furthermore, another previous analysis [51] for granular gases without an interstitial fluid showed that $\Delta T/L$ has a much weaker effect on the emergence non-Newtonian behavior.

Thus, let us analyze the class of solutions for steady flows where this $\Delta T/L$ is the only active source of spatial gradients; i.e., $a = 0$, $\gamma = 0$ and $\Delta T/L \lesssim 1$

$$\frac{\partial \hat{u}_y}{\partial \hat{l}} = 0, \tag{20}$$

$$\frac{\partial^2 \hat{T}}{\partial \hat{l}^2} = 0. \tag{21}$$

Moreover, in the $\alpha = 1$ limit this state corresponds to the classic Fourier flow of a molecular gas. In order to obtain $\gamma = 0$ (with $a = 0$), and taking into account (19) we find

$$\hat{\xi}(\hat{l})^2 \hat{T}(\hat{l})^{-1/2} = \zeta^*(\alpha) \quad \Rightarrow \quad \hat{\xi}(\hat{l})^2 = \zeta^*(\alpha)\hat{T}(\hat{l})^{1/2}. \tag{22}$$

Equation (20) implies that flow velocity $\hat{u}$ is constant respect to $\hat{l}$, which, from the definition of l leads to $\hat{u}(\hat{y})$ also constant (and to all physical effects, flow velocity may be considered as null). Moreover, Equation (21) implies that temperature $\hat{T}$ is linear in $\hat{l}$, which (since $\sqrt{T/T_r}\partial/\partial y = \partial/\partial l$) yields $\hat{y} \propto \hat{l}^{3/2}$ and thus $\hat{T}(\hat{y}) \propto \hat{y}^{2/3}$; i.e., the stochastic force generates a Fourier-like temperature profile [37]. At a physical level, this makes sense if we consider that the granular gas is fluidized by a low viscosity interstitial fluid that is heated from the boundaries. It is easy to deduce from (8), (10) and (19) that $\gamma = 0$ implies that inelastic cooling is balanced by volume energy input and thus, that heat flux is uniform throughout the system.

Therefore our base steady states are (not previously reported) granular flows with uniform heat flux. Summarizing, these NS steady states have the following properties: (i) $P_{xy} = \eta(\alpha)\partial u_x/\partial y$,

$q_y = -\lambda \partial T / \partial y - \mu \partial n / \partial y$, (ii) constant hydrostatic pressure $\hat{p}$, (iii) null normal stress differences ($\hat{P}_{ii} = \hat{P}_{jj} = p$) and null $\hat{q}_x$ [48], (iv) constant $\hat{q}_y$, and (v) linear $\hat{T}(\hat{l})$ profiles, or equivalently, $\hat{T}(\hat{y}) \propto \hat{y}^{3/2}$ [37]. Properties (i) to (iii) are fulfilled by all Couette-Fourier steady flows at NS order [47] and conditions (iv) and (v) are fulfilled by all Couette-Fourier steady flows with uniform heat flux [37].

We need to remark that the relation in (22) between fluctuating force intensity and temperature is not a hypothesis but a consequence of the experimental fact that there is local energy balance. On the other hand, there is an essential difference of this new flow respect to previously studied granular flows with uniform heat flux (the simple shear flow included [36,54]).

Very surprisingly, we can show that contrary to other steady granular flows, the kind of state defined by (22) can be accurately described in the context of Navier-Stokes equations. In order to prove this feature, we have simulated the system described by Equations (20)–(22) by means of the DSMC method (more details on simulations are given in the Appendix A). We have checked that the temperature profiles obtained under steady state in effect follow the behaviour $\hat{T}(\hat{y}) \propto y^{2/3}$, inherent to gas steady flows with uniform heat flux [37]. Results can be seen in Figure 1, for which the agreement with the NS theory prediction is excellent, independently of the inelasticity value.

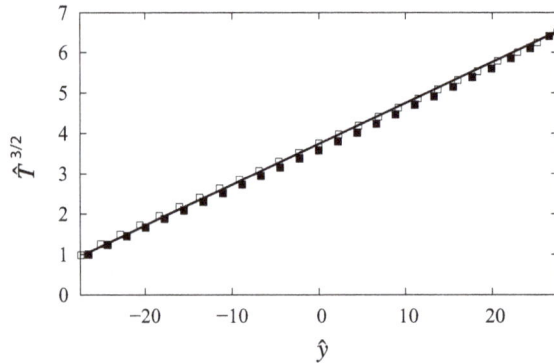

Figure 1. DSMC simulation data ($d = 3$, spheres) for temperature profile for two different values of inelasticity: $\alpha = 0.9$ (solid symbols) and $\alpha = 0.7$ (open symbols) and with the same boundary condition for wall temperature difference ($T_+/T_- = 5$, $\Delta L = 15\bar{\lambda}$) with $\bar{\lambda} = \sqrt{2}\pi\bar{n}\sigma^{d-1}$, $\bar{n}$ being the average density in the system. There is excellent agreement with the theoretical prediction of $\hat{T}(\hat{y}) \propto y^{2/3}$ (solid line stands for the NS theoretical profile), for both α values.

4. Weakly Sheared Steady States

We discuss here weakly sheared states, in order to capture the gradual the eventual appearance of non-Newtonian behavior. Thus, starting out of the reference base states in the previous subsection; i.e., still with condition (22) being fulfilled, we analyze now sheared states ($a \neq 0$), for which we have from (19),

$$-\gamma = -\frac{(d-1)}{(d+2)\beta^*(\alpha)}\eta^*(\alpha)a^2 \neq 0. \tag{23}$$

For this type of flow we have also performed DSMC simulations. Relation (23) helps us understand that an increase in shear rate a (one of the relevant Knudsen numbers) implies an increase in γ (one of the other Knudsen numbers). Thus, for increasing shear we should gradually depart from NS hydrodynamics as both a and $|\gamma|^{1/2}$ increase. However, relation (23) is valid only as long as NS hydrodynamics applies. For non-Newtonian regime, the relation between the temperature curvature γ, local shear rate a and coefficient of restitution α would be more intricate and in general it must be solved numerically [51]. A non-linear theory is beyond the scope of the present work, but, fortunately, we may use results from DSMC to analyze deviations from NS hydrodynamics.

The results we obtained are quite surprising (see Table 1). In effect, we can see that similar values of local shear rate *a* tend to yield greater values of the other Knudsen number $|\gamma|$ for more elastic gases than for more inelastic gases. This means that, contrary to what would be expected, the Knudsen numbers in this type of flow tend globally to be smaller for higher inelasticities, which could indicate that departures from NS hydrodynamics are more important, for similar shear rates, for more elastic gases. However, we will confirm this point in the following Section 5, by measuring from DSMC simulations the set of transport coefficients and rheological properties and comparing with NS theory predictions.

Table 1. DSMC measurements of Knudsen numbers *a* and $|\gamma|^{-1/2}$, for steady states described by (20)–(22), with $T_+/T_- = 5$, $\Delta L = 15\bar{\lambda}$ (i.e., $\Delta T/\Delta L = (4/15)((T_+ - T_-)/\bar{\lambda})$, and ΔU defined as $\Delta U \equiv (U_+ - U_-)/\sqrt{T_-/m}$.

α	$\Delta U = 5$ a	$\Delta U = 5$ $\|\gamma\|^{-1/2}$	$\Delta U = 3$ a	$\Delta U = 3$ $\|\gamma\|^{-1/2}$
0.99	0.173	0.058	0.115	0.036
0.9	0.191	0.034	0.120	0.018
0.8	0.193	0.027	0.119	0.017
0.7	0.193	0.024	0.119	$\lesssim 1 \times 10^{-2}$
0.6	0.188	$\lesssim 1 \times 10^{-2}$	0.115	$\lesssim 1 \times 10^{-2}$
0.5	0.187	$\lesssim 1 \times 10^{-2}$	0.113	$\lesssim 1 \times 10^{-2}$

We also performed an additional series with varying $\Delta T/\Delta L$ (not shown in figures) and checked that the measured transport coefficients do not depend on its value, analogously to the result in a previous work on granular flows with uniform heat flux [37].

5. Transport Coefficients and Rheology

Results in the preceding section suggest that: (a) our steady states are well described by NS hydrodynamics, even for strong inelasticities, as long as shearing from the boundaries is not too strong; (b) deviations from NS hydrodynamics are stronger for more elastic gases. Both observations would be contrary to what occurs in steady granular flows (like the simple shear flow [36], just to put one example).

Let us analyze in more detail to what extent the reference (no-shear) steady states described in Section 3 and the sheared states described in Section 4 are strictly NS flows or not, by analyzing the relevant transport properties. This analysis will be done as a function of the degree of inelasticity in grains collisions and intensity of shearing. In order to analyze non-Newtonian behavior, we need to define the cross thermal ϕ^* conductivity transport coefficient ϕ^* and the reduced normal stresses, defined by

$$q_x = \phi^* \lambda_0 \partial T/\partial y, \quad \theta_i = P_{ii}/p, \tag{24}$$

where λ_0 is the thermal conductivity for a gas with elastic particle collisions [49]. For hydrodynamics at NS order, $q_x, \phi^* = 0$ and $\theta_i = P_{ii}/p = 1$. The degree of divergence from these numerical values quantifies eventual departures from NS hydrodynamics. The appearance of $q_x \neq 0$ and $\theta_i \neq 1$ occurs with the emergence in the fluxes of terms of second order in the gradients (Burnett order). For our geometry, the only remaining second order terms are of form $(\partial T/\partial y)(\partial u_x/\partial y), \partial^2 u_x/\partial y^2$ in the heat flux and of the form $\partial^2 T/\partial y^2, (\partial T/\partial y)^2, (\partial u_x/\partial y)^2$ in the stress tensor (moment flux) respectively (the rest of the gradient second order terms in the fluxes are null for the geometry of our system) [48,49].

In Figure 2, we present the theory-simulation comparison for reduced viscosity $\eta^*(\alpha)$ and effective thermal conductivity $\beta^*(\alpha)$, for different shearing levels. For sufficiently low shearing, the agreement between theory and simulation is very good for the range of inelasticities represented here. In the limit of no shear, the agreement for the thermal conductivity is excellent from $\alpha = 1$ down to $\alpha = 0.7$. The viscosity, however, cannot be measured on the reference (null shear) steady states. Of course, this limitation is not intrinsic of the granular gas since the same happens for classic fluids. Nevertheless, a relatively good viscosity measurement can be obtained in the series with lowest shear used in the simulations ($a \simeq 0.019$), for which the agreement at high inelasticities between theory and simulation is actually slightly better than for the thermal conductivity measured at $a = 0$. For both coefficients, and at high inelasticities ($\alpha \simeq 0.5$), the agreement is improved when the zeroth order distribution function in the Chapman-Enskog method [49] is not the Maxwellian but an approximation to the solution of the homogeneous state for the heated granular gas (slightly different from the Maxwellian [55]). This approximation improve (represented with dashed lines in Figure 2) consists in taking into account the first term in an expansion in Sonine polynomials around the Maxwellian (for more references addressing the Sonine polynomials expansion method in kinetic theory, see for instance [48,49,56]).

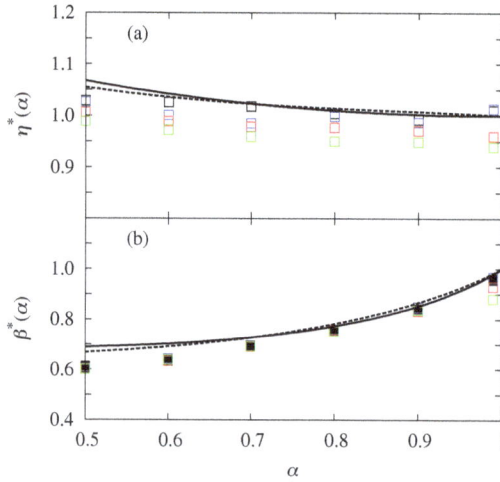

Figure 2. (a) Viscosity vs. coefficient of restitution α. (b) Effective thermal conductivity vs. α. In both panels, lines represent the results for the theoretical NS coefficient (solid for standard Sonine approximation and dashed for improved Sonine approximation) whereas the symbols stand for DSMC simulations. Several DSMC series have been represented, with increasing relative wall velocity. Solid symbols represent the no-shear case and open symbols are the sheared states for: $\Delta U = 0.25$ with $a \simeq 0.019$ (black), $\Delta U = 1$ with $a \simeq 0.04$ (blue), $\Delta U = 3$ with $a \simeq 0.10$ (red), $\Delta U = 5$ with $a \simeq 0.18$ (green).

We report in Figure 3a the reduced normal stresses $\theta_x \equiv P_{xx}/p, \theta_y \equiv P_{yy}/p$ measured in the simulations, as a function of the coefficient of restitution α and for several series with different shearing intensities. As we see, deviations from the NS behavior ($\theta_i \neq 1$) are small (less than 5%) even at moderately high values of shear (close to $a = 0.2$). In this case, the coefficient of restitution seems not to have an important impact on the behavior of both θ_x and θ_y. In Figure 3b we represent the value of the cross conductivity coefficient ϕ^*. As we can see, the deviations from linear behavior (i.e., from $\phi^*(\alpha) = 0$) are more significant here: up to a maximum of approximately 40% for the highest shear rate series represented here. However, as expected, the NS behavior is recovered for small shear rate (always less than 6.5% for the lowest non-null shear rate series). It is remarkable that, again, departures from NS behavior are stronger in the quasi-elastic limit than for more inelastic gases (i.e., more inelastic flows show less significantly non-Newtonian behavior here). This is consistent with the non-trivial observation in the previous section that $|\gamma|$ increases as we approach the quasi-elastic limit. As we already noticed, a higher $|\gamma|$ implies higher Knudsen number, and thus, our observations are self-consistent and not an artifact of the theory. The NS behavior ($\theta_i = 1$, $\phi^* = 0$) is strictly recovered for the reference states ($a = 0$), for all values of inelasticity.

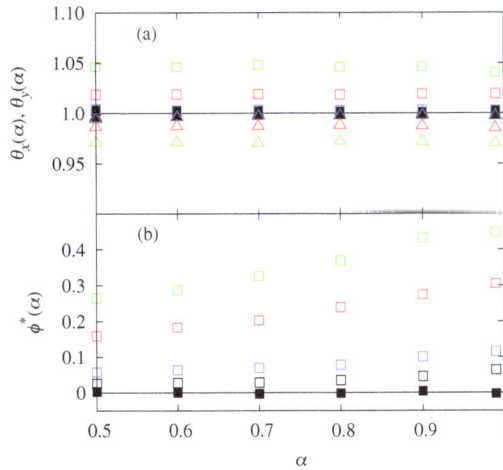

Figure 3. (**a**) DSMC data for reduced normal stresses $\theta_x \equiv P_{xx}/p, \theta_y \equiv P_{yy}/p$ vs. coefficient of restitution α. Square symbols stand for θ_x and triangles stand for θ_y. (**b**) DSMC data for thermal cross conductivity vs. α. In both panels, solid symbols represent the no-shear case and open symbols are the sheared states for: $\Delta U = 1$ (black) with $a \simeq 0.04$, $\Delta U = 3$ with $a \simeq 0.10$ (blue), $\Delta U = 5$ with $a \simeq 0.18$ (red).

6. Conclusions

In summary, our results on the transport properties of our system strongly indicate that NS hydrodynamics may apply for steady flows of inert particles in a heterogeneous medium, as long as the relevant Knudsen number scales imposed by the boundary conditions in the problem do keep small. This is in no way different to the condition that is required to a molecular gas in order to obey hydrodynamics at NS order [53].

More generally, the fact that this result is independent of the degree of inelasticity supports the hypothesis that inelasticity by itself does not cause necessarily the breakdown of scale separation in granular gases such as it appears in the Homogeneous Cooling State. A recent work on the aging time to hydrodynamics for the homogeneous cooling state of the granular gas also supports this hypothesis [57]. The dynamics is actually a bit more complex and the energy source necessary to fluidize the system may cancel out the gradients inherently produced by inelasticity, bringing the granular gas back to the realm of NS hydrodynamics. In fact, theory-simulation comparisons seem to indicate that a part of the disagreement between theory and simulation can be ruled out when we consider an improve in the approximation to the reference distribution function in the Chapman-Enskog method (when a Sonine polynomial expansion to first order [48,56] for the homogeneous cooling state [58] is used instead of the Maxwellian [55]) and thus, the relatively small disagreement would not be due to an eventual failure of hydrodynamics but to the inherent weakness of the perturbative solution used in the Chapman-Enskog procedure [18,49] for calculation of the transport coefficients.

Therefore, there is no a priori reason for concern respect to the validity of hydrodynamics for granular gases. This would depend essentially on the specific properties and geometry of the flow of interest, not just the inelasticity.

Considering a granular gas fluidized by an active fluid should be regarded, not as an artifact, but as a natural situation since, in the first place, rapid granular flows are possible just because an energy input is able to compensate for inelastic cooling (otherwise, clustering instabilities and total freezing of the dynamics occur [19]). However, an interstitial fluid is a very frequent configuration for granular dynamics problems present in nature [59]. Applications to biology problems are therefore promising. Moreover, it has been experimentally proved that this type of fluctuating force like the one we use to model the active fluid reaches local equilibrium with a granular particle [35]. Based on these observations, our model for energy input from the active fluid should be closer, in most cases, to experimental situations [60] where, interestingly, particle clustering instabilities also may occur.

In addition, the results in the present work have helped us to produce further results, since we have already applied NS hydrodynamics with a non-uniform stochastic force also to problems and applications like granular segregation [61], where excellent agreement is found between granular impurity segregation criteria resulting from NS theory and computer simulations. Finally, we plan to extend this work by developing the corresponding non-Newtonian hydrodynamic theory and also by calculating the linear stability criteria of theses base flows. On the other hand, this work could be relevant for the study of transport in crowded active enviroments [7] where the time scales of the active fluids is faster than the inert particles relaxation time. The existence of a well defined NS hydrodynamics in heterogeneous active bath [8], is also an important result in order to explain biological systems. Another possible future work is to study the effect of sorting of active swimmers with inert particles in heterogeneous active media. Also the study of the behaviour of particles, mean square displacement, mobility, diffusion coefficient depending on the media. A study of the hydrodynamics of more complex forms of the volume force, for example with time dependency, such as for a traveling wave [9], is also a promising line of research.

Acknowledgments: Support of the Spanish Government through Grant FIS2010-12587 (Francisco Vega Reyes), MTM2014-56948-C2-2-P (Antonio Lasanta) is acknowledged. Partial support from FEDER funds and by Junta de Extremadura through Grant No. GRU10158 is also acknowledged (Francisco Vega Reyes). Antonio Lasanta thanks the hospitality of Universidad de Extremadura where during a stay this work was done.

Author Contributions: Francisco Vega Reyes and Antonio Lasanta conceived the research and wrote the paper. Francisco Vega Reyes performed the simulations and the theory

Conflicts of Interest: The authors declare no conflicts of interest.

Appendix A. Computer Simulations

As we said, we used for this work data from the direct simulation Monte Carlo method (DSMC) [62] of the kinetic Equation (6). The standard DSMC algorithm consists of two basic steps: collisions (particle-particle collisions and, in this case, particle-boundaries collisions) and free drift [46]. The boundaries are modeled as hard walls, similarly to previous works. In the presence of a volume force, there is an additional step that accounts for the action of the fluctuating volume force. Implementation of this step, is described for instance in [32,63]. However, for this work, we consider a more general situation where the noise intensity $\xi^2(y)$ is space-dependent. More specifically, the fluctuating force intensity has a profile obeying equation (22); i.e., of the form $\xi(y)^2 \propto T(y)^{1/2}$. For the solution to be self-consistent, the temperature profile should be obtained, as we explained, from the differential Equation (21), that is the same differential equation that the 'LTu' profiles obey [37]. For this reason, we have extracted simulation temperature profiles $T(y)_{\text{LTu}}$ from LTu states obtained in a previous work [37]. We introduced them in condition (22), which defines a force intensity profile that is maintained constant in time during the simulation: $\xi(y)^2 \propto T(y)_{\text{LTu}}^{1/2}$. The initial temperature of the granular gas was however always set to the lower wall temperature (constant): $T_0(y) = T_-$. Once the simulation starts, and independently of the value of inelasticity, we always obtained after a transient a steady temperature profile $T(y)$ with the same form of a Fourier flow, as expected (see Figure 1). We have considered in all simulation figures presented in this work a system with $T_+/T_- = 5$ and $\Delta L \equiv L = 15\bar{\lambda}$ ($\bar{\lambda} = \sqrt{2}\pi\bar{n}\sigma^{d-1}$, $\bar{n}$ being the average density in the system). We also performed additional series with different $T_+/T_- = 2, 10$ (not shown in figures), and checked that with this variation we obtained the same results.

Appendix B. Navier-Stokes Transport Coefficients

The Navier-Stokes (NS) transport coefficients that we use in this work have been calculated by Garzó and Montanero in a previous work [52]. Their expressions need to be inserted in Equations (19) and (22) in order to complete the analytical solution of the hydrodynamic profiles, that result from the differential Equations (15) and (16) for sheared states and (20) and (21) for reference states (no shear). We write these transport coefficients in this appendix too, for the sake of completion. The reduced viscosity $\eta^*(\alpha)$, and reduced heat flux coefficients $\kappa^*(\alpha), \mu^*(\alpha)$ are [52]

$$\eta^*(\alpha) = \frac{1}{\nu_\eta^*}, \quad \kappa^*(\alpha) = \frac{d-1}{d}\frac{1+c}{\nu_\kappa^*},$$

$$\mu^*(\alpha) = \frac{d-1}{2d}\frac{c}{\nu_\mu^*}. \tag{A1}$$

Here, $\nu_\eta^*, \nu_\kappa^*, \nu_\mu^*$ are the collisional frequencies associated to the transport coefficients and their expressions were calculated by Brey and Cubero, together with the expression for the reduced cooling rate $\zeta^*(\alpha)$ [64]

$$\zeta^*(\alpha) = \frac{d+2}{4d}(1-\alpha^2)\left(1+\frac{3}{32}c\right), \tag{A2}$$

$$\nu_\eta^* = \frac{3}{4d}(1-\alpha+\frac{2}{3}d)(1+\alpha)\left(1-\frac{c}{64}\right), \tag{A3}$$

$$\nu_\kappa^* = \nu_\mu^* = \frac{1+\alpha}{d}\left(\frac{d-1}{2}+\frac{3}{16}(d+8)(1-\alpha)\right)$$

$$+\frac{1+\alpha}{d}\left(\frac{4+5d-3(4-d)\alpha}{1024}c\right). \tag{A4}$$

The coefficient c in (A1)–(A4) was calculated by van Noije and Ernst [58] and has the expression

$$c = \frac{32(1-\alpha)(1-2\alpha^2)}{73+56d-3\alpha(35+8d)+30(1-\alpha)\alpha^2}. \tag{A5}$$

As we already mentioned, the additional (dashed) theoretical curves in Figure 2 correspond to the transport coefficients calculated with an improved approximation to the reference distribution function in the Chapman-Enskog method. In effect, in this improved procedure the reference distribution function an approximate form of the homogeneous state. This approximate form consists in an expansion in Sonine polynomials, where we only retain the first non-Maxwellian contribution. For more reference on this alternative procedure, please refer to [55] or [65].

References

1. Marchetti, M.C.; Joanny, J.F.; Ramaswamy, S.; Liverpoo, T.B.; Prost, J.; Rao, M.; Aditi Simha, R. Hydrodynamics of soft active matter. *Rev. Mod. Phys.* **2013**, *85*, 1143–1189.
2. Bertin, E.; Droz, M.; Grégoire, G. Hydrodynamic equations for self-propelled particles: Microscopic derivation and stability analysis. *J. Phys. A* **2009**, *42*, 445001.
3. Bertin, E.; Droz, M.; Grégoire, G. Boltzmann and hydrodynamic description for self-propelled particles. *Phys. Rev. E* **2006**, *74*, 22101.
4. Chou, Y.; Wolfe, R.; Ihle, T. Kinetic theory for systems of self-propelled particles with metric-free interactions. *Phys. Rev. E* **2012**, *86*, 21120.
5. Ihle, T. Chapman-Enskog expansion for the Vicsek model of self-propelled particles. *J. Stat. Mech.* **2016**, doi:10.1088/1742-5468/2016/08/083205.
6. Vicsek, T.; Czirók, A.; Ben-Jacob, E.; Cohen, I.; Shochet, O. Novel Type of Phase Transition in a System of Self-Driven Particles. *Phys. Rev. Lett.* **1995**, *75*, 1226–1229.
7. Bechinger, C.; Leonardo, R.D.; Lowen, H.; Reichhardt, C.; Volpe, G. Active particles in complex and crowded enviroments. *Rev. Mod. Phys.* **2016**, *88*, 45006.

8. Argun, A.; Moradi, A.R.; Pinçe, E.; Bagci, G.B.; Imparato, A.; Volpe, G. Non-Boltzmann stationary distribution and nonequilibrium relations in active baths. *Phys. Rev. E* **2016**, *94*, 62150.

9. Sándor, C.; Libàl, A.; Reichhardt, C.; Reichhardt, C.J.O. Collective transport for active matter run-and-tumble disk systems on a travellin-wave substrate. *Phys. Rev. E* **2017**, *95*, 12607.

10. Sándor, C.; Libàl, A.; Reichhardt, C.; Reichhardt, C.J.O. Dynamic phases of active matter systems with quenched disorder. *Phys. Rev. E* **2017**, *95*, 32606.

11. Sándor, C.; Libàl, A.; Reichhardt, C.; Reichhardt, C.J.O. Dewetting and spreading transitions for active matter on random pinning substrate. *J. Chem. Phys.* **2017**, *146*, 204903.

12. Angelini, T.E.; Hanezzo, E.; Trepat, X.; Marquez, M.; Fredberg, J.J.; Weitz, D.A. Glass-like dynamics of collective cell migration. *Proc. Nat. Acad. Sci. U.S.A.* **2011**, *108*, 4714–4719.

13. Malinverno, C.; Corallino, S.; Giavazzi, F.; Bergert, M.; Li, Q.; Leoni, M.; Disanza, A.; Frittoli, E.; Oldani, A.; Martini, E.; et al. Endocytic reawakening of motility in jammed epithelia. *Nat. Mater.* **2017**, *16*, 587–596.

14. Garcia, S.; Hannezo, E.; Elgeti, J.; Joanny, J.F.; Silberzan, P.; Gov, N.S. Physics of active jamming during collective cellular motion in a monolayer. *Proc. Natl. Acad. Sci. U.S.A.* **2015**, *112*, 15314–15319.

15. Bi, D.; Lopez, J.H.; Schwarz, J.M.; Manning, M.L. A density-independent glass transition in biological tissues. *Nat. Phys.* **2015**, *11*, 1074–1079.

16. Malgaretti, P.; Pagonabarraga, I.; Rubi, M.J. Rectification and Non-Gaussian Diffusion in Heterogeneous Media. *Entropy* **2016**, *18*, 349.

17. Dufty, J.W. Kinetic theory and hydrodynamics for a low density granular gas. *Adv. Complex Syst.* **2001**, *4*, 397–406.

18. Brilliantov, N.V.; Pöschel, T. *Kinetic Theory of Granular Gases*; Oxford University Press: Oxford, UK, 2004.

19. Goldhirsch, I. Rapid granular flows. *Annu. Rev. Fluid Mech.* **2003**, *35*, 267–293.

20. De Bruyn, J. Unifying Liquid and Granular Flow. *Physics* **2011**, *4*, doi:10.1103/Physics.4.86.

21. Batchelor, G.K. *The Theory of Homogeneous Turbulence*; Cambridge University Press: Cambridge, UK, 1953.

22. Haff, P.K. Grain flow as a fluid-mechanical phenomenon. *J. Fluid Mech.* **1983**, *134*, 401–430.

23. Puglisi, A.; Gnoli, A.; Gradenigo, G.; Sarracino, A.; Villamaina, D. Structure factors in granular experiments with homogeneous fluidization. *J. Chem. Phys.* **2012**, *136*, 14704.

24. Puglisi, A. *Transport and Fluctuations in Granular Fluids*; Springer: Cham, Switzerland, 2014.

25. Jaeger, H.M.; Nagel, S.R. Physics of the Granular State. *Science* **1992**, *255*, 1523–1531.

26. Geminard, J.C.; Losert, W.; Gollub, J.P. Frictional mechanics of wet granular material. *Phys. Rev. E* **1999**, *59*, 5881–5890.

27. Williams, D.R.M.; MacKintosh, F.C. Driven granular media in one dimension: Correlations and equation of state. *Phys. Rev. E* **1996**, *54*, R9–R12.

28. Cafiero, R.; Luding, S.; Herrmann, H.J. Two-dimensional granular gas of inelastic spheres with multiplicative driving. *Phys. Rev. Lett.* **2000**, *84*, 6014–6017.

29. Visco, P.; Puglisi, A.; Barrat, A.; Trizac, E.; van Wijland, F. Fluctuations of Power Injection in Randomly Driven Granular Gases. *J. Stat. Phys.* **2006**, *125*, 533–568.

30. Villamaina, D.; Puglisi, A.; Vulpiani, A. The fluctuation-dissipation relation in sub-diffusive systems: The case of granular single-file diffusion. *J. Stat. Mech.* **2008**, doi:10.1088/1742-5468/2008/10/L10001.

31. Fiege, A.; Aspelmeier, T.; Zippelius, A. Long-Time Tails and Cage Effect in Driven Granular Fluids. *Phys. Rev. Lett.* **2009**, *102*, 98001.

32. Gradenigo, G.; Sarracino, A.; Villamaina, D.; Puglisi, A. Fluctuating hydrodynamics and correlation lengths in a driven granular fluid. *J. Stat. Mech.* **2011**, *2011*, doi:10.1088/1742-5468/2011/08/P08017.

33. Losert, W.; Bocquet, L.; Lubensky, T.C.; Gollub, J.P. Particle Dynamics in Sheared Granular Matter. *Phys. Rev. Lett.* **2000**, *85*, 1428–1431.

34. Yu, H.; Chin, M.; Yuan, T.; Bian, H.; Remer, L.A.; Prospero, J.M.; Omar, A.; Winker, D.; Yang, Y.; Zhang, Y.; et al. The fertilizing role of African dust in the Amazon rainforest: A first multiyear assessment based on data from Cloud-Aerosol Lidar and Infrared Pathfinder Satellite Observations. *Geophy. Res. Lett.* **2015**, *42*, 1984–1991.

35. Ojha, R.P.; Lemieux, P.A.; Dixon, P.K.; Liu, A.J.; Durian, D.J. Statistical mechanics of a gas-fluidized particle. *Nature* **2004**, *427*, 521–523.

36. Campbell, C.S. Rapid Granular Flows. *Annu. Rev. Fluid Mech.* **1990**, *22*, 57–92.

37. Vega Reyes, F.; Santos, A.; Garzó, V. Non-Newtonian granular hydrodynamics: What do the inelastic simple shear flow and the elastic Fourier flow have in common? *Phys. Rev. Lett.* **2010**, *104*, 28001.

38. Foerster, S.F.; Louge, M.Y.; Chang, H.; Alla, K. Measurements of the collision properties of Small Spheres. *Phys. Fluids* **1994**, *6*, 1108–1115.

39. Grasselli, Y.; Bossis, G.; Morini, R. Translational and rotational temperatures of a 2D vibrated granular gas in microgravity. *Eur. Phys. J. E* **2015**, *38*, doi:10.1140/epje/i2015-15008-5.

40. Van Kampen, N.G. *Stochastic Processes in Physics and Chemistry*; Elsevier: Amsterdam, The Netherlands, 1992.

41. Marini-Bettolo-Marconi, U.; Tarazona, P.; Cecconi, F. Theory of thermostatted inhomogeneous granular fluids: A self-consistent density functional description. *J. Chem. Phys.* **2007**, *126*, 164904.

42. Sarracino, A.; Villamaina, D.; Constantini, G.; Puglisi, A. Granular Brownian motion. *J. Stat. Mech.* **2010**, doi:10.1088/1742-5468/2010/04/P04013.

43. Garzó, V.; Chamorro, M.G.; Vega Reyes, F. Transport properties for driven granular fluids in situations close to homogeneous steady states. *Phys. Rev. E* **2013**, *87*, 32201.

44. Marconi, U.M.B.; Puglisi, A.; Rondonic, L.; Vulpiani, A. Fluctuation-dissipation: Response theory in statistical physics. *Phys. Rep.* **2008**, *461*, 111–195.

45. Brey, J.J.; Dufty, J.W.; Kim, C.S.; Santos, A. Hydrodynamics for granular flow at low density. *Phys. Rev. E* **1998**, *58*, 4638–4653.

46. Bird, G.I. *Molecular Gas Dynamics and the Direct Simulation of Gas Flows*; Oxford University Press: Oxford, UK, 1994.

47. Vega Reyes, F.; Urbach, J.S. Steady base states for Navier–Stokes granular hydrodynamics with boundary heating and shear. *J. Fluid Mech.* **2009**, *636*, 279–293.

48. Burnett, D. Velocity Distribution in a non-uniform gas. *Proc. Lond. Math.* **1935**, *s2-39*, 385–430.

49. Chapman, C.; Cowling, T.G. *The Mathematical Theory of Non-Uniform Gases*, 3rd ed.; Cambridge University Press: Cambridge, UK, 1970.

50. Abramowitz, M.; Stegun, I.A. *Handbook of Mathematical Functions*; Dover Publications: New York, NY, USA, 1965.

51. Vega Reyes, F.; Santos, A.; Garzó, V. Steady base states for non-Newtonian granular hydrodynamics. *J. Fluid Mech.* **2013**, *719*, 431–464.

52. Garzó, V.; Montanero, J.M. Transport coefficients of a heated granular gas. *Phys. A* **2002**, *313*, 336–356.

53. Boyd, I.D.; Chen, G.; Candler, G.V. Predicting failure of the continuum fluid equations in transitional hypersonic flows. *Phys. Fluids* **1995**, *7*, 210–219.

54. Campbell, C.S. The stress tensor for simple shear flows of a granular material. *J. Fluid Mech.* **1989**, *203*, 449–473.

55. Garzó, V.; Santos, A.; Montanero, J.M. Modified Sonine approximation for the Navier-Stokes transport coefficients of a granular gas. *Phys. A* **2007**, *376*, 94–107.

56. Truesdell, C. *Mathematical Aspects of the Kinetic Theory of Gases*; Universidade Federal do Rio de Janeiro: Rio de Janeiro, Brazil, 1973. (In Portuguese)

57. Vega Reyes, F.; Santos, A.; Kremer, G.M. Role of roughness on the hydrodynamic homogeneous base state of inelastic spheres. *Phys. Rev. E* **2014**, *89*, 20202.

58. Van Noije, T.; Ernst, M. Velocity distributions in homogeneous granular fluids: Velocity distributions in homogeneous granular fluids: The free and the heated case. *Gran. Matt.* **1998**, *1*, 57–64.

59. Bagnold, R.A. *The Physics of Blown Sand and Desert Dunes*; Dover Publications: Mineola, New York, USA, 1954.

60. Mognetti, B.M.; Šarić, A.; Angioletti-Uberti, S.; Cacciuto, A.; Valeriani, C.; Frenkel, D. Living Clusters and Crystals from Low-Density Suspensions of Active Colloids. *Phys. Rev. Lett.* **2013**, *111*, 245702.

61. Vega Reyes, F.; Garzó, V.; Khalil, N. Hydrodynamic granular segregation induced by boundary heating and shear. *Phys. Rev. E* **2014**, *89*, 52206.

62. Pöschel, T.; Schwager, T. *Computational Granular Dynamics: Models and Algorithms*; Springer: Berlin/Heidelberg, Germany, 2005.

63. Montanero, J.M.; Santos, A. Computer simulation of uniformly heated granular fluids. *Gran. Matt.* **2000**, *2*, 53–64.

64. Brey, J.J.; Cubero, D. Hydrodynamic Transport Coefficients of Granular Gases. In *Granular Gases*; Pöschel, T., Luding, S., Eds.; Lecture Notes in Physics; Springer: Berlin/Heidelberg, Germany, 2001; p. 59.

65. Garzó, V.; Vega Reyes, F.; Montanero, J.M. Modified Sonine approximation for granular binary mixtures. *J. Fluid Mech.* **2009**, *623*, 387–411.

Article

Participation Ratio for Constraint-Driven Condensation with Superextensive Mass

Giacomo Gradenigo * and Eric Bertin

Laboratoire Interdisciplinaire de Physique (LIPHY), Université Grenoble Alpes and CNRS, F-38000 Grenoble, France; eric.bertin@univ-grenoble-alpes.fr
* Correspondence: ggradenigo@gmail.com

Received: 30 August 2017; Accepted: 22 September 2017; Published: 26 September 2017

Abstract: Broadly distributed random variables with a power-law distribution $f(m) \sim m^{-(1+\alpha)}$ are known to generate condensation effects. This means that, when the exponent α lies in a certain interval, the largest variable in a sum of N (independent and identically distributed) terms is for large N of the same order as the sum itself. In particular, when the distribution has infinite mean ($0 < \alpha < 1$) one finds unconstrained condensation, whereas for $\alpha > 1$ *constrained* condensation takes places fixing the total mass to a large enough value $M = \sum_{i=1}^{N} m_i > M_c$. In both cases, a standard indicator of the condensation phenomenon is the participation ratio $Y_k = \langle \sum_i m_i^k / (\sum_i m_i)^k \rangle$ ($k > 1$), which takes a finite value for $N \to \infty$ when condensation occurs. To better understand the connection between constrained and unconstrained condensation, we study here the situation when the total mass is fixed to a superextensive value $M \sim N^{1+\delta}$ ($\delta > 0$), hence interpolating between the unconstrained condensation case (where the typical value of the total mass scales as $M \sim N^{1/\alpha}$ for $\alpha < 1$) and the extensive constrained mass. In particular we show that for exponents $\alpha < 1$ a condensate phase for values $\delta > \delta_c = 1/\alpha - 1$ is separated from a homogeneous phase at $\delta < \delta_c$ from a transition line, $\delta = \delta_c$, where a *weak* condensation phenomenon takes place. We focus on the evaluation of the participation ratio as a generic indicator of condensation, also recalling or presenting results in the standard cases of unconstrained mass and of fixed extensive mass.

Keywords: large deviations; condensation phenomenon; ensemble inequivalence; canonical ensemble

1. Introduction

In the context of the sum of a large number of positive random variables, an interesting phenomenon occurs when a single variable carries a finite fraction of the sum [1]. Such a phenomenon has been put forward for instance in the context of the glass transition [2,3]. In the framework of particle or mass transport models [4–13], where the sum of the random variables is fixed to a constant value due to a conservation law of the underlying dynamics, this phenomenon has been called "condensation". This condensation phenomenon has since then been reported in different contexts like in extreme value statistics [14], and in the sample variance of exponentially distributed random variables as well as for conditioned random-walks [1,15,16]. A similar mechanism is also at the basis of the condensation observed in the non-equilibrium dynamics of non-interacting field-theoretical models [17–20]. A more general type of condensation, induced by interaction, has also been put forward [21], but in the following we shall focus on cases without interaction, apart from a possible constraint on the total mass.

As mentioned above, standard condensation results from the presence of a constraint fixing the sum of the random variables to a given value. However, the fact that a single random variable carries a finite fraction of the sum is also observed for fat-tailed random variables with infinite mean—a phenomenon sometimes called the Noah effect [22,23]. The goal of the present paper is to present a comparative study of these two scenarios, that we shall respectively denote as *constrained condensation* and *unconstrained condensation*. Note that the term "condensation" is commonly used in the literature to describe the constrained case, but we shall extend its use to the unconstrained case, to emphasize possible analogies between the two scenarios. Considering the set of N random variables m_i with joint distribution $P(m_1, \ldots, m_N)$, unconstrained condensation takes place when the sum $M = \sum_{i=1}^{N} m_i$, in the limit $N \to \infty$, is dominated by few terms, i.e., a number of terms of order $\mathcal{O}(1)$. This happens for instance to the sum of N independent and identically distributed (iid) Levy-type random variables, with probability density $f(m)$ decaying asymptotically as a power law, $f(m) \sim m^{-(1+\alpha)}$, with an exponent $0 < \alpha < 1$. This unconstrained condensation effect (sometimes also referred to as "localization" [24] depending on the context) is often characterized by the participation ratio Y_k [3,25], defined as

$$Y_k = \left\langle \frac{\sum_{i=1}^{N} m_i^k}{\left(\sum_{i=1}^{N} m_i\right)^k} \right\rangle \tag{1}$$

where $k > 1$ is a real number, and where the brackets indicate an average over the m_i's. For broadly distributed random variables it can be shown, with the calculation presented in [25] and briefly recalled here in Section 2, that there is a critical value $\alpha_c = 1$ for the exponent of the power-law distribution such that for $\alpha > 1$ the asymptotic value of the participation ratio is zero, $\lim_{N \to \infty} Y_k = 0$, whereas for a broad enough tail, $0 < \alpha < 1$, one has $\lim_{N \to \infty} Y_k > 0$ for any value $k > 1$. The average in Equation (1) is computed with respect to the probability distribution $P(m_1, \ldots, m_N) = \prod_{i=1}^{N} f(m_i)$. The participation ratio is therefore the "order parameter" for condensation in the sum of random variables. It is in fact easy to see from Equation (1) that when all the random variables contribute "democratically" to the sum, namely when each of them is of order $m_i \sim 1/N$, then the asymptotic behaviour of the participation ratio is $Y_k \sim 1/N^{k-1}$, which goes to zero when $N \to \infty$. In contrast, if the sum is dominated by few terms of order $m_i \sim N$, asymptotically one has $Y_k \sim 1$.

As a physical example, the relevance of participation ratios to unveil unconstrained condensation was also shown for the phase-space condensation taking place in the glass phase of the Random Energy Model (REM) [3,25]. The REM is a system with 2^N configurations, where each configuration i has the Boltzmann weight $e^{-\beta E_i}$, and the energies E_i are iid random variables, usually assumed to have a Gaussian distribution with a variance proportional to N. The random variables with respect to which condensation takes place in the glass phase are the probabilities $z_i(\beta) = e^{-\beta E_i}$ of the different configurations. The corresponding participation ratio takes the same form as Equation (1), simply replacing m_i by $z_i(\beta)$. It has been shown [3,25] that for values of the inverse temperature $\beta > \beta_c$, where β_c is the critical value for the glass transition, the value of the sum $\mathcal{Z} = \sum_{i=1}^{2^N} z_i(\beta)$ is dominated, in the limit $N \to \infty$, by $\mathcal{O}(1)$ terms: in this case the asymptotic value of Y_k is finite. In particular one can prove that for an inverse temperature $\beta > \beta_c$ the participation ratio of the REM has precisely the same form as for the sum of iid Levy random variables m_i with distribution $p(m_i) = m_i^{1+\beta_c/\beta}$ (the exponent $\alpha = \beta_c/\beta$ is thus proportional to temperature).

In the above cases unconstrained condensation occurs for $\alpha < 1$, that is when the first moment of the power-law distribution is infinite. The situation is different, though, when one considers power-law distributed random variables with a fixed total sum $M = \sum_{i=1}^{N} m_i$, a case which we refer to as *constraint-driven* condensation, or simply *constrained condensation*. Such a phenomenon, which is also related to the large of heavy-tailed sums (see, e.g., [26]), is found for instance in the stationary distribution of the discrete Zero Range Process and its continuous variables generalization [7,8,10]. The latter is represented by a lattice with N sites, each carrying a continuous mass m_i, endowed with some total-mass conserving dynamical rules. For this model the stationary distribution is:

$$P(m_1, \ldots, m_N | \rho) = \frac{1}{\mathcal{Z}_N(\rho)} \prod_{i=1}^{N} f(m_i) \, \delta \left[\rho N - \sum_{i=1}^{N} m_i \right] \tag{2}$$

213

where ρ is the average density fixed by the initial total mass $M = \rho N$, and where

$$\mathcal{Z}_N(\rho) = \int_0^\infty dm_1 \dots dm_N \prod_{i=1}^N f(m_i)\, \delta \left[\rho N - \sum_{i=1}^N m_i \right] \tag{3}$$

is a normalization constant (or partition function). In mass transport models the shape of the distribution $f(m_i)$ depends on the dynamical rules, and has typically a power-law tail,

$$f_\alpha(m) \approx \frac{A}{m^{1+\alpha}} \tag{4}$$

In [7,8,10] it has been shown that, in the presence of a constraint on the total value of the mass, constrained condensation never takes place for exponents of the local power-law distribution in the interval $0 < \alpha < 1$, while on the contrary when $\alpha > 1$ there exists a critical value ρ_c such that for $\rho > \rho_c$ the system is in the condensed phase. It thus turns out that constraining the random variables to have a fixed sum deeply modifies their statistical properties in this case—while naive intuition based on elementary statistical physics, like the equivalence of ensembles, may suggest that fixing the sum may not make an important difference. It is also worth emphasizing that there is a significant difference between constrained and unconstrained condensation. In the unconstrained case, a few variables carry a finite fraction of the sum, while in the constrained case, only a single variable takes a macroscopic fraction of the sum (note that the situation may be different, though, in the presence of correlations between the variables [21]). We thus see that the "condensation" phenomenon we define here as a non-vanishing value of the participation ratio in the infinite N limit is a weak notion of condensation, which is more general than the standard condensation reported in the constrained case. In particular, this weak condensation effect does not imply the existence of a proper condensate, that is a "bump" in the tail of the marginal distribution $p(m)$ with a vanishing relative width. The bump may have a non-vanishing relative width, or may even not exist, the distribution $p(m)$ being monotonously decreasing in this case (see [8] for an exactly solvable example). When relevant, we shall emphasize this specific character of the condensation by using the term "weak condensation".

The goal of the present work is to understand the relation between these two cases, which differ only by the presence or absence of a constraint on the total mass, but yield opposite ranges of values of α for the existence of condensation. To better grasp the nature of this difference, we study here the case where the total mass is fixed to a superextensive value $M \sim N^{1+\delta}$, with $\delta > 0$, thus extending some of the results presented in [8]. The choice of a superextensive mass is motivated by the fact that in the unconstrained case, the total mass $M = \sum_{i=1}^N m_i$, being the sum of iid broadly distributed variables, typically scales superextensively, as $N^{1/\alpha}$, for $\alpha < 1$. This suggests that the case of a superextensive fixed mass may be closer to the unconstrained case, and that the value $1 + \delta = 1/\alpha$ may play a specific role. This will be confirmed by the detailed calculations presented in Section 4. Yet, before dealing with the superextensive mass case, we will first recall in Section 2 how to compute the participation ratios in the case of unconstrained condensation, and present in Section 3 a simplified evaluation of the participation ratio in the case of constrained condensation with an extensive fixed mass.

2. Unconstrained Condensation

In the unconstrained case, where the masses are simply independent and identically distributed random variables with a broad distribution, the evaluation of the average participation ratio Y_k is well-know and has been performed using different methods [3,25]. We sketch here the derivation of Y_k using the auxiliary integral method put forward in [25].

Noting that

$$\frac{1}{\left(\sum_{i=1}^N m_i \right)^k} = \frac{1}{\Gamma(k)} \int_0^\infty dt\, t^{k-1} \exp\left(-t \sum_{i=1}^N m_i \right) \tag{5}$$

one obtains, using the property that the random variables m_i are independent and identically distributed,

$$Y_k = \frac{N}{\Gamma(k)} \int_0^\infty dt\, t^{k-1} \langle e^{-tm} \rangle_\alpha^{N-1} \langle m^k e^{-tm} \rangle_\alpha \tag{6}$$

where the brackets $\langle \dots \rangle_\alpha$ indicate an average over a single variable m with distribution given in Equation (4); Γ is the Euler Gamma function, defined as $\Gamma(k) = \int_0^\infty dt\, t^{k-1} e^{-t}$. For large N, the factor $\langle e^{-tm} \rangle_\alpha^{N-1}$ in Equation (6) takes very small values except if t is small, in which case $\langle e^{-tm} \rangle_\alpha$ is close to 1. Using a simple change of variable, one finds for $0 < \alpha < 1$ [25]

$$1 - \langle e^{-tm} \rangle_\alpha \approx a\, t^\alpha \qquad (t \to 0) \tag{7}$$

with $a = A\Gamma(1-\alpha)/\alpha$, so that for $N \to \infty$,

$$\langle e^{-tm} \rangle_\alpha^{N-1} \approx e^{-Nat^\alpha} \tag{8}$$

again for small t. In a similar way, one also obtains for $k > \alpha$ [25]

$$\langle m^k e^{-tm} \rangle_\alpha \approx A\Gamma(k - \alpha)\, t^{-(k-\alpha)} \qquad (t \to 0) \tag{9}$$

One thus has for large N

$$Y_k \approx \frac{NA\Gamma(k - \alpha)}{\Gamma(k)} \int_0^\infty dt\, t^{\alpha-1} e^{-Nat^\alpha} \tag{10}$$

Using now the change of variable $v = Nat^\alpha$, the last integral can be expressed in terms of the Gamma function, eventually leading, in the limit $N \to \infty$, to [3,25],

$$Y_k = \frac{\Gamma(k - \alpha)}{\Gamma(k)\Gamma(1 - \alpha)} \qquad (0 < \alpha < 1) \tag{11}$$

The participation ratio Y_k is thus non-zero for $0 < \alpha < 1$, and goes to zero linearly when $\alpha \to 1$. A similar calculation in the case $\alpha > 1$ yields $Y_k = 0$ in the limit $N \to \infty$. Hence condensation occurs for $0 < \alpha < 1$ in the unconstrained case. As we shall see below, the opposite situation occurs in the constrained case.

3. Constrained Condensation

We now turn to the computation of the participation ratio Y_k when the total mass in the system is constrained to have the extensive value $M = \rho N$, as a function of the exponent α and of the density ρ. Evaluating Y_k as defined in Equation (1) by averaging over the constrained probability distribution given in Equation (2), the denominator is a constant and can be factored out of the average, yielding the simple result:

$$Y_k = \frac{1}{\rho^k N^{k-1}} \langle m^k \rangle = \frac{1}{\rho^k N^{k-1}} \int_0^\infty dm\, p(m)\, m^k \tag{12}$$

where the marginal distribution $p(m)$ is defined as

$$p(m) = f_\alpha(m) \frac{Z_{N-1}\left(\rho - \frac{m}{N}\right)}{Z_N(\rho)} \tag{13}$$

Before discussing what happens for the range of exponents α where constrained condensation takes place, let us briefly explain why for $\alpha < 1$ the presence of the constraint removes the condensation and the participation ratio in Equation (12) vanish when $N \to \infty$.

3.1. $\alpha < 1$: Absence of Condensation

The first important issue to clarify is why the condensation taking place in the unconstrained case for values of the power-law exponent α (see Equation (4)) in the range $0 < \alpha < 1$, then disappear when a constraint on the total mass value is applied. Why the constraint forces the system to stay in the homogeneous phase? To answer this question, it is useful to recall the expression of the partition function of the model in terms of its inverse Laplace transform:

$$Z_N(\rho) = \frac{1}{2\pi i} \int_{s_0 - i\infty}^{s_0 + i\infty} ds \, \exp\{N[\log g_\alpha(s) + \rho s]\} \tag{14}$$

where

$$g_\alpha(s) = \int_0^\infty dm f_\alpha(m) e^{-sm} \tag{15}$$

The values of ρ in the homogeneous "fluid" phase are those for which the integral in Equation (14) can be solved with the saddle-point method. In contrast, constrained condensation occurs for all values of ρ such that the saddle-point equation

$$\rho = -\frac{g'_\alpha(s)}{g_\alpha(s)} = \frac{\int_0^\infty dm \, f_\alpha(m) \, m \, e^{-sm}}{\int_0^\infty dm \, f_\alpha(m) \, e^{-sm}} = \frac{\langle m \, e^{-sm} \rangle_\alpha}{\langle e^{-sm} \rangle_\alpha} \tag{16}$$

admits no solution on the real axis. It can be checked by inspection that the function $h(s) = \log g_\alpha(s) + \rho s$ has a branch cut in the complex s plane coinciding with the negative part of the real axis. The domain over which s can be varied to look for a solution of the saddle point equation is the positive semiaxis $[0, +\infty]$. When increasing s, the function $\langle m \, e^{-sm} \rangle_\alpha / \langle e^{-sm} \rangle_\alpha$ monotonically decreases from its value $\langle m \rangle_\alpha$ reached for $s \to 0$ to 0 for $s \to \infty$.

At this point we just need to recall that for $0 < \alpha < 1$ one has $\langle m \rangle_\alpha = \infty$. This means that it is possible to find a value s^* which is a solution of Equation (16) for any given value of ρ. Hence the integral representation of the partition function in Equation (14) can always be treated in the saddle-point approximation and condensation, which is related to the breaking of the saddle-point approximation, never occurs. By exploiting the saddle-point approximation for the partition function it is then not difficult to compute explicitly the expression in Equation (13), which reads

$$p(m) \sim \frac{e^{-s^* m}}{m^{1+\alpha}} \tag{17}$$

where $s^* < \infty$ is the solution of the saddle point equation. The vanishing of the participation ratio Y_k then follows easily, according to its expression in Equation (12), from the presence of the exponential cutoff in $p(m)$ (Equation (17)):

$$Y_k = \frac{\langle m^k \rangle}{\rho^k N^{k-1}} \tag{18}$$

where $\langle m^k \rangle = \int_0^\infty dm \, m^k \, p(m)$ is a function which does not depend on N in the large N limit, so that

$$\lim_{N \to \infty} \frac{\langle m^k \rangle}{\rho^k N^{k-1}} = 0 \tag{19}$$

We have therefore seen that for any $\alpha < 1$, if one constrains the system to have an extensive mass $M = \rho N$, the participation ratio vanishes in the thermodynamic limit: $\lim_{N \to \infty} Y_k = 0$, as expected since no condensation occurs in this case.

3.2. $\alpha > 1$: Homogeneous Phase at Low Density ($\rho < \rho_c$)

As soon as the exponent α of the single-variable distribution is increased above $\alpha_c = 1$, namely as soon as the first moment of the distribution $f_\alpha(m)$ becomes finite, a condensed phase appears at finite ρ_c. For any given value of α the critical density ρ_c is the maximal density for which the saddle point equation Equation (16) has a solution. As we have already noticed in the previous section, the maximum value which can be attained by the term on the right of Equation (16) is $\langle m \rangle_\alpha$, so that for all values $\rho > \rho_c = \langle m \rangle_\alpha$ the saddle-point approximation breaks down and one has condensation [7,8]. On the contrary, for $\alpha > 1$ and $\rho < \rho_c$ the system is still in the homogeneous phase and one can compute the marginal distribution $\rho(m)$ according to its definition in Equation (13) by using the saddle-point approximation. The result is also in this case

$$\rho(m) \sim f_\alpha(m)\, e^{-m/m_0} \tag{20}$$

and we know from [7,8] that the characteristic mass m_0 diverges when the density ρ tends to ρ_c as $m_0 \sim (\rho - \rho_c)^{-1}$ for $\alpha > 2$ and as $m_0 \sim (\rho - \rho_c)^{-1/(\alpha-1)}$ for $1 < \alpha < 2$.

3.3. $\alpha > 1$: Condensed Phase at High Density ($\rho > \rho_c$)

Let us now study what happens in the condensed phase. For $\alpha > 1$ and $\rho > \rho_c$, one observes for large N a coexistence between a homogeneous fluid phase carrying a total mass approximately equal to $N\rho_c$ and a condensate of mass $M_{\text{cond}} \approx (\rho - \rho_c)N$. The marginal distribution $p(m)$ can be approximately written as [14]

$$p(m) \approx f(m) + p_{\text{cond}}(m, \rho, N) \tag{21}$$

where $p_{\text{cond}}(m, \rho, N)$ is the mass distribution of the condensate, normalized according to $\int_0^\infty p_{\text{cond}}(m, \rho, N) = 1/N$ to account for the fact that the condensate is present on a single site. It has been shown [7,8] that for $\alpha > 2$, the distribution $p_{\text{cond}}(m, \rho, N)$ is Gaussian, with a width proportional to $N^{1/2}$. In other words, the condensate exhibits normal fluctuations. In contrast, for $1 < \alpha < 2$, the distribution $p_{\text{cond}}(m, \rho, N)$ has a broader, non-Gaussian shape, with a typical scale of fluctuation $\sim N^{1/\alpha}$ [7,8]. For all values of $\alpha > 1$, however, the relative fluctuations of M_{cond} vanish in the large N limit:

$$\frac{M_{\text{cond}} - \overline{M}_{\text{cond}}}{M_{\text{cond}}} \begin{array}{ll} \sim \ N^{-(\alpha-1)/\alpha} & \text{if } \ 1 < \alpha < 2 \\ \sim \ N^{-1/2} & \text{if } \ \alpha > 2 \end{array} \tag{22}$$

with $\overline{M}_{\text{cond}} = (\rho - \rho_c)N$. Hence in order to compute the large N behavior of moments of the distribution $p(m)$, one can further approximate $p(m)$ as

$$p(m) \approx f(m) + \frac{1}{N}\, \delta\left(\sum_{i=1}^N m_i - \overline{M}_{\text{cond}}\right) \tag{23}$$

Note that more accurate expressions of the distribution $p(m)$ can be found in [8].

From Equation (23), the moment $\langle m^k \rangle$ is evaluated as

$$\langle m^k \rangle \approx \int_0^{\rho_c N} dm\, m^k f(m) + \frac{\overline{M}_{\text{cond}}^k}{N} \tag{24}$$

The integral in Equation (24), corresponding to the fluid phase contribution to the moment, has a different scaling with N depending on the values of k and α. If $k < \alpha$, the integral converges to a finite limit when N goes to infinity. On the contrary, when $k > \alpha$, the integral diverges with N and scales as $N^{k-\alpha}$.

The participation ratio reads $Y_k = \langle m^k \rangle / (\rho^k N^{k-1})$, so that the contribution of the fluid phase to the participation ratio scales as $1/N^{k-1}$ for $k < \alpha$, and as $1/N^{\alpha-1}$ for $k > \alpha$; in both cases, this contribution vanishes for $N \to \infty$, when $k > 1$ and $\alpha > 1$. The remaining contribution, resulting from the condensate, simply leads to

$$Y_k = \left(\frac{\rho - \rho_c}{\rho}\right)^k \tag{25}$$

where we recall that $\rho > \rho_c$ in the condensed phase. Hence Y_k goes to zero at the onset of condensation ($\rho \to \rho_c$), so that the transition can be thought as continuous if one considers the participation ratio as an order parameter. In the opposite limit $\rho \to \infty$, the participation ratio goes to 1, indicating a full condensation.

A phase diagram in the (α, ρ)-plane summarizing the results of this section for the case of constrained condensation with extensive mass is shown in Figure 1.

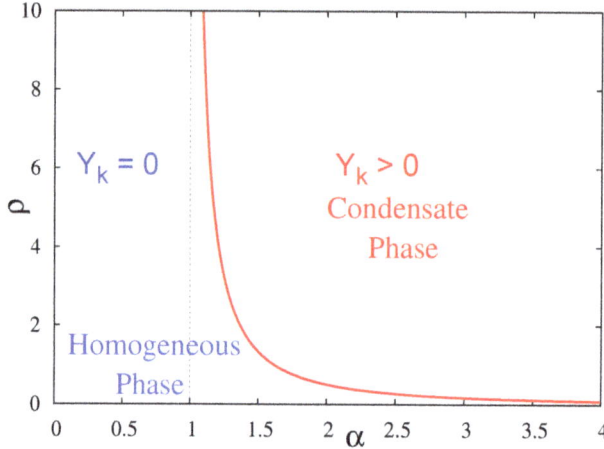

Figure 1. Phase diagram for the values of the participation rations in the (α, ρ) plane in the presence of an extensive constraint on the total value of the mass $\sum_{i=1}^{N} m_i = \rho N$. The (red) continuous line marks the separation of the condensed phase ($\lim_{N \to \infty} Y_k > 0$) from the homogeneous phase ($\lim_{N \to \infty} Y_k = 0$). The vertical dotted (black) line marks the critical value $\alpha_c = 1$ where the critical density ρ_c for condensation diverges.

4. Constraint to a Superextensive Total Mass $M = \tilde{\rho} N^{1+\delta}$

As explained in the introduction, the unconstrained condensation occurs for $\alpha < 1$, while the constraint-driven condensation occurs at $\alpha > 1$ (and at high enough density). Given that the typical total mass in the unconstrained case is superextensive for $\alpha < 1$, it is of interest to study condensation effects in the more general case of a fixed superextensive total mass $M = \tilde{\rho} N^{1+\delta}$, with $\delta > 0$ and $\tilde{\rho}$ a parameter which generalizes the usual notion of density. The joint probability distribution $p(m_1, \ldots, m_N | \tilde{\rho})$ reads in this case

$$p(m_1, \ldots, m_N | \tilde{\rho}) = \frac{1}{\mathcal{Z}_{N,\delta}(\tilde{\rho})} \prod_{i=1}^{N} f(m_i) \delta \left[\sum_{i=1}^{N} m_i - \tilde{\rho} N^{1+\delta} \right] \tag{26}$$

where $\mathcal{Z}_{N,\delta}(\tilde{\rho})$ is a normalization factor (see Equation (3)). We wish to determine for which values of α and $\tilde{\rho}$ condensation occurs in this case, using as an order parameter for condensation the participation ratio $Y_k = N \langle m^k \rangle / M^k$, which reads in the present case as

$$Y_k = \frac{\langle m^k \rangle}{\tilde{\rho}^k N^{k-1+k\delta}} = \frac{1}{\tilde{\rho}^k N^{k-1+k\delta}} \int_0^{\infty} dm \, p(m) \, m^k \tag{27}$$

The expression of the marginal distribution $p(m)$ in the case of the superextensive total mass is given below in Equation (41).

In the following paragraphs, we first use in Section 4.1 the integral representation of the partition function $\mathcal{Z}_N(\tilde{\rho})$ in order to get indications on the phase diagram in the (α, δ) plane. This preliminary analysis will suggest the existence of a transition line, that will be confirmed in Sections 4.2–4.4 by an explicit determination of the marginal distribution $p(m)$ and the participation ratio Y_k respectively below, above and on the transition line.

4.1. Preliminary Analysis of the Phase Diagram

We start by expressing the partition function $\mathcal{Z}_{N,\delta}(\tilde{\rho})$ as an integral representation in terms of its inverse Laplace transform. The Laplace transform $\hat{\mathcal{Z}}_{N,\delta}(s)$ of $\mathcal{Z}_{N,\delta}(\tilde{\rho})$ is expressed as

$$\hat{\mathcal{Z}}_{N,\delta}(s) \equiv \int_0^\infty d\tilde{\rho}\, e^{-s\tilde{\rho}}\, \mathcal{Z}_{N,\delta}(\tilde{\rho}) = \frac{1}{N^{1+\delta}}\, g_\alpha \left(\frac{s}{N^{1+\delta}} \right)^N \tag{28}$$

where $g_\alpha(s)$ is defined in Equation (15). After a simple change of variable, the inverse Laplace representation of the partition function $\mathcal{Z}_N(\tilde{\rho})$ reads

$$\mathcal{Z}_{N,\delta}(\tilde{\rho}) = \frac{1}{2\pi i} \int_{s_0-i\infty}^{s_0+i\infty} ds\, \exp\{N[\log g_\alpha(s) + \tilde{\rho}N^\delta s]\} \tag{29}$$

The value of s_0, although arbitrary, can be conveniently chosen to be the saddle-point value of the argument of the exponential in Equation (29), when a saddle-point $s_0 > 0$ exists (this is due to the presence of a branch-cut singularity on the negative real axis, as discussed in the previous section). When no saddle-point exists, the equivalence between canonical and grand-canonical ensemble breaks down, and condensation is expected to occur. A saddle-point solution of the integral in Equation (29) should satisfy the following equation,

$$\tilde{\rho}N^\delta = -\frac{g_\alpha'(s)}{g_\alpha(s)} = \frac{\langle m\, e^{-sm} \rangle_\alpha}{\langle e^{-sm} \rangle_\alpha} \tag{30}$$

Note that this approach is heuristic, since the saddle-point should in principle not depend on N. However, the N-dependence is not a problem when testing the existence of a saddle-point. If it exists, the saddle-point evaluation of the integral then requires a change of variable for (some power of) N to appear only as a global prefactor in the argument of the exponential.

For $\alpha > 1$, we know from the results of Section 3 that the saddle-point Equation (30) has a solution only if $\tilde{\rho}N^\delta < \rho_c$. This condition is never satisfied for $\delta > 0$ and $N \to \infty$, so that no saddle-point exists and condensation occurs for any value of $\tilde{\rho} > 0$ when $\alpha > 1$.

The situation is thus quite similar to the extensive mass case: there is a homogeneous phase carrying a total mass $N\rho_c$ which coexists with a superextensive condensate with a mass $M_c = \tilde{\rho}N^{1+\delta} - N\rho_c \approx \tilde{\rho}N^{1+\delta}$, so that the condensate carries a fraction of the total mass equal to one in the limit $N \to \infty$. It follows that the participation ratio $Y_k = 1$ in this limit.

In contrast, for $\alpha < 1$, the function $g_\alpha'(s)/g_\alpha(s)$ spans the whole positive real axis, and the saddle-point Equation (30) always has a solution $s_0(N)$, which goes to 0 when $N \to \infty$. One then has to factor out the N-dependence through an appropriate change of variable, and to check whether a saddle-point evaluation of the integral can be made. For $s \to 0$, one has $g_\alpha'(s)/g_\alpha(s) \approx a\alpha s^{\alpha-1}$, so that $s_0 \sim N^{-\delta/(1-\alpha)}$. Using the change of variable $s = zN^{-\delta/(1-\alpha)}$ in the integral appearing in Equation (30), the argument of the exponential can be rewritten as

$$N^{1-\alpha\delta/(1-\alpha)}(-az^\alpha + \tilde{\rho}z) \tag{31}$$

with $a = A\Gamma(1-\alpha)/\alpha$, and where we have used the small-s expansion $g_\alpha(s) \approx 1 - as^\alpha$, valid for $0 < \alpha < 1$. The saddle-point evaluation of the integral is valid only if the N-dependent prefactor diverges, meaning that $1 - \alpha\delta/(1-\alpha) > 0$, or equivalently

$$\delta < \delta_c = \frac{1-\alpha}{\alpha} \tag{32}$$

Hence for $\alpha < 1$ and $\delta < \delta_c$, a saddle-point evaluation of the partition function is possible, and the equivalence between canonical and grand-canonical ensembles holds: the system is in the homogeneous phase, and no condensation occurs. For $\delta \geq \delta_c$, the saddle-point evaluation of the partition function is no longer possible, which suggests that the equivalence of ensembles breaks down. This is an indication that condensation may occur. We show through explicit calculations in Sections 4.3 and 4.4 that condensation occurs when $\delta \geq \delta_c$, in the sense that the participation ratio Y_k takes a nonzero value in the infinite N limit.

Before proceeding to a detailed characterization of this condensation, let us briefly comment on the value of δ_c. For $\delta = \delta_c$, the total mass in the system scales as $M \sim N^{1/\alpha}$, and this scaling precisely corresponds to the typical value of the total mass present in the unconstrained case (see Section 2), as already noticed in [8]. Hence $\delta < \delta_c$ corresponds to imposing a total mass much smaller than the "natural" unconstrained mass, while for $\delta > \delta_c$ one imposes a mass much larger than the typical unconstrained mass, leading to condensation. In this sense, the situation is similar to that of the extensive mass case for $\alpha > 1$, where condensation occurs when a mass M larger that the unconstrained mass $N\rho_c$ is imposed, $M > N\rho_c$. In Figure 2 we present the phase diagram of the model for the case of a constraint to a superextensive total mass, which is a phase diagram in the plane (α, δ). Two observations are in order for this phase diagram. First, as will be explained in detail in in Sections 4.3 and 4.4, the presence of a condensed phase never depends on the value of the parameter $\tilde{\rho}$, apart on the transition line $\delta_c(\alpha)$. Second, on the transition line $\delta_c(\alpha)$ (see Figure 2), the system is in the condensed phase (see Section 4.4), at variance with the behaviour on the critical line $\rho_c(\alpha)$ for the case of extensive mass (see the phase diagram in Figure 1), in which case the system is *not* in the condensed phase on the transition line.

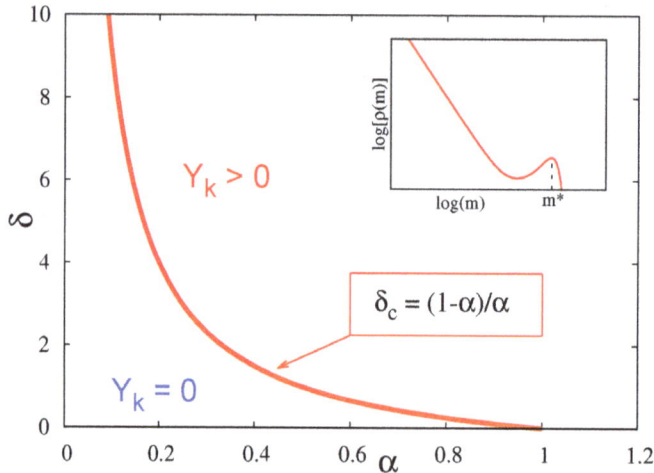

Figure 2. *Main*: Phase diagram for the values of the participation ratio in the (α, δ) plane in the presence of a super-extensive constraint on the total value of the mass $\sum_{i=1}^{N} m_i = \tilde{\rho} N^{1+\delta}$. The (red) continuous line marks the separation of the condensed phase ($\lim_{N\to\infty} Y_k > 0$) and the homogeneous phase ($\lim_{N\to\infty} Y_k = 0$). *Inset*: schematic representation of the marginal probability distribution of the local mass, $p(m)$, in the presence of condensation, namely in the whole region $\delta \geq \delta_c$.

In the forthcoming sections, we evaluate the distribution $p(m)$ and the participation ratio Y_k for $\alpha < 1$ in the three cases $\delta < \delta_c$ (Section 4.2), $\delta > \delta_c$ (Section 4.3) and $\delta = \delta_c$ (Section 4.4) respectively.

4.2. Case $\delta < \delta_c$: Homogeneous Phase

As we have seen above, the system remains homogeneous for $\alpha < 1$ and $\delta < \delta_c$, and the equivalence between canonical and grand-canonical ensembles holds. One can thus more conveniently perform calculations in the grand-canonical ensemble, with a chemical potential μ_N which depends on N, and which will be determined below as a function of the total mass. The single-mass distribution $p(m)$ simply reads

$$p(m) \approx f(m)\,e^{-\mu_N m} \qquad (0 < m < M) \tag{33}$$

where we have neglected the correction to the normalization factor, as the latter remains very close to 1 since μ_N is very small.

Note that the distribution $p(m)$ monotonously decays at large m as

$$p(m) \approx \frac{A}{m^{1+\alpha}}\,e^{-\mu_N m} \tag{34}$$

as is typical for a homogeneous phase. The k-th moment of this distribution is obtained for $k > \alpha$ as

$$\begin{aligned}
\langle m^k \rangle &= \int_0^M dm\, m^k p(m) \\
&\approx A \int_0^\infty dm\, m^{k-1-\alpha} e^{-\mu_N m} = \frac{A\,\Gamma(k-\alpha)}{\mu_N^{k-\alpha}}
\end{aligned} \tag{35}$$

The value of μ_N is then determined from the condition $\langle m \rangle = M/N = \tilde{\rho}N^\delta$, yielding

$$\mu_N = \left[\frac{A}{\tilde{\rho}}\Gamma(1-\alpha)\right]^{1/(1-\alpha)} N^{-\delta/(1-\alpha)} \tag{36}$$

The participation ratio Y_k is then given by

$$Y_k = K\tilde{\rho}^{(k-1)/\delta_c}\,N^{-(k-1)(\delta_c-\delta)/\delta_c} \tag{37}$$

with

$$K = A^{1-\omega}\frac{\Gamma(k-\alpha)}{\Gamma(1-\alpha)^\omega}, \qquad \omega = \frac{k-\alpha}{1-\alpha} \tag{38}$$

and where δ_c is defined in Equation (32). One thus obtains that $Y_1 = 1$ as it should, and that for $k > 1$, $Y_k \to 0$ when $N \to \infty$, which confirms the absence of condensation for $\delta < \delta_c$. Yet, it is interesting to note that the decay of Y_k becomes slower when increasing δ, and becomes approximately logarithmic in N when $(\delta_c - \delta)/\delta_c \ll 1$.

4.3. Case $\delta > \delta_c$: Condensed Phase

When $\delta > \delta_c$ (and $\alpha < 1$), the partition function $Z_{N,\delta}(\tilde{\rho})$ can no longer be evaluated by a saddle-point method, and equivalence of ensembles breaks down, so that one has to work in the canonical ensemble. From Equation (28), the Laplace transform $\hat{Z}_{N,\delta}(s)$ reads in the large N limit, using the small-s behavior $g_\alpha(s) \approx 1 - as^\alpha$,

$$\hat{Z}_{N,\delta}(s) = N^{-1-\delta}\left(1 - \frac{as^\alpha}{N^\nu}\right) \tag{39}$$

with $\nu = \alpha(1+\delta) - 1$.

By assuming a scaling function $G(x)$ which satisfies the normalization condition $\int_0^\infty G(x)\,dx = 1$ and which has the asymptotic behaviour $G(x) \approx A/x^{1+\alpha}$ for $x \to \infty$, one can guess the following expression for the partition function $Z_{N,\delta}(\tilde{\rho})$ in direct space:

$$Z_{N,\delta}(\tilde{\rho}) \approx N^{-1-\delta+\nu/\alpha}G(\tilde{\rho}N^{\nu/\alpha}) \tag{40}$$

It is then not difficult to check that the expression in Equation (40) is (asymptotically in N) the correct one: the expansion for small s of its Laplace transform corresponds precisely to the expression of $\hat{Z}_{N,\delta}(s)$ in Equation (39).

From the knowledge of $Z_{N,\delta}(\tilde{\rho})$ one can then compute the distribution $p(m)$, which reads as

$$
\begin{aligned}
p(m) &= \frac{f(m)}{Z_{N,\delta}(\tilde{\rho})} Z_{N-1,\delta}\left(\tilde{\rho} - \frac{m}{(N-1)^{1+\delta}}\right) \\
&\approx \frac{f(m)}{G(\tilde{\rho}N^{\delta-\delta_c})} G\left(\tilde{\rho}N^{\delta-\delta_c} - \frac{m}{N^{1/\alpha}}\right)
\end{aligned}
\tag{41}
$$

and has a non-monotonous shape, as seen by evaluating $p(m)$ in the regime $m \sim xM$ with $0 < x < 1$, which leads to

$$
p(xM) \approx \frac{A}{N^{(1+\delta)(1+\alpha)}\tilde{\rho}^{1+\alpha}[x(1-x)]^{1+\alpha}}
\tag{42}
$$

Note that the divergences at $x = 0$ and $x = 1$ are regularized for values of x such that $x \sim M^{-1}$ and $1 - x \sim M^{-1}$ respectively. The non-monotonic shape of $p(xM)$, which is schematically represented in the inset of Figure 2, is a strong similarity that the constrained condensation for $\alpha < 1$ (and superextensive total mass) bears with the constrained condensation for $\alpha > 1$ (and extensive total mass). At the same time such a non-monotonic shape of $p(m)$ is a remarkable qualitative difference with the case of unconstrained condensation found for the same range of the exponent, $\alpha < 1$, in which case the local mass distribution decays monotonously at large values as $p(m) \sim 1/m^{1+\alpha}$. Interestingly, the expression Equation (41) of $p(m)$ can be rewritten as

$$
p(m) \approx \frac{f(m)}{G(\tilde{\rho}N^{\delta-\delta_c})} G\left(\frac{M-m}{N^{1/\alpha}}\right)
\tag{43}
$$

with $M = \tilde{\rho}N^{1+\delta}$, which shows that the "bump" occuring for $m \approx M$ has a width $\sim N^{1/\alpha}$. Hence its relative width scales as $N^{1+\delta}/N^{1/\alpha} = 1/N^{\delta-\delta_c}$ and goes to zero when $N \to \infty$ for $\delta > \delta_c$. In this case it is therefore legitimate to call the bump a condensate, because it has a well-defined mass $M_{\text{cond}} \sim M$.

To complete the analysis, let us compute the participation ratio Y_k. In the large N limit, the moment $\langle m^k \rangle = \int_0^M dm\, m^k p(m)$ can be computed as, using the change of variable $v = \tilde{\rho}N^{\delta-\delta_c} - m/N^{1/\alpha}$,

$$
\langle m^k \rangle \approx \frac{M^k}{N} \int_0^{M/N^{1/\alpha}} dv \left(1 - \frac{vN^{1/\alpha}}{M}\right)^{k-1-\alpha} G(v)
\tag{44}
$$

where we have used the asymptotic (large argument) behavior of $f(m) \sim 1/m^{1+\alpha}$ and $G(x) \sim 1/x^{1+\alpha}$. It is then easy to show (see Appendix A) that the integral in Equation (44) tends to $\int_0^\infty dv\, G(v) = 1$ when $N \to \infty$. One thus simply gets

$$
\langle m^k \rangle \approx \frac{M^k}{N}
\tag{45}
$$

which, using $Y_k = N\langle m^k \rangle/M^k$, immediately leads to the conclusion that $Y_k = 1$ in the limit $N \to \infty$. Hence, as anticipated above, a strong condensation occurs for $\delta > \delta_c$ and $\alpha < 1$, in the sense that the condensate carries almost all the mass present in the system.

4.4. Case $\delta = \delta_c$: Marginal Condensed Phase

For $\alpha < 1$ and $\delta = \delta_c$, the Laplace transform of the partition function reads for large N as

$$
\hat{Z}_{N,\delta_c}(s) = N^{-1/\alpha} e^{-as^\alpha}
\tag{46}
$$

and from the expression in Equation (46) the partition function is obtained as

$$
Z_{N,\delta_c}(\tilde{\rho}) = N^{-1/\alpha} H(\tilde{\rho})
\tag{47}
$$

where the function $H(\tilde{\rho})$ is independent of N and is defined by its Laplace transform,

$$\int_0^\infty d\tilde{\rho}\, e^{-s\tilde{\rho}} H(\tilde{\rho}) = e^{-as^\alpha} \tag{48}$$

(H is actually a one-sided Lévy distribution). The small s behavior $e^{-as^\alpha} \approx 1 - as^\alpha$ implies the large $\tilde{\rho}$ behavior

$$H(\tilde{\rho}) \approx \frac{A}{\tilde{\rho}^{1+\alpha}} \tag{49}$$

where again A is defined from the large m behavior $f(m) \approx A/m^{1+\alpha}$.

The distribution $p(m)$ is given for large N by

$$p(m) = \frac{f(m)}{H(\tilde{\rho})} H\left(\frac{M-m}{N^{1/\alpha}}\right) \tag{50}$$

with $M = \tilde{\rho} N^{1/\alpha}$. It is interesting to evaluate $p(m)$ in the regime where $m \sim xM$ with $0 < x < 1$, which leads to

$$p(xM) \approx \frac{A}{N^{1+1/\alpha} \tilde{\rho}^{1+\alpha} H(\tilde{\rho})} \frac{H\left(\tilde{\rho}(1-x)\right)}{x^{1+\alpha}} \tag{51}$$

For large enough $\tilde{\rho}$, the shape of $p(m)$ is not monotonous, since the large $\tilde{\rho}$ expansion of Equation (51) yields

$$p(xM) \approx \frac{A}{N^{1+1/\alpha} \tilde{\rho}^{1+\alpha} [x(1-x)]^{1+\alpha}} \tag{52}$$

with a regularization of the divergence appearing at $x = 1$ for $1 - x \sim \tilde{\rho}^{-1}$, and of the divergence at $x = 0$ for $x \sim M^{-1}$. So here again, a bump appears in the distribution, but its width scales as $N^{1/\alpha}$ as seen from Equation (50), so that the relative width remains of the order of one. Following [8], one may call this bump a "pseudo-condensate".

The above argument on the existence of the bump in the distribution $p(m)$ was based on a large $\tilde{\rho}$ limit. The explicit example studied in [8] indeed shows that the bump may disappear below a certain value of $\tilde{\rho}$.

We now turn to the evaluation of the moment $\langle m^k \rangle$. Using the change of variable $v = \tilde{\rho} - m/N^{1/\alpha}$, as well as the asymptotic (large argument) behaviors of $f(m)$ and $H(\tilde{\rho})$, the moment $\langle m^k \rangle$ can be evaluated as

$$\langle m^k \rangle = \tilde{\rho}^{1+\alpha} N^{-1+k/\alpha} \int_0^{\tilde{\rho}} dv\, (\tilde{\rho} - v)^{k-1-\alpha} H(v) \tag{53}$$

It follows that $Y_k = N \langle m^k \rangle / M^k$ is given by

$$Y_k = \frac{1}{\tilde{\rho}^{k-1-\alpha}} \int_0^{\tilde{\rho}} dv\, (\tilde{\rho} - v)^{k-1-\alpha} H(v) \tag{54}$$

which is one of the main results of this paper. Note that the convergence of the integral at the upper bound implies $k - \alpha > 0$. Note also that the integral in Equation (54) is a convolution, which in some cases may be conveniently evaluated using a Laplace transform, given that $H(\tilde{\rho})$ is known through its Laplace transform. For a numerical evaluation of Y_k, one may thus compute analytically the Laplace transform of the integral in Equation (54), yielding

$$\mathcal{L}\left(\tilde{\rho}^{k-1-\alpha} Y_k(\tilde{\rho})\right) = \frac{\Gamma(k-\alpha)}{s^{k-\alpha}} e^{-as^\alpha} \tag{55}$$

and perform numerically the inverse Laplace transform.

The Laplace transform approach is also convenient to determine analytically the small $\tilde{\rho}$ behavior of Y_k, since the inverse Laplace transform can be evaluated through a saddle-point calculation in this limit. One finds

$$Y_k \approx B\,\tilde{\rho}^\lambda\,e^{-c/\tilde{\rho}^{\alpha/(1-\alpha)}} \tag{56}$$

with parameters λ, c and B given by

$$\lambda = \frac{\alpha(2k - 2\alpha - 1)}{2(1 - \alpha)} \tag{57}$$

$$c = \frac{1 - \alpha}{\alpha}\,(\alpha a)^{1/(1-\alpha)} \tag{58}$$

$$B = \frac{\Gamma(k - \alpha)}{\sqrt{2\pi(1 - \alpha)}}\,(\alpha a)^{\frac{1+2\alpha-2k}{2-2\alpha}} \tag{59}$$

More detailed calculations on the derivation of Equation (56) are reported in Appendix A.

In the large $\tilde{\rho}$ limit, it is easy to show that Y_k goes to 1, following a procedure similar to the one used in the case $\delta > \delta_c$. It is of interest to compute the first correction in $\tilde{\rho}$ (see Appendix A), and one finds

$$Y_k \approx 1 - \frac{B'}{\tilde{\rho}^\alpha} \tag{60}$$

with

$$B' = \frac{A}{\alpha}\,\frac{\Gamma(1 - \alpha)\,\Gamma(k - \alpha)}{\Gamma(k - 2\alpha)} \tag{61}$$

Note in particular that if $\alpha \ll 1$, the convergence of Y_k to 1 is very slow.

In summary, one has in the case $\delta = \delta_c$ a non-standard, weak condensation effect, which does not correspond to the genuine condensation effect reported in the literature [4–13]. Here, the weak condensation effect simply means that the participation ratio takes a nonzero value in the infinite size limit, indicating that a few random variables carry a finite fraction of the sum. However, as mentioned above, there is no well-defined condensate that would coexist with a fluid phase. Depending on the generalized density $\tilde{\rho}$, the marginal distribution $p(m)$ either decreases monotonously, or has a bump which corresponds only to a pseudo-condensate, since the relative width of the bump remains of the order of one, see Equation (50). In addition, the line $\delta = \delta_c$ does not correspond to a well-defined transition line in the (α, δ) plane, in the sense that the state of the system continuously depends on the generalized density $\tilde{\rho}$, as shown by the expression of the participation ratio Y_k given in Equation (54).

5. Conclusions

The general motivation of this work was to better understand the connection between condensation in the unconstrained case and in the constrained case with extensive mass, because condensation occurs on opposite ranges of the exponent α (which defines the power-law decay of the unconstrained probability distribution), respectively $\alpha < 1$ and $\alpha > 1$. To this aim, we have studied condensation in the case where the total mass is constrained to a superextensive value $M = \tilde{\rho}N^{1+\delta}$, where $\delta > 0$, motivated by the fact that the typical scaling of the total mass is also superextensive, $M \sim N^{1/\alpha}$, when condensation takes place in the unconstrained case, which happens for $\alpha < 1$.

We indeed found that the case of a fixed superextensive total mass interpolates in a sense between the case with a fixed extensive mass and the unconstrained case: condensation is found for values of the power law exponent in the interval $0 < \alpha < 1$, as in the case of *unconstrained* condensation, but with qualitative features more similar to the case of *constrained* condensation with extensive mass: for $\delta > \delta_c$ (and for $\delta = \delta_c$ at large enough $\tilde{\rho}$) the marginal distribution $p(m)$ of the local mass has a secondary peak related to the condensate fraction, at variance with unconstrained case where $p(m)$ decays monotonously for increasing values of m.

The inclusion in the problem of the new parameter δ, which characterizes the superextensive scaling $M \sim N^{1+\delta}$ of the total mass, allowed us to draw the two-dimensional (α, δ) phase diagram shown in Figure 2. At variance with the two models usually studied in the literature, where condensation takes place *either* for $\alpha < 1$, without the constraint, *or* for $\alpha > 1$, with constrained extensive mass, in the case of a constrained superextensive mass condensation is found *both* for $\alpha < 1$ (when $\delta > \delta_c$) *and* for $\alpha > 1$ (when $\delta > 0$).

More in detail, we have shown that as soon as $\delta > 0$, constrained condensation occurs for any $\alpha > 1$, irrespective of the value of the generalized density $\tilde{\rho}$, when the system is constrained to have a superextensive value of the mass. This case is qualitatively similar to the case of an extensive mass $M = \rho N$ with a large density ρ. For $\alpha < 1$, a weak form of condensation occurs if $\delta = \delta_c(\alpha)$, in the sense that the participation ratio takes a nonzero value in the infinite N limit. Here, $1 + \delta_c = 1/\alpha$ is precisely the scaling exponent of the mass in the unconstrained case. When $\delta = \delta_c$, only a pseudo-condensate with non-vanishing relative fluctuations appears, or the distribution $p(m)$ may even decay monotonously. This is confirmed by the expression Equation (54) of Y_k ($\alpha < 1$ and constraint to superextensive mass with $\delta = \delta_c$), which differs from Equation (25) obtained in the case of $\alpha > 1$ and a fixed extensive mass. The situation is thus different from standard condensation, but the nonzero asymptotic value of the participation ratio indicates that some non-trivial phenomenon (that we call weak condensation) takes place. On the contrary, when $\delta > \delta_c$ condensation takes the form of a bump with vanishing relative width in the marginal distribution $p(m)$. It thus shares similarities with the standard condensation phenomenon.

To conclude, we note that the qualitative idea that condensation occurs when one imposes a total mass larger than the "natural" mass the system would have in the unconstrained case remains valid: this is always the case for $\alpha > 1$ (both for extensive and superextensive constraints), but it is also the case to some extent for $\alpha < 1$, where condensation is present for $\delta > \delta_c$. Yet, one has to be aware that the notion of "natural mass" is not firmly grounded in this case, and is just a heuristic concept associated to a typical scaling $M \sim N^{1/\alpha}$ with the system size N. One further subtlety is whether condensation occurs or not on the transition line. For the constrained case with an extensive mass, condensation does not occur at the critical density $\rho = \rho_c$. In contrast, a weak form of condensation occurs at $\delta = \delta_c$ (see Section 4.4), which may suggest a discontinuous condensation transition as a function of δ. However, for $\delta = \delta_c$ the condensation properties actually depend on the generalized density $\tilde{\rho}$, see Equation (54), so that this weak condensation is actually continuous (in the sense that Y_k goes to zero when $\tilde{\rho} \to 0$) if one looks on a finer scale in terms of $\tilde{\rho}$.

Acknowledgments: G.G. acknowledges Financial support from ERC Grant No. ADG20110209.

Author Contributions: Both authors designed the project, performed calculations and wrote the paper.

Conflicts of Interest: The authors declare no conflict of interest.

Appendix A. Participation Ratio for $M = \tilde{\rho} N^{1+\delta}$

In this appendix, we provide some technical details on the evaluation of participation ratios for $\alpha > 1$, in the cases $\delta > \delta_c$ and $\delta = \delta_c$.

Appendix A.1. Case $\delta > \delta_c$

Considering the case $\alpha < 1$ and $\delta > \delta_c$, we wish here to justify the approximation made to go from Equation (44) to Equation (45) in the evaluation of $\langle m^k \rangle$. Considering the integral appearing in Equation (44) as well as its approximation, one can write, setting $V_0 = M/N^{1/\alpha}$

$$\int_0^{V_0} dv\, G(v) - \int_0^{V_0} dv\, G(v) \left(1 - \frac{v}{V_0}\right)^{k-1-\alpha}$$

$$= V_0 \int_0^1 du\, G(V_0 u) \left[1 - (1-u)^{k-1-\alpha}\right] \tag{A1}$$

$$\approx \frac{A}{V_0^\alpha} \int_0^1 \frac{du}{u^{1+\alpha}} \left[1 - (1-u)^{k-1-\alpha}\right]$$

where we have used the change of variable $v = V_0 u$ as well as the asymptotic behavior of the function $G(v)$. Assuming $k - \alpha > 0$, the last integral in Equation (A1) converges, so that the difference of the two integrals in the lhs of Equation (A1) indeed converges to 0 when $N \to \infty$, since $V_0 \to \infty$ in this limit.

Appendix A.2. Case $\delta = \delta_c$

We discuss here the asymptotic, small $\tilde{\rho}$ and large $\tilde{\rho}$, behavior of the participation ratio Y_k in the case $\delta = \delta_c$.

Let us start by the small $\tilde{\rho}$ regime. As discussed in the main text, Y_k can be obtained by an inverse Laplace transform, see Equation (55). Introducing $\psi(\tilde{\rho}) = \tilde{\rho}^{k-1-\alpha} Y_k(\tilde{\rho})$, the Laplace transform $\hat{\psi}(s)$ is given by

$$\hat{\psi}(s) = \frac{\Gamma(k-\alpha)}{s^{k-\alpha}} e^{-as^\alpha} \tag{A2}$$

Taking the inverse Laplace transform, one has

$$\psi(\tilde{\rho}) = \frac{1}{2\pi i} \int_{s_0 - i\infty}^{s_0 + i\infty} ds\, \frac{\Gamma(k-\alpha)}{s^{k-\alpha}} e^{-as^\alpha + s\tilde{\rho}} \tag{A3}$$

In order to see whether this integral can be performed through a saddle-point evaluation, we note that balancing the two terms in the argument of the exponential leads to $as^\alpha \sim s\tilde{\rho}$, which results in $s \sim \tilde{\rho}^{-1/(1-\alpha)}$, eventually leading to $-as^\alpha + s\tilde{\rho} \sim \tilde{\rho}^{-\alpha/(1-\alpha)}$ (note that the algebraic prefactor in front of the exponential does not change the location of the saddle-point). The argument can then be made sharper using the change of variable $s = z/\tilde{\rho}^{1/(1-\alpha)}$, yielding

$$-as^\alpha + s\tilde{\rho} = \tilde{\rho}^{-\alpha/(1-\alpha)}(-az^\alpha + z) \tag{A4}$$

In this form, a diverging prefactor is obtained when $\tilde{\rho} \to 0$, so that a saddle-point calculation can indeed be performed in this limit. Defining $\phi(z) = -az^\alpha + z$, the saddle-point z_0 is obtained for $\phi'(z_0) = 0$, yielding $z_0 = (\alpha a)^{1/(1-\alpha)}$. Choosing $s_0 = z_0/\tilde{\rho}^{1/(1-\alpha)}$ in the integral Equation (A3), and setting $z = z_0 + iy$, one can write in the small $\tilde{\rho}$ limit

$$\psi(\tilde{\rho}) \approx \frac{\Gamma(k-\alpha)}{2\pi s^{k-\alpha}} \int_{-\infty}^\infty dy\, \exp\left(\tilde{\rho}^{\frac{\alpha}{\alpha-1}} \left[\phi(z_0) - \phi''(z_0)\frac{y^2}{2}\right]\right)$$

$$\approx \frac{\Gamma(k-\alpha)}{z_0^{k-\alpha}\sqrt{2\pi\phi''(z_0)}} \tilde{\rho}^\lambda \exp\left(\phi(z_0)\tilde{\rho}^{-\frac{\alpha}{1-\alpha}}\right) \tag{A5}$$

where λ is given in Equation (57), thus recovering Equation (56).

We now turn to the computation in the large $\tilde{\rho}$ limit. We have seen that the inverse Laplace transform cannot be computed through a saddle-point evaluation in this limit. We thus come back to Equation (54) and rewrite it as

$$Y_k = \int_0^{\tilde{\rho}} dv \left(1 - \frac{v}{\tilde{\rho}}\right)^{k-1-\alpha} H(v) \tag{A6}$$

Since the integral $\int_0^\infty dv \, H(v)$ converges, one can approximate for large $\tilde{\rho}$ the factor $(1 - v/\tilde{\rho})^{k-1-\alpha}$ in Equation (A6) by 1, assuming $k - 1 - \alpha > 0$. Hence, again for $\tilde{\rho} \to \infty$,

$$Y_k \approx \int_0^{\tilde{\rho}} dv \, H(v) \to \int_0^\infty dv \, H(v) = 1 \tag{A7}$$

The correction to $Y_k = 1$ can be computed as follows:

$$1 - Y_k = \int_{\tilde{\rho}}^\infty dv \, H(v) + \int_0^{\tilde{\rho}} dv \, H(v) \left[1 - \left(1 - \frac{v}{\tilde{\rho}}\right)^{k-1-\alpha} \right] \tag{A8}$$

The first integral in Equation (A8) is easily evaluated for large $\tilde{\rho}$ as

$$\int_0^{\tilde{\rho}} dv \, H(v) \approx \frac{A}{\alpha \tilde{\rho}^\alpha} \tag{A9}$$

The second integral in Equation (A8) can be rewritten with the change of variable $v = \tilde{\rho} u$ as

$$\tilde{\rho} \int_0^1 du \, H(\tilde{\rho} u) \left[1 - (1 - u)^{k-1-\alpha} \right] \approx \frac{A}{\tilde{\rho}^\alpha} \int_0^1 \frac{du}{u^{1+\alpha}} \left[1 - (1 - u)^{k-1-\alpha} \right] \equiv \frac{A \, I}{\tilde{\rho}^\alpha} \tag{A10}$$

where we have denoted as I the last integral, and where we have used the asymptotic behavior of $H(\rho)$ given in Equation (49). Assuming $k - 1 - \alpha > 0$ (which is consistent since k is in most cases of interest an integer >1), the integral I can be computed through an integration by part, leading to

$$I = -\frac{1}{\alpha} + \frac{k - 1 - \alpha}{\alpha} \frac{\Gamma(1 - \alpha)\Gamma(k - 1 - \alpha)}{\Gamma(k - 2\alpha)} \tag{A11}$$

where we have also used the standard result

$$\int_0^1 du \, u^{\mu-1}(1 - u)^{\nu-1} = \frac{\Gamma(\mu)\Gamma(\nu)}{\Gamma(\mu + \nu)} \tag{A12}$$

for $\mu, \nu > 0$. Then combining Equations (A8)–(A11), one eventually obtains Equation (60).

References

1. Szavits-Nossan, J.; Evans, M.R.; Majumdar, S.N. Conditioned random walks and interaction-driven condensation. *J. Phys. A Math. Theor.* **2016**, *50*, 024005.
2. Mézard, M.; Parisi, G.; Virasoro, M. *Spin Glass Theory and Beyond*; World Scientific: Singapore, 1987.
3. Bouchaud, J.P.; Mézard, M. Universality classes for extreme-value statistics. *J. Phys. A Math. Gen.* **1997**, *30*, 7997.
4. Bialas, P.; Burda, Z.; Johnston, D. Condensation in the Backgammon model. *Nucl. Phys. B* **1997**, *493*, 505–516.
5. Majumdar, S.N.; Krishnamurthy, S.; Barma, M. Nonequilibrium Phase Transitions in Models of Aggregation, Adsorption, and Dissociation. *Phys. Rev. Lett.* **1998**, *81*, 3691.
6. Grosskinsky, S.; Schütz, G.M.; Spohn, H. Condensation in the zero range process: Stationary and dynamical properties. *J. Stat. Phys.* **2003**, *113*, 389–410.
7. Majumdar, S.N.; Evans, M.R.; Zia, R.K.P. Nature of the Condensate in Mass Transport Models. *Phys. Rev. Lett.* **2005**, *94*, 180601.
8. Evans, M.R.; Majumdar, S.N.; Zia, R.K.P. Canonical Analysis of Condensation in Factorised Steady States. *J. Stat. Phys.* **2006**, *123*, 357–390.
9. Evans, M.R.; Hanney, T. Nonequilibrium statistical mechanics of the Zero-Range Process and related models. *J. Phys. A Math. Gen.* **2005**, *38*, R195.

10. Majumdar, S.N. Real-space Condensation in Stochastic Mass Transport Models. In *Exact Methods in Low-Dimensional Statistical Physics and Quantum Computing*; Les Houches Lecture Notes for the Summer School; Jacobsen, J., Ouvry, S., Pasquier, V., Serban, D., Cugliandolo, L.F., Eds.; Oxford University Press: Oxford, UK, 2008.

11. Hirschberg, O.; Mukamel, D.; Schütz, G.M. Condensation in Temporally Correlated Zero-Range Dynamics. *Phys. Rev. Lett.* **2009**, *103*, 090602.

12. Whitehouse, J.; Costa, A.; Blythe, R.A.; Evans, M.R. Maintenance of order in a moving strong condensate. *J. Stat. Mech.* **2014**, P11029, doi:10.1088/1742-5468/2014/11/P11029.

13. Evans, M.R.; Waclaw, B. Condensation in stochastic mass transport models: Beyond the zero-range process. *J. Phys. A Math. Theor.* **2014**, *47*, 095001.

14. Evans, M.R.; Majumdar, S.N. Condensation and extreme value statistics. *J. Stat. Mech.* **2008**, P05004, doi:10.1088/1742-5468/2008/05/P05004.

15. Szavits-Nossan, J.; Evans, M.R.; Majumdar, S.N. Constraint-Driven Condensation in Large Fluctuations of Linear Statistics. *Phys. Rev. Lett.* **2014**, *112*, 020602.

16. Szavits-Nossan, J.; Evans, M.R.; Majumdar, S.N. Condensation Transition in Joint Large Deviations of Linear Statistics. *J. Phys. A Math. Theor.* **2014**, *47*, 455004.

17. Zannetti, M.; Corberi, F.; Gonnella, G. Condensation of Fluctuations in and out of Equilibrium. *Phys. Rev. E* **2014**, *90*, 012143.

18. Corberi, F.; Gonnella, G.; Piscitelli, A. Singular behavior of fluctuations in a relaxation process. *J. Non-Cryst. Solids* **2015**, *407*, 51–56.

19. Zannetti, M. The Grand Canonical catastrophe as an istance of condensation of fluctuations. *Europhys. Lett.* **2015**, *111*, 20004.

20. Crisanti, A.; Sarracino, A.; Zannetti, M. Heat fluctuations of Brownian oscillators in nonstationary processes: Fluctuation theorem and condensation transition. *Phys. Rev. E* **2017**, *95*, 052138.

21. Evans, M.R.; Hanney, T.; Majumdar, S.N. Interaction driven real-space condensation. *Phys. Rev. Lett.* **2006**, *97*, 010602.

22. Mandelbrot, B.B.; Wallis, J.R. Noah, Joseph, and operational hydrology. *Water Resour. Res.* **1968**, *4*, 909–918.

23. Magdziarz, M. Fractional Ornstein-Uhlenbeck processes. Joseph effect in models with infinite variance. *Physica A* **2008**, *387*, 123–133.

24. Bertin, E.M.; Bouchaud, J.P. Subdiffusion and localization in the one dimensional trap model. *Phys. Rev. E* **2003**, *67*, 026128.

25. Derrida, B. From random walks to spin glasses. *Physica D* **1997**, *107*, 186–198.

26. Mikosch, T.; Nagaev, A.V. Large deviations of heavy-tailed sums with applications in insurance. *Extremes* **1998**, *1*, 81–110.

entropy

MDPI

Article

Far-From-Equilibrium Time Evolution between Two Gamma Distributions

Eun-jin Kim [1,*]**, Lucille-Marie Tenkès** [1,2]**, Rainer Hollerbach** [3]**and Ovidiu Radulescu** [4]

[1] School of Mathematics and Statistics, University of Sheffield, Sheffield S3 7RH, UK;
lucillemarie.tenkes@gmail.com
[2] ENSTA ParisTech Université Paris-Saclay, 828 boulevard des Maréchaux, 91120 Palaiseau, France
[3] Department of Applied Mathematics, University of Leeds, Leeds LS2 9JT, UK; R.Hollerbach@leeds.ac.uk
[4] DIMNP-UMR 5235 CNRS, Université de Montpellier, Place Eugène Bataillon, 34095 Montpellier, France;
ovidiu.radulescu@umontpellier.fr
* Correspondence: e.kim@shef.ac.uk; Tel.: +44-114-222-3876

Received: 18 August 2017; Accepted: 21 September 2017; Published: 22 September 2017

Abstract: Many systems in nature and laboratories are far from equilibrium and exhibit significant fluctuations, invalidating the key assumptions of small fluctuations and short memory time in or near equilibrium. A full knowledge of Probability Distribution Functions (PDFs), especially time-dependent PDFs, becomes essential in understanding far-from-equilibrium processes. We consider a stochastic logistic model with multiplicative noise, which has gamma distributions as stationary PDFs. We numerically solve the transient relaxation problem and show that as the strength of the stochastic noise increases, the time-dependent PDFs increasingly deviate from gamma distributions. For sufficiently strong noise, a transition occurs whereby the PDF never reaches a stationary state, but instead, forms a peak that becomes ever more narrowly concentrated at the origin. The addition of an arbitrarily small amount of additive noise regularizes these solutions and re-establishes the existence of stationary solutions. In addition to diagnostic quantities such as mean value, standard deviation, skewness and kurtosis, the transitions between different solutions are analysed in terms of entropy and information length, the total number of statistically-distinguishable states that a system passes through in time.

Keywords: non-equilibrium statistical mechanics; gamma distribution; stochastic processes; Fokker-Planck equation; fluctuations and noise

1. Introduction

In classical statistical mechanics, the Gaussian (or normal) distribution and mean-field type theories based on such distributions have been widely used to describe equilibrium or near equilibrium phenomena. The ubiquity of the Gaussian distribution stems from the central limit theorem that random variables governed by different distributions tend to follow the Gaussian distribution in the limit of a large sample size [1–3]. In such a limit, fluctuations are small and have a short correlation time, and mean values and variance completely describe all different moments, greatly facilitating analysis.

Many systems in nature and laboratories are however far from equilibrium, exhibiting significant fluctuations. Examples are found not only in turbulence in astrophysical and laboratory plasmas, but also in forest fires, the stock market and biological ecosystems [4–23]. Specifically, anomalous (much larger than average values) transport associated with large fluctuations in fusion plasmas can degrade the confinement, potentially even terminating fusion operation [6]. Tornadoes are rare, large amplitude events, but can cause very substantial damage when they do occur. In biology, the pioneering work of Delbrück on bacteriophages showed that viruses replicate in strongly fluctuating bursts [24]. The fluctuations of the burst amplitudes were explained mathematically by stochastic autocatalytic reaction models first introduced in [25]. Delbrück's autocatalytic models predict discrete negative-binomial distributions, which can be well approximated by gamma distributions when the average number of particles is large. Furthermore, gene expression and protein productions, which used to be thought of as smooth processes, have also been observed to occur in bursts, leading to negative binomial and gamma distributed protein copy numbers (e.g., [19–23]). Such rare events of large amplitude (called intermittency) can dominate the entire transport even if they occur infrequently [8,26]. They thus invalidate the assumption of small fluctuations with short correlation time, making mean value and variances meaningless. Therefore, to understand the dynamics of a system far from equilibrium, it is crucial to have a full knowledge of Probability Distribution Functions (PDFs), including time-dependent PDFs [27].

The consequences of strong fluctuations in far-from-equilibrium systems are multiple. In physics, far-from-equilibrium fluctuations produce dissipative patterns, shift or wipe out phase transitions, etc. In economics, finance and actuarial science, strong fluctuations are important issues of risk evaluation. In biology, strong fluctuations generate phenotypic heterogeneity that helps multicellular organisms or microbial populations to adapt to changes of the environment by so-called "bet-hedging" strategies. In such a strategy, only a part of the cell population adapts upon emergence of new environmental conditions. The remaining part retains the memory of the old conditions and is thus already adapted if environmental conditions revert to previous ones [28]. Exceptional behaviour can also rescue cell subpopulations from drug-induced lethal conditions, thus generating drug resistance [29]. In particular, because of the skewness and exponential tail of the gamma distribution, gamma-distributed populations contain a significant proportion of individuals with an exceptionally high phenotypic variation. Bet-hedging being a dynamic phenomenon, it is important, for biological studies, to be able to predict not only steady-state, but also time-dependent distributions.

Obtaining good quality PDFs is often very challenging, as it requires a sufficiently large number of simulations or observations. Therefore, a PDF is usually constructed by averaging data from a long time series and is thus stationary (independent of time). Unfortunately, such stationary PDFs miss crucial information about the dynamics/evolution of non-equilibrium processes (e.g., tumour evolution). Theoretical prediction of time-dependent PDFs has proven to be no less challenging due to the limitation in our understanding of nonlinear stochastic dynamical systems, as well as the complexity in the required analysis.

Spectral analysis, for example, using theoretical tools similar to those used in quantum mechanics (e.g., raising and lower operators) is useful (e.g., [1]), but the summation of all eigenfunctions is necessary for time-dependent PDFs far from equilibrium. Various different methodologies have also been developed to obtain approximate PDFs, such as the variational principle, the rate equation method or the moment method [30–35]. In particular, the rate equation method [31,32] assumes that the form of a time-dependent PDF during the relaxation is similar to that of the stationary PDF and thus approximates a time-dependent PDF during transient relaxation by a PDF having the same functional form as a stationary PDF, but with time-varying parameters.

In this work, we show that this assumption is not always appropriate. We consider a stochastic logistic model with multiplicative noise. We show that for fixed parameter values, the stationary PDFs are always gamma distributions (e.g., [36,37]), one of the most popular distributions used in fitting experimental data. However, we find numerically that the time-dependent PDFs in transitioning from one set of parameter values to another are significantly different from gamma distributions, especially for strong stochastic noise. For sufficiently strong multiplicative noise, it is necessary to introduce additive noise, as well, to obtain stationary distributions at all. We note that in inferential statistics, gamma distributions facilitate Bayesian model learning from data, as a gamma distribution is a conjugate prior to many likelihood functions. It is therefore interesting to test whether models with stationary gamma distributions also have time-dependent gamma distributions.

2. Stochastic Logistic Model

We consider the logistic growth with a multiplicative noise given by the following Langevin equation:

$$\frac{dx}{dt} = (\gamma + \xi)x - \epsilon x^2, \tag{1}$$

where x is a random variable and ξ is a stochastic forcing, which for simplicity, can be taken as a short-correlated random forcing as follows:

$$\langle \xi(t)\xi(t') \rangle = 2D\delta(t - t'). \tag{2}$$

In Equation (2), the angular brackets represent the average over ξ, $\langle \xi \rangle = 0$, and D is the strength of the forcing. γ is the control parameter in the positive feedback, representing the growth rate of x, while ϵ represents the efficiency in self-regulation by a negative feedback. $\gamma x - \epsilon x^2$ can be considered as the gradient of the potential V as $\gamma x - \epsilon x^2 = -\frac{\partial V}{\partial x}$, where $V = -\frac{\gamma}{2}x^2 + \frac{\epsilon}{3}x^3$. Thus, V has its minimum value at $x = \frac{\gamma}{\epsilon}$. When $\xi = 0$, $x = \frac{\gamma}{\epsilon}$ (the carrying capacity) is a stable equilibrium point since $\partial_{xx}V\big|_{x=\gamma/\epsilon} = \gamma > 0$; $x = 0$ is an unstable equilibrium point since $\partial_{xx}V\big|_{x=0} = -\gamma < 0$.

The multiplicative noise in Equation (1) shows that the linear growth rate contains the stochastic part ξ. This model is entirely phenomenological, and x can be interpreted as the size of a critical physical phenomenon (vortex, tornado, etc.), stock market, number of biological cells, viruses and proteins. It is reminiscent of Delbrück's autocatalytic processes [25], but is different from these in many ways, the most important being the lack of discreteness and the possibility of reaching a steady-state due to the finite capacity of logistic growth. We will show in the following that in spite of these differences, our model is capable of producing large fluctuations.

By using the Stratonovich calculus [2,3,38], we can obtain the following Fokker–Planck equation for the PDF $p(x,t)$ (see Appendix A for details):

$$\frac{\partial}{\partial t}p(x,t) = -\frac{\partial}{\partial x}\Big[(\gamma x - \epsilon x^2)p(x,t)\Big] + D\frac{\partial}{\partial x}\left[x\frac{\partial}{\partial x}\big[x\,p(x,t)\big]\right] \tag{3}$$

corresponding to the Langevin Equation (1). By setting $\partial_t p = 0$, we can analytically solve for the stationary PDFs as:

$$p(x) = \frac{b^a}{\Gamma(a)}x^{a-1}e^{-bx}, \tag{4}$$

which is the well-known gamma distribution. The two parameters a and b are given by $a = \gamma/D$ and $b = \epsilon/D$. The mean value and variance of the gamma distribution are found to be:

$$\langle x \rangle = \frac{a}{b} = \frac{\gamma}{\epsilon}, \qquad \text{Var}(x) = \sigma^2 = \langle(x - \langle x \rangle)^2\rangle = \frac{a}{b^2} = \frac{\gamma D}{\epsilon^2}, \tag{5}$$

where $\sigma = \sqrt{\text{Var}(x)}$ is the standard deviation. We recognise $\langle x \rangle$ as the carrying capacity for a deterministic system with $\xi = 0$. It is useful to note that $\langle x \rangle$ is given by the linear growth rate scaled by ϵ, while $\text{Var}(x)$ is given by the product of the linear growth rate and the diffusion coefficient, each scaled by ϵ. That is, the effect of stochasticity should be measured relative to the linear growth rate.

Therefore, the case of small fluctuations is modelled by values of D small compared with γ and ϵ. In such a limit, a and b are large, making $\sqrt{\text{Var}(x)} \ll \langle x \rangle$ in Equation (5). That is, the width of the PDF is much smaller than its mean value. In this limit, Equation (4) reduces to a Gaussian distribution. To show this, we express Equation (4) in the following form:

$$p \equiv \frac{b^a}{\Gamma(a)}e^{-f(x)}, \tag{6}$$

where $f(x) = bx - (a - 1) \ln x$. For large b, we expand $f(x)$ around the stationary point $x = x_p$ where $\partial_x f(x) = 0 = b - (a - 1)/x$ up to the second order in $x - x_p$ to find:

$$x_p = \frac{a - 1}{b} \sim \frac{a}{b}, \qquad f(x = x_p) \sim a \left(1 - \ln \frac{a}{b}\right), \tag{7}$$

$$f(x) \sim f(x_p) + \frac{1}{2}(x - x_p)^2 \partial_{xx} f(x)\Big|_{x=x_p} = a \left(1 - \ln \frac{a}{b}\right) + \frac{b^2}{2a}\left(x - \frac{a}{b}\right)^2. \tag{8}$$

Here, $a \gg 1$ was used. Using Equation (8) in Equation (6) then gives us:

$$p \propto \exp\left[-\frac{b^2}{2a}\left(x - \frac{a}{b}\right)^2\right] \propto \exp\left[-\beta(x - \langle x \rangle)^2\right], \tag{9}$$

which is a Gaussian PDF with mean value $\langle x \rangle$. Here, $\beta = 1/\mathrm{Var}(x)$ is the inverse temperature, and the variance $\mathrm{Var}(x)$ is given by Equation (5). Therefore, for a sufficiently small D, the gamma distribution is approximated as a Gaussian PDF, which is consistent with the central limit theorem as small D corresponds to small fluctuations and large system size. See also [39] for a different derivation.

As D increases, the Gaussian approximation becomes increasingly less valid. Indeed, even the gamma distribution becomes invalid asymptotically, when $t \to \infty$, if $D > \gamma$; according to Equation (4), having $a < 1$ yields $\lim_{x \to 0} p = \lim_{x \to 0} \frac{\partial p}{\partial x} = \infty$. However, from the full time-dependent Fokker–Planck Equation (3), one finds that if the initial condition satisfies $p = 0$ at $x = 0$, then $p(x = 0)$ will remain zero for all later times. As we will see, the resolution to this seeming paradox is that no stationary distribution is ever reached for $D > \gamma$, but instead, the peak approaches ever closer to $x = 0$, without ever reaching it.

If we are interested in obtaining stationary solutions even when $D > \gamma$, one way to achieve that is to return to the original Langevin Equation (1), but now include a further additive stochastic noise η:

$$\frac{dx}{dt} = (\gamma + \xi)x - \epsilon x^2 + \eta, \tag{10}$$

where ξ and η are uncorrelated and η satisfies $\langle \eta(t)\eta(t') \rangle = 2Q\delta(t - t')$. The new version of the Fokker–Planck Equation (3) then becomes:

$$\frac{\partial}{\partial t} p = -\frac{\partial}{\partial x}\left[(\gamma x - \epsilon x^2)p\right] + D \frac{\partial}{\partial x}\left[x \frac{\partial}{\partial x}[x p]\right] + Q \frac{\partial^2}{\partial x^2} p, \tag{11}$$

which has stationary solutions given by:

$$\ln p(x) = \int \frac{(\gamma - D)x - \epsilon x^2}{Dx^2 + Q} dx. \tag{12}$$

This integral can be evaluated analytically, but the final form is not particularly illuminating. The only point to note is that for non-zero Q, the denominator is never zero, even for $x \to 0$, which avoids any possible singularities at the origin. For $\gamma > D$ and $Q \ll D$, the solutions are also essentially indistinguishable from the previous gamma distribution (4). The only significant effect of including η therefore is to avoid the previous difficulties at the origin when $D > \gamma$.

As we have seen, both Fokker–Planck Equations (3) and (11) can be solved exactly for their stationary solutions. This is unfortunately not the case regarding time-dependent solutions, where no closed-form analytic solutions exist (see Appendix B for the extent to which analytic progress can be made). We therefore developed finite-difference codes, second-order accurate in both space and time. Most aspects of the numerics are standard and similar to previous work [40–42]. The only point that requires discussion are the boundary conditions. As noted above, for (3), the equation itself states that $p = 0$ at $x = 0$ is the appropriate boundary condition, provided only that the initial condition also satisfies this. In contrast, for (11), the appropriate boundary condition is $\frac{\partial}{\partial x} p = 0$ at $x = 0$. To derive this boundary condition for (11), we simply integrate (11) over the range $x = [0, \infty]$ and require that the total probability should always remain one, so that $\frac{d}{dt} \int p \, dx = 0$. Regarding the outer boundary, choosing some moderately large outer value for x and then imposing $p = 0$ there was sufficient. Resolutions up to 10^6 grid points were used, and results were carefully checked to ensure they were independent of the grid size, time step and precise choice of outer boundary.

3. Diagnostics

Once the time-dependent solutions are computed, we can analyse them using a number of diagnostics. First, we can evaluate the mean value $\langle x \rangle$ and standard deviation σ from (5). Next, to explore the extent to which the time-dependent PDFs differ from gamma distributions, we can simply compare them with 'equivalent' gamma distributions and compute the difference. That is, given $\langle x \rangle$ and σ, the gamma distribution p_{equiv} having the same mean and variance would have as its two parameters $a = \langle x \rangle^2 / \sigma^2$ and $b = \langle x \rangle / \sigma^2$. With these values, we define:

$$\text{Difference} = \int |p - p_{\text{equiv}}| \, dx \tag{13}$$

to measure how different the actual time-dependent PDF is from its equivalent gamma distribution.

Two other familiar quantities often useful in analysing PDFs are the skewness and kurtosis, defined by:

$$\text{Skewness} = \frac{\langle (x - \langle x \rangle)^3 \rangle}{\sigma^3}, \qquad \text{Kurtosis} = \frac{\langle (x - \langle x \rangle)^4 \rangle}{\sigma^4} - 3. \tag{14}$$

Skewness measures the extent to which a PDF is asymmetric about its peak, whereas kurtosis measures how concentrated a PDF is in the peak versus the tails, relative to a Gaussian having the same variance (the -3 is included in the definition of the kurtosis to ensure that a Gaussian would yield zero). For gamma distributions, one finds analytically that the skewness is $2\sqrt{D/\gamma}$, and the kurtosis is $6D/\gamma$. Comparing the skewness and kurtosis of the time-dependent PDFs with these formulas is therefore another useful way of quantifying how similar or different they are from gamma distributions.

Another quantity that can be useful is the so-called differential entropy as a measure of order versus disorder (as entropy always is):

$$S = - \int p \ln p \, dx. \tag{15}$$

In particular, we expect S to be small for localised PDFs and large for spread out ones (e.g., [40–43]). For unimodal PDFs as the ones studied here, entropy and standard deviation are typically comparably good measures of localization, but for bimodal peaks, entropy can be significantly better [42]. For the gamma distribution in Equation (4), the differential entropy can be shown to be given by:

$$S = a - \ln b + \ln (\Gamma(a)) + (1 - a)\psi(a), \tag{16}$$

where $\psi(a) = \frac{d \ln (\Gamma(x))}{dx} \Big|_{x=a}$ is the double gamma function.

Our final diagnostic quantity is what is known as information length. Unlike all the previous diagnostics, which are simply evaluated at any instant in time, but otherwise do not involve t, information length is a Lagrangian quantity, explicitly concerned with the full time-history of the evolution of a given PDF. It is thus ideally suited to understanding time-dependent PDFs. Very briefly, we begin by defining:

$$\mathcal{E} \equiv \frac{1}{[\tau(t)]^2} = \int \frac{1}{p(x,t)} \left[\frac{\partial p(x,t)}{\partial t} \right]^2 dx. \tag{17}$$

Note how τ has units of time and quantifies the correlation time over which the PDF changes, thereby serving as a time unit in statistical space. Alternatively, $1/\tau$ quantifies the (average) rate of change of information in time. $\mathcal{E}$ is due to the change in either width (variance) of the PDF or the mean value, which are determined by γ, D and ϵ for the gamma distribution (e.g., see Equation (4)). In the standard Brownian motion, the mean value is zero so that $\mathcal{E}$ is due to the change in the variance of a PDF.

The total change in information between initial and final times, zero and t respectively, is then defined by measuring the total elapsed time in units of τ as:

$$\mathcal{L}(t) = \int_0^t \frac{dt_1}{\tau(t_1)} = \int_0^t \sqrt{\int dx \frac{1}{p(x,t_1)} \left[\frac{\partial p(x,t_1)}{\partial t_1}\right]^2} \, dt_1. \tag{18}$$

This information length $\mathcal{L}$ measures the total number of statistically distinguishable states that a system evolves through, thereby establishing a distance between the initial and final PDFs in the statistical space. Note that $\mathcal{L}$ by construction is a continuous variable and thus measures the total "number" of statistically-different states as a continuous number. See also [40–48] for further applications and theoretical background of $\mathcal{E}$ and $\mathcal{L}$.

4. Results

4.1. $\gamma > D$

We start with the case $\gamma > D$, where Equation (3) has stationary solutions, given by (4). Keeping ϵ and D fixed, we then switch γ back and forth between two values, in the following sense: Take the gamma distribution (4) corresponding to one value; call it γ_1; and use that as the initial condition to solve (3) with the other value; call it γ_2. We then interchange γ_1 and γ_2 to complete the pair of "inward" and "outward" processes. Such a pair can be thought of as an order/disorder phase transition [40,41], caused for example by cyclically adjusting temperature in an experiment. This protocol is also inspired from adaptation of a biological system. During adaptation, a model parameter can be abruptly changed in response to the change of environmental conditions, for instance a particle replication parameter γ, but the resulting changes can be extremely heterogeneous in the population.

Figure 1 shows the result of switching γ between $\gamma_1 = 0.5$ and $\gamma_2 = 0.05$, for fixed $\epsilon = 1$ and $D = 0.02$ (one of the three parameters ϵ, γ and D can of course always be kept fixed by rescaling the entire equation, so throughout this entire section, we keep $\epsilon = 1$ fixed and focus on how the various quantities depend on γ and D). We immediately see that the inward and outward processes behave differently. When γ is decreased and the peak therefore moves inward, the PDF is relatively narrow, and the peak amplitude is monotonically increasing. When γ is switched from 0.05 back to 0.5, the PDF is much broader, and the peak amplitude in the intermediate stages is less than either the initial or final gamma distributions.

Figure 2 shows how $\langle x \rangle$, $\mathcal{E}$ and $\mathcal{L}$ vary as functions of time, for the three values $D = 0.01$, 0.02, 0.04. For $\langle x \rangle$, the movement from 0.5 to 0.05 is somewhat slower than the reverse process; but, both processes occur on a similar time-scale, and both are essentially independent of D. This is in contrast with other Fokker–Planck systems where the magnitude of the diffusion coefficient can have a very strong influence on the equilibration time-scales [40,41].

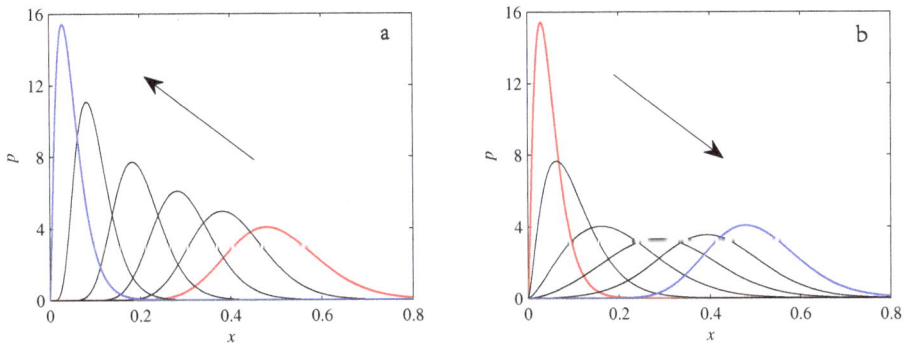

Figure 1. (**a**) shows the result of switching $\gamma = 0.5 \to 0.05$; (**b**) $\gamma = 0.05 \to 0.5$, both at fixed $\epsilon = 1$ and $D = 0.02$. The initial (red) and final (blue) gamma distributions are shown as heavy lines. The four intermediate lines are when the time-dependent solutions have $\langle x \rangle = 0.1, 0.2, 0.3, 0.4$. The arrows are a reminder of the direction of motion, inward on the left and outward on the right.

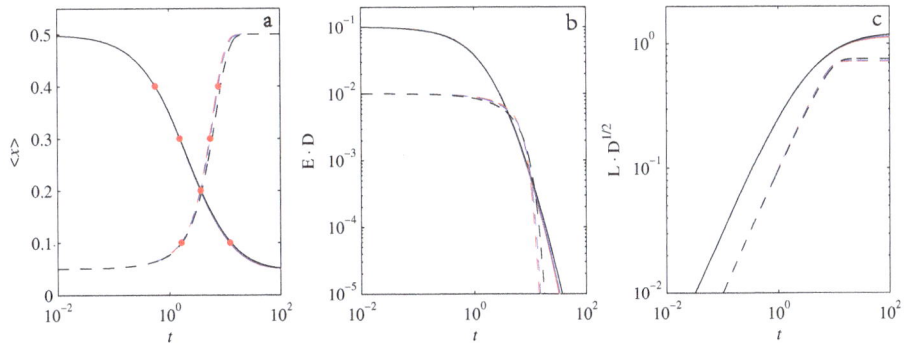

Figure 2. (**a**) shows $\langle x \rangle$ as a function of time; (**b**) shows $\mathcal{E} \cdot D$ (to indicate the $\mathcal{E} \sim D^{-1}$ scaling); (**c**) shows $\mathcal{L} \cdot D^{1/2}$ (to indicate the $\mathcal{L} \sim D^{-1/2}$ scaling). Solid lines denote $\gamma = 0.5 \to 0.05$, dashed lines the reverse. Each solid or dashed "line" is in fact three, occasionally barely distinguishable, lines with $D = 0.01, 0.02, 0.04$. The dots on the $\langle x \rangle$ curves correspond to the PDFs shown in Figure 1.

For $\mathcal{E}$ and $\mathcal{L}$, the equilibration is again somewhat slower for $\gamma = 0.5 \to 0.05$ than the reverse. We can further identify clear scalings $\mathcal{E} \sim D^{-1}$ and $\mathcal{L} \sim D^{-1/2}$. Finally, $\mathcal{L}$ is greater for $\gamma = 0.5 \to 0.05$ than the reverse. These results are all understandable in terms of the interpretation of $\mathcal{L}$ as the number of statistically-distinguishable states that the PDF evolves through: First, we recall from Figure 1 that $\gamma = 0.5 \to 0.05$ had consistently narrower PDFs than the reverse. Narrower PDFs means more distinguishable states, hence larger $\mathcal{L}$ for $\gamma = 0.5 \to 0.05$ than the reverse. The $\mathcal{L} \sim D^{-1/2}$ scaling has the same explanation; smaller D yields narrower PDFs, hence larger $\mathcal{L}$.

The first panel in Figure 3 shows the previous quantities $\langle x \rangle$ and $\mathcal{L} \cdot D^{1/2}$, but now plotted against each other rather than separately against time. The behaviour is exactly as one might expect, with $\mathcal{L}$ growing more or less linearly with distance from the initial position. The right panel in Figure 3 shows the entropy (15), again as a function of $\langle x \rangle$ rather than time, to emphasize the cyclic nature of the two processes. The significance is indeed as claimed above, with more localized PDFs having smaller entropy values. Note how $\gamma = 0.5 \to 0.05$, which had the narrower PDFs, has lower entropy values than the reverse process. Note also how reducing D by a factor of two, thereby making the PDFs narrower, causes the entire cyclic pattern to shift downward by an essentially constant amount.

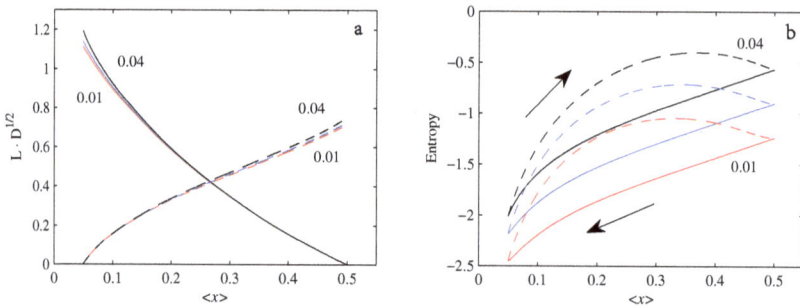

Figure 3. (**a**) shows $\mathcal{L} \cdot D^{1/2}$ and (**b**) entropy, both as functions of $\langle x \rangle$. Solid lines denote $\gamma = 0.5 \to 0.05$, dashed lines the reverse. Numbers besides curves indicate $D = 0.01, 0.02, 0.04$. The arrows on the entropy plot are a reminder of the direction of inward/outward motion.

Figure 4 shows how the standard deviation, skewness and kurtosis behave, again, as functions of $\langle x \rangle$ throughout the two processes. The heavy green lines also show the behaviour that would be expected if the time-dependent PDFs were always gamma distributions throughout their evolution. That is, if gamma distributions have $\langle x \rangle = \gamma$, $\sigma = \sqrt{\gamma D}$, skewness $= 2\sqrt{D/\gamma}$ and kurtosis $= 6D/\gamma$ (setting $\epsilon = 1$), then expressed as functions of $\langle x \rangle$, we would have $(\sigma / D^{1/2}) = \sqrt{\langle x \rangle}$, (skewness $/ \sqrt{D}) = 2/\sqrt{\langle x \rangle}$ and (kurtosis $/D) = 6/\langle x \rangle$. As we can see, the $\gamma = 0.5 \to 0.05$ process follows these functional relationships reasonably well (especially for skewness and kurtosis), but for $\gamma = 0.05 \to 0.5$, all three quantities deviate substantially.

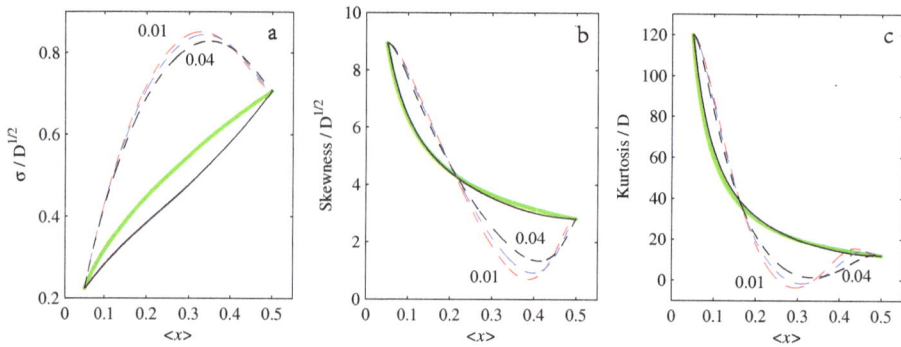

Figure 4. (**a**) $\sigma / D^{1/2}$, (**b**) (skewness $/D^{1/2}$) and (**c**) (kurtosis $/D$), as functions of $\langle x \rangle$. Solid lines denote $\gamma = 0.5 \to 0.05$, dashed lines the reverse. Numbers besides curves indicate $D = 0.01, 0.02, 0.04$. The heavy green curves are $\sqrt{\langle x \rangle}$, $2/\sqrt{\langle x \rangle}$ and $6/\langle x \rangle$, respectively, and indicate the behaviour expected for exact gamma distributions.

Further evidence of significant deviations from gamma distribution behaviour is seen in Figure 5, showing the difference (13) directly. As expected from Figure 4, $\gamma = 0.05 \to 0.5$ has a much greater difference than $\gamma = 0.5 \to 0.05$. The second and third panels show how the PDFs compare with the equivalent gamma distributions having the same $\langle x \rangle$ and σ values as the actual PDFs at that instant. The differences are clearly visible, especially for $\gamma = 0.05 \to 0.5$, but also for $\gamma = 0.5 \to 0.05$.

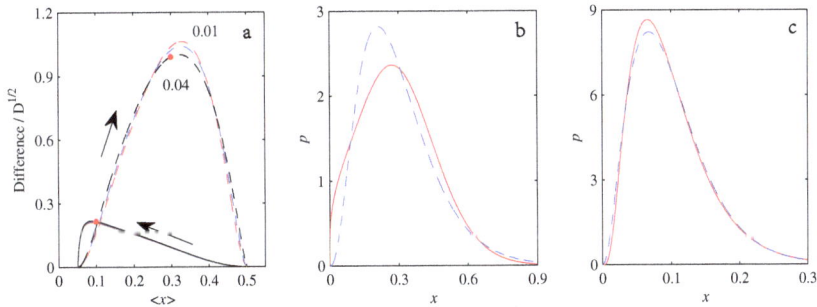

Figure 5. (**a**) shows the difference (13) between the actual PDF and the equivalent gamma distribution, as functions of $\langle x \rangle$. Solid lines denote $\gamma = 0.5 \rightarrow 0.05$, dashed lines the reverse, with arrows also indicating the direction of motion. The dots at $\langle x \rangle = 0.3$ for $\gamma = 0.05 \rightarrow 0.5$ and $\langle x \rangle = 0.1$ for $\gamma = 0.5 \rightarrow 0.05$, correspond to the other two panels: (**b**) compares the $\gamma = 0.05 \rightarrow 0.5$ PDF with its equivalent gamma distribution; (**c**) compares the $\gamma = 0.5 \rightarrow 0.05$ PDF with its equivalent gamma distribution. The actual PDFs in each case are solid (red), and the equivalent gamma distributions are dashed (blue). $D = 0.04$ for both sets.

4.2. $D > \gamma$

We next consider the case $D > \gamma$, where we demonstrated above that stationary solutions cannot exist at all, because the time-dependent PDF can only ever get closer and closer to the gamma distribution singularity at the origin, but can never actually achieve it. To explore what does happen in this case then, we simply repeat the above procedure, except that there is now only an 'inward' process, and no reverse. That is, instead of $\gamma = 0.5 \rightarrow 0.05$, let us consider $\gamma = 0.5 \rightarrow 0$ (Throughout this section, we will also take $D = 10^{-3}$, to facilitate comparison with results in the next section. For $\gamma = 0$, of course, any D is greater than γ.).

Figure 6 shows the resulting PDFs and how they approach ever closer to the origin, but never actually achieve the x^{-1} blow-up that would be implied by Equation (4) for $a = \gamma/D = 0$. The peak amplitude simply increases indefinitely, as $t^{1/2}$. The widths correspondingly also decrease; the apparent increase is an illusion caused by the logarithmic scale for x. The dashed lines also show the equivalent gamma distributions, as before. Note how the difference becomes increasingly noticeable; in line with the fact that the equivalent gamma distribution is tending toward its singular behaviour as $\langle x \rangle$ decreases, but the actual PDFs must always have $p(0) = 0$.

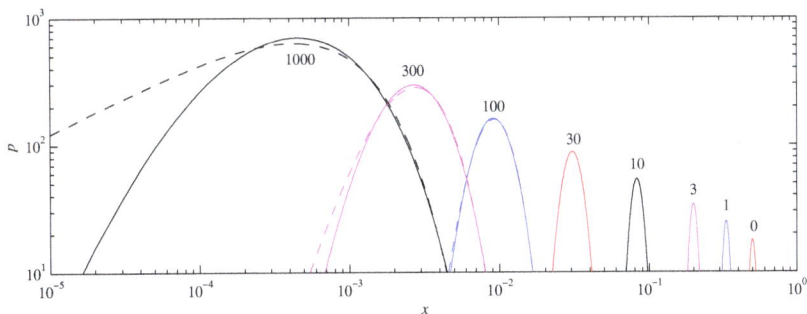

Figure 6. The initial condition is a gamma distribution with $\gamma = 0.5$, $\epsilon = 1$ and $D = 10^{-3}$; γ is then switched to zero, and the solution is evolved according to Equation (3). Numbers besides curves indicate time, from the initial condition at $t = 0$ to the final time 1000. The dashed curves indicate the equivalent gamma distributions having the same $\langle x \rangle$ and σ.

Figure 7 is the equivalent of Figure 2 and directly compares $\gamma = 0.5 \to 0$ here with the previous $\gamma = 0.5 \to 0.05$. We see that $\langle x \rangle$ starts out very similarly, but instead of equilibrating to 0.05, it now tends to zero as t^{-1}. $\mathcal{E}$ again starts out similarly, but ultimately tends to zero much more slowly, as t^{-3} instead of exponentially. This t^{-3} scaling for $\mathcal{E}$ has an interesting consequence for $\mathcal{L}$, namely that $\mathcal{L}$ does saturate to a finite value $\mathcal{L}_\infty$ (since $\int t^{-3/2} \, dt$ remains bounded for $t \to \infty$) even though the PDF itself never settles to a stationary state.

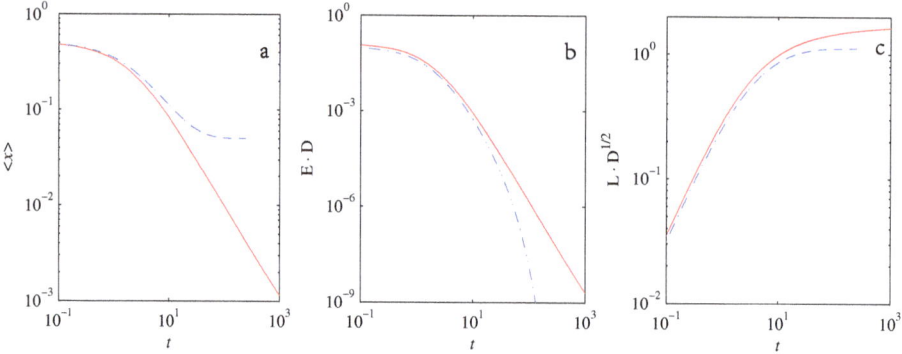

Figure 7. As in Figure 2, (**a**) shows $\langle x \rangle$; (**b**) shows $\mathcal{E} \cdot D$; and (**c**) $\mathcal{L} \cdot D^{1/2}$. Solid lines denote $\gamma = 0.5 \to 0$ for $D = 10^{-3}$, dashed lines the previous $\gamma = 0.5 \to 0.05$ for $D = 0.01$. Note how the scalings of $\mathcal{E}$ and $\mathcal{L}$ with D are still preserved even when D is changed by a factor of 10.

Figure 8 shows entropy, σ, skewness and kurtosis, so some of the results as in Figures 3 and 4. Entropy and σ are again both good measures of how narrow the PDF is, becoming ever smaller as the peak moves toward the origin. Skewness and kurtosis seem to follow the expected gamma distribution relationship extremely well, even though we saw before in Figure 6 that the PDFs are actually different from gamma distributions. As $\langle x \rangle \to 0$, both skewness and kurtosis thus become indefinitely large.

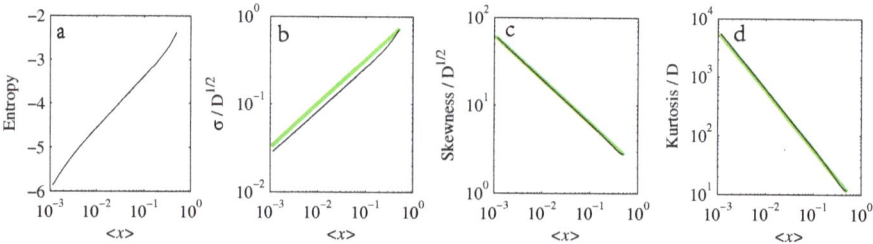

Figure 8. (**a**) Entropy, (**b**) $\sigma / D^{1/2}$, (**c**) (skewness $/ D^{1/2}$) and (**d**) (kurtosis $/D$), as functions of $\langle x \rangle$, for the $\gamma = 0.5 \to 0$ calculation from Figure 6. The heavy green curves in the last three panels are $\sqrt{\langle x \rangle}$, $2/\sqrt{\langle x \rangle}$ and $6/\langle x \rangle$, respectively, and indicate the behaviour expected for exact gamma distributions.

4.3. $Q \neq 0$

Finally, we turn to the Fokker–Planck Equation (11) with additive noise included and use it to explore the two questions that could not be addressed otherwise. First, how does a process like $\gamma = 0.5 \to 0$ then equilibrate to a stationary solution? Second, what does the reverse process $\gamma = 0 \to 0.5$ look like?

We will keep $D = 10^{-3}$ and $Q = 10^{-5}$ fixed throughout this section. Since the effective diffusion coefficients in (11) are Dx^2 and Q (recall also the denominator of Equation (12)), this means that Q is dominant only within $x \leq 0.1$; any stationary solutions with peaks much beyond that are effectively pure gamma distributions.

Figure 9 shows the same type of inward/outward process as before in Figure 1, only now switching γ between 0.5 and 0.1. Comparing with Figure 1, we see that the dynamics are very similar, just with all the peaks considerably narrower, which is to be expected if $D = 10^{-3}$ rather than 0.02. The only other point to note is how the final peak in the left panel is lower than the previous peak at $\langle x \rangle = 0.2$, which is different from Figure 1, where $\gamma = 0.5 \to 0.05$ had peaks monotonically increasing throughout the entire evolution. The reason for the final peak here decreasing slightly is precisely the influence of Q in this region; if this peak is now seeing just as much diffusion from Q as from D, it is not surprising that it spreads out somewhat more and is correspondingly somewhat lower than a pure gamma distribution would be.

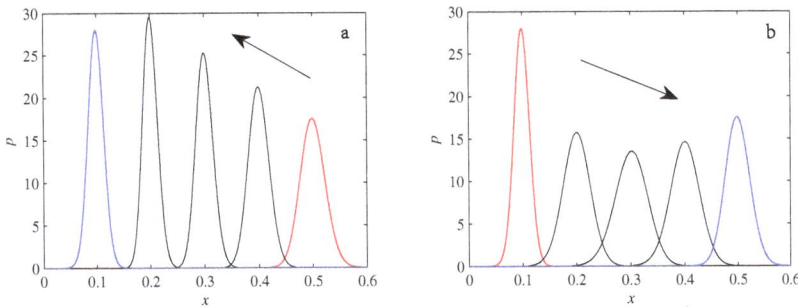

Figure 9. (**a**) shows the result of switching $\gamma = 0.5 \to 0.1$, (**b**) $\gamma = 0.1 \to 0.5$, both at fixed $\epsilon = 1$, $D = 10^{-3}$ and $Q = 10^{-5}$. The initial (red) and final (blue) gamma distributions are shown as heavy lines. The three intermediate lines are when the time-dependent solutions have $\langle x \rangle = 0.2$, 0.3, 0.4. $\mathcal{L}_\infty = 25$ on the left and 16 on the right.

Figure 10 shows the fundamentally new case, namely switching γ between 0.5 and zero. The inward process $\gamma = 0.5 \to 0$ is again very similar to either Figure 1 or 9. The only difference from Figure 6 is that the process does actually equilibrate to a stationary solution now, as given by Equation (12). The reverse process $\gamma = 0 \to 0.5$ is rather different though. The initial central peak now broadens far more than previously seen in Figures 1 and 9.

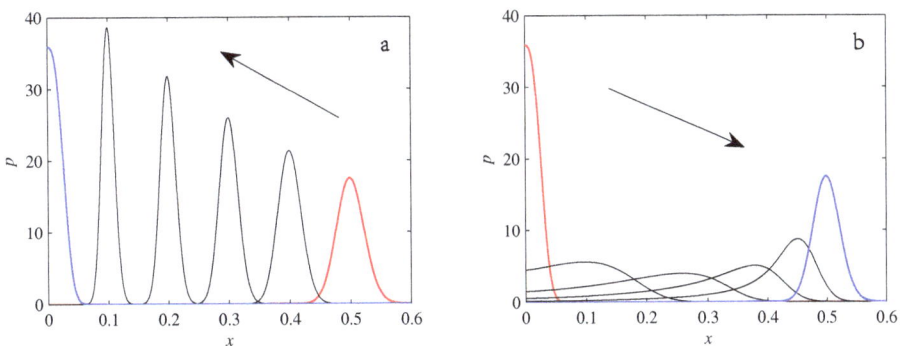

Figure 10. (**a**) the result of switching $\gamma = 0.5 \to 0$, (**b**) $\gamma = 0 \to 0.5$, both at fixed $\epsilon = 1$, $D = 10^{-3}$ and $Q = 10^{-5}$. The initial (red) and final (blue) gamma distributions are shown as heavy lines. The four intermediate lines are when the time-dependent solutions have $\langle x \rangle = 0.1$, 0.2, 0.3, 0.4. $\mathcal{L}_\infty = 35$ on the left and 9.5 on the right.

One interesting consequence of this extreme broadening for $\gamma = 0 \rightarrow 0.5$ is on the total information length $\mathcal{L}_\infty$. In Figure 9, these values are 25 and 16, respectively, whereas in Figure 10, they are 35 and 9.5. That is, in both cases, decreasing γ yields larger $\mathcal{L}_\infty$ values than increasing γ does, consistent with the peaks being narrower, hence passing through more statistically-distinguishable states. Next, comparing $\mathcal{L}_\infty = 25$ for $\gamma = 0.5 \rightarrow 0.1$ versus 35 for $\gamma = 0.5 \rightarrow 0$, this is exactly as one might expect: having the peak travel somewhat further yields extra information length. However, comparing $\mathcal{L}_\infty = 16$ for $\gamma = 0.1 \rightarrow 0.5$ versus 9.5 for $\gamma = 0 \rightarrow 0.5$ is puzzling then. The peak has further to travel, but accomplishes it with less information length. The reason is precisely this extreme broadening, which substantially reduces the number of distinguishable states along the way. See also [40,41], where the same effect was studied for Gaussian PDFs and values of D as small as 10^{-7}, leading to fundamentally different scalings of $\mathcal{L}_\infty$ with D for inward and outward processes.

Returning to the central question of this paper, namely how close the time-dependent PDFs are to gamma distributions, the results for Figure 9 are similar to the previous ones. In particular, we recall that before in Figure 5, we had the difference scaling as $D^{1/2}$, so a smaller D here means a smaller difference. These results are approaching the small D regime where gamma distributions become very close to Gaussians anyway, which generally remain close to Gaussian as they move.

However, for the $\gamma = 0 \rightarrow 0.5$ process in Figure 10, the intermediate stages do not look much like gamma distributions (the final equilibrium is indistinguishable from a gamma distribution though, consistent with Q being completely negligible for these values of x). For the intermediate stages, these were found to be so different from gamma distributions that attempting to fit a gamma distribution having the same $\langle x \rangle$ and σ made little sense; this extreme broadening and long tail trailing behind the peak meant that both $\langle x \rangle$ and σ were too different from the normal expectation that they should be measures of 'peak' and 'width'.

Instead, we simply asked the question, which values of a and b would minimize the quantity $\int |p - p_{\text{bf}}| \, dx$, where p is the time-dependent PDF to be fitted and p_{bf} is the best-fit gamma distribution. Unlike our previous difference formula, this does not yield simple analytic formulas for the a and b to choose, but is numerically still straightforward to implement. Figure 11 shows the results, for two of the intermediate stages in the $\gamma = 0 \rightarrow 0.5$ process. We can see that the fit is rather poor, indicating that these PDFs are significantly different from gamma distributions.

This misfit is also not caused by the inclusion of Q; if this or any similar central peak is evolved for either small or zero Q in the Fokker–Planck equation, the result is always similar to here. As explained also in [40,41], the dynamics of how central peaks move away from the origin is simply different from how peaks already away from the origin move, regardless of whether the final states are Gaussians as in [40,41] or gamma distributions as here.

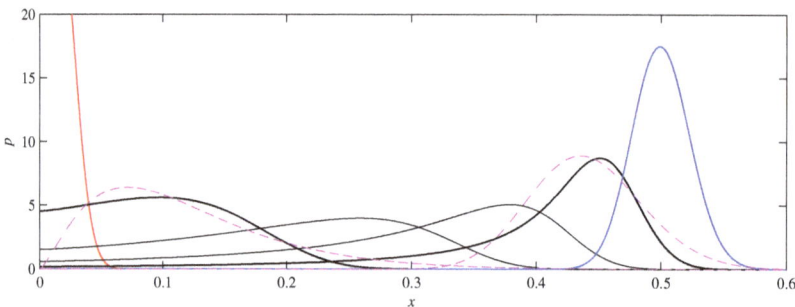

Figure 11. The $\gamma = 0 \rightarrow 0.5$ process as in Figure 10, but now shown in more detail. The dashed (magenta) curves are the gamma distributions that best fit the two thicker curves at intermediate times. Note how even a "best-fit" is a rather poor approximation to the actual PDFs.

5. Conclusions

Gamma distributions are among the most popular choices for modelling a broad range of experimentally-determined PDFs. It is often assumed that time-dependent PDFs can then simply be modelled as gamma distributions with time-varying parameters a and b. In this work, we have demonstrated that one should be cautious with such an approach. By numerically solving the full time-dependent Fokker–Planck equation, we found that there are three sets of circumstances where the PDFs can differ significantly from gamma distributions:

- If $D < \gamma$, so that stationary solutions exist, but D is also sufficiently close to γ that a gamma distribution differs significantly from a Gaussian, then the time-dependent PDFs will also differ significantly from gamma distributions.
- If $D > \gamma$, stationary gamma distributions do not exist at all. Instead, peaks move ever closer to the origin and in the process increasingly differ from gamma distributions.
- If the initial condition is a peak right on the origin—either as a result of adding additive noise to produce stationary solutions even for $D > \gamma$, or simply as an arbitrary initial condition—then any evolution away from the origin will differ significantly from gamma distributions. Unlike the previous two items, which become more pronounced for larger D, this effect is most clearly visible for smaller D, where the mismatch between the naturally narrower peaks and the extreme broadening seen in Figure 11 becomes increasingly significant.

In summary, our results show that a simple Langevin equation model mimics the strong fluctuations of far-from-equilibrium systems. This model has gamma distributions as steady-state solutions, but the time-dependent solutions can deviate considerably from this law. This makes tasks such as Bayesian and frequentist inference of the model from data more complicated. On the other hand, the model shows complex asymptotic dynamics with situations when a steady state is reached or not, different from the one of the deterministic logistic model that invariably evolves to the maximum capacity. The studied model is general enough and can be applied to many practical situations in biology, economics, finance and physics. Future work will apply some of these ideas to fitting actual data.

Author Contributions: Eun-jin Kim and Rainer Hollerbach conceived the basic mathematical ideas; Ovidiu Radulescu provided background in biology and other areas; Eun-jin Kim conducted the analytical derivations; Lucille-Marie Tenkès and Rainer Hollerbach conducted the numerical calculations; all authors were involved in writing the paper. All the authors have read and approved the final manuscript.

Conflicts of Interest: The authors declare no conflict of interest.

Appendix A. Derivation of the Fokker–Planck Equations

In order to derive the Fokker–Planck Equation (3) from the Langevin Equation (1), it is useful to introduce a generating function Z:

$$Z = e^{i\lambda x(t)}. \tag{A1}$$

Then, by definition of 'average', the average of Z is related to the PDF, $p(x, t)$, as:

$$\langle Z \rangle = \int dx\, Z\, p(x, t) = \int dx\, e^{i\lambda x(t)}\, p(x, t). \tag{A2}$$

Thus, we see that $\langle Z \rangle$ is the Fourier transform of $p(x, t)$. The inverse Fourier transform of $\langle Z \rangle$ then gives $p(x, t)$:

$$p(x, t) = \frac{1}{2\pi} \int d\lambda\, e^{-i\lambda x}\, \langle Z \rangle. \tag{A3}$$

We note that Equation (A3) can be written as:

$$p(x, t) = \left\langle \frac{1}{2\pi} \int d\lambda\, e^{i\lambda(x - x(t))} \right\rangle = \langle \delta(x - x(t)) \rangle, \tag{A4}$$

which is another form of $p(x, t)$. To obtain the equation for $p(x, t)$, we first derive the equation for $\langle x \rangle$ and then take the inverse Fourier transform as summarised in the following.

We differentiate Z with respect to time t and use Equation (1) to obtain:

$$\partial_t Z = i\lambda \partial_t x Z = i\lambda(\gamma x - \epsilon x^2 + \xi(t)x)Z = \lambda \left[\gamma \partial_\lambda + i\epsilon \partial_{\lambda\lambda} + \xi \partial_\lambda\right] Z, \tag{A5}$$

where $xZ = -i\partial_\lambda Z$ was used. The formal solution to Equation (A5) is:

$$Z(t) = \lambda \int dt_1 \left[\gamma \partial_\lambda + i\epsilon \partial_{\lambda\lambda} + \xi(t_1)\partial_\lambda\right] Z(t_1). \tag{A6}$$

The average of Equation (A5) gives:

$$\partial_t \langle Z \rangle = \lambda(\gamma \partial_\lambda + i\epsilon \partial_{\lambda\lambda})\langle Z \rangle + \lambda \langle \xi(t)\partial_\lambda Z(t) \rangle. \tag{A7}$$

To find $\langle \xi(t)\partial_\lambda Z(t) \rangle$, we use Equation (A6) iteratively as follows:

$$
\begin{aligned}
\langle \xi(t)\partial_\lambda Z \rangle &= \left\langle \xi(t)\partial_\lambda \left[\lambda \int dt_1 \left[\gamma \partial_\lambda + i\epsilon \partial_{\lambda\lambda} + \xi(t_1)\partial_\lambda\right] Z(t_1) \right] \right\rangle \\
&= \langle \xi(t) \rangle \partial_\lambda \left[\lambda \int dt_1 \left[\gamma \partial_\lambda + i\epsilon \partial_{\lambda\lambda}\right] \langle Z(t_1) \rangle \right] + \partial_\lambda \left[\lambda \int dt_1 \langle \xi(t)\xi(t_1) \rangle \partial_\lambda \langle Z(t_1) \rangle \right] \\
&= \partial_\lambda \left[\lambda \left[D\partial_\lambda \langle Z(t) \rangle \right] \right].
\end{aligned}
\tag{A8}
$$

Here, we used the independence of $\xi(t)$ and $Z(t_1)$ for $t_1 < t$, $\langle \xi(t)Z(t_1) \rangle = \langle \xi(t) \rangle \langle Z(t_1) \rangle = 0$, together with Equation (2), $\int_0^t dt_1 \, \delta(t - t_1) = 1/2$, and $\langle \xi \rangle = 0$. By substituting Equation (A8) into Equation (A7), we obtain:

$$\partial_t \langle Z \rangle = \lambda(\gamma \partial_\lambda + i\epsilon \partial_{\lambda\lambda})\langle Z \rangle + \lambda \partial_\lambda \left[\lambda \left[D\partial_\lambda \langle Z(t) \rangle \right] \right]. \tag{A9}$$

The inverse Fourier transform of Equation (A9) then gives us:

$$\frac{\partial}{\partial t} p(x, t) = -\frac{\partial}{\partial x}\left[(\gamma x - \epsilon x^2)p(x, t) \right] + D\frac{\partial}{\partial x}\left[x\frac{\partial}{\partial x}\left[x\, p(x, t) \right] \right] \tag{A10}$$

which is Equation (3). Specifically, the inverse Fourier transforms of the first and last terms in Equation (A9) are shown explicitly in the following:

$$\frac{1}{2\pi} \int d\lambda \, e^{-i\lambda x} \langle \partial_t Z \rangle = \partial_t \left[\frac{1}{2\pi} \int d\lambda \, e^{-i\lambda x} \langle Z \rangle \right] = \frac{\partial}{\partial t} p(x, t), \tag{A11}$$

$$\frac{D}{2\pi} \int d\lambda \, e^{-i\lambda x} \lambda \partial_\lambda \left[\lambda \partial_\lambda \langle Z \rangle \right] = D\frac{\partial}{\partial x}\left[x\frac{\partial}{\partial x}\left[x\, p(x, t) \right] \right], \tag{A12}$$

where integration by parts was used twice in obtaining Equation (A12). The additional $Q\partial_{xx}p$ term in the Fokker–Planck Equation (11) can be derived in the same way.

Appendix B. Time-Dependent Analytical Solutions of Equation (3)

We begin by making the change of variables $y = 1/x$ in Equation (1) to obtain:

$$\frac{dy}{dt} = -(\gamma + \xi)y + \epsilon. \tag{A13}$$

By using the Stratonovich calculus [2,3,38], the solution to Equation (A13) is found as:

$$y(t) = y_0 e^{-(\gamma t + B(t))} + \epsilon e^{-(\gamma t + B(t))} \int_0^t dt_1 e^{(\gamma t_1 + B(t_1))}, \tag{A14}$$

where $y_0 = y(t = 0)$ and $B(t) = \int_0^t dt_1 \, \xi(t_1)$ is the Brownian motion. Therefore,

$$x(t) = \frac{x_0 e^{\gamma t + B(t)}}{1 + \epsilon x_0 \int_0^t dt_1 e^{(\gamma t_1 + B(t_1))}}, \tag{A15}$$

where $x_0 = x(t = 0)$. In Equation (A15), $e^{B(t)}$ is the geometric Brownian motion while $e^{-\gamma t - B(t)}$ is the geometric Brownian motion with a drift (e.g., [2]). The time integral of the latter is used in understanding stochastic processes in financial mathematics and many other areas [49,50]. In particular, in the long time limit, its PDF can be shown to be a gamma distribution. However, this PDF of x is not particularly useful as it involves complicated summations and integrals that cannot be evaluated in closed form [49,50].

References

1. Risken, H. *The Fokker-Planck Equation: Methods of Solution and Applications*; Springer: Berlin, Germany, 1996.
2. Klebaner, F. *Introduction to Stochastic Calculus with Applications*; Imperial College Press: London, Britain, 2012.
3. Gardiner, C. *Stochastic Methods*, 4th ed.; Springer: Berlin, Germany, 2008.
4. Saw, E.-W.; Kuzzay, D.; Faranda, D.; Guittonneau, A.; Daviaud, F.; Wiertel-Gasquet, C.; Padilla, V.; Dubrulle, B. Experimental characterization of extreme events of inertial dissipation in a turbulent swirling flow. *Nat. Commun.* **2016**, *7*, 12466, doi:10.1038/ncomms12466.
5. Kim, E.; Diamond, P.H. On intermittency in drift wave turbulence: Structure of the probability distribution function. *Phys. Rev. Lett.* **2002**, *88*, 225002.
6. Kim, E.; Diamond, P.H. Zonal flows and transient dynamics of the L-H transition. *Phys. Rev. Lett.* **2003**, *90*, 185006.
7. Kim, E. Consistent theory of turbulent transport in two dimensional magnetohydrodynamics. *Phys. Rev. Lett.* **2006**, *96*, 084504.
8. Kim, E.; Anderson, J. Structure-based statistical theory of intermittency. *Phys. Plasmas* **2008**, *15*, 114506.
9. Newton, A.P.L.; Kim, E.; Liu, H.-L. On the self-organizing process of large scale shear flows. *Phys. Plasmas* **2013**, *20*, 092306.
10. Srinivasan, K.; Young, W.R. Zonostrophic instability. *J. Atmos. Sci.* **2012**, *69*, 1633–1656.
11. Sayanagi, K.M.; Showman, A.P.; Dowling, T.E. The emergence of multiple robust zonal jets from freely evolving, three-dimensional stratified geostrophic turbulence with applications to Jupiter. *J. Atmos. Sci.* **2008**, *65*, 3947–3962.
12. Tsuchiya, M.; Giuliani, A.; Hashimoto, M.; Erenpreisa J.; Yoshikawa, K. Emergent self-organized criticality in gene expression dynamics: Temporal development of global phase transition revealed in a cancer cell line. *PLoS ONE* **2015**, *10*, e0128565.
13. Tang, C.; Bak, P. Mean field theory of self-organized critical phenomena. *J. Stat. Phys.* **1988**, *51*, 797–802.
14. Jensen, H.J. *Self-Organized Criticality: Emergent Complex Behavior in Physical and Biological Systems*; Cambridge University Press: Cambridge, UK, 1998.
15. Pruessner, G. *Self-Organised Criticality*; Cambridge University Press: Cambridge, UK, 2012.
16. Longo, G.; Montévil, M. From physics to biology by extending criticality and symmetry breaking. *Prog. Biophys. Mol. Biol.* **2011**, *106*, 340–347.
17. Flynn, S.W.; Zhao, H.C.; Green, J.R. Measuring disorder in irreversible decay processes. *J. Chem. Phys.* **2014**, *141*, 104107.
18. Nichols, J.W.; Flynn, S.W.; Green, J.R. Order and disorder in irreversible decay processes. *J. Chem. Phys.* **2015**, *142*, 064113.
19. Ferguson, M.L.; Le Coq, D.; Jules, M.; Aymerich, S.; Radulescu, O.; Declerck, N.; Royer, C.A. Reconciling molecular regulatory mechanisms with noise patterns of bacterial metabolic promoters in induced and repressed states. *Proc. Natl. Acad. Sci. USA* **2012**, *109*, 155–160.
20. Shahrezaei, V.; Swain, P.S. Analytical distributions for stochastic gene expression. *Proc. Natl. Acad. Sci. USA* **2008**, *105*, 17256.
21. Thomas, R.; Torre, L.; Chang, X.; Mehrotra, S. Validation and characterization of DNA microarray gene expression data distribution and associated moments. *BMC Bioinform.* **2010**, *11*, 576.
22. Iyer-Biswas, S.; Hayot, F.; Jayaprakash, C. Stochasticity of gene products from transcriptional pulsing. *Phys. Rev. E* **2009**, *79*, 031911.

23. Elgart, V.; Jia, T.; Fenley, A.T.; Kulkarni, R. Connecting protein and mRNA burst distributions for stochastic models of gene expression. *Phys. Biol.* **2011**, *8*, 046001.

24. Delbrück, M. The burst size distribution in the growth of bacterial viruses (bacteriophages). *J. Bacteriol.* **1945**, *50*, 131.

25. Delbrück, M. Statistical fluctuations in autocatalytic reactions. *J. Chem. Phys.* **1940**, *8*, 120–124.

26. Kim, E.; Liu, H.; Anderson, J. Probability distribution function for self-organization of shear flows. *Phys. Plasmas* **2009**, *16*, 052304.

27. Kim, E.; Hollerbach, R. Time-dependent probability density function in cubic stochastic processes. *Phys. Rev. E* **2016**, *94*, 052118.

28. De Jong, I.G.; Haccou, P.; Kuipers, O.P. Bet hedging or not? A guide to proper classification of microbial survival strategies. *Bioessays* **2011**, *33*, 215–223.

29. Balaban, N.; Merrin, J.; Chait, R.; Kowalik, L.; Leibler, S. Bacterial persistence as a phenotypic switch. *Science* **2004**, *305*, 1622–1625.

30. Glansdorff, P.; Prigogine, I. Thermodynamic theory of structure, stability and fluctuations. *Am. J. Phys.* **1973**, *41*, 147–148.

31. Suzuki, M. Microscopic theory of formation of macroscopic order. *Phys. Lett. A* **1980**, *75*, 331–332.

32. Suzuki, M. The variational theory and rate equation method with applications to relaxation near the instability point. *Phys. A Stat. Mech. Appl.* **1981**, *105*, 631–641.

33. Langer, J.S.; Baron, M.; Miller, H.D. New computational method in the theory of spinodal decomposition. *Phys. Rev. A* **1975**, *11*, 1417.

34. Saito, Y. Relaxation in a bistable system. *J. Phys. Soc. Jpn.* **1976**, *61*, 388–393.

35. Hasegawa, H. Variational approach in studies with Fokker-Planck equations. *Prog. Theor. Phys.* **1977**, *58*, 128–146.

36. Dennis, B.; Costantino, R.F. Analysis of steady-state populations with the gamma abundance model: Application to Tribolium. *Ecology* **1988**, *69*, 1200–1213.

37. Liao, H.-Y.; Ai, B.-Q.; Hu, L. Effects of multiplicative colored noise on bacteria growth. *Braz. J. Phys.* **2007**, *37*, 1125–1128.

38. Wong, E.; Zakai, M. On the convergence of ordinary integrals to stochastic integrals. *Ann. Math. Stat.* **1960**, *36*, 1560–1564.

39. Bagui, S.C.; Mehra, K.L. Convergence of binomial, Poisson, negative-binomial, and gamma to normal distribution: Moment generating functions technique. *Am. J. Math. Stat.* **2016**, *6*, 115–121.

40. Kim, E.; Hollerbach, R. Geometric structure and information change in phase transitions. *Phys. Rev. E* **2017**, *95*, 062107.

41. Hollerbach, R.; Kim, E. Information geometry of non-equilibrium processes in a bistable system with a cubic damping. *Entropy* **2017**, *19*, 268.

42. Tenkès, L.-M.; Hollerbach, R.; Kim, E. Time-dependent probability density functions and information geometry in stochastic logistic and Gompertz models. *arXiv* **2017**, arXiv:1708.02789.

43. Frieden, B.R. *Physics from Fisher Information*; Cambridge University Press: Cambridge, Britain, 2000.

44. Wootters, W.K. Statistical distance and Hilbert space. *Phys. Rev. D* **1981**, *23*, 357.

45. Nicholson, S.B.; Kim, E. Investigation of the statistical distance to reach stationary distributions. *Phys. Lett. A* **2015**, *379*, 83–88.

46. Nicholson, S.B.; Kim, E. Structures in sound: Analysis of classical music using the information length. *Entropy* **2016**, *18*, 258, doi: 10.3390/e18070258.

47. Heseltine, J.; Kim, E. Novel mapping in a non-equilibrium stochastic process. *J. Phys. A* **2016**, *49*, 175002.

48. Kim, E.; Lee, U.; Heseltine, J.; Hollerbach, R. Geometric structure and geodesic motion in a solvable model of non-equilibrium stochastic process. *Phys. Rev. E* **2016**, *93*, 062127.

49. Bertoin, J.; Yor, M. Exponential functionals of Lévy processes. *Prob. Surv.* **2005**, *2*, 191–212.

50. Matsumoto, H.; Yor, M. Exponential functionals of Brownian motion, I: Probability laws at fixed time. *Prob. Surv.* **2005**, *2*, 312–347.

entropy

MDPI

Article

Parameterization of Coarse-Grained Molecular Interactions through Potential of Mean Force Calculations and Cluster Expansion Techniques

Anastasios Tsourtis [1,*], Vagelis Harmandaris [1,2,*] and Dimitrios Tsagkarogiannis [3,*]

[1] Department of Mathematics and Applied Mathematics, University of Crete, Heraklion 70013, Greece
[2] Institute of Applied and Computational Mathematics (IACM), Foundation for Research and Technology Hellas (FORTH), Heraklion 70013, Greece
[3] Department of Mathematics, University of Sussex, Brighton BN1 9QH, UK
* Correspondence: tsourtis@uoc.gr (A.T.); harman@uoc.gr (V.H.); D.Tsagkarogiannis@sussex.ac.uk (D.T.);
    Tel.: +30-2810-393768 (A.T.); +30-2810-393735 (V.H.); +44-1273-876824 (D.T.)

Received: 24 May 2017; Accepted: 24 July 2017; Published: 1 August 2017

Abstract: We present a systematic coarse-graining (CG) strategy for many particle molecular systems based on cluster expansion techniques. We construct a hierarchy of coarse-grained Hamiltonians with interaction potentials consisting of two, three and higher body interactions. In this way, the suggested model becomes computationally tractable, since no information from long n-body (bulk) simulations is required in order to develop it, while retaining the fluctuations at the coarse-grained level. The accuracy of the derived cluster expansion based on interatomic potentials is examined over a range of various temperatures and densities and compared to direct computation of the pair potential of mean force. The comparison of the coarse-grained simulations is done on the basis of the structural properties, against detailed all-atom data. On the other hand, by construction, the approximate coarse-grained models retain, in principle, the thermodynamic properties of the atomistic model without the need for any further parameter fitting. We give specific examples for methane and ethane molecules in which the coarse-grained variable is the centre of mass of the molecule. We investigate different temperature (T) and density (ρ) regimes, and we examine differences between the methane and ethane systems. Results show that the cluster expansion formalism can be used in order to provide accurate effective pair and three-body CG potentials at high T and low ρ regimes. In the liquid regime, the three-body effective CG potentials give a small improvement over the typical pair CG ones; however, in order to get significantly better results, one needs to consider even higher order terms.

Keywords: cluster expansions; PMF calculations; systematic coarse-graining; three-body effective potential

1. Introduction

The theoretical study of complex molecular systems is a very intense research area due to both basic scientific questions and technological applications [1]. A main challenge in this field is to provide a direct quantitative link between the chemical structure at the molecular level and measurable macroscopic quantities over a broad range of length and time scales. Such knowledge would be especially important for the tailored design of materials with the desired properties, over an enormous range of possible applications in nano-, bio-technology, food science, drug industry, cosmetics, etc.

A common characteristic of all complex fluids is that they exhibit multiple length and time scales. Therefore, simulation methods across scales are required in order to study such systems. On the all-atom-level description, classical atomistic models have successfully been used in order to quantitatively predict the properties of molecular systems over a considerable range of different thermodynamic conditions [1–4]. However, due to the broad spectrum of characteristic lengths and times involved in complex molecular systems, it is desirable to reduce the required computational cost by describing the system through a small number of degrees of freedom. Thus, coarse-grained (CG) models have been used in order to increase the length and time scales accessible by simulations [1,3,5–22].

From a mathematical point of view, coarse-graining is a sub-field of dimensionality reduction; there are several statistical methods for the reduction of the degrees of freedom under consideration in a deterministic or stochastic model, such as principal component analysis, polynomial chaos and diffusion maps [4,20]. Here, we focus our discussion on CG methods based on a combination of recent computational methods and old theoretical tools from statistical mechanics. Such CG models, which are developed by lumping groups of atoms into CG particles and deriving the effective CG interaction potentials directly from more detailed (microscopic) simulations, are capable of predicting quantitatively the properties of specific molecular systems (see, for example, [5–9,11–13,15,17–19,23–25] and the references therein).

The most important part in all systematic CG models, based on detailed atomistic data, is to develop rigorous all-atom to CG methodologies that allow, as accurate as possible, estimation of the CG effective interaction. With such approaches, the combination of atomistic and hierarchical CG models could allow the study of specific molecular systems without adjustable parameters and by that become truly predictive [11,14,15]. There exists a variety of methods that construct a reduced CG model that approximates the properties of molecular systems based on statistical mechanics. For example:

(a) In structural, or correlation-based, methods, the main goal is to find effective CG potentials that reproduce the pair radial distribution function $g(r)$, and the distribution functions of bonded degrees of freedom (e.g., bonds, angles, dihedrals) for CG systems with intramolecular interaction potential [6,7,9,10,21,22]. The CG effective interactions in such methods are obtained using the direct Boltzmann inversion, or reversible work, method [10,26–28], or iterative techniques, such as the iterative Boltzmann inversion (IBI) [7,29] and the inverse Monte Carlo (IMC); or inverse Newton approach [22,30].

(b) Force matching (FM) or multi-scale CG (MSCG) methods [5,14,16,31–33] comprise a mean least squares problem that considers as the observable function the total force acting on a coarse bead.

(c) The relative entropy (RE) [8,18,34] method employs the minimization of the relative entropy, or Kullback–Leibler divergence, between the microscopic Gibbs measure μ and μ^θ, representing approximations to the exact coarse space Gibbs measure. In this case, the microscopic probability distribution can be thought of as the observable. The minimization of the relative entropy is performed through Newton–Raphson approaches and/or stochastic optimization techniques [19,35].

In practice, all of the above numerical methods are employed to approximate a many body potential of mean force (PMF), U_{PMF}, describing the equilibrium distribution of CG particles observed in simulations of atomically-detailed models. Besides the above numerical parametrization schemes, more analytical approaches have also been developed for the approximation of the CG effective interaction, based on traditional liquid state theory and on pair correlation functions [36–43]. In these methods, in order to account for the many-body contributions (from the higher order terms), a closure scheme is performed. However, although these methods do perform well in some cases, there is no rigorous quantification of the closure error in the corresponding diagrammatic expansions; see [44,45]. The closures take into account some diagrams from all orders in the expansions (and hence, can potentially give good approximations in the liquid state), but the sum of the neglected terms (i.e., the error) has not been quantified. The quantification of these terms is a challenging open problem, both from a theoretical point of view due to the complicated combinatorial structure, as well as from the computational one since it involves many-body interactions. Hence, a common idea in all methods is to reduce this complexity. We also refer to [46], where simplifications are produced by considering the infinite dilution limit.

Here, we discuss an approach for estimating U_{PMF}, and the corresponding effective CG non-bonded potential, based on cluster expansion methods. Such methods originate from the works of Mayer and collaborators [47] in the 1940s. In the 1960s, numerous approximate expansions had been further developed [44,48,49] for the study of the liquid state. We also refer to the later textbook [50]. Later, with the advancement of powerful computational machines, the main focus has been directed on improving the computational methods such as Monte Carlo and molecular dynamics. However, the latter are mostly bulk calculations, and they become quite slow for large systems. Reducing the degrees of freedom by coarse-graining has been a key strategy to construct more efficient methods, but with many open questions with respect to the error estimation, transferability and adaptivity of the suggested methods. Based on recent developments of the mathematical theory of expansion methods in the canonical ensemble [51], our purpose is to combine the two approaches and obtain powerful computational methods, whose error compared to the target atomistic calculations can be quantified via rigorous estimates. In principle, the validity of these methods is limited to the gas regime. Here, we examine the accuracy of these methods in different state points. This attempt consists of the following: a priori error estimation of the approximate schemes depending on the different regimes, a posteriori error validation of the method from the coarse-grained data and the design of related adaptive methods. We note that our suggested approach is a bottom up methodology, in the sense that we directly sample the CG effective potential based on atomistic simulations of only a few particles, instead of matching average forces or equilibrium quantities of larger n-body (bulk) systems, as in (a)–(c). By construction, the effective potential is independent of the density ρ, but naturally, it depends on the number of hierarchy terms in each computation, as well as on the temperature. Furthermore, in each regime, different expansions might be appropriate. Note that density-dependent CG effective interaction potentials, which are often used in the literature, are a non-systematic way to account for the many-body interactions [52].

In previous years, we have developed CG models, based on cluster expansions, for lattice systems, obtaining higher order schemes and a posteriori error estimates [53], for both short- and long-range interactions [54] and designing adaptive methods [55] and investigating possible strategies for reconstruction of the atomistic information [56]. This is very much in the spirit of the polymer science literature [10,11,57], and in this paper, we get closer by considering off-lattice models. The proposed approach is based on typical schemes that are based on isolated molecules [26,27,58]. Here, we extend such approaches using cluster expansion tools for deriving CG effective potentials. We start from the typical two-body (pair) effective interaction, but some results can be extended to many-body interactions, as well. We also present a detailed theoretical investigation about the effect of higher order terms in obtaining CG effective interaction potentials for realistic molecular systems. In the present work, we focus on two prototypical examples of molecular systems, one with spherical symmetry (methane) and one without (ethane). In future work, we will extend these results to longer molecules, not necessarily with spherical symmetry. We show some first results from the implementation of three-body terms on the effective CG potential; a more detailed work on the higher order terms will be given in a forthcoming work [59].

The structure of the paper is as follows: In Section 2, we introduce the atomistic molecular system and its coarse-graining via the definition of the CG map, the n-body distribution function and the corresponding n-body potential of mean force. The cluster expansion-based formulation of the CG effective interaction is presented in Section 3. Details about the model systems (methane and ethane) and the simulation considered here are discussed in Section 4. Results are presented in Section 5; we briefly discuss the three-body case in Section 6. Finally, we close with Section 7 summarizing the results of this work.

2. Molecular Models

2.1. Atomistic and "Exact" Coarse-Grained Description

Here, we give a short description of the molecular model in the microscopic (all-atom) and mesoscopic (coarse-grained) scale. Assume a system of N (classical) atoms (or molecules) in a box $\Lambda(\ell) := (-\frac{\ell}{2}, \frac{\ell}{2}]^d \subset \mathbb{R}^d$ (for some $\ell > 0$), at temperature T. We will also denote the box by Λ when we do not need to make explicit the dependence on ℓ. We consider a configuration $\mathbf{q} \equiv \{q_1, \ldots, q_N\}$ of N atoms, where q_i is the position of the i-th atom. The particles interact via a pair potential $V : \mathbb{R}^d \to \mathbb{R} \cup \{\infty\}$, for which we assume the standard conditions of stability and temperedness; namely, that there exists a constant $B \geq 0$ such that: $\sum_{1 \leq i < j \leq N} V(q_i - q_j) \geq -BN$ for all N and all $q_1, \ldots, q_N$ and that $C(\beta) := \int_{\mathbb{R}^d} |e^{-\beta V(r)} - 1| dr < \infty$, where $\beta = \frac{1}{k_B T}$ and k_B is the Boltzmann's constant.

The canonical partition function of the system is given by:

$$Z_{\beta,\Lambda,N} := \frac{1}{N!} \int_{\Lambda^N} dq_1 \ldots dq_N \, e^{-\beta H_\Lambda(\mathbf{p},\mathbf{q})}, \tag{1}$$

where H_Λ is the Hamiltonian (total energy) of the system confined in a domain Λ:

$$H_\Lambda(\mathbf{p},\mathbf{q}) := \sum_{i=0}^{N} \frac{p_i^2}{2m} + U(\mathbf{q}). \tag{2}$$

By $U(\mathbf{q})$, we denote the total potential energy of the system, which for pair type potentials is:

$$U(\mathbf{q}) := \sum_{1 \leq i < j \leq N} V(q_i - q_j), \tag{3}$$

where, for simplicity, we assume periodic boundary conditions on Λ. Integrating over the momenta in (1), we get:

$$Z_{\beta,\Lambda,N} = \frac{\lambda^N}{N!} \int_{\Lambda^N} dq_1 \ldots dq_N \, e^{-\beta U(\mathbf{q})} =: \lambda^N Z_{\beta,\Lambda,N}^U, \tag{4}$$

where $\lambda := (\frac{2m\pi}{\beta})^{d/2}$. In the sequel, for simplicity, we will consider $\lambda = 1$ and identify $Z_{\beta,\Lambda,N} \equiv Z_{\beta,\Lambda,N}^U$. Fixing the positions q_1 and q_2 of two particles, we define the two-point correlation function:

$$\rho_{N,\Lambda}^{(2),at}(q_1, q_2) := \frac{1}{(N-2)!} \int dq_3 \ldots dq_N \frac{1}{Z_{\beta,\Lambda,N}} e^{-\beta U(\mathbf{q})}. \tag{5}$$

It is easy to see that in the thermodynamic limit, the leading order is ρ^2, where $\rho = \frac{N}{|\Lambda|}$, and $|\Lambda|$ is the volume of the box Λ. Thus, it is common to define the following order one quantity $g(r) := \frac{1}{\rho^2} \rho_{N,\Lambda}^{(2),at}(q_1, q_2)$, for $r = |q_1 - q_2|$. More generally, for $n \leq N$, we define the n-body version:

$$g^{(n)}(q_1, \ldots, q_n) = \frac{1}{(N-n)! \rho^n} \int_{\Lambda^{N-n}} dq_{n+1} \ldots dq_N \frac{1}{Z_{\beta,\Lambda,N}} e^{-\beta U(\mathbf{q})}, \tag{6}$$

and from that, the order n potential of mean force (PMF), $U_{\text{PMF}}(q_1, \ldots, q_n)$ [60,61], given by:

$$U_{\text{PMF}}(q_1, \ldots, q_n) := -\frac{1}{\beta} \log g^{(n)}(q_1, \ldots, q_n). \tag{7}$$

We define the coarse-graining map $T : (\mathbb{R}^d)^N \to (\mathbb{R}^d)^M$ on the microscopic state space, given by $T : \mathbf{q} \mapsto T(\mathbf{q}) \equiv (T_1(\mathbf{q}), \ldots, T_M(\mathbf{q})) \in \mathbb{R}^M$, which determines the M ($M < N$) CG degrees of freedom as a function of the atomic configuration $\mathbf{q}$. We call "CG particles" the elements of the coarse space with positions $\mathbf{r} \equiv \{r_1, \ldots, r_M\}$. We further define the effective CG potential energy by:

$$U_{\text{eff}}(r_1, \ldots, r_M) := -\frac{1}{\beta} \log \int_{\{Tq=r\}} dq_1 \ldots dq_N \, e^{-\beta U(\mathbf{q})}, \tag{8}$$

where the integral is over all atomistic configurations that correspond to a specific CG one using the coarse-graining map. Note that U_{eff} is in practice equivalent, up to a constant, to the (conditional) PMF. In the example we will deal with later, the configuration $\mathbf{r}$ will represent the centres of mass of groups of atomistic particles. This coarse-graining gives rise to a series of multi-body effective potentials of one, two, up to M-body interactions, which are unknown functions of the CG configuration. Note also that by the construction of the CG potential in (8), the partition function is the same:

$$Z_{\beta,\Lambda,N} = \int dr_1 \ldots dr_M \int_{\{Tq=r\}} d\mathbf{q} e^{-\beta U(\mathbf{q})} = \int dr_1 \ldots dr_M e^{-\beta U_{\text{eff}}(r_1,\ldots,r_M)} =: Z^{cg}_{\beta,\Lambda,M}. \tag{9}$$

The main purpose of this article is to give a systematic way (via the cluster expansion method) of constructing controlled approximations of U_{eff} that can be efficiently computed, and at the same time, we have a quantification of the corresponding error for both "structural" and "thermodynamic" quantities. By structural, we refer to $g(r)$, while by thermodynamic to the pressure and the free energy. Note that both depend on the partition function, but they can also be related [61] to each other as follows:

$$\beta p = \rho - \frac{\beta}{6}\rho^2 \int_0^\infty ru'(r)g(r)4\pi r^2 dr, \tag{10}$$

for the general case of pair-interaction potentials $u(r)$.

2.2. Coarse-Grained Approximations

As mentioned above, there are several methods in the literature that give approximations to the effective (CG) interaction potential U_{eff} as defined in (8). Below, we list some of them without the claim of being exhaustive:

(a) The correlation-based (e.g., DBI, IBI and IMC) methods that use the pair radial distribution function $g(r)$, related to the two-body potential of mean force for the intermolecular interaction potential, as well as distribution functions of bonded degrees of freedom (e.g., bonds, angles, dihedrals) for CG systems with intramolecular interaction potential [6,7,9,10,21,22]. These methods will be further discussed below.

(b) Force matching (FM) methods [5,16,31] in which the observable function is the average force acting on a CG particle. The CG potential is then determined from atomistic force information through a least-square minimization principle, to variationally project the force corresponding to the potential of mean force onto a force that is defined by the form of the approximate potential.

(c) Relative entropy (RE)-type [8,18,19] methods that produce optimal CG potential parameters by minimizing the relative entropy, Kullback–Leibler divergence between the atomistic and the CG Gibbs measures sampled by the atomistic model.

In addition to the above numerical methods, analytical works for the estimation of the effective CG interaction, based on integral equation theory, have also been developed for polymers and polymer blends [40,42]. A brief review and categorization of parametrization methods at equilibrium is given in [17,62].

The correlation-based iterative methods use the fact that for a pair interaction $u(r)$, by plugging the virial expansion of p in powers of ρ into (10) and comparing the orders of ρ, one obtains that [61]:

$$g(r) = e^{-\beta u(r)}\gamma(r), \quad \text{where} \quad \gamma(r) = 1 + c_1(r)\rho + c_2(r)\rho^2 + \ldots \tag{11}$$

Given the atomistic "target" $g(r)$ from a free (i.e., without constraints) atomistic run, by inverting (11) and neglecting the higher order terms of $\gamma(r)$, one can obtain a first candidate for a pair coarse-grained potential $u(r)$. Then, one calculates the $g(r)$ that corresponds to the first candidate and by iterating this procedure eventually obtains the desired two-body coarse-grained potential. This scheme is based on the theorem that if there exists a pair CG interaction that reproduces the target $g(r)$, this is unique [63]. However, this is only an approximation (accounting for the neglected terms of order ρ and higher in the expansion of $\gamma(r)$) since we know that the "true" CG interaction potential should be multi-body, as a result of integrating atomistic degrees of freedom. Hence, having agreement on $g(r)$, within numerical accuracy, does not secure proper thermodynamic behaviour, and several methods have been employed towards this direction; see, for example, [7,40,52] and the references within.

In order to maintain the correct thermodynamic properties, our approach in this paper is based on cluster expanding (8) with respect to some small, but finite parameter ϵ depending on the regime in which we are interested. For technical reasons we will focus here on the low density-high temperature regime. However, the cluster expansion is a perturbative method that can in principle be applied to other regimes once the appropriate target system is identified. For example, we refer to [53] where similar perturbations around mean field models for lattice systems have been implemented. In that case, the validity of the expansion can go beyond the usual high temperature regime. Extending this result to off-lattice systems, as is the case here, is more complicated, and we leave it for future investigation. As will be explained in detail in the next section, the resulting cluster expansion provides us with a hierarchy of terms. Uniformly in $\mathbf{r} \equiv r_1, \dots, r_M$, we have:

$$
\begin{aligned}
U_{\text{eff}}(\mathbf{r}) = U^{(2)}(\mathbf{r}) + O(\epsilon^2), \quad &\text{or} \quad U_{\text{eff}}(\mathbf{r}) = U^{(2)}(\mathbf{r}) + U^{(3)}(\mathbf{r}) + O(\epsilon^3), \\
U^{(2)}(\mathbf{r}) := \sum_{i,j} W^{(2)}(r_i, r_j), \quad &U^{(3)}(\mathbf{r}) := \sum_{i,j,k} W^{(3)}(r_i, r_j, r_k), \quad \text{etc.,}
\end{aligned}
\tag{12}
$$

together with the corresponding error estimates. Note that a precise choice of the error quantity ϵ, depending on the density and the temperature, will be given in Section 3.

The above terms can in principle be calculated independently via fast atomistic simulations of 2, 3, etc., molecules, in the spirit of the conditional reversible work (CRW) method [13,27,28,58]. In more detail, the effective non-bonded (two-body) CG potential can be computed as follows:

(a) One method is by fixing the distance $r_{1,2} := r_1 - r_2$ between two molecules and performing molecular dynamics with such forces that maintain the fixed distance $r_{1,2}$. In this way, we sample atomistic potential energy (and forces) over the constrained phase space and obtain the conditional partition function as the integral in (8). Alternatively, by integration of the constrained force $(-\int_{r_{min}}^{r_{cut}} \langle f \rangle_{r_{1,2}=R} dR)$, the two-body effective potential can be obtained. These are the $W^{(2),\text{full},U}$ and $W^{(2),\text{full},F}$ terms, respectively. Note that we have not used any kind of fitting or projection over a basis as in [64,65]; the data are in tabulated form.

(b) Upon inverting $g(r)$ in (11) for two isolated molecules, the two-body effective potential can be directly obtained, since for such a system, $\gamma(r) = 1$; i.e., the low ρ regime. This method is only used for comparison with (a)as it uses the $g(r)$.

Here, we examine both of the above methods; see Section 5.1. Note also that the validity of cluster expansion provides rigorous expansions for $g(r)$, the pressure and the other relevant quantities. Hence, with this approach, we can have a priori estimation of the errors made in (11). Another benefit of the cluster expansion is that the error terms can be written in terms of the coarse-grained quantities allowing for a posteriori error estimates and the design of adaptive methods [55]; see also the discussion in Section 7.

2.3. Thermodynamic Consistency

As already mentioned, several coarse-graining strategies lack thermodynamic consistency; see also the discussion by Louis [46] and Guenza [40], where thermodynamically-consistent theories have been developed based on the standard closures of the liquid state theory. However, as the error in these closures has not been rigorously quantified, we first investigate here a much simpler case;

namely, what is the contribution of the three-body terms in approximating thermodynamic quantities in the regimes where the cluster expansion can be valid. Hence, defining the coarse-graining as in (12), we demonstrate below that these approximations directly imply similar approximations for thermodynamic quantities, such as the free energy and the pressure, and we also comment on the isothermal compressibility. This is a direct consequence of the validity of the cluster expansions. Thus, from (12), by considering only the two-body contribution, for the finite volume free energy $f_{\beta,\Lambda}(N)$, we have that:

$$f_{\beta,\Lambda}(N) = -\frac{1}{\beta|\Lambda|} \log Z_{\beta,\Lambda,N} = \int d\mathbf{r} e^{\beta U_{\text{eff}}(\iota)} = -\frac{1}{\beta|\Lambda|} \log \int d\mathbf{r} e^{-\beta U^{(\tau)}(r)} + O(\epsilon^2), \tag{13}$$

where the error is uniform in N and $|\Lambda|$. Thus, the approximation $U^{(2)}$ of the CG Hamiltonian implies a good approximation of the free energy. By adding the next order contribution of the three-body terms $U^{(3)}$, we can further improve the error in computing the free energy and obtain that:

$$-\frac{1}{\beta|\Lambda|} \log Z_{\beta,\Lambda,N} = -\frac{1}{\beta|\Lambda|} \log \int d\mathbf{r} e^{-\beta(U^{(2)}(r) + U^{(3)}(r))} + O(\epsilon^3). \tag{14}$$

Note once again that these estimates are valid only in the low density regime where these expansions are justified.

Similarly, in order to construct approximations for the thermodynamic pressure (at finite volume) $p_{\beta,\Lambda}(z)$ as a function of the activity z, we work in the grand-canonical ensemble and obtain:

$$p_{\beta,\Lambda}(z) = \frac{1}{\beta|\Lambda|} \log \sum_{N \geq 0} z^N Z_{\beta,\Lambda,N} = \frac{1}{\beta|\Lambda|} \log \sum_{N \geq 0} z^N \int d\mathbf{r} e^{-\beta U^{(2)}(r)} + O(\epsilon^2). \tag{15}$$

Both quantities have limits given by absolutely convergent series with respect to $\rho = N/|\Lambda|$ for the first and z or ρ for the second.

In the same spirit, good approximations to the atomistic (full) partition function can give controlled approximations for other thermodynamic quantities, as well. For example, the isothermal compressibility can be related to the variance in the number of particles, which, in turn, can be expressed in terms of the derivative of the density with respect to the chemical potential (in the grand canonical ensemble). Since the density is again expressed in terms of the derivative of the logarithm of the partition function, our suggested method is also applicable here. All further manipulations needed to arrive at an expansion for the isothermal compressibility; namely, taking derivatives, as well as inverting series, can be made rigorous in the framework of the cluster expansion, but a full analysis of this and similar cases of other thermodynamic quantities is beyond the scope of the present work, so we just refer to [45] for an analysis in this direction.

3. Cluster Expansion

The cluster expansion method originates from the work of Mayer and collaborators (see [47] for an early review) and consists of expanding the logarithm of the partition function in an absolutely convergent series of an appropriately chosen small, but finite parameter. Here, we will adapt this method to obtain an expansion of the conditional partition function and consequently the desired many-body PMF.

For the purpose of this article, we assume that the CG map T is a product $T = \otimes_{i=1}^{M} T^i$ creating M groups of $l_1, \ldots, l_M$ particles each (e.g., centre of mass mapping). We index the particles in the i-th group of the coarse-grained variable r_i by $k_1^i, \ldots, k_{l_i}^i$. We also denote them by $\mathbf{q}^i := (q_{k_1^i}, \ldots, q_{k_{l_i}^i})$, for $i = 1, \ldots, M$. Then, (8) can be written as:

$$U_{\text{eff}}(r_1, \ldots, r_M) := -\frac{1}{\beta} \log \prod_{i=1}^{M} \lambda^i(\{T^i \mathbf{q}^i = r_i\}) - \frac{1}{\beta} \log \int \prod_{i=1}^{M} \mu(d\mathbf{q}^i; r_i) e^{-\beta U(\mathbf{q})}, \tag{16}$$

where, for simplicity, we have introduced the normalized conditional measure, related to the specific CG map:

$$\mu(d\mathbf{q}^i; r_i) := \frac{1}{l_i!} dq_{k_1^i} \ldots dq_{k_{l_i}^i} \frac{\mathbf{1}_{\{T^i \mathbf{q}^i = r_i\}}}{\lambda^i(\{T^i \mathbf{q}^i = r_i\})}, \tag{17}$$

and by λ^i, we denote the measure $\frac{1}{l_i!} dq_{k_1^i} \ldots dq_{k_{l_i}^i}$.

To perform a cluster expansion in the second term of (16), we rewrite the interaction potential as follows:

$$U(\mathbf{q}) = \sum_{i<j} \tilde{V}(\mathbf{q}^i, \mathbf{q}^j), \quad \text{where}$$

$$\tilde{V}(\mathbf{q}^i, \mathbf{q}^j) := \sum_{m=1}^{l_i} \sum_{m'=1}^{l_j} V(|q_{k_m^i} - q_{k_{m'}^j}|). \tag{18}$$

Then, letting $f_{i,j}(\mathbf{q}^i, \mathbf{q}^j) := e^{-\beta \tilde{V}(\mathbf{q}^i, \mathbf{q}^j)} - 1$, we have:

$$e^{-\beta U(\mathbf{q})} = \prod_{i<j} \left(1 + e^{-\beta \tilde{V}(\mathbf{q}^i, \mathbf{q}^j)} - 1\right) = \sum_{\substack{V_1, \ldots, V_m \\ |V_i| \geq 2, V_i \subset \{1, \ldots, N\}}} \prod_{l=1}^{m} \sum_{g \in \mathcal{C}_{V_i}} \prod_{\{i,j\} \in E(g)} f_{i,j}(\mathbf{q}^i, \mathbf{q}^j), \tag{19}$$

where for $V \subset \{1, \ldots, N\}$, we denote by $\mathcal{C}_V$ the set of connected graphs on the set of vertices with labels in V. Furthermore, for $g \in \mathcal{C}_V$, we denote by $E(g)$ the set of its edges.

Since μ in (17) is a normalized measure, from (16), we obtain:

$$U_{\text{eff}}(r_1, \ldots, r_M) = -\frac{1}{\beta} \log \prod_{i=1}^{M} \lambda^i(\{T^i \mathbf{q}^i = r_i\}) - \frac{1}{\beta} \log \sum_{\substack{V_1, \ldots, V_m \\ |V_i| \geq 2, V_i \subset \{1, \ldots, N\}}} \prod_{l=1}^{m} \zeta(V_i)$$

$$= -\frac{1}{\beta} \log \prod_{i=1}^{M} \lambda^i(\{T^i \mathbf{q}^i = r_i\}) - \frac{1}{\beta} \sum_{V \subset \{1, \ldots, N\}} \zeta(V) + \frac{1}{\beta} \sum_{\substack{V, V': \\ V \cap V' = \varnothing}} \zeta(V) \zeta(V') + \ldots, \tag{20}$$

where:

$$\zeta(V) := \int \sum_{g \in \mathcal{C}_V} \prod_{\{i,j\} \in E(g)} f_{i,j}(\mathbf{q}^i, \mathbf{q}^j) \prod_{i \in V} \mu(d\mathbf{q}^i; r_i) \tag{21}$$

is a function over the atomistic details of the system. Note that the above expression involves a sum over all possible pairs, triplets, etc., which is a convergent series for the values of the density $\rho = \frac{N}{|\Lambda|}$ and of the inverse temperature β such that $\rho \bar{C}_\beta(r) < c_0$, where $\bar{C}_\beta(r) := \int |f_{i,j}(\mathbf{q}^i, \mathbf{q}^j)| \mu(d\mathbf{q}^i; 0) \mu(d\mathbf{q}^i; r)$ and c_0 is a known small positive constant [51]. Here, we do not give the full proof, which can be easily obtained by a slight modification of the one in [51], but in order to motivate the error estimates in (12), we note that the sum over triplets will give a contribution of the order $\binom{N}{3}$#{trees on 3 labels} $\frac{1}{|\Lambda|^3} \bar{C}_\beta(r)^2 \sim \rho(\rho \bar{C}_\beta(r))^2$, i.e., here, $\epsilon \sim \rho \bar{C}_\beta(r)$. If we simplify the sum in (20), one can obtain [51] Expansion (12), where:

$$W^{(2)}(r_1, r_2) := -\frac{1}{\beta} \int \mu(d\mathbf{q}^1; r_1) \, \mu(d\mathbf{q}^2; r_2) \, f_{1,2}(\mathbf{q}^1, \mathbf{q}^2) \tag{22}$$

and:

$$W^{(3)}(r_1, r_2, r_3) := -\frac{1}{\beta} \int \mu(d\mathbf{q}^1; r_1) \, \mu(d\mathbf{q}^2; r_2) \, \mu(d\mathbf{q}^3; r_3) \, f_{1,2}(\mathbf{q}^1, \mathbf{q}^2) \, f_{2,3}(\mathbf{q}^2, \mathbf{q}^3) \, f_{3,1}(\mathbf{q}^3, \mathbf{q}^1). \tag{23}$$

Recall also the definition of $f_{i,j}$ in (19).

3.1. Full Calculation of the PMF

Notice that the potentials $W^{(2)}$ and $W^{(3)}$ in (22) and (23), respectively, have been expressed via the Mayer functions $f_{i,j}$. However, the full effective interaction potential between two CG particles can be directly defined as the (conditional) two-body PMF given by:

$$W^{(2),\text{full}}(r_1, r_2) := -\frac{1}{\beta} \log \int \mu(d\mathbf{q}^1; r_1)\, \mu(d\mathbf{q}^2; r_2)\, e^{-\beta \tilde{V}(\mathbf{q}^1, \mathbf{q}^2)}. \tag{24}$$

By adding and subtracting one and expanding the logarithm, we can relate it to (22):

$$
\begin{aligned}
-\beta W^{(2),\text{full}}(r_1, r_2) &= \log \int \mu(d\mathbf{q}^1; r_1)\mu(d\mathbf{q}^2; r_2)e^{-\beta\tilde{V}(\mathbf{q}^1,\mathbf{q}^2)} \\
&= \log(1 + \int \mu(d\mathbf{q}^1; r_1)\mu(d\mathbf{q}^2; r_2)f_{1,2}(\mathbf{q}^1,\mathbf{q}^2)) \\
&= \int \mu(d\mathbf{q}^1; r_1)\,\mu(d\mathbf{q}^2; r_2)f_{1,2}(\mathbf{q}^1,\mathbf{q}^2) - \tfrac{1}{2}\left(\int \mu(d\mathbf{q}^1; r_1)\,\mu(d\mathbf{q}^2; r_2)f_{1,2}(\mathbf{q}^1,\mathbf{q}^2)\right)^2 + \dots
\end{aligned} \tag{25}
$$

Higher order terms in the above equation are expected to be less/more important in high/low temperature. Similarly, for three CG degrees of freedom r_1, r_2, r_3, the full PMF is given by:

$$W^{(3),\text{full}}(r_1, r_2, r_3) := -\frac{1}{\beta} \log \int \mu(d\mathbf{q}^1; r_1)\, \mu(d\mathbf{q}^2; r_2)\mu(d\mathbf{q}^3; r_3)\, e^{-\beta\sum_{1\le i<j\le 3} \tilde{V}(\mathbf{q}^i,\mathbf{q}^j)}. \tag{26}$$

By adding and subtracting one, we can relate it to (22) and (23) (in the following, we simplify notation by not explicitly showing the dependence on the atomistic configuration and neglecting the normalized conditional measure):

$$
\begin{aligned}
e^{-\beta W^{(3),\text{full}}} &= \int e^{-(V_{12}+V_{13}+V_{23})} \\
&= 1 + \int f_{12} + \int f_{13} + \int f_{23} + \int f_{12}f_{13} + \int f_{13}f_{23} + \int f_{12}f_{23} + \int f_{12}f_{23}f_{13},
\end{aligned} \tag{27}
$$

which implies that:

$$
\begin{aligned}
W^{(3),\text{full}} = -\frac{1}{\beta}\Bigg(&\int f_{12} + \int f_{13} + \int f_{23} + \int f_{12}f_{23}f_{13} + \int f_{12}f_{13} + \int f_{13}f_{23} + \int f_{12}f_{23} - \\
&\left[\int f_{12} \int f_{13} + \int f_{13} \int f_{23} + \int f_{12} \int f_{23}\right]\Bigg) + \dots
\end{aligned} \tag{28}
$$

In principle, we can rewrite (12) with respect to $W^{(2),\text{full}}$ and $W^{(3),\text{full}}$. Note, however, that both of these terms contain the coarse-grained two-body interactions; hence, in order to avoid double-counting, when we use both, we have to appropriately subtract the two-body contributions. For some related results, see also the discussion about Figure 11.

Note finally that in the proposed approach, the hierarchy of CG effective potential (two-body, three-body, etc.) terms is not fitted or included via some uncontrolled closure; they can be computed, through a conditional sampling of a few (two, three, etc.) CG particles. Furthermore, our work distinguishes from those using the standard liquid state theory also by the fact that we do not use any simplified models (PRISM, thread model, etc.), but integrate over the atomistic degrees of freedom.

4. Model and Simulations

4.1. The Model

A main goal of this work, as mentioned before, is to examine the parameterization of a coarse-grained model using the cluster expansion formalism described above for simple realistic molecular systems; in this work, we study methane and ethane in various density and temperature regimes. We refer the reader to the Supplementary Material for the atomistic models, as well as their coarse-graining through the cluster expansion formalism. In order to clarify the notation, we refer to Figure 1; for the three-body clustering case, each set is composed by exactly three interacting CG molecules (dots connected with solid lines). The centres of mass (COM) of the set i is located at r_i and associated with $\mu(d\mathbf{q}^i, r_i)$.

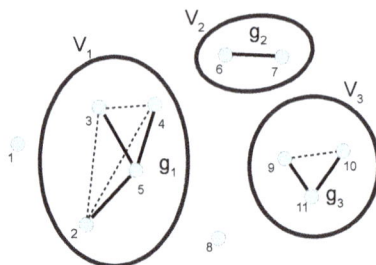

Figure 1. Visualization of the partition in (19) for non-intersecting sets $V_1 = \{2, 3, 4, 5\}$, $V_2 = \{6, 7\}$, $V_3 = \{9, 10, 11\}$ in each of which we display by solid lines the connected graphs $g_i \in \mathcal{C}_{V_i}$, $i = 1, 2, 3$.

4.2. Simulations

The simplest system to simulate is the one with only two interacting methane, or ethane, molecules in a vacuum. This is a reference system for which the many-body PMF is equal to the two-body one. In addition, we have also simulated the corresponding systems in various densities, as well as in the liquid regime. The atomistic and CG model methane systems were studied through molecular dynamics and Langevin dynamics (LD) simulations. All simulations were conducted in the NVT ensemble. For the MD simulations, the Nose–Hoover thermostat was used. Langevin dynamics models a Hamiltonian system that is coupled with a thermostat [66]. The thermostat serves as a reservoir of energy. The densities of both the liquid methane and ethane systems were chosen as the average values of NPT runs at atmospheric pressure. NVT equilibration and production runs of a few *ns* followed, and the sizes of the systems were 512 CH_4 and 500 CH_3–CH_3 molecules. We note here that the BBK integrator used for Langevin dynamics exhibits pressure fluctuations of the order of ± 40 atm in the liquid phase, whereas temperature fluctuations have small variance, and the system is driven to the target temperature much faster than with conventional MD. Long production atomistic and CG simulations have been performed, from 10 ns up to 100 ns, depending on the system. The error bars (computed via the standard deviation) in all reported data are about 1–2% of the actual values. Details about the simulation parameters are shown in Table 1.

Table 1. Simulation parameters for CH_4.

System	N (Molecules)	T/K	Simulation Time/ns
CH_4, CH_3–CH_3	2, 3	100–900	10–20
CH_4	512	80, 100, 120, 300, 900	100
CH_3–CH_3	500	150, 300, 650	100

In order to compute the hierarchy of CG effective non-bonded potential terms, discussed above, different simulation runs have been performed, which are discussed below.

4.2.1. Constrained Runs

The first method that we use in order to estimate the effective CG potential is by constraining the intermolecular distance between two molecules, $r = r_{1,2}$, in order to compute the constrained partition function (9). We call it the "constrained run" of two methane, or ethane, molecules, and special care had to be taken in order to avoid long sampling of the low probability short distances. This notion is similar to the free energy calculation methods in which CG degrees of freedom are constrained at fixed distances. Technically, we pin the centres of mass (COM) of each CG particle in space, and on every step throughout the stochastic (Langevin) dynamics trajectory, we subtract the total force acting on each COM. Hence, we allow the atoms to move, resulting in rotations, but not translations of the CG degrees of freedom OM. During these runs, the constraint forces are recorded. The mean value $\langle F \rangle_{r_{12}=r}$ is calculated in the same manner, and we get $W^{(2),\text{full, F}}(r)$, from $f = -\nabla W$, as explained in Section 2.2. Both $W^{(2),\text{full, F}}(r)$ and $W^{(2),\text{full, U}}(r)$ are based on the same trajectory.

The constrained run technique described above accelerates the sampling for short distances, but there is a caveat; the ensemble average at very short distances (left part of the potential well) is strongly affected by the geometric arrangement of specific atoms between the two molecules, and the system might be trapped in the minimum of energy. For example, the two CH_4 molecules are oriented according to the highly repulsive forces and rotate around the axis connecting the two COMs. Due to this specific reason, we utilized stochastic (Langevin) dynamics in order to better explore the subspace of the phase space, as a random kick breaks this alignment. We determine the minimum amount of steps needed for the ensemble average to converge, in a semi-empirical manner upon inspection of the error-bars.

4.2.2. Geometric Direct Computation of PMF

In order to further accelerate the sampling and alleviate the noise problems at high energy regions, which might become catastrophic in the case of the non-symmetric CH_3–CH_3 model, we have also calculated the two-body PMF (constraint partition function) directly, through "full sampling" of all possible configurations using a geometrical method proper for rigid bodies. In more detail, the geometric averaged constrained two-body effective potential $W^{(2),geom}(r)$ given in (24) is obtained by rotating the two (methane or ethane) molecules around their COMs, through their Eulerian angles and taking account of all of the possible (up to a degree of angle discretization) orientations. The main idea is to cover every possible (discretized) orientation and associate it with a corresponding weight. The Euler angles proved to be the easiest way to implement this; each possible orientation is calculated via a rotation matrix using three (Euler) angles in spherical coordinates.

The above way of sampling is more accurate (less noisy) than constrained canonical sampling and considerably faster. In addition, the nature of the computations allows massive parallelization of the procedure. We used a ZYZ rotation with $d\phi = d\psi = d\theta = \pi/20$ for CH_4 and simple spherical coordinate sampling with $d\phi = \pi/20, d\theta = \pi/45$ for CH_3–CH_3 (as it is diagonally symmetric in the united atom description). Note, however, that in this case, the molecules are treated as rigid bodies; i.e., bond lengths and bond angles are kept fixed; essentially, it is assumed that intra-molecular degrees of freedom do not affect the intermolecular (non-bonded potential) ones. However, for this system, there is no considerable difference in the resulting non-bonded effective potential. The advantage of this method is that we avoid long (and more expensive) molecular simulations of the canonical ensemble, which might also get trapped in local minima and inadequately sample the phase space. We should also state that this method is very similar to the one used by McCoy and Curro in order to develop a CH_4 united-atom model from all-atom configurations [58].

All atomistic and coarse-grained simulations have been performed using a home-made simulation package that has been extensively used in the past [62,67], whereas all analysis has been executed through home-made codes in MATLAB and Python. In addition, long simulation runs have been performed resulting in error bars that are of the order of the width of the lines in all figures presented below. Characteristic snapshots from two-body, three-body and bulk simulations are shown in Figure 2.

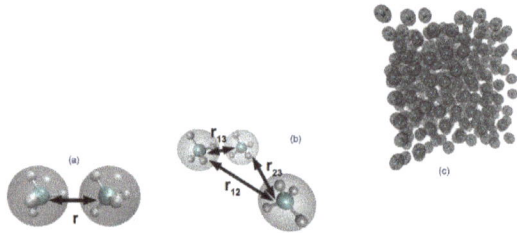

Figure 2. Snapshot of model systems in atomistic and coarse-grained description. (**a,b**) Two and three methanes used for the estimation of the coarse-graining (CG) effective potential from isolated molecules; (**c**) bulk methane liquid.

5. Results

5.1. Calculation of the Effective Two-Body CG Potential

First, we present data related to the calculation of the two-body potential of mean force for the ideal system of two (isolated) molecules. For such a system, the conditional M-body CG PMF is a two-body one. In Figure 3a,b, we provide data for the CG effective interaction between two methane and ethane molecules, through the following methods:

(a) A calculation of the PMF using the constraint force approach, $W^{(2),\text{full},f}$, as described in Section 4.2.1. Alternatively, through the same set of atomistic configurations, the two-body PMF, $W^{(2),\text{full},u}$, can be directly calculated through Equation (24).

(b) A direct calculation of the PMF, $W^{(2),geom}$, using a geometrical approach as described in Section 4.2.2.

(c) DBI method: The CG effective potential, $W^{(2),g(r)}$, is obtained by inverting the pair (radial) correlation function, $g(r)$, computed through a stochastic LD run with only two methane (or ethane) molecules freely moving in the simulation box. The pair correlation function, $g(r)$, of the two methane molecules is also shown in Figure 3a.

The first two of the above methods refer to the direct calculation of the constrained partition function (8) with constrained forces and canonical sampling, while the third uses the "direct Boltzmann inversion" approach. All of the above data correspond to temperatures in which both methane and ethane are liquid at atmospheric pressure (values of $-k_B T$ are also shown in Figure 3).

First, for the case of the two methane molecules (Figure 3a), we see very good agreement between the different methods. As expected, slightly more noisy is the $W^{(2),\text{full},U}(r_{12})$ curve as fluctuations in the $\langle e^{-\beta V(q)} \rangle$ term for a given r_{12} distance in Equation (24) are difficult to cancel out. The small probability configurations in high potential energy regimes have a large impact on the average containing the exponent; hence, the corresponding plot is not as smooth as the others are. In addition, as previously mentioned, $W^{(2),\text{full},F}$ comes from the same trajectory (run), but the integration of the $\langle f \rangle_{r_{12}}$ from r_{cutoff} up to r_{12} washes out any non-smoothness. Note that for the same system, recently CG effective potentials based on IBI, force matching and relative entropy methods have been derived and compared against each other [62].

Second, for the case of the two ethane molecules (Figure 3b), we see a good, but not perfect, agreement between the different sets of data, especially in the regions of high potential energy (short distances). This is not surprising if we consider that high energy data from any simulation technique that samples the canonical ensemble exhibit large error bars, due to difficulties in sampling. The latter is more important for the ethane compared to the methane case due to its molecular structure; indeed, the atomistic structure of methane approximates much better the spherical structure of CG particles than ethane. The only method that provides a "full", within the numerical discretization, sampling at any distance is the geometric one; however, as discussed before (see Section 4), such a method neglects the bond lengths and bond angle fluctuations.

Next, we also examine an alternative method for the computation of the effective CG potential, by calculating the approximate terms from the cluster expansion approach. For the latter, we use the data from the constraint runs of two methane molecules integrated over all atomistic degrees of freedom, as given in Formula (22). In Figure 4a,b, we demonstrate the PMF through cluster expansions and the effect of higher order terms as shown in Equation (25), of the two isolated molecules, for CH_4 and CH_3–CH_3, respectively. As discussed in Section 3, cluster expansion is expected to be more accurate at high temperatures and/or lower densities. For this reason, we examine both systems at higher temperatures, than of the data shown in Figure 3; values of $-k_B T$ are shown with full lines.

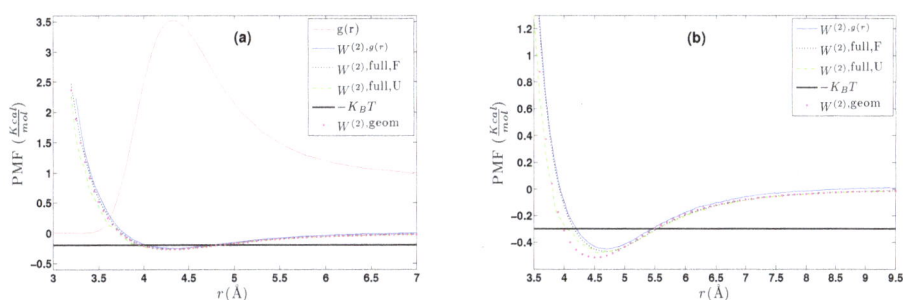

Figure 3. Representation of the two-body PMF, for two isolated molecules, as a function of distance r, through different approximations: geometric averaging, (constrained) force matching and inversion of $g(r)$. (**a**) CH_4 at $T = 100$ K; (**b**) CH_3–CH_3 at $T = 150$ K. For the methane, the corresponding $g(r)$ curve is also shown.

Both systems show the same behaviour. First, it is clear that the agreement between $W^{(2)}$ and the (more accurate) $W^{(2),full}$ is very good only at long distances (not surprisingly, since the logarithm expansion holds for every β, as $\bar{V} \rightarrow 0$ in $C(\beta)$), whereas there are strong discrepancies in the regions where the potential is minimum, as well as in the high energy regions (short distances); Second, it is evident that adding terms up to the second order with respect to β, we obtain a better approximation of $W^{(2),full}$. All of the above suggest that geometric averaging is the most accurate and computationally less expensive.

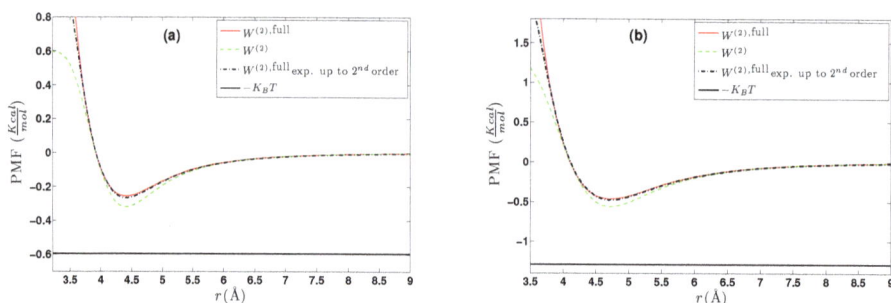

Figure 4. Relation of the PMF through cluster expansions and energy averaging at high temperatures, i.e., $W^{(2)}(r_1, r_2)$ and $W^{(2),full}(r_1, r_2)$, through expansion over β for (**a**) CH_4 at $T = 300$ K; (**b**) CH_3–CH_3 at $T = 650$ K. As expected from the analytic form and the relation between the two formulas, $W^{(2)}$ and $W^{(2),full}$ tend to converge to the same effective potential.

5.1.1. Effect of Temperature-Density

Next, we further examine the dependence of the PMF, for the two isolated methanes, on the temperature, by studying the system at $T = 80$ K , 100 K, 120 K, 300 K and 900 K. In more detail, in Figure 5a,b, we compare the difference between $W^{(2)}$ and $W^{(2),full}$ at different temperatures. As discussed in Section 3, the cluster expansion method is valid only in the high temperature regime. This is directly observed in Figure 5a; at high temperatures, $W^{(2)}$ is very close to $W^{(2),full}$, which is exact for the system consisting of two molecules. Note the small differences at short distances, which, as also discussed in the previous subsection, are even smaller if higher order terms are included in the calculation of $W^{(2)}$; see also Figure 4.

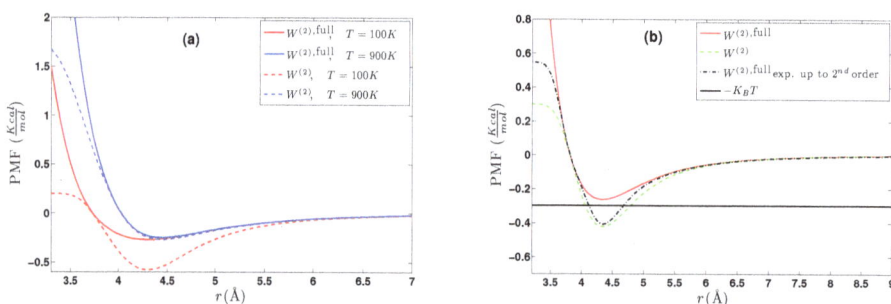

Figure 5. (**a**) PMF through cluster expansions, using (22) and (25) for different temperatures for the CH_4 model; (**b**) PMF through cluster expansions and energy averaging, i.e., $W^{(2)}(r_1, r_2)$ and $W^{(2),full}(r_1, r_2)$ through expansion over β for CH_4 at $T = 150$ K. The expansion is not valid at this temperature.

On the contrary, at low temperatures, there is a strong discrepancy around the potential well, as shown in Figure 5b. In fact, for values of r close to the potential well and for rather high values of β, the contribution to the integral $C(\beta)$ is large, and the latter can exceed one, rendering the expansion in (25) not valid. In Figure 5b, we see that the term (22) is not small, so the expansion (25) is not valid. The case for ethane is qualitatively similar.

For completeness, we also plot the potential of mean force at different temperatures for the system of two CH_4 molecules; see Figure 6. In principle, Equation (22) is a calculation of free energy; hence, it incorporates the temperature of the system, and thus, both approximations to the exact two-body PMF, $W^{(2)}$ and $W^{(2),full}$, are not transferable. Indeed, we observe slight differences in the

CG effective interactions (free energies) for the various temperatures, which become larger for the highest temperature.

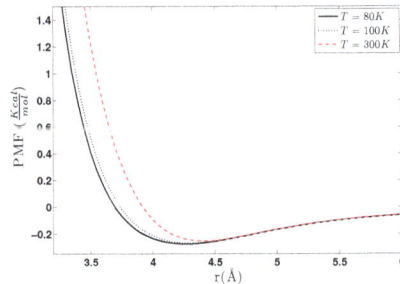

Figure 6. Potential of mean force at different temperatures (geometric averaging). Two CH_4 molecules at $T = 80$ K, 100 K, 300 K.

5.2. Bulk CG CH_4 Runs Using a Pair Potential

In the next stage, we quantitatively examine the accuracy of the effective CG interaction potential (approximation of the two-body PMF), in the liquid state based on structural properties like $g(r)$. Here, we use the different CG models (approximated pair CG interaction potentials) derived above, to predict the properties of the bulk CG methane and ethane liquids. In all cases, we compare with structural data obtained from the reference all-atom bulk system, projected on the CG description.

In Figure 7a,b, we assess the discrepancy between the CG (projected) pair distribution function, $g(r)$, taken from an atomistic run and the one obtained from the corresponding CG run based on $W^{(2),\text{full}}$, as already seen in Figure 3 of methane and ethane, respectively. Note that $g(r)$ is directly related to the effective CG potentials ($N = 2$ in Equation (6)).

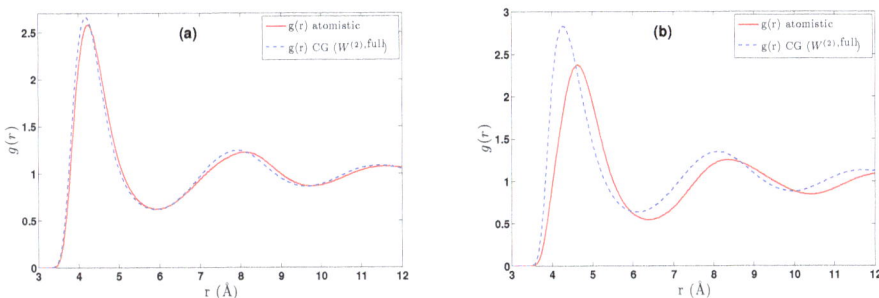

Figure 7. RDF from atomistic and CG using pair potential, $W^{(2)}$, for (**a**) CH_4 at $T = 80$ K and (**b**) CH_3–CH_3 at $T = 150$ K. Spherical CG approximation to the non-symmetric ethane molecule induces discrepancy and implies that there is more room for improvement.

It is clear that for methane (Figure 7a), the CG model with the $W^{(2),\text{full}}$ potential gives a g(r) very close to the one derived from the analysis of the all-atom data. This is not surprising if we consider that for most molecular systems, small differences in the interaction potential lead to even smaller differences in the obtained pair correlation function. Interestingly, the CG model with the $W^{(2)}$ is also in good agreement with the reference one, despite the small differences in the CG interaction potential discussed above (see Figures 4a and 8b). As expected, the difference comes from the missing higher order terms of Equation (12).

The fact that the CG effective potential, which is derived from two isolated methane molecules, gives a very good estimate for the methane structure in the liquid state is not surprising if we consider the geometrical structure of methane, which is rather close to the spherical one. On the contrary, for the case of ethane (Figure 7b), predictions of $g(r)$ using pair CG potential are much different compared to the atomistic one, especially for the short distances. Even larger differences would be expected for more complex systems with long-range interactions, such as water [62]. Similar is the case also for the other temperatures ($T = 80$ K) studied in this work (data not shown here).

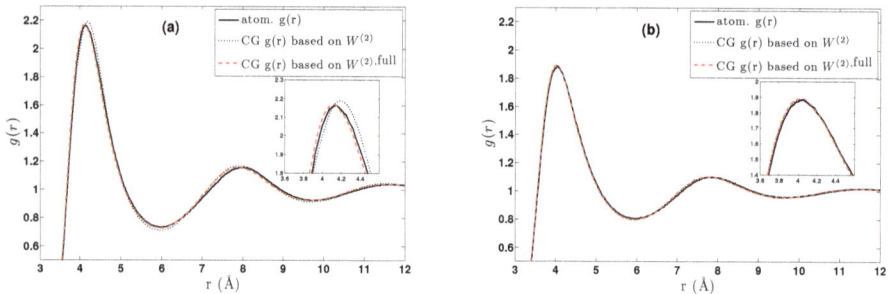

Figure 8. RDF of methane from atomistic data and CG models using pair potential at different temperatures: (**a**) $T = 300$ K; (**b**) $T = 900$ K. In both cases, the density is $\rho_1 = 0.3799 \frac{g}{cm^3}$.

5.2.1. Effect of Temperature-Density

We further study the structural behaviour of the CG systems at different state points; i.e., temperature/density conditions, compared to the atomistic ones. First, we examine the temperature effect by simulating the systems discussed above (see Figure 7) at higher temperatures; however, keeping the same density. In Figure 8a,b, we present the RDF of methane from atomistic and CG runs using pair potential at $T = 300$ K and $T = 900$ K respectively.

It is clear that the analysis of the CG runs using the $W^{(2),full}$ potential gives a pair distribution function $g(r)$ close to the atomistic one for both (high) temperatures, similar to the case of the $T = 80$ K shown before. In addition, the CG model with the $W^{(2)}$ potential is in very good agreement with the atomistic data at high temperature (Figure 8b), whereas there are small discrepancies at lower temperatures (Figure 8a), in particular at the maximum of $g(r)$. This is shown in the inset of Figure 8a,b. Note also that at this high temperature, the incorporation of the higher order terms in $W^{(2)}$ leads to very similar potential as the $W^{(2),full}$ (see also Figure 4a) and consequently to very accurate structural $g(r)$ data, as well.

Next, we examine the structural behaviour of the CG systems at different densities. In Figure 9a, we present the $g(r)$ from atomistic and CG runs using pair potential at different densities ($\rho_1 = 0.3799 \frac{g}{cm^3}$ and $\rho_2 = 0.0395 \frac{g}{cm^3}$ and $T = 300$ K and $T = 900$ K). There is apparent discrepancy from the reference (atomistic) system in both densities in agreement with the data discussed above in Figure 8a.

For the case of higher temperature data ($T = 900$ K) and the same densities, as shown in Figure 9b, the pair distribution function, $g(r)$, obtained from the CG model with the $W^{(2)}$ effective interaction is very close to the data derived from the $W^{(2),full}$ one, and in very good agreement with the reference, all-atom, data. This is not surprising since, as discussed before, at high temperatures, the cluster expansion is expected to be more accurate, since cluster expansions hold for high T and low ρ.

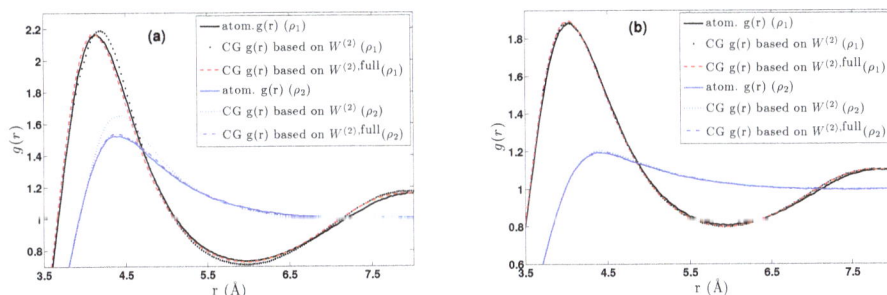

Figure 9. RDF of methane from atomistic and CG using pair potential at different densities $\rho_1 > \rho_2$. (a) $T = 300K$; (b) $T = 900$ K. For this model, the pair approximation is sufficient, and in low ρ, high T conditions, $W^{(2)}$ converges to the reference $g(r)$.

Overall, the higher the temperature, the better the agreement in the $g(r)$ derived from the CG models using any of $W^{(2)}$ and $W^{(2),\text{full}}$. These data are in better agreement with the atomistic data, as well.

6. Effective Three-Body Potential

In the last part of this work, we briefly discuss the direct computation of the three-body effective CG potential and its implementation in a (stochastic) dynamic simulation. More results about the three-body terms will be presented in a future work [59].

6.1. Calculation of the Effective Three-Body Potential

In the following, we present data for the three-body potential of mean force estimated from simulation runs and geometric computations involving three isolated molecules. We have two suggestions for the three-body PMF: (a) Formula (23) derived from the cluster expansion formalism, which is valid for rather high temperatures; (b) another one based on the McCoy–Curro scheme given in Formula (26). Here, we present data using the latter formula, a detailed comparison of the three-body effective potentials, $W^{(3),\text{full}}$ and $W^{(3)}$, using Formulas (23), and (26) will be given elsewhere [59].

Similarly to the two-body potential, the corresponding calculations can be performed by constrained molecular dynamics (or any other method that performs canonical sampling). For this, one needs to calculate the derivative of the three-body potential with respect to some distance. However, as previously stated, deterministic MD simulations of a constrained system might easily get trapped in local energy minima, so we utilized stochastic dynamics for the three-body case. In addition, rare events (high energy, low probability configurations) induce noise to the data, despite long equilibration (burn-in) periods or stronger heat-bath coupling in the simulations. Although smoothing could in principle have been applied, it would wash-out important information needed upon derivation with respect to positions ($f = -\nabla_\mathbf{q} W^{(3)}$). Therefore, we choose here to present results from the "direct" geometric averaging approach. The total calculations are one order of magnitude more than the two-body ones (all possible orientations of the two molecules for one of the third one), so special care was given to spatial symmetries.

The new effective three-body potential, $W^{(3),\text{full}}$, is naturally a function of three intermolecular distances: r_{12}, r_{13}, r_{23}. The discretization of the COMs in space is on top of the angular discretization mentioned in Section 4.2.2 and relates to the above three distances. The investigation of $W^{(3),\text{full}}$ for all possible distances is beyond the scope of this article. Here, we only study some characteristic cases, showing $W^{(3),\text{full}}$ data as a function of distance r_{23} for fixed r_{12} and r_{13}, comparing always with the sum of the corresponding two-body terms. In more detail, in Figure 10a–d, we present simulations based on the effective three body potential $W^{(3),\text{full}}$ and the sum $\sum W^{(2),\text{full}}$ (geometric averaging) for CH_4 at $T = 80$ K for different COM distances (): (a) $r_{12} = 3.9$, $r_{13} = 3.9$; (b) $r_{12} = 4.0$ $r_{13} = 4.0$; (c) $r_{12} = 4.3$,

$r_{13} = 4.0$; (d) $r_{12} = 3.8$, $r_{13} = 5.64$. At smaller distances, the potential of the triplet deviates from the sum of the three pairwise potentials, and this is where improvement in accuracy can be obtained. As shown in Figure 10, improvement is needed for close distances around the (three-dimensional) well. We used a three-dimensional cubic polynomial to fit the potential data (conjugate gradient method), which means that 20 constants should be determined. A lower order polynomial cannot capture the curvature of the forces upon differentiation. The benefit of this fitting methodology (over partial derivatives for instance) is the analytical solution of the forces with respect to any of r_{12}, r_{13}, r_{23} in contrast to tabulated data that induce some small error.

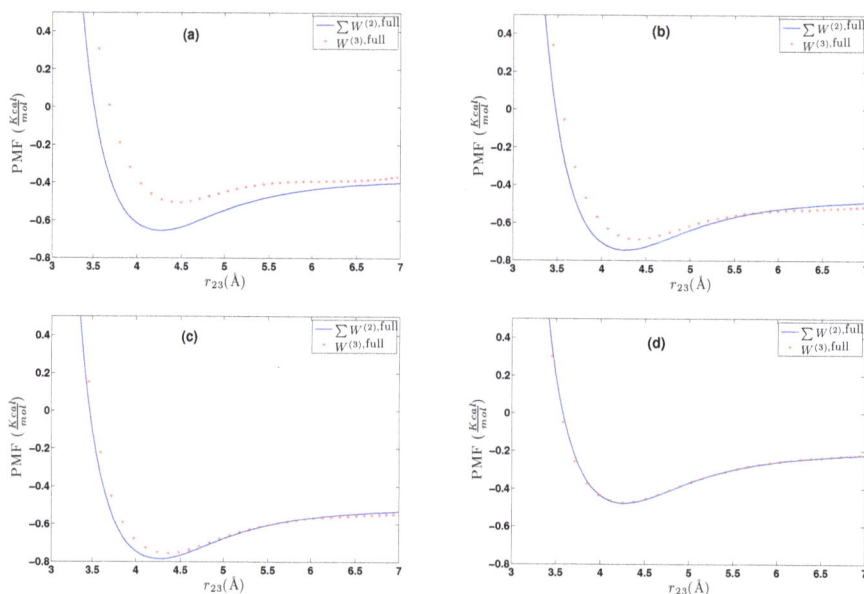

Figure 10. Effective potential comparison between the $W^{(3),full}$ three-body and $\sum W^{(2),full}$ simulations (geometric averaging) for CH_4 at $T = 80$ K for different fixed centre of mass (COM) distances. (a) $r_{12} = 3.9, r_{13} = 3.9$; (b) $r_{12} = 4.0, r_{13} = 4.0$; (c) $r_{12} = 4.3, r_{13} = 4.0$; (d) $r_{12} = 3.8, r_{13} = 5.64$.

Overall, there are clear differences between the three-body PMF, $W^{(3),full}$, and the sum of three two-body interactions, $\sum W^{(2),full}$, at short r_{12}, r_{13} and r_{23} distances. On the contrary, for larger distances, the sum of two-body interactions seems to represent the full three-body PMF very accurately. This is a clear indication of the rather short range of the three-body terms. Based on the above data, the range of the three-body terms for this system (methane at $T = 80$ K) is: $r_{12} \in [3.8 : 4.1], r_{13} \in [3.8 : 4.1]$ and $r_{23} \in [3.8 : 5]$; hence, the maximum distance for which three-body terms were considered is $r_{cut-off,3} = 5$. In practice, we need to identify all possible triplets within $r_{cut-off,3}$. Naturally, by including higher-order terms, the computational cost has increased, as well. More information about the numerical implementation of the three-body CG effective potential and its computational efficiency will be given elsewhere [59]. We should state here that in order to keep the temperature constant (in the BBK algorithm), due to the extra three-body terms in the CG force field, a larger coupling constant value for the heat bath was required.

6.2. CG Runs with the Effective Three-Body Potential

Next, we examine the effect of the three-body term on the CG model by performing bulk CG stochastic dynamics simulations using the new CG model with the three-body terms described above.

In this case, we incorporate the two-body CG effective potential described before for distances larger than $r_{\text{cut-off,3}}$, whereas we use $W^{(3),\text{full}}$ for triplets with all distances (r_{12}, r_{13} and r_{23}) below $r_{\text{cut-off,3}}$. In practice, we compute all possible pair interactions, and for the triplets with distances in the above defined range, we "correct" by adding the difference between $W^{(3),\text{full}}$ and the corresponding sum of $W^{(2),\text{full}}$; i.e., the difference between the datasets shown in Figure 10a–d.

Results on the pair distribution function, $g(r)$, for bulk (liquid) methane at $T = 80$ K are shown in Figure 11. In this graph, data from the atomistic MD runs (projected on the CG description), the CG model involving only pair CC potentials and the new CG model that also involves three-body terms are shown. First, it is clear that $g(r)$ data derived from the CG model that involves only pair CG interactions deviate, compared to the reference all-atom data; Second, the incorporation of the three-body terms in the effective CG potential slightly improves the prediction of the $g(r)$, mainly in the first maximum regime.

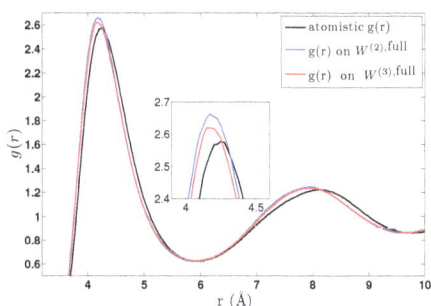

Figure 11. RDF from atomistic and CG using pair, $W^{(2),\text{full}}$, and three-body, $W^{(3),\text{full}}$, potential for CH_4 ($T = 80$ K). The three-dimensional cubic polynomial was used for the fitting.

7. Discussion and Conclusions

In recent years, we have experienced an enormous increase of computational power due to both hardware improvements and clever CPU-architecture. However, atomistic simulations of large complex molecular systems are still out of reach in particular when long computational times are desirable. A generic strategy in order to improve the efficiency of the computational methods is to reduce the dimensionality (degrees of freedom) by considering systematic coarse-grained models. There have been many suggestions on how to compute the relevant CG effective interactions in such models; a main issue here is that even if at the microscopic (atomistic) level there are only pair interactions, after coarse-graining, a multi-body effective potential (many-body PMF) is derived, which for realistic molecular complex systems cannot be calculated. Therefore, a common trend has been to approximate them by an "effective" pair potential by comparing the pair correlation function $g(r)$. This seems reasonable since given the correlation function, one can solve the "inverse problem" [68] and find an interaction to which it corresponds. However, this is an uncontrolled approximation without thermodynamic consistency.

Instead, here, we suggest to explicitly compute the constrained configuration integral over all atomistic configurations that correspond to a given coarse-grained state and from that suggest approximations with a quantifiable error. This is similar to the virial expansion where one needs to integrate over all positions of particles that correspond to a fixed density, and it is based on the recent development of establishing the cluster expansion in the canonical ensemble [51]; see also [45,61] for the corresponding (in the canonical ensemble) expansions for the correlation functions and the Ornstein–Zernike equation. See also the references therein for a more comprehensive list of the literature on the diagrammatic expansions. The main drawback that limits the applicability of these expansions is that they are rigorously valid only in the gas phase. To extend them to the liquid state is an outstanding problem, and even several successful closures like the Percus–Yevick are not rigorously justified. Therefore, there is the need for further developing these methods and relating them to computational strategies. Hence, in this paper, working in the rigorous framework of the cluster expansion, we seek quantifiable coarse-grained approximations, which we compute using molecular simulations. As a first test, we presented a detailed investigation of the proposed methodology to derive CG potentials for methane and ethane molecular systems. Each CG variable corresponds to the centre of mass for each molecule. Below, we summarize our main findings:

(a) The hierarchy of the cluster expansion formalism allowed us to systematically define the CG effective interaction as a sum of pair, triplets, etc., interactions. Then, CG effective potentials can be computed as they arise from the cluster expansion. Note, that for this estimation, no information from long simulations of n-body (bulk) systems is required.

(b) The two-body coarse-grained potentials can be efficiently computed via the cluster expansion giving comparable results with the existing methods, such as the conditional reversible work. In addition, we present a more efficient direct geometric computation of the constrained partition function, which also alleviates sampling noise issues. No basis function is needed in any of these methods.

(c) The obtained pair CG potentials were used to model methane and ethane systems in various regimes. The derived $g(r)$ data were compared against the all-atom ones. Clear differences between methane and ethane systems were observed; for the (almost spherical) methane, pair CG potentials seem to be a very good approximation, whereas much larger differences between CG and atomistic distribution functions were observed for ethane.

(d) We further investigated different temperature and density regimes and in particular cases where the two-body approximations are not good enough compared to the atomistic simulations. In the latter case, we considered the next term in the cluster expansion, namely the three-body effective potentials, and we found that they give a small improvement over the pair ones in the liquid state.

Finally, note that the proposed approach becomes computationally efficient, since the hierarchy of the effective non-boned CG potentials terms is developed by the conditional sampling of a few CG particles, and no information from long n-body (bulk) simulations is required. However, for dense systems and in the liquid regime, numerical estimation of higher, than the three-body, order terms is required; this is a major computational challenge.

Overall, we conjecture that the cluster expansion formalism can be used in order to provide accurate effective pair and three-body CG potentials at high T and low ρ regimes. In order to get significantly better results in the liquid regime, one needs to consider even higher order terms, which are in general more expensive to compute and more difficult to treat. Furthermore, expansions in the correlation functions can also be considered, as was already suggested in one of the original works of Hiroike and Morita [48]. Finally, another future goal is to extend this investigation to larger molecules (e.g., polymeric chains) that involve intra-molecular CG effective interactions, as well, and to systems with long range (e.g., Coulombic) interactions.

Supplementary Materials: The following are available online at www.mdpi.com/1099-4300/19/8/395/s1: More details about (a) the model; (b) the two-body and (c) the three-body simulation algorithms.

Acknowledgments: We acknowledge support by the European Union (ESF) and Greek national funds through the Operational Program "Education and Lifelong Learning" of the NSRF-Research Funding Programs: THALES and ARISTEIA II.

Author Contributions: Dimitrios Tsagkarogiannis and Vagelis Harmandaris conceived and designed the research work; Anastasios Tsourtis performed the simulations; All authors contributed equally in writing the paper.

Conflicts of Interest: The authors declare no conflict of interest.

References

1. Frenkel, D.; Smit, B. *Understanding Molecular Simulation: From Algorithms to Applications*, 2nd ed.; Academic Press: New York, NY, USA, 2002.

2. Allen, M.P.; Tildesley, D.J. *Computer Simulation of Liquids*; Oxford University Press: Oxford, UK, 1987.

3. Harmandaris, V.A.; Mavrantzas, V.G.; Theodorou, D.; Kröger, M.; Ramírez, J.; Öttinger, H.C.; Vlassopoulos, D. Dynamic crossover from Rouse to entangled polymer melt regime: Signals from long, detailed atomistic molecular dynamics simulations, supported by rheological experiments. *Macromolecules* **2003**, *36*, 1376–1387.

4. Kotelyanskii, M.; Theodorou, D.N. *Simulation Methods for Polymers*; Taylor & Francis: Abingdon, UK, 2004.

5. Izvekov, S.; Voth, G.A. Multiscale coarse-graining of liquid-state systems. *J. Chem. Phys.* **2005**, *123*, 134105.

6. Tschöp, W.; Kremer, K.; Hahn, O.; Batoulis, J.; Bürger, T. Simulation of polymer melts. I. Coarse-graining procedure for polycarbonates. *Acta Polym.* **1998**, *49*, 61–74.

7. Müller-Plathe, F. Coarse-Graining in Polymer Simulation: From the Atomistic to the Mesoscopic Scale and Back. *ChemPhysChem* **2002**, *3*, 754–769.

8. Shell, M.S. The relative entropy is fundamental to multiscale and inverse thermodynamic problems. *J. Chem. Phys.* **2008**, *129*, 144108.

9. Briels, W.J.; Akkermans, R.L.C. Coarse-grained interactions in polymer melts: A variational approach. *J. Chem. Phys.* **2001**, *115*, 6210.

10. Harmandaris, V.A.; Adhikari, N.P.; van der Vegt, N.F.A.; Kremer, K. Hierarchical Modeling of Polystyrene: From Atomistic to Coarse-Grained Simulations. *Macromolecules* **2006**, *39*, 6708.

11. Harmandaris, V.A.; Kremer, K. Dynamics of Polystyrene Melts through Hierarchical Multiscale Simulations. *Macromolecules* **2009**, *42*, 791.

12. Harmandaris, V.A.; Kremer, K. Predicting polymer dynamics at multiple length and time scales. *Soft Matter* **2009**, *5*, 3920.

13. Johnston, K.; Harmandaris, V. Hierarchical simulations of hybrid polymer–solid materials. *Soft Matter* **2013**, *9*, 6696–6710.

14. Noid, W.G.; Chu, J.W.; Ayton, G.S.; Krishna, V.; Izvekov, S.; Voth, G.A.; Das, A.; Andersen, H.C. The multiscale coarse-graining method. I. A rigorous bridge between atomistic and coarse-grained models. *J. Chem. Phys.* **2008**, *128*, 4114.

15. Lu, L.; Izvekov, S.; Das, A.; Andersen, H.C.; Voth, G.A. Efficient, Regularized, and Scalable Algorithms for Multiscale Coarse-Graining. *J. Chem. Theor. Comput.* **2010**, *6*, 954–965.

16. Rudzinski, J.F.; Noid, W.G. Coarse-graining entropy, forces, and structures. *J. Chem. Phys.* **2011**, *135*, 214101.

17. Noid, W.G. Perspective: Coarse-grained models for biomolecular systems. *J. Chem. Phys.* **2013**, *139*, 090901.

18. Chaimovich, A.; Shell, M.S. Anomalous waterlike behaviour in spherically-symmetric water models optimized with the relative entropy. *Phys. Chem. Chem. Phys.* **2009**, *11*, 1901–1915.

19. Bilionis, I.; Zabaras, N. A stochastic optimization approach to coarse-graining using a relative-entropy framework. *J. Chem. Phys.* **2013**, *138*, 044313.

20. Coifman, R.R.; Kevrekidis, I.G.; Lafon, S.; Maggioni, M.; Nadler, B. Diffusion Maps, Reduction Coordinates, and Low Dimensional Representation of Stochastic Systems. *Multiscale Model. Simul.* **2008**, *7*, 842–864.

21. Soper, A.K. Empirical potential Monte Carlo simulation of fluid structure. *Chem. Phys.* **1996**, *202*, 295–306.

22. Lyubartsev, A.P.; Laaksonen, A. On the Reduction of Molecular Degrees of Freedom in Computer Simulations. In *Novel Methods in Soft Matter Simulations*; Karttunen, M., Lukkarinen, A., Vattulainen, I., Eds.; Springer: Berlin, Germany, 2004; Volume 640, pp. 219–244.

23. Harmandaris, V.A. Quantitative study of equilibrium and non-equilibrium polymer dynamics through systematic hierarchical coarse-graining simulations. *Korea Aust. Rheol. J.* **2014**, *26*, 15–28.

24. Espanol, P.; Zuniga, I. Obtaining fully dynamic coarse-grained models from MD. *Phys. Chem. Chem. Phys.* **2011**, *13*, 10538–10545.

25. Padding, J.T.; Briels, W.J. Uncrossability constraints in mesoscopic polymer melt simulations: Non-Rouse behaviour of $C_{120}H_{242}$. *J. Chem. Phys.* **2001**, *115*, 2846–2859.

26. Deichmann, G.; Marcon, V.; van der Vegt, N.F.A. Bottom-up derivation of conservative and dissipative interactions for coarse-grained molecular liquids with the conditional reversible work method. *J. Chem. Phys.* **2014**, *141*, 224109.

27. Fritz, D.; Harmandaris, V.; Kremer, K.; van der Vegt, N. Coarse-grained polymer melts based on isolated atomistic chains: Simulation of polystyrene of different tacticities. *Macromolecules* **2009**, *42*, 7579–7588.
28. Brini, E.; Algaer, E.A.; Ganguly, P.; Li, C.; Rodriguez-Ropero, F.; van der Vegt, N.F.A. Systematic Coarse-Graining Methods for Soft Matter Simulations—A Review. *Soft Matter* **2013**, *9*, 2108–2119.
29. Reith, D.; Pütz, M.; Müller-Plathe, F. Deriving effective mesoscale potentials from atomistic simulations. *J. Comput. Chem.* **2003**, *24*, 1624–1636.
30. Lyubartsev, A.P.; Laaksonen, A. Calculation of effective interaction potentials from radial distribution functions: A reverse Monte Carlo approach. *Phys. Rev. E* **1995**, *52*, 3730–3737.
31. Izvekov, S.; Voth, G.A. Effective force field for liquid hydrogen fluoride from ab initio molecular dynamics simulation using the force-matching method. *J. Phys. Chem. B* **2005**, *109*, 6573–6586.
32. Noid, W.G.; Liu, P.; Wang, Y.; Chu, J.; Ayton, G.S.; Izvekov, S.; Andersen, H.C.; Voth, G.A. The multiscale coarse-graining method. II. Numerical implementation for coarse-grained molecular models. *J. Chem. Phys.* **2008**, *128*, 244115.
33. Lu, L.; Dama, J.F.; Voth, G.A. Fitting coarse-grained distribution functions through an iterative force-matching method. *J. Chem. Phys.* **2013**, *139*, 121906.
34. Katsoulakis, M.A.; Plechac, P. Information-theoretic tools for parametrized coarse-graining of non-equilibrium extended systems. *J. Chem. Phys.* **2013**, *139*, 4852–4863.
35. Chaimovich, A.; Shell, M.S. Coarse-graining errors and numerical optimization using a relative entropy framework. *J. Chem. Phys.* **2011**, *134*, 094112.
36. Cho, H.M.; Chu, J.W. Inversion of radial distribution functions to pair forces by solving the Yvon–Born–Green equation iteratively. *J. Chem. Phys.* **2009**, *131*, 134107.
37. Noid, W.G.; Chu, J.; Ayton, G.S.; Voth, G.A. Multiscale coarse-graining and structural correlations: Connections to liquid-state theory. *J. Phys. Chem. B* **2007**, *111*, 4116–4127.
38. Mullinax, J.W.; Noid, W.G. Generalized Yvon–Born–Green theory for molecular systems. *Phys. Rev. Lett.* **2009**, *103*, 198104.
39. Mullinax, J.W.; Noid, W.G. Generalized Yvon–Born–Green theory for determining coarse-grained interaction potentials. *J. Phys. Chem. C* **2010**, *114*, 5661–5674.
40. McCarty, J.; Clark, A.J.; Copperman, J.; Guenza, M.G. An analytical coarse-graining method which preserves the free energy, structural correlations, and thermodynamic state of polymer melts from the atomistic to the mesoscale. *J. Chem. Phys.* **2014**, *140*, 204913.
41. Li, Z.; Bian, X.; Li, X.; Karniadakis, G.E. Incorporation of memory effects in coarse-grained modeling via the Mori-Zwanzig formalism. *J. Chem. Phys.* **2015**, *143*, 243128.
42. McCarty, J.; Clark, A.J.; Lyubimov, I.Y.; Guenza, M.G. Thermodynamic Consistency between Analytic Integral Equation Theory and Coarse-Grained Molecular Dynamics Simulations of Homopolymer Melts. *Macromolecules* **2012**, *45*, 8482–8493.
43. Clark, A.J.; McCarty, J.; Guenza, M.G. Effective potentials for representing polymers in melts as chains of interacting soft particles. *J. Chem. Phys.* **2013**, *139*, 124906.
44. Stell, G. The Percus-Yevick equation for the radial distribution function of a fluid. *Physica* **1963**, *29*, 517–534.
45. Kuna, T.; Tsagkarogiannis, D. Convergence of density expansions of correlation functions and the Ornstein-Zernike equation. *arXiv* **2016**, arXiv:1611.01716.
46. Bolhuis, P.G.; Louis, A.A.; Hansen, J.P. Many-body interactions and correlations in coarse-grained descriptions of polymer solutions. *Phys. Rev. E* **2001**, *64*, 021801.
47. Mayer, J.E.; Mayer, M.G. *Statistical Mechanics*; John Wiley & Sons: Hoboken, NJ, USA, 1940.
48. Morita, T.; Hiroike, K. The statistical mechanics of condensing systems. III. *Prog. Theor. Phys.* **1961**, *25*, 537.
49. Frisch, H.; Lebowitz, J. *Equilibrium Theory of Classical Fluids*; W.A. Benjamin: New York, NY, USA, 1964.
50. Hansen, J.P.; McDonald, I.R. *Theory of Simple Lipquids*; Academic Press: London, UK, 1986.
51. Pulvirenti, E.; Tsagkarogiannis, D. Cluster expansion in the canonical ensemble. *Commun. Math. Phys.* **2012**, *316*, 289–306.
52. Louis, A.A. Beware of density dependent pair potentials. *J. Phys. Condens. Matter* **2002**, *14*, 9187–9206.
53. Katsoulakis, M.; Plecháč, P.; Rey-Bellet, L.; Tsagkarogiannis, D. Coarse-graining schemes and a posteriori error estimates for stochastic lattice systems. *ESAIM Math. Model. Numer. Anal.* **2007**, *41*, 627–660.
54. Katsoulakis, M.; Plecháč, P.; Rey-Bellet, L.; Tsagkarogiannis, D. Coarse-graining schemes for stochastic lattice systems with short and long range interactions. *Math. Comput.* **2014**, *83*, 1757–1793.

55. Katsoulakis, M.; Plecháč, P.; Rey-Bellet, L.; Tsagkarogiannis, D. Mathematical strategies and error quantification in coarse-graining of extended systems. *J. Non Newton. Fluid Mech.* **2008**, *152*, 101–112.

56. Trashorras, J.; Tsagkarogiannis, D. Reconstruction schemes for coarse-grained stochastic lattice systems. *SIAM J. Numer. Anal.* **2010**, *48*, 1647–1677.

57. Kremer, K.; Müller-Plathe, F. Multiscale problems in polymer science: Simulation approaches. *MRS Bull.* **2001**, *26*, 205–210.

58. McCoy, J.D.; Curro, J.G. Mapping of Explicit Atom onto United Atom Potentials. *Macromolecules* **1998**, *31*, 9352–9368.

59. Tsourtis, A.; Harmandaris, V.; Tsagkarogiannis, D. Effective coarse-grained interactions: The role of three-body terms through cluster expansions, under preparation.

60. Kirkwood, J.G. Statistical Mechanics of Fluid Mixtures. *J. Chem. Phys.* **1935**, *3*, 300–313.

61. McQuarrie, D.A. *Statistical Mechanics*; University Science Books: Sausalito, CA, USA, 2000.

62. Kalligiannaki, E.; Chazirakis, A.; Tsourtis, A.; Katsoulakis, M.; Plecháč, P.; Harmandaris, V. Parametrizing coarse grained models for molecular systems at equilibrium. *Eur. Phys. J.* **2016**, *225*, 1347–1372.

63. Henderson, R. A uniqueness theorem for fluid pair correlation functions. *Phys. Lett. A* **1974**, *49*, 197–198.

64. Larini, L.; Lu, L.; Voth, G.A. The multiscale coarse-graining method. VI. Implementation of three-body coarse-grained potentials. *J. Chem. Phys.* **2010**, *132*, 164107.

65. Das, A.; Andersen, H.C. The multiscale coarse-graining method. IX. A general method for construction of three body coarse-grained force fields. *J. Chem. Phys.* **2012**, *136*, 194114.

66. Lelièvre, T.; Rousset, M.; Stoltz, G. *Free Energy Computations: A Mathematical Perspective*; Imperial College Press: London, UK, 2010.

67. Tsourtis, A.; Pantazis,Y.; Katsoulakis, M.A.; Harmandaris, V. Parametric sensitivity analysis for stochastic molecular systems using information theoretic metrics. *J. Chem. Phys.* **2015**, *143*, 014116.

68. Kuna, T.; Lebowitz, J.; Speer, E. Realizability of point processes. *J. Stat. Phys.* **2007**, *129*, 417–439.

entropy

MDPI

Review

Fluctuations, Finite-Size Effects and the Thermodynamic Limit in Computer Simulations: Revisiting the Spatial Block Analysis Method

Maziar Heidari [1], Kurt Kremer [1], Raffaello Potestio [1,2,3] and Robinson Cortes-Huerto [1,*]

[1]   Max Planck Institute for Polymer Research, Ackermannweg 10, 55128 Mainz, Germany;
     heidari@mpip-mainz.mpg.de (M.H.); kremer@mpip-mainz.mpg.de (K.K.); raffaello.potestio@unitn.it (R.P.)
[2]   Physics Department, University of Trento, via Sommarive 14, I-38123 Trento, Italy
[3]   INFN-TIFPA, Trento Institute for Fundamental Physics and Applications, I-38123 Trento, Italy
*   Correspondence: corteshu@mpip-mainz.mpg.de; Tel.: +49-6131-379-148

Received: 1 March 2018; Accepted: 22 March 2018; Published: 24 March 2018

Abstract: The spatial block analysis (SBA) method has been introduced to efficiently extrapolate thermodynamic quantities from finite-size computer simulations of a large variety of physical systems. In the particular case of simple liquids and liquid mixtures, by subdividing the simulation box into blocks of increasing size and calculating volume-dependent fluctuations of the number of particles, it is possible to extrapolate the bulk isothermal compressibility and Kirkwood–Buff integrals in the thermodynamic limit. Only by explicitly including finite-size effects, ubiquitous in computer simulations, into the SBA method, the extrapolation to the thermodynamic limit can be achieved. In this review, we discuss two of these finite-size effects in the context of the SBA method due to (i) the statistical ensemble and (ii) the finite integration domains used in computer simulations. To illustrate the method, we consider prototypical liquids and liquid mixtures described by truncated and shifted Lennard–Jones (TSLJ) potentials. Furthermore, we show some of the most recent developments of the SBA method, in particular its use to calculate chemical potentials of liquids in a wide range of density/concentration conditions.

Keywords: computer simulations; finite-size effects; calculation of free energies; thermodynamic limit

---

1. Introduction

In the last few decades, computational studies of soft matter have gained ground in the no-man's land between purely theoretical studies and experimental investigations. Arguably, this success is due to the use of statistical mechanics relations between macroscopic thermodynamic properties and microscopic components and interactions of a physical system in the thermodynamic limit (TL) [1,2]. However, and apart from a few examples [3–5], computer simulations are mainly constrained to consider closed systems with a finite and usually small number of particles N_0. These limitations introduce spurious finite-size effects, apparent in the simulation results, that in spite of the current computing capabilities are still the subject of intense investigations [6–13].

A meaningful comparison between computer simulations of finite systems and experimental results has been always a difficult task. In principle, it is possible to extrapolate the simulation data to the quantities of interest in the thermodynamic limit by considering systems of increasing size and performing simulations for each of them. The SBA method has been proposed as a more efficient alternative where only one system is examined and then subdivided into blocks of different size from which the data are extracted. The method is rather general since it was originally proposed to study the critical behavior of Ising systems [14,15] and then extended to study liquids [16–21] and even the elastic constants of model solids [22].

In this paper, we examine the SBA method focusing on the extrapolation of bulk thermodynamic properties of simple liquids. We use prototypical liquids and mixtures described by truncated and shifted Lennard–Jones (TSLJ) potentials to discuss the original ideas [16,17] and explore the background [20,21,23–26] for the most recent developments [6,7,9] of the method. The simple examples presented here, in addition to the results available in the literature [6], suggest that the method is suitable for the calculation of trends in the chemical potential of complex liquids in a wide range of density/concentration conditions.

The paper is organized as follows: In Section 2, we introduce the relevant finite-size effects present in standard computer simulations. In Section 3, we introduce the finite-size integral equations for liquids and illustrate the procedure to extrapolate thermodynamic quantities. In Section 4, we discuss the extension of the block analysis method to liquid mixtures. We conclude the paper in Section 5.

2. Boundary and Ensemble Finite-Size Effects

Statistical mechanics establishes the connection between macroscopic thermodynamic properties and the microscopic components and interactions of a physical system. An interesting example of this relation is provided by the compressibility equation that identifies the density fluctuations of a system in the grand canonical ensemble with the bulk isothermal compressibility κ_T [27]. In the thermodynamic limit (TL), the isothermal compressibility of a homogeneous system is related to the fluctuations of the number of particles via the expression [1]:

$$\chi_T^\infty = \frac{\langle N^2 \rangle - \langle N \rangle^2}{\langle N \rangle},$$

(1)

with $\langle N \rangle$ the average number of particles contained in a volume V of the fluid. The reduced isothermal compressibility $\chi_T^\infty = \rho k_B T \kappa_T$ is the ratio between the bulk isothermal compressibility of the system, κ_T, and the isothermal compressibility of the ideal gas $(\rho k_B T)^{-1}$ with $\rho = \langle N \rangle / V$.

Various finite-size effects can be included in the block analysis aiming at extrapolating interesting thermodynamic quantities. In practice, let us consider a system of N_0 particles where the simulation box of volume $V_0 = L_0^3$ is divided into subdomains of volume $V = L^3$, as illustrated in Figure 1. By evaluating the fluctuations of the number of particles in these subdomains, it is possible to obtain the distribution $P_{L,L_0}(N)$ of the number of particles, with k-moments given by [25]:

$$\langle N^k \rangle_{L,L_0} = \sum_{N=0}^{N_0} N^k P_{L,L_0}(N).$$

(2)

The second moment of the distribution is related to the reduced isothermal compressibility of the finite system $\chi_T(L, L_0)$ [14,16,17,25]:

$$\chi_T(L, L_0) = \frac{\langle N^2 \rangle_{L,L_0} - \langle N \rangle_{L,L_0}^2}{\langle N \rangle_{L,L_0}}.$$

(3)

The finite-size reduced isothermal compressibility, $\chi_T(L, L_0)$, can be extrapolated to the reduced isothermal compressibility in the TL, χ_T^∞, taking the limits $L, L_0 \to \infty$. Originally [17,25], by applying periodic boundary conditions (PBCs) to the total linear size L_0 and taking into account volumes such that $L \gg \zeta$ with ζ the correlation length of the system, it has been proposed that the difference between $\chi_T(L, L_0)$ and χ_T^∞ is related to boundary effects associated with the finite-size of the subdomains. This difference takes the form [16,17]:

$$\chi_T(L, L_0 \to \infty) = \chi_T^\infty + \frac{c}{L} + O\left(\frac{1}{L^2}\right),$$

(4)

with c a constant. Recently, Equation (4) has been obtained [28] using arguments based on the thermodynamics of small systems [29,30], underpinning the consistency of the result.

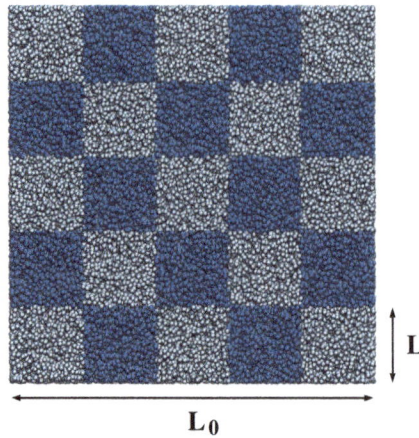

Figure 1. Snapshot of the simulation box for a system of particles interacting via a TSLJ potential at density $\rho\sigma^3 = 0.1$ and temperature $k_B T = 1.2\epsilon$. In this particular example, a box of linear size L_0 has been subdivided into blocks of linear dimension $L = L_0/5$ as indicated by the different color shades. The figure has been rendered with the Visual Molecular Dynamics (VMD) program [31].

To investigate this expression, we consider a liquid system whose potential energy is described by a 12–6 Lennard–Jones potential truncated, with cutoff radius $r_c/\sigma = 2^{1/6}$, and shifted. The parameters ϵ, σ and m, define the units of energy, length and mass, respectively. All the results are expressed in LJ units with time $\sigma(m/\epsilon)^{1/2}$, temperature ϵ/k_B and pressure ϵ/σ^3. Various system sizes, namely $N_0 = 10^4, 10^5$ and 10^6, are considered, and the density is fixed at $\rho\sigma^3 = 0.864$, thus defining the linear size of the simulation box L_0. The systems are equilibrated at $k_B T = 1.2\epsilon$, enforced with a Langevin thermostat with damping coefficient $\gamma(\sigma(m/\epsilon)^{1/2}) = 1.0$, for 2×10^6 molecular dynamics (MD) steps using a time step of $\delta t/(\sigma(m/\epsilon)^{1/2}) = 10^{-3}$. Production runs span 10^6 MD steps. All the simulations have been performed with the ESPResSo++ [32] simulation package.

To use the block analysis method, we compute the fluctuations of the number of particles. In particular, we choose domains of size $1 < L/\sigma < L_0/\sigma$ to scan continuously the fluctuations as a function of domain size. To increase the amount of statistics, we use 100 randomly-positioned subdomains per simulation frame.

In Figure 2, we report $\chi_T(L, L_0)$ as a function of σ/L. The linear behavior predicted in Equation (4) is apparent for $L \ll L_0$. There are evident deviations from the linear behavior, which are not included in Equation (4), since this equation has been obtained for a system in the grand canonical ensemble. As a matter of fact, the deviations from linearity are mainly related to the fixed size of the system because when $L \to L_0$, $\chi_T(L_0, L_0) = 0$, that is, the fluctuations of the number of particles for a closed system are equal to zero. In principle, the isothermal compressibility in the TL can be extracted by extrapolating a line to the y-axis, i.e., $\sigma/L \to 0$, and determining the y-intercept. This procedure, however, might lead to ambiguous and strongly-size-dependent results as suggested by the same plot.

From the previous discussion, Equation (4) satisfactorily describes the boundary size effects present in a system described in the grand canonical ensemble. However, ensemble size effects, i.e., the fact that we are computing quantities defined in the grand canonical ensemble using information obtained from a system in a canonical ensemble, are important even in cases where the size of the system might appear to be enormous ($L_0/\sigma = 105$ for $N_0 = 10^6$ where $\zeta/\sigma \approx 10$).

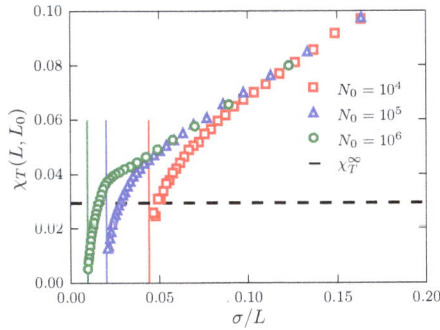

Figure 2. Fluctuations of the number of particles $\chi_T(L, L_0)$ as a function of σ/L for systems described by a TSLJ potential with $r_c/\sigma = 2^{1/6}$. Data corresponding to system sizes $N_0 = 10^4, 10^5$ and 10^6 are presented using red squares, blue triangles and green circles, respectively. The vertical lines indicate the limit σ/L_0 at which fluctuations become zero. The black horizontal dashed line indicates the value $\chi_T^\infty = \rho k_B T \kappa_T = 0.0295$ with κ_T the bulk compressibility obtained with the method described in [6].

It is thus clear that the isothermal compressibility of a finite-size system in the TL, i.e., $L, L_0 \to \infty$ with $\rho = N_0/L_0^3$, should equate to the bulk isothermal compressibility κ_T. An elegant analysis using probabilistic arguments for the ideal gas case [26,33] shows that the finite-size reduced isothermal compressibility can be written as:

$$\chi_T(L, L_0) = \chi_T^\infty \left(1 - \left(\frac{L}{L_0}\right)^3\right). \tag{5}$$

In spite of the simplicity of the system chosen in this study, it cannot be identified with the ideal gas. However, at very low densities and temperature $k_B T = 1.2\epsilon$, the system behaves more like a real gas, and a meaningful trend could be identified. Therefore, to investigate Equation (5), we consider the density range $\rho\sigma^3 = 0.1, \cdots, 1.0$ for systems of size $N_0 = 10^5$ particles. Results are presented in Figure 3 for the cases $\rho\sigma^3 = 0.1, 0.2$ and 0.3. The three datasets follow the theoretical prediction in Equation (5) with deviations from this behavior for $L \ll L_0$, thus indicating the signature of boundary finite-size effects. As expected, the data presented also suggest that upon increasing density, the deviations from the ideal gas behavior become more evident, as can be seen in the case $\rho\sigma^3 = 0.3$.

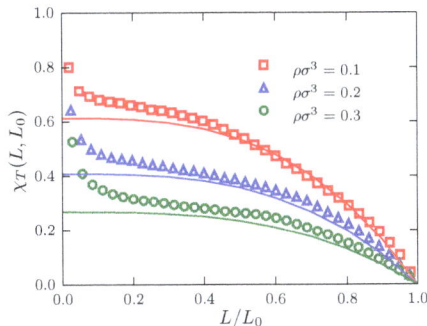

Figure 3. Fluctuations of the number of particles $\chi_T(L, L_0)$ as a function of the ratio L/L_0 for systems described by a TSLJ potential with $r_c/\sigma = 2^{1/6}$. Results corresponding to systems of $N_0 = 10^5$ particles with densities $\rho\sigma^3 = 0.1, 0.2$ and 0.3 are presented using red squares, blue triangles and green circles, respectively. The theoretical prediction presented in the text is plotted using the corresponding value for χ_T^∞, obtained as described in [6], and solid-line curves with the same color code.

This is also seen in Figure 4, where for a system with density $\rho\sigma^3 = 0.864$, the deviations from the ideal gas case are much more evident. As a matter of fact, even for the largest size considered ($N_0 = 10^6$), it is not possible to convincingly reproduce the ideal gas behavior.

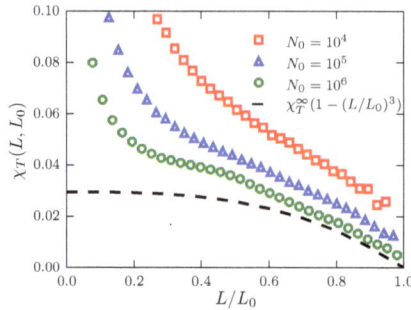

Figure 4. Fluctuations of the number of particles $\chi_T(L, L_0)$ as a function of the ratio L/L_0 for systems described by a TSLJ potential with $r_c/\sigma = 2^{1/6}$. Results corresponding to sizes $N_0 = 10^4, 10^5$ and 10^6, with density $\rho\sigma^3 = 0.864$, using red squares, blue triangles and green circles, respectively. The theoretical prediction presented in the text is plotted as the black dashed curve using $\chi_T^\infty = 0.0295$.

Nonetheless, one intuitively could imagine that the following expression:

$$\chi_T(L, L_0) = \chi_T^\infty \left(1 - \left(\frac{L}{L_0}\right)^3\right) + \frac{c}{L} + O\left(\frac{1}{L^2}\right), \tag{6}$$

captures the two finite-size effects, ensemble and boundary [25]. By neglecting the $O(1/L^2)$ terms, defining $\lambda = L/L_0$ and multiplying everything by λ, we obtain:

$$\lambda\chi_T(\lambda) = \lambda\chi_T^\infty \left(1 - \lambda^3\right) + \frac{c}{L_0}. \tag{7}$$

Equation (7) is more convenient to analyze because in the limit $\lambda \to 0$, provided that $\zeta < L < L_0$, λ^3 is negligible, and this expression can be approximated to a linear function in λ with slope χ_T^∞ and y-intercept equal to c/L_0. In particular, we use a simple linear regression in the interval $0.0 < \lambda < 0.3$, with the fluctuations data for $N_0 = 10^5$, to find $\chi_T^\infty = 0.0295(5)$ and $c = 0.415(5)\sigma$. Results of the scaled fluctuations $\lambda\chi_T(\lambda)$ minus c/L_0 are presented in Figure 5, where the intensive character of the constant c becomes clear. By replacing the calculated values χ_T^∞ and c in Equation (7), we obtain the black curve that superimposes on the simulation data in the full range $0 < \lambda < 1$.

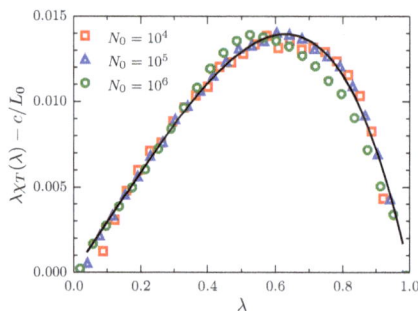

Figure 5. Scaled fluctuations of the number of particles $\lambda\chi_T(L, L_0)$, minus c/L_0, versus the ratio $\lambda = L/L_0$ for systems described by a TSLJ potential with $r_c/\sigma = 2^{1/6}$. Results corresponding to sizes $N_0 = 10^4, 10^5$ and 10^6, with density $\rho\sigma^3 = 0.864$, using red squares, blue triangles and green circles, respectively. The theoretical prediction Equation (7) presented in the text is plotted as the black solid curve using $\chi_T^\infty = 0.0295$ and $c = 0.415\sigma$.

In addition to the explicit finite-size effects discussed above, there is another type of effect related to the periodicity of the simulation box. This is the case of implicit finite-size effects that appear due to anisotropies in the pair correlation function of the system, generated by the use of PBCs [34,35]. These effects, extremely important for small simulation setups, appear as oscillations in $\lambda\chi_T(\lambda)$ for $\lambda \approx 1$ caused by short range interactions between the system and its nearest neighbor images. However, given the large sizes of the systems considered here, implicit finite-size effects can be safely ignored in the present discussion.

With the trajectories of the system with $N_0 = 10^5$ particles in the density interval $0.1 < \rho\sigma^3 < 1.0$, we compute the scaled fluctuations $\lambda\chi_T(\lambda)$ and determine, as before, the ratio $\chi_T^\infty = \kappa_T/\kappa_T^{IG}$ as a function of the density, with $\kappa_T^{IG} = (\rho k_B T)^{-1}$ the isothermal compressibility of the ideal gas (see Figure 6). As expected for this system at $k_B T = 1.2\epsilon$, a monotonically-decreasing behavior is observed since the system becomes less compressible as the density increases.

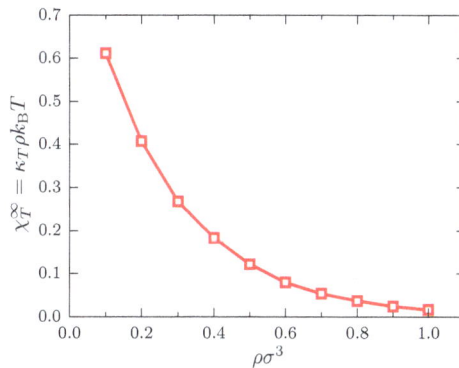

Figure 6. Ratio $\chi_T^\infty = \kappa_T/\kappa_T^{IG}$ at $k_B T = 1.2\epsilon$ as a function of the density for systems described by a TSLJ potential with $r_c/\sigma = 2^{1/6}$, with $\kappa_T^{IG} = (\rho k_B T)^{-1}$ the isothermal compressibility of the ideal gas. The red curve is a guide to the eye.

The isothermal compressibility as a function of the density allows one to investigate more interesting thermodynamic properties, as has been recently demonstrated [6,7]. For example, the isothermal compressibility can be written as:

$$\kappa_T = \frac{1}{\rho^2} \frac{\partial\rho}{\partial\mu}\bigg|_T, \tag{8}$$

which can be rearranged, in terms of the chemical potential μ, as:

$$\delta\mu = \int_{\rho_0}^{\rho} \frac{d\rho'}{\rho'^2 \kappa_T} \tag{9}$$

with $\delta\mu = \mu - \mu_0$ and μ_0 the chemical potential of the system at the reference density ρ_0. In practice, one usually is interested in the excess chemical potential (In this context, the word excess should be replaced with residual. The residual chemical potential is the difference between the chemical potential of the target system and that of an ideal gas at the same density, temperature and composition. We misuse the expression excess chemical potential to match the modern literature.):

$$\delta\mu^{ex} = \delta\mu - k_B T \ln\rho, \tag{10}$$

obtained by subtracting from $\delta\mu$ the density-dependent part of the chemical potential of the ideal gas.

To validate the results obtained using Equation (10), it is necessary to use a different computational method to evaluate μ_0. For that purpose, any computational method aiming at calculating chemical potentials could be used. In particular, we use the spatially-resolved thermodynamic integration (SPARTIAN) method [36], recently implemented by us. In SPARTIAN, the target system, described with atomistic resolution, is embedded in a reservoir of ideal gas particles. An interface between the two subdomains is defined such that molecules are free to diffuse, adapting their resolution on the fly. A uniform density across the simulation box is guaranteed by applying a single-molecule external potential that is identified with the difference in chemical potential between the two resolutions, i.e., the excess chemical potential of the target system. This method has been validated by calculating excess chemical potentials for Lennard–Jones liquids, mixtures, as well as for simple point-charge (SPC) and extended simple point-charge (SPC/E) water models and aqueous sodium chloride solutions, all in good agreement with state-of-the-art computational methods.

For the comparison, we consider the same system at the same temperature with densities $\rho\sigma^3 = 0.2$, 0.4, 0.6, 0.8 and 1.0. Results for the excess chemical potential as a function of the density are presented in Figure 7 where the value of $\rho_0\sigma = 0.6$ has been used as the reference value. Once $\delta\mu^{ex}$ is rescaled, it becomes clear that the agreement between the two methods is remarkable. This result suggests that the simple calculation of the fluctuations of the number of particles, used in combination with Equation (7), provides us with an efficient and accurate method to compute the chemical potential of simple liquids, which can be extended to more complex fluids [6].

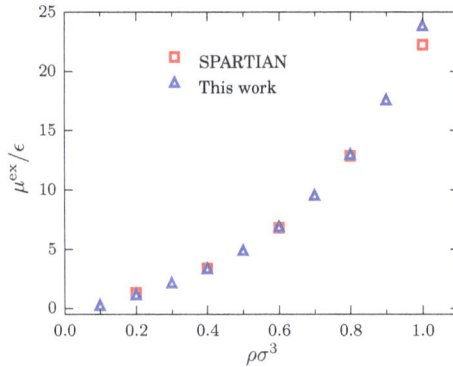

Figure 7. Excess chemical potential μ^{ex}/ϵ at $k_{\mathrm{B}}T = 1.2\epsilon$ as a function of the density for systems described by a TSLJ potential with $r_c/\sigma = 2^{1/6}$. Red squares indicate the data obtained with the spatially-resolved thermodynamic integration (SPARTIAN) method [36], and the blue triangles are the data points obtained with the method outlined in the text.

In this section, Equation (7) has been introduced in a rather intuitive manner. However, the presented results suggest that it encompasses the relevant finite-size effects of the system and allows one to compute bulk thermodynamic quantities. In the following section, we derive Equation (7) more rigorously and explore, using a different example, its range of validity.

3. Finite-Size Ornstein–Zernike Integral Equation

Fluctuations of the number of particles are related to the local structure of a liquid. Let us consider a molecular liquid of average density ρ at temperature T in equilibrium with a reservoir of particles, i.e., an open system. The fluctuations of the number of molecules are related to the local structure of the liquid via the Ornstein–Zernike integral equation [1,37]:

$$\frac{\Delta^2(N)}{\langle N \rangle} = 1 + \frac{\rho}{V} \int_V \int_V [g^\circ(\mathbf{r}_1, \mathbf{r}_2) - 1] \, d\mathbf{r}_1 \, d\mathbf{r}_2 , \tag{11}$$

where $\Delta^2(N)/\langle N \rangle$ are the fluctuations of the number of particles, $\Delta^2(N) = \langle N^2 \rangle - \langle N \rangle^2$ and $g^o(\mathbf{r}_1, \mathbf{r}_2)$ is the pair correlation function of the open system and $\mathbf{r}_1, \mathbf{r}_2$ the position vectors of a pair of fluid particles. To solve the integral in Equation (11), one assumes that the fluid is homogeneous, isotropic and that the system is in the thermodynamic limit (TL), i.e., $V \to \infty$, $\langle N \rangle \to \infty$ with $\rho = \langle N \rangle / V =$ constant. An infinite, homogeneous and isotropic system is translationally invariant; therefore, we rewrite Equation (11) as [1]:

$$\chi_T^\infty - \frac{\Delta^2(N)}{\langle N \rangle} = 1 + 4\pi\rho \int_0^\infty (g^o(r) - 1)\, r^2\, dr, \tag{12}$$

with $\chi_T^\infty = \rho k_B T \kappa_T$, κ_T being the isothermal compressibility of the bulk system. We have replaced $g^o(\mathbf{r}_1, \mathbf{r}_2)$ with $g^o(r)$ the radial distribution function (RDF) of the open system, with $r = |\mathbf{r}_2 - \mathbf{r}_1|$.

An alternative version of the OZ integral equation for finite systems has been introduced [25]. For a finite system with total volume V_0 with PBCs we have:

$$\chi_T(V, V_0) = \frac{\Delta^2(N; V, V_0)}{\langle N \rangle_{V, V_0}} = 1 + \frac{\rho}{V} \int_V \int_V [g^c(r_{12}) - 1]\, d\mathbf{r}_1\, d\mathbf{r}_2, \tag{13}$$

where $g^c(r_{12})$, $r_{12} = |\mathbf{r}_2 - \mathbf{r}_1|$, is the pair correlation function of the closed system with total number of particles N_0, and $\Delta^2(N; V, V_0) = \langle N^2 \rangle_{V, V_0} - \langle N \rangle^2_{V, V_0}$. The fluctuations of the number of particles thus depend on both subdomain and simulation box volumes.

For a single component fluid of density ρ at temperature T with fixed number of particles N_0 and volume V_0, its RDF can be written in terms of an expansion around N_0 as [23–26,33]:

$$g^c(r) = g^o(r) - \frac{\chi_T^\infty}{N_0}. \tag{14}$$

As a matter of fact, the expansion includes terms that depend on the partial derivative of $g^o(r)$ with respect to the density. However, we anticipate here that for the present analysis, their contribution is negligible [6]. By replacing $g^c(r)$ in the integral on the r.h.s of Equation (13), we obtain:

$$\frac{\rho}{V} \int_V \int_V (g^c(r_{12}) - 1)\, d\mathbf{r}_1\, d\mathbf{r}_2 = I_{V,V} - \frac{V}{V_0} \chi_T^\infty, \tag{15}$$

where:

$$I_{V,V} = \frac{\rho}{V} \int_V \int_V (g^o(r_{12}) - 1)\, d\mathbf{r}_1\, d\mathbf{r}_2, \tag{16}$$

and we use that $\rho = N_0 / V_0$.

Next, we include explicitly the second finite-size effect, i.e., the fact that the volume V is finite and embedded into a finite volume V_0 with PBCs. For this, we rewrite $I_{V,V}$ as [17]:

$$I_{V,V_0-V} = I_{V,V_0} - I_{V,V},$$

with:

$$I_{V,V_0} = \frac{\rho}{V} \int_V \int_{V_0} (g^o(r_{12}) - 1)\, d\mathbf{r}_1\, d\mathbf{r}_2$$

$$I_{V,V_0-V} = \frac{\rho}{V} \int_V \int_{V_0-V} (g^o(r_{12}) - 1)\, d\mathbf{r}_1\, d\mathbf{r}_2.$$

As pointed out by Rovere, Heermann and Binder [17], the two integrals $I_{V,V}$ and I_{V,V_0} are equal when $\mathbf{r}_1$ and $\mathbf{r}_2$ are both within the volume V. When $r_{12} > \zeta$, the integrand $(g^o(r_{12}) - 1) = 0$, and it does not contribute to the integrals. Close to the boundary of the subdomain V, for $r_{12} < \zeta$, and in particular when $\mathbf{r}_1$ lies inside and $\mathbf{r}_2$ outside the volume V, there are contributions missing in $I_{V,V}$, which are present in I_{V,V_0}. Therefore, the difference between the two integrals $I_{V,V_0-V} = I_{V,V_0} - I_{V,V}$ must be proportional to the surface volume ratio of the subdomain V [17], i.e.,

$$I_{V,V_0-V} = \frac{c_1}{L} + \left(\frac{c_2}{L}\right)^2 + O\left(\frac{1}{L^3}\right), \tag{17}$$

with c_1, c_2 proportionality constants with units of length that, at this point, we assume to be intensive.

To compute I_{V,V_0}, we require that $\zeta < L < L_0$. Since we assume PBCs, the system is translationally invariant. Hence, upon applying the transformation $\mathbf{r}_{12} \to \mathbf{r} = \mathbf{r}_2 - \mathbf{r}_1$, the expression:

$$I_{V,V_0} = \rho \int_{V_0} (g^o(r) - 1)\, d\mathbf{r} = \chi_T^\infty - 1 \tag{18}$$

is obtained, where we assume that $g^o(r > \zeta) = 1$, thus ignoring fluctuations of the RDF beyond the volume V. By combining these two results, we obtain:

$$I_{V,V} = \chi_T^\infty - 1 + \frac{c_1}{L} + \left(\frac{c_2}{L}\right)^2, \tag{19}$$

and by including this result in Equation (15), we arrive at the following expression:

$$\frac{\rho}{V} \int_V \int_V (g^c(r_{12}) - 1)\, d\mathbf{r}_1\, d\mathbf{r}_2 = \chi_T^\infty \left(1 - \left(\frac{L}{L_0}\right)^3\right) - 1 + \frac{c_1}{L} + \left(\frac{c_2}{L}\right)^2. \tag{20}$$

Finally, this expression becomes:

$$\chi_T(L, L_0) = \chi_T^\infty \left(1 - \left(\frac{L}{L_0}\right)^3\right) + \frac{c_1}{L} + \left(\frac{c_2}{L}\right)^2, \tag{21}$$

and by defining $\lambda = L/L_0$, we write:

$$\lambda \chi_T(\lambda) = \lambda \chi_T^\infty \left(1 - \lambda^3\right) + \frac{c_1}{L_0} + \left(\frac{c_2}{L_0}\right)^2 \frac{1}{\lambda}. \tag{22}$$

Equations (7) and (22) differ in the $c_2^2/L_0^2\lambda$ term that appears from considering the boundary finite-size effects. One possible scenario in which this difference might play a role is in the case of simulations near critical conditions where the correlation length of the system tends to infinity.

To test this expression, we perform simulations of systems with potential energy described by the truncated, at $r_c/\sigma = 2.5$, and shifted 12–6 Lennard–Jones potential. We consider systems with $N_0 = 24,000$ particles, with densities spanning the range $0.05 < \rho\sigma^3 < 0.70$. Two temperatures were considered, $k_BT = 2.00\epsilon$ and 1.15ϵ. The critical point of this system has been reported at $\rho_c\sigma^3 = 0.319$ and $k_BT_c = 1.086\epsilon$ [38].

We report the reduced fluctuations $\lambda\chi_T(\lambda)$ as a function of λ for $\rho\sigma^3 = 0.3$ in Figure 8. In the case $k_BT = 2.00\epsilon$, the effect of the λ^{-1} term in Equation (22) is negligible, and a linear approximation in the region $\lambda < 0.3$ seems to be well justified. However, for the case close to the critical point, i.e., $k_BT = 1.15\epsilon$, the effect of this term is evident and should be included in the extrapolation to χ_T^∞.

Finally, upon extrapolating to χ_T^∞, an interesting behavior is observed for the bulk isothermal compressibility κ_T as a function of density (Figure 9). In the case $k_BT = 2.00\epsilon$, as expected, a monotonically-decreasing behavior with increasing density is observed. More interestingly, in the case $k_BT = 1.15\epsilon$, the monotonically-decreasing behavior is interrupted by a singularity in the isothermal compressibility in the vicinity of the critical density. This cusp in the curve is expected since the isothermal compressibility of a fluid at the critical point is infinite.

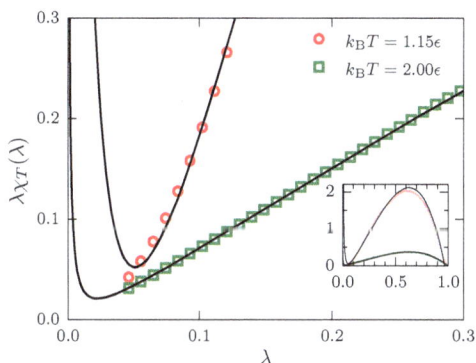

Figure 8. Reduced fluctuations as a function of λ for systems described by a TSLJ potential with $r_c/\sigma = 2.5$ with density $\rho\sigma^3 = 0.3$ at temperatures $k_BT = 2.00\epsilon$ and 1.15ϵ. For the latter case, it is apparent that the contribution proportional to λ^{-1} is not negligible. The inset shows the full range $0 < \lambda < 1$. The black curves are the result of fitting the data to Equation (22).

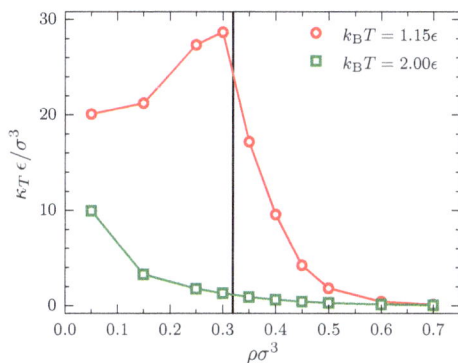

Figure 9. Bulk isothermal compressibility κ_T as a function of the density ρ at $k_BT = 1.15\epsilon$ (red circles) and $k_BT = 2.00\epsilon$ (green squares) for systems described by a TSLJ potential with $r_c/\sigma = 2.5$. The vertical black line indicates the location of the critical density $\rho\sigma^3 = 0.319$ [38].

The use of finite-size integral equations is general enough to admit generalizations of other systems of interest. In the next section, we describe one of such possible extensions: the study of binary mixtures.

4. Mixtures

Kirkwood–Buff (KB) theory [39] is arguably the most successful framework to investigate the properties of liquid mixtures that relates the local structure of a system to density fluctuations in the grand canonical ensemble. These quantities are in turn related to equilibrium thermodynamic quantities such as the compressibility, the partial molar volumes and the derivatives of the chemical potentials [2]. Formulated more than sixty years ago, KB enjoys renewed interest in the computational soft-matter and statistical physics communities [6,7,9–13]. Recent works have shown promising applications related to solvation of biomolecules [40] and potential uses to compute multicomponent diffusion in liquids [41] and to study complex phenomena such as self-assembly of proteins [42] and polymer conformation in complex mixtures [4,43].

For a multicomponent fluid of species i, j in equilibrium at temperature T, the Kirkwood–Buff integral (KBI) is defined as:

$$G_{ij}^o = V \left(\frac{\langle N_i N_j \rangle - \langle N_i \rangle \langle N_j \rangle}{\langle N_i \rangle \langle N_j \rangle} - \frac{\delta_{ij}}{\langle N_i \rangle} \right) = \frac{1}{V} \int_V \int_V [g_{ij}^o(\mathbf{r}_{12}) - 1] \, d\mathbf{r}_1 \, d\mathbf{r}_2, \tag{23}$$

with δ_{ij} the Kronecker delta. The superscript (o) indicates that this definition holds for an open system, i.e., a system in the grand canonical ensemble. In practice, we compute fluctuations of the number of particles in a subdomain of volume V embedded in a reservoir whose size goes to infinity. Thus, $\langle N_i \rangle$ is the average number of i-particles inside V, or $\rho_i = \langle N_i \rangle / V$. $g_{ij}^o(\mathbf{r}_{12})$ is the multicomponent radial distribution function (RDF) of the infinite system, with $\mathbf{r}_{12} = \mathbf{r}_2 - \mathbf{r}_1$.

Let us recall that in computer simulations one considers systems with total fixed number of particles N_0 and volume V_0 with PBCs. In this case, we have [35]:

$$G_{ij}(L, L_0) = V \left(\frac{\langle N_i N_j \rangle' - \langle N_i \rangle' \langle N_j \rangle'}{\langle N_i \rangle' \langle N_j \rangle'} - \frac{\delta_{ij}}{\langle N_i \rangle'} \right) = \frac{1}{V} \int_V \int_V [g_{ij}^c(\mathbf{r}_{12}) - 1] \, d\mathbf{r}_1 \, d\mathbf{r}_2. \tag{24}$$

The finite-size KBI $G_{ij}(L, L_0)$ is evaluated by computing fluctuations of the number of particles in finite subdomains of volume V inside a simulation box of volume V_0. The average number of i-particles $\langle N_i \rangle' \equiv \langle N_i \rangle_{V,V_0}$ depends on both subdomain and simulation box volumes. Moreover, the integral on the r.h.s. of Equation (24) should be evaluated for the RDF of the finite system $g_{ij}^c(\mathbf{r}_{12})$ with volume V_0 by using a finite integration domain V.

As has been done for the single component case, we include in this example both, ensemble and boundary, finite-size effects. For the former, the following correction has been suggested [44]:

$$g_{ij}^c(r) = g_{ij}^o(r) - \frac{1}{V_0} \left(\frac{\delta_{ij}}{\rho_i} + G_{ij}^\infty \right), \tag{25}$$

based on the asymptotic limit $g_{ij}^c(r \gg \zeta) = 1 - (\delta_{ij}/\rho_i + G_{ij}^\infty)/V_0$ discussed in [2]. As expected, when the total volume $V_0 \to \infty$, we recover $g_{ij}^c(r) = g_{ij}^o(r)$. By including Equation (25) in the integral on the r.h.s. of Equation (24) and evaluating the finite-size integral as for the single component case, we finally obtain:

$$\lambda G_{ij}(\lambda) = \lambda G_{ij}^\infty \left(1 - \lambda^3 \right) - \lambda^4 \frac{\delta_{ij}}{\rho_i} + \frac{\alpha_{ij}}{L_0}, \tag{26}$$

with $\lambda \equiv L/L_0$ and α_{ij} an intensive parameter with units of length. In the limit $L_0 \to \infty$, the following expression is obtained:

$$G_{ij}(L, L_0 \to \infty) = G_{ij}^\infty + \frac{\alpha_{ij}}{L}, \tag{27}$$

that describes the finite-size effects on the KBIs for a system in the grand canonical ensemble. Consistent with this limiting case in Equation (26), Equation (27) has been obtained from the thermodynamics of small systems [45,46].

For the investigation of Equation (26), we perform simulations for binary mixtures (A, B) of Lennard–Jones (LJ) fluids. We use a purely repulsive 12-6 LJ potential truncated and shifted with cutoff radius $2^{1/6}\sigma$. The potential parameters are chosen as $\sigma_{AA} = \sigma_{BB} = \sigma_{AB} = \sigma$, and $\epsilon_{AA} = 1.2\epsilon$, $\epsilon_{BB} = 1.0\epsilon$ with $\epsilon_{AB} = (\epsilon_{AA} + \epsilon_{BB})/2 = 1.1\epsilon$. All the results are expressed in LJ units with energy ϵ, length σ, mass $m_A = m_B = m$, time $\sigma(m/\epsilon)^{1/2}$, temperature ϵ/k_B and pressure ϵ/σ^3. As before, simulations are carried out using ESPResSo++ [32] with a time step of $\delta t/(\sigma(m/\epsilon)^{1/2}) = 10^{-3}$. Constant temperature $k_B T = 1.2\epsilon$ is enforced through a Langevin thermostat with damping coefficient $\gamma(\sigma(m/\epsilon)^{1/2}) = 1.0$. The size of the system is $N_0 = 23,328$ in the range of mole fractions of A-molecules $x_A = 0.1, \cdots, 1.0$. The pressure is fixed at $P\sigma^3/\epsilon = 9.8$ by adjusting the number density of the system at values around $\rho\sigma^3 \approx 0.86$ (or $L_0/\sigma \approx 30$). We perform equilibration runs of 64×10^6 MD steps and production runs of 2×10^6 MD steps. To compute $G_{ij}(\lambda)$, we select 800 frames per trajectory and for each frame identify 1000 randomly-positioned subdomains with linear sizes ranging from $2 < L/\sigma < L_0/\sigma$.

In Figure 10, results for finite-size KBIs are presented for four mole fractions, namely (a) $x_A = 0.20$; (b) $x_A = 0.30$; (c) $x_A = 0.50$ and (d) $x_A = 0.80$. Plots of G_{AB} (green circles) tend to zero when $\lambda \rightarrow 1$, as indicated by the horizontal green lines. By contrast, $G_{AA} \rightarrow 1/\rho_A$ (indicated by horizontal red lines) when $\lambda \rightarrow 1$. The region $\lambda < 3$, indicated by vertical black lines, is where simple linear regression is used to find G_{ij}^∞ and α_{ij}. By replacing such values in Equation (26), we obtained the black curves that, in all cases, superimpose on the simulation data for the full interval $0 < \lambda < 1$.

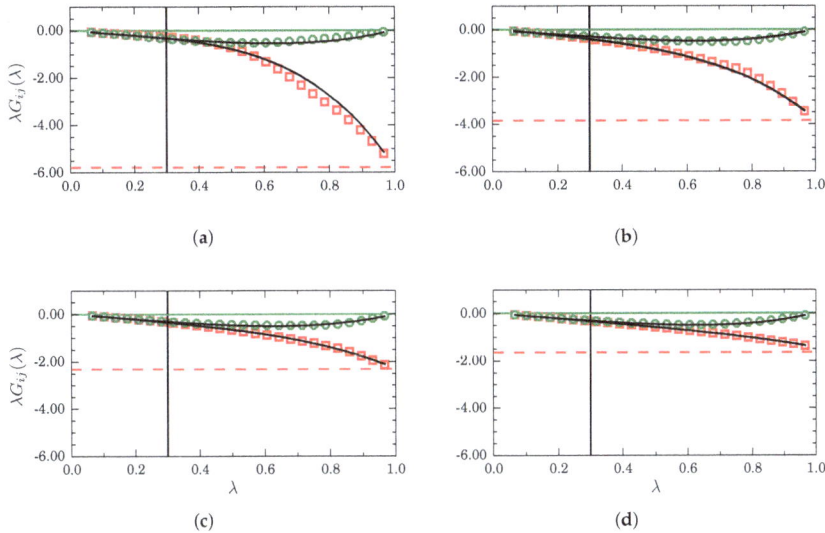

Figure 10. Scaled finite-size Kirkwood–Buff integrals $\lambda G_{ij}(\lambda)$ as a function of λ for different mole fractions: **(a)** $x_A = 0.20$; **(b)** $x_A = 0.30$; **(c)** $x_A = 0.50$ and **(d)** $x_A = 0.80$, for mixtures described by a TSLJ potential with $r_c/\sigma = 2^{1/6}$. For clarity, only the cases G_{AA} (red squares) and G_{AB} (green circles) are plotted. In the asymptotic case $\lambda \rightarrow 1$, $G_{AB} \rightarrow 0$ and $G_{AA} \rightarrow 1/\rho_A$, as indicated by the horizontal green and red lines, respectively. The black curves correspond to Equation (26) with G_{ij}^∞ and α_{ij} obtained from a simple regression analysis in the interval $\lambda < 0.3$.

The bulk KBIs are related to various thermodynamic quantities. For example, the isothermal compressibility is given by [39]:

$$\kappa_T = \frac{1 + \rho_A G_{AA} + \rho_B G_{BB} + \rho_A \rho_B (G_{AA} G_{BB} - G_{AB}^2)}{k_B T (\rho_A + \rho_B + \rho_A \rho_B (G_{AA} + G_{BB} - 2G_{AB}))}, \tag{28}$$

with $\rho_{A,B}$ the number density of the corresponding species.

Results for the isothermal compressibility obtained from the G_{ij}^∞ values are presented in Figure 11. Single component cases corresponding to systems composed by only type-A and type-B particles are indicated by the horizontal black lines. As expected, the system composed by strongly interacting particles, i.e., the type-A, has a lower compressibility. The behavior of the isothermal compressibility is nearly ideal since it follows closely the relation $\kappa_T = (1 - x_A)\kappa_T^B + x_A \kappa_T^A$, with $\kappa_T^A \epsilon/\sigma^3 = 0.012(1)$ and $\kappa_T^B \epsilon/\sigma^3 = 0.0281(8)$, as indicated by the solid black line.

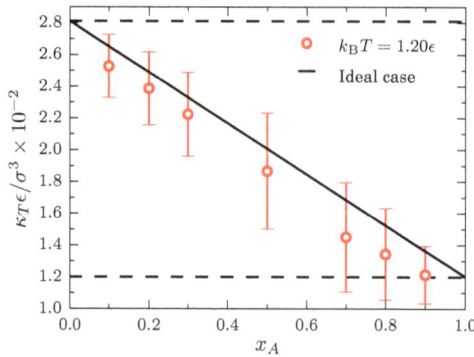

Figure 11. Isothermal compressibility at $k_B T = 1.20\epsilon$ and $P\sigma^3/\epsilon = 9.8$ as a function of the mole fraction of type-A particles x_A for mixtures described by a TSLJ potential with $r_c/\sigma = 2^{1/6}$. The horizontal black lines indicate the compressibility for a pure system of type-A particles $\kappa_T^A \epsilon/\sigma^3 = 0.012(1)$ and for a pure system of type-B particles $\kappa_T^B \epsilon/\sigma^3 = 0.0281(8)$. The red line is a guide to the eye. The ideal case corresponds to $\kappa_T = (1 - x_A)\kappa_T^B + x_A \kappa_T^A$.

Finally, the extrapolated KBIs have been used to compute the derivative of the chemical potential of type-A particles with respect to the number density ρ_A using the expression [39]:

$$\frac{1}{k_B T}\left(\frac{\partial \mu_A}{\partial \rho_A}\right)_{P,T} = \frac{1}{\rho_A} + \frac{G_{AB} - G_{AA}}{1 + \rho_A(G_{AA} - G_{AB})}, \tag{29}$$

that, as has been done for the single component case, can be integrated to obtain [6]:

$$\delta\mu_A = k_B T \int_{\rho_A^0}^{\rho_A} \left[\frac{1}{\rho_A'} + \frac{G_{AB} - G_{AA}}{1 + \rho_A'(G_{AA} - G_{AB})}\right] d\rho_A'. \tag{30}$$

This is the chemical potential shifted by a reference chemical potential computed at density ρ_A^0 [4,43]. By removing the density and concentration terms of the chemical potential of an ideal mixture, the excess chemical potential can be written as:

$$\delta\mu_A^{ex} = \delta\mu_A - k_B T \ln(x_A) - k_B T \ln(\rho_A). \tag{31}$$

We compare the results obtained using Equations (30) and (31) with the results obtained with the SPARTIAN method [36] and use the excess chemical potential result from $x_A = 0.3$ to find the reference value. We present the results in Figure 12 where a good agreement between the two datasets is apparent. To conclude this section, it has been shown that the block analysis method constitutes a robust strategy to compute chemical potentials of liquids and mixtures in a wide range of density/concentration conditions.

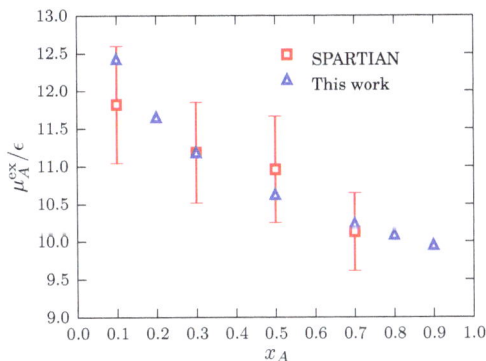

Figure 12. Excess chemical potential of type-A particles as a function of the mole fraction x_A for mixtures described by a TSLJ potential with $r_c/\sigma = 2^{1/6}$ at $k_B T = 1.2\epsilon$ and $P\sigma^3/\epsilon = 9.8$. Data points obtained with the method in [36], in particular for $x_A = 0.3$, are used as a reference for the data points obtained with Equations (30) and (31).

5. Summary and Concluding Remarks

In general, a direct comparison between a real system and a finite-size simulation is prevented by the fixed and relatively small number of particles used in the latter. As has been encoded in the title of the paper, the spatial block analysis method employs a clever combination of finite-size effects, ensemble and boundary, and density fluctuations to extrapolate bulk isothermal compressibilities and Kirkwood–Buff integrals in the thermodynamic limit.

In this work, we have illustrated with prototypical Lennard–Jones liquids and liquid mixtures the working mechanisms of the method. Upon identifying the relevant finite-size effects and assessing their impact on the simulation results, we have intuitively introduced an analytical expression connecting the fluctuations measured in a small subdomain of the simulation box with the bulk isothermal compressibility for a single component fluid.

Subsequently, the same analytical expression has been rigorously obtained from a finite-size version of the Ornstein–Zernike integral equation. Using a challenging system close to critical point conditions, we have tested the range of validity of the method and obtained results in line with theoretical expectations.

Then, for a multicomponent system, we have applied the same protocol to the Kirkwood–Buff integrals. Using the corresponding analytical expression, it is possible to obtain the Kirkwood–Buff integrals in the thermodynamic limit. In both single and multicomponent systems, the method allows one to compute the chemical potential of a liquid/mixture for a wide range of density/concentration conditions, provided a single reference chemical potential has been determined using a different computational method. These results contribute to establishing the spatial block analysis method as a powerful tool to investigate systems where the accurate computation of the chemical potential is of paramount importance.

Acknowledgments: We thank Debashish Mukherji and Roberto Menichetti for many stimulating discussions. We also thank Claudio Perego and Nancy C. Forero Martinez for a critical reading of the manuscript. Maziar Heidari, Kurt Kremer and Raffaello Potestio gratefully acknowledge funding from SFB-TRR146 of the German Research Foundation (DFG). Robinson Cortes-Huerto gratefully acknowledge the Alexander von Humboldt Foundation for financial support.

Author Contributions: Robinson Cortes-Huerto conceived the study. Kurt Kremer, Raffaello Potestio and Robinson Cortes-Huerto planned the computer simulations. Maziar Heidari and Robinson Cortes-Huerto carried out the numerical work. All authors discussed the results, helped with their interpretation and contributed to the final manuscript. All authors have read and approved the final manuscript.

Conflicts of Interest: The authors declare no conflict of interest.

Abbreviations

The following abbreviations are used in this manuscript:

SBA Spatial block analysis
TSLJ Truncated and shifted Lennard–Jones
MD Molecular dynamics
TL Thermodynamic limit
PBCs Periodic boundary conditions
OZ Ornstein–Zernike
KB Kirkwood–Buff

References

1. Hansen, J.P.; McDonald, I.R. *Theory of Simple Liquids*, 3rd ed.; Academic Press: Cambridge, MA, USA, 2006.
2. Ben-Naim, A. *Molecular Theory of Solutions*; Oxford University Press: Oxford, UK, 2006.
3. Adams, D. Chemical potential of hard-sphere fluids by Monte Carlo methods. *Mol. Phys.* **1974**, *28*, 1241–1252.
4. Mukherji, D.; Kremer, K. Coil-Globule-Coil Transition of PNIPAm in Aqueous Methanol: Coupling All-Atom Simulations to Semi-Grand Canonical Coarse-Grained Reservoir. *Macromolecules* **2013**, *46*, 9158–9163.
5. Wang, H.; Hartmann, C.; Schütte, C.; Site, L.D. Grand-Canonical-like Molecular-Dynamics Simulations by Using an Adaptive-Resolution Technique. *Phys. Rev. X* **2013**, *3*, 011018.
6. Heidari, M.; Kremer, K.; Potestio, R.; Cortes-Huerto, R. Finite-size integral equations in the theory of liquids and the thermodynamic limit in computer simulations. Submitted.
7. Galata, A.A.; Anogiannakis, S.D.; Theodorou, D.N. Thermodynamic analysis of Lennard–Jones binary mixtures using Kirkwood–Buff theory. *Fluid Phase Equilibria* **2017**, in press.
8. Site, L.D.; Ciccotti, G.; Hartmann, C. Partitioning a macroscopic system into independent subsystems. *J. Stat. Mech. Theory Exp.* **2017**, 083201.
9. Strom, B.A.; Simon, J.M.; Schnell, S.K.; Kjelstrup, S.; He, J.; Bedeaux, D. Size and shape effects on the thermodynamic properties of nanoscale volumes of water. *Phys. Chem. Chem. Phys.* **2017**, *19*, 9016–9027.
10. Dawass, N.; Krüger, P.; Schnell, S.K.; Bedeaux, D.; Kjelstrup, S.; Simon, J.M.; Vlugt, T.J.H. Finite-size effects of Kirkwood–Buff integrals from molecular simulations. *Mol. Simul.* **2018**, *44*, 599–612.
11. Milzetti, J.; Nayar, D.; van der Vegt, N.F.A. Convergence of Kirkwood–Buff Integrals of Ideal and Nonideal Aqueous Solutions Using Molecular Dynamics Simulations. *J. Phys. Chem. B* **2018**, doi:10.1021/acs.jpcb.7b11831.
12. Rogers, D.M. Extension of Kirkwood–Buff theory to the canonical ensemble. *J. Chem. Phys.* **2018**, *148*, 054102.
13. Dawass, N.; Krüger, P.; Simon, J.M.; Vlugt, T.J.H. Kirkwood–Buff integrals of finite systems: Shape effects. *Mol. Phys.* **2018**, doi:10.1080/00268976.2018.1434908.
14. Binder, K. Finite size scaling analysis of ising model block distribution functions. *Z. Phys. B* **1981**, *43*, 119–140.
15. Binder, K. Critical Properties from Monte Carlo Coarse Graining and Renormalization. *Phys. Rev. Lett.* **1981**, *47*, 693–696.
16. Rovere, M.; Hermann, D.W.; Binder, K. Block Density Distribution Function Analysis of Two-Dimensional Lennard–Jones Fluids. *EPL* **1988**, *6*, 585.
17. Rovere, M.; Heermann, D.W.; Binder, K. The gas-liquid transition of the two-dimensional Lennard–Jones fluid. *J. Phys. Condens. Matter* **1990**, *2*, 7009–7032.
18. Rovere, M.; Nielaba, P.; Binder, K. Simulation studies of gas-liquid transitions in two dimensions via a subsystem-block-density distribution analysis. *Z. Phys. B* **1993**, *90*, 215–228.
19. Weber, H.; Marx, D.; Binder, K. Melting transition in two dimensions: A finite-size scaling analysis of bond-orientational order in hard disks. *Phys. Rev. B* **1995**, *51*, 14636–14651.
20. Román, F.L.; White, J.A.; Velasco, S. Block analysis method in off-lattice fluids. *EPL* **1998**, *42*, 371.
21. Salacuse, J. Particle fluctuations within sub-regions of an N-particle, three-dimensional fluid: Finite-size effects and compressibility. *Phys. A* **2008**, *387*, 3073–3083.

22. Sengupta, S.; Nielaba, P.; Rao, M.; Binder, K. Elastic constants from microscopic strain fluctuations. *Phys. Rev. E* **2000**, *61*, 1072–1080.

23. Lebowitz, J.L.; Percus, J.K. Long-Range Correlations in a Closed System with Applications to Nonuniform Fluids. *Phys. Rev.* **1961**, *122*, 1675–1691.

24. Salacuse, J.J.; Denton, A.R.; Egelstaff, P.A. Finite-size effects in molecular dynamics simulations: Static structure factor and compressibility. I. Theoretical method. *Phys. Rev. E* **1996**, *53*, 2382–2389.

25. Román, F.L.; White, J.A.; Velasco, S. Fluctuations in an equilibrium hard-disk fluid: Explicit size effects. *J. Chem. Phys.* **1997**, *107*, 4635.

26. Villamaina, D.; Trizac, E. Thinking outside the box. Fluctuations and finite size effects. *Eur. J. Phys.* **2014**, *35*, 035011.

27. Rowlinson, J.S. The equation of state of dense systems. *Rep. Prog. Phys.* **1965**, *28*, 169.

28. Schnell, S.K.; Vlugt, T.J.; Simon, J.M.; Bedeaux, D.; Kjelstrup, S. Thermodynamics of a small system in a μT reservoir. *Chem. Phys. Lett.* **2011**, *504*, 199–201.

29. Lebowitz, J.L.; Percus, J.K. Thermodynamic Properties of Small Systems. *Phys. Rev.* **1961**, *124*, 1673–1681.

30. Hill, T.L. *Thermodynamics of Small Systems*; Courier Corporation: Dover, UK, 1963.

31. Humphrey, W.; Dalke, A.; Schulten, K. VMD—Visual Molecular Dynamics. *J. Mol. Graph.* **1996**, *14*, 33–38.

32. Halverson, J.D.; Brandes, T.; Lenz, O.; Arnold, A.; Bevc, S.; Starchenko, V.; Kremer, K.; Stuehn, T.; Reith, D. ESPResSo++: A modern multiscale simulation package for soft matter systems. *Comput. Phys. Commun.* **2013**, *184*, 1129–1149.

33. Román, F.L.; González, A.; White, J.A.; Velasco, S. Fluctuations in the number of particles of the ideal gas: A simple example of explicit finite-size effects. *Am. J. Phys.* **1999**, *67*, 1149–1151.

34. Román, F.L.; White, J.A.; González, A.; Velasco, S. Fluctuations in a small hard-disk system: Implicit finite size effects. *J. Chem. Phys.* **1999**, *110*, 9821.

35. Román, F.; White, J.; González, A.; Velasco, S. *Theory and Simulation of Hard-Sphere Fluids and Related Systems*; Chapter Ensemble Effects in Small Systems; Springer: Berlin/Heidelberg, Germany, 2008; pp. 343–381.

36. Heidari, M.; Kremer, K.; Cortes-Huerto, R.; Potestio, R. Spatially Resolved Thermodynamic Integration: An Efficient Method to Compute Chemical Potentials of Dense Fluids. *ArXiv* **2018**, arXiv:cond-mat.soft/1802.08045.

37. Ornstein, L.S.; Zernike, F. Accidental deviations of density and opalescence at the critical point of a single substance. *Proc. Akad. Sci. (Amsterdam)* **1914**, *17*, 793–806.

38. Thol, M.; Rutkai, G.; Span, R.; Vrabec, J.; Lustig, R. Equation of State for the Lennard–Jones Truncated and Shifted Model Fluid. *Int. J. Thermophys.* **2015**, *36*, 25–43.

39. Kirkwood, J.G.; Buff, F.P. The Statistical Mechanical Theory of Solutions. I. *J. Chem. Phys.* **1951**, *19*, 774–777.

40. Pierce, V.; Kang, M.; Aburi, M.; Weerasinghe, S.; Smith, P.E. Recent Applications of Kirkwood–Buff Theory to Biological Systems. *Cell Biochem. Biophys.* **2008**, *50*, 1–22.

41. Kjelstrup, S.; Schnell, S.K.; Vlugt, T.J.H.; Simon, J.M.; Bardow, A.; Bedeaux, D.; Trinh, T. Bridging scales with thermodynamics: From nano to macro. *Adv. Nat. Sci. Nanosci. Nanotechnol.* **2014**, *5*, 023002.

42. Ben-Naim, A. Theoretical aspects of self-assembly of proteins: A Kirkwood–Buff-theory approach. *J. Chem. Phys.* **2013**, *138*, 224906.

43. Mukherji, D.; Marques, C.M.; Stuehn, T.; Kremer, K. Depleted depletion drives polymer swelling in poor solvent mixtures. *Nat. Commun.* **2017**, *8*, 1374.

44. Cortes-Huerto, R.; Kremer, K.; Potestio, R. Communication: Kirkwood–Buff integrals in the thermodynamic limit from small-sized molecular dynamics simulations. *J. Chem. Phys.* **2016**, *145*, 141103.

45. Schnell, S.K.; Liu, X.; Simon, J.; Bardow, A.; Bedeaux, D.; Vlugt, T.J.H.; Kjelstrup, S. Calculating Thermodynamic Properties from Fluctuations at Small Scales. *J. Phys. Chem. B* **2011**, *115*, 10911.

46. Ganguly, P.; van der Vegt, N.F.A. Convergence of Sampling Kirkwood–Buff Integrals of Aqueous Solutions with Molecular Dynamics Simulations. *J. Chem. Theory Comput.* **2013**, *9*, 1347–1355.

entropy

MDPI

Article

Thermodynamics of Small Magnetic Particles

Eugenio E. Vogel [1,2,*], Patricio Vargas [2,3], Gonzalo Saravia [1], Julio Valdes [1],
Antonio Jose Ramirez-Pastor [4] and Paulo M. Centres [4]

[1]   Department of Physics, Universidad de La Frontera, Francisco Salazar 01145, Temuco, Chile;
      gonzalo.saravia@gmail.com (G.S.); julio.valdes@ufrontera.cl (J.V.)
[2]   Center for the Development of Nanoscience and Nanotechnology, 9170124 Santiago, Chile;
      patricio.vargas@usm.cl
[3]   Department of Physics, Universidad Técnica Federico Santa María, Avenida España 1680, Valparaíso, Chile
[4]   Departamento de Física, Instituto de Física Aplicada, Universidad Nacional de San Luis-CONICET,
      Ejército de Los Andes 950, San Luis D5700BWS, Argentina; antorami@unsl.edu.ar (A.J.R.-P.);
      pcentres@gmail.com (P.M.C.)
*   Correspondence: eugenio.vogel@ufrontera.cl; Tel.: +56-45-232-5316

Received: 2 August 2017; Accepted: 13 September 2017; Published: 15 September 2017

Abstract: In the present paper, we discuss the interpretation of some of the results of the thermodynamics in the case of very small systems. Most of the usual statistical physics is done for systems with a huge number of elements in what is called the thermodynamic limit, but not all of the approximations done for those conditions can be extended to all properties in the case of objects with less than a thousand elements. The starting point is the Ising model in two dimensions (2D) where an analytic solution exits, which allows validating the numerical techniques used in the present article. From there on, we introduce several variations bearing in mind the small systems such as the nanoscopic or even subnanoscopic particles, which are nowadays produced for several applications. Magnetization is the main property investigated aimed for two singular possible devices. The size of the systems (number of magnetic sites) is decreased so as to appreciate the departure from the results valid in the thermodynamic limit; periodic boundary conditions are eliminated to approach the reality of small particles; 1D, 2D and 3D systems are examined to appreciate the differences established by dimensionality is this small world; upon diluting the lattices, the effect of coordination number (bonding) is also explored; since the 2D Ising model is equivalent to the clock model with $q = 2$ degrees of freedom, we combine previous results with the supplementary degrees of freedom coming from the variation of q up to $q = 20$. Most of the previous results are numeric; however, for the case of a very small system, we obtain the exact partition function to compare with the conclusions coming from our numerical results. Conclusions can be summarized in the following way: the laws of thermodynamics remain the same, but the interpretation of the results, averages and numerical treatments need special care for systems with less than about a thousand constituents, and this might need to be adapted for different properties or devices.

Keywords: small systems; thermodynamics; magnetization; Ising model; Potts model; dilution

1. Introduction

Recent developments in experimental techniques and refinements of synthesis methods allow for direct measurements of very small systems [1–6]. Examples of such systems include magnetic domains in ferromagnets, which are typically smaller than 300 nm, quantum dots and solid-like clusters that are important in the relaxation of glassy systems and whose dimensions are a few nanometers. Due to the small size of such systems and the high surface/volume ratio, the usual thermodynamic properties of macroscopic systems do not apply in the usual sense, and the analysis of fundamental principles of statistical thermodynamics needs to be reconsidered [7–10]. Phase transitions, specific heat, thermal conductivity, magnetization and other phenomena need careful examination of the corresponding thermodynamic variables used to characterize them. Ergodicity, fluctuations and error bars are among the aspects that are greatly affected by the small size of the systems.

The increasing interest in the theory of small systems is not limited to physics, but also extends to biological and chemical systems, such as clusters [11,12], thin films [13] and biological molecular machines with sizes under 100 nm [14]. This broad picture calls attention to the extension of the mathematical treatments valid in the (usually called) thermodynamic limit (TL) to systems formed by a small number of elements.

In 1962, Hill [15,16] developed a formalism for describing the thermodynamics of small systems, where thermodynamic properties such as enthalpy and Gibbs energy are no longer extensive. The pioneer work of Hill established the fundamental laws that govern finite-size effects in thermodynamics. Later, Hill and Chamberlin [17–20] introduced the term nanothermodynamics to denote the study of systems far away from the TL. Despite these results, there are still many open questions regarding the thermodynamic behavior and possible applications of very small systems.

Magnetic nanoparticles [21,22] are also examples of small systems whose properties are influenced by finite size effects. The competition between core and surface effects determines the state of the particle, which can go from a paramagnetic state to a ferromagnetic state. The temperature characterizing this transition is the well-known Curie temperature (T_C) in the TL. Experimental studies for different materials show that the T_C in nanoparticles is not a unique property of the material (as in the bulk), since it is strongly size dependent and decreases as the characteristic dimension of the particle, $V^{1/3}$ (where V is the volume of the particle), is reduced [23–25].

In a recent work [26], it has been shown that for tiny systems, not only surface sites are special, but also some inner sites might not be equivalent to each other. This fact can have important consequences on the thermodynamic properties of small nanoparticles (up to sizes of the order of ≈ 10 nm). In the magnetic systems described below, this effect mainly introduces dispersion in the exchange coupling constants, which in turn may affect the values of the thermodynamical parameters. These considerations tend to lose importance as the nanoparticle size increases. Moreover, they do not affect the overall underlying physics, but impose limitations in the interpretation of the magnitudes defined in the TL.

In the present paper, we would like to answer the following questions: How stable and reliable are devices built of subnanoscopic magnetic particles? How do the representative properties of these devices depend on the size, dimensionality, coordination and internal degrees of freedom among other variables? We will consider two examples offered by very small magnetic particles responding to temperature changes; such particles could be components of devices.

Let us first consider the magnetic switch shown in Figure 1a where a tiny nanomagnet (nm) sticks to a ferromagnetic material (F) under a sensitive temperature T^* (a way to avoid calling it the critical temperature). To overcome the difficulties of defining temperature for very small isolated systems we will assume that the devices and everything that is connected to them are immersed in a thermal bath at temperature T. Changes in temperature are smooth and slow, so there is time to equilibrate. As temperature raises, nm can decrease its magnetization under a critical value, which will weaken the attraction to F; then an elastic restoring force (like the spring in the figure) brings nm to its natural position away from F. This magnetic switch could activate an alarm that goes off over a prefixed temperature, which ideally is precisely T^*. Let us assume that for this system, mechanical relaxation times are longer than electronic ones; then, such a device is not sensitive to magnetization reversals, so it is basically controlled by the absolute value of the magnetization, namely $|m(T)|$.

Figure 1b presents a tiny rod-like nanomagnet, which can rotate about its center on the plane of the figure. A stronger fixed magnet (NS) with high T_C orientates the tiny magnet. When the temperature goes over a sensitive temperature T^{**}, a magnetization reversal happens in the small magnet. Such polarity reversion triggers a rotation of the small particle disconnecting the terminals that are connected when the conducting element attached to the "black" part of the particle closes the circuit.

This device is strongly dependent on the polarity of the nanoparticle, namely on the sign of $m(T)$. Magnetization reversals are usually quite fast as compared to most mechanical processes in materials. Some orders of magnitude for the switching time (defined as the time in which the magnetization reverses polarity) have been proposed at about 200 ps for small Co systems [27]. The switching time could be larger depending on the system size, temperature and magnetic anisotropies, but a few tenths of ns have been reported for other small systems in the absence of an external magnetic field [28].

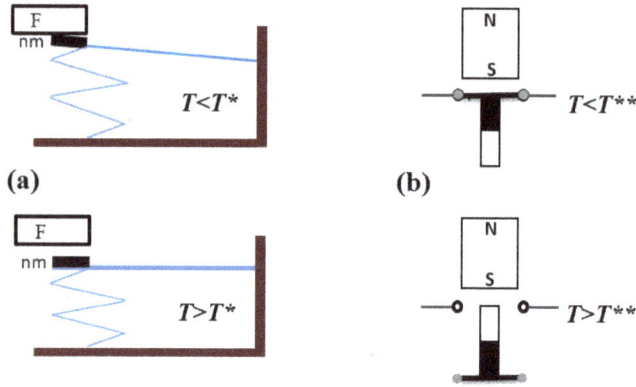

Figure 1. Magnetic switches. (**a**) Top: Under T^* the magnetic nanoparticle (nm) sticks to the ferromagnetic material (F) even against a restoring elastic force represented by a zig-zag line. Bottom: As temperature T goes over T^*, nm loses enough magnetization, so the elastic force dominates switching off the contact; (**b**) Top: Under T^{**}, a weak magnet labeled NS, with high enough T_C, attracts the north pole (painted black) of a rod-like nanomagnet made of a material with lower T_C; the black part has a conducting element that closes a circuit. Bottom: Over T^{**}, an internal magnetization reversal occurs in the nanomagnet; the north pole is now in the sector painted white; the rod rotates with respect to an axis going through the center of the rod and perpendicular to the figure; in this new position, the circuit is switched off.

We shall come back to these examples later on since this is the kind of system we have in mind, rather than statistical averages over different samples or instances. At this stage, we point to the gross picture, namely the tendency of properties, and not to material-related or geometrical features controlling a particular device. To focus the discussion, we turn our attention to two general and well-known models whose properties can be reexamined in this new scenario. We refer to the Ising model and to the clock model (related to the better-known Potts model).

The Ising model plays a central role in the study of phase transitions in magnetic systems (transitions from paramagnet to ferromagnet, order-disorder transitions, etc.) [29–37]. Ising [38] gave an exact solution to the one-dimensional lattice problem in 1925. All other cases are expressed in terms of series solution [39,40], except for the special case of two-dimensional lattices in zero external field, which was exactly solved by Onsager [41] in 1944. Close approximate solutions in dimensions higher than one can be obtained, and the two most important of these are the mean-field approach [42] and the quasi-chemical approximation [42,43].

In the last few years, the Ising model has been adopted as a prototype for studying the magnetic properties of systems far away from the TL [44–46]. In [44], Velásquez et al. studied the magnetic properties and pseudocritical behavior of ferromagnetic pure and metallic nanoparticles in the presence of atomic disorder, dilution and competing interactions. Using variational and Monte Carlo methods, the authors reported the particle size dependence of the ordering temperature for pure and random diluted particles.

Bertoldi et al. [45] solved analytically the Ising model for small clusters in the microcanonical formalism, using the mean-field theory and considering the presence of surfaces. In agreement with the work of Velásquez et al. [44], the authors found that the critical temperature decreases proportionally to the inverse of the radius of the cluster. Recently [46], the Ising model on a simple-cubic lattice was used to analyze the main concepts of nanothermodynamics applied to Monte Carlo simulations.

Finally, we increase the internal degrees of freedom at each site upon generalizing the Ising model to the discretized XY model in what is called the clock model which is a well-known example of the famous Berezinskii–Kosterlitz–Thouless (BKT) transition [47,48]. In this way, we will be able to appreciate what is the role of the increase of the configuration space due to the characteristics of the constituents of small systems and not only the fact that they interact among themselves.

This paper is organized as follows: In Section 2, we present the system and the methodology based both on theoretical formulations and on computer simulations. In Section 3, we present the results stressing those points where the analysis concerning the TL is not applicable for very small particles; the discussion and relations among results are done along with the presentation of the results given by plots and particular numerical values. In Section 4, we list and comment on the main conclusions.

2. Model and Methods

2.1. General Definitions

Let us begin by considering the Ising model on a square lattice of dimensions $L \times L = N$, where local magnetic moment or "spin" S_i at site i can point either along the positive ($+1$) or negative direction (-1) of a given axis. The isotropic Hamiltonian for such a system can be written as:

$$H = -J \sum_{i>j}^{N} S_i S_j, \tag{1}$$

where J is the exchange interaction to nearest neighbors; the sum extends to all pairs through the lattice, which is indicated by the symbol $i > j$ under the summation symbol. The magnetization $m(T,t)$ of the system at any temperature T and time t is simply obtained by just adding the spins in an algebraic manner, namely,

$$m(T,t) = \frac{1}{N} \sum_{i=1}^{N} S_i, \tag{2}$$

where we have normalized it over the number of sites N, which is equivalent to normalizing it with respect to the saturation magnetization.

Large macroscopic systems are modeled by making use of periodic boundary conditions (PBC) where surface states are ignored; in this approximation, Lars Onsager [41] found an analytic solution for the magnetization of the system in 2 dimensions (2D):

$$
\begin{aligned}
m_O(T) &= (1 - \sinh^{-4}(2J/k_B T))^{1/8} \; ; \; T < T_C \\
&= 0 \; ; \; T \geq T_C
\end{aligned}
\tag{3}
$$

where k_B is the Boltzmann constant, which we can take as 1.0 from now on (this is equivalent to measuring energy and J in T units). The critical temperature T_C can be readily obtained at the point where the magnetization vanishes:

$$\left[\sinh\left(\frac{2J}{T_C}\right)\right]^2 = 1, \tag{4}$$

Then, the exact numeric solution for this problem for PBC can be easily evaluated, yielding:

$$T_C = 2.269185...J \approx 2.27, \tag{5}$$

where *J* has been omitted, as will be done from now on.

The reason to have chosen the 2D Ising model to begin with is precisely this result, namely the existence of an analytic solution for the transition temperature in the TL. We can use it as a reference point to compare with other results and simulations where different conditions will be varied.

2.2. Exact Theoretical Approach for a Small System

Next, we consider a very small Ising square lattice, $L = 4$, where we can calculate everything exactly. The partition function can be expressed as:

$$Z(T) = \sum_{n=1}^{\lambda} c_n e^{-E_n/T} , \tag{6}$$

where the coefficients c_n give all of the possible spin configurations compatible with energy E_n according to the Ising Hamiltonian; λ is the number of energy levels.

We can write the exact partition functions $Z_P(T)$ and $Z_F(T)$ for PBC and free boundary conditions (FBC), respectively:

$$Z_P(T) = \sum_{n=1}^{15} c_{Pn} e^{-E_n/T} , \tag{7}$$

$$Z_F(T) = \sum_{n=1}^{23} c_{Fn} e^{-E_n/T} , \tag{8}$$

where coefficients c_{Pn} and c_{Fn} are to be obtained from Tables 1 and 2, respectively. These tables list the corresponding degeneracies $p(n,k)$ (PBC) and $f(n,k)$ (FBC) for the set of states sharing energy E_n given by Equation (1) and magnetization $M_{n,k}$:

$$M_{n,k} = 18 - 2k, \tag{9}$$

which runs between -16 and $+16$ at intervals of 2, as shown by the second row in the tables. The coefficients in the expansions given by Equations (7) and (8) can be obtained as:

$$c_{Pn} = \sum_{k=1}^{17} p(n,k), \tag{10}$$

$$c_{Fn} = \sum_{k=1}^{17} f(n,k), \tag{11}$$

where we take into consideration the symmetry in Tables 1 and 2, namely $p(n, 18 - k) = p(n,k)$ and $f(n, 18 - k) = f(n,k)$ with $k = 1, 2, ...8$, and

$$M_{n,k} = -M_{n,18-k} , \quad k = 1, 2, ...8. \tag{12}$$

It should be noticed that Tables 1 and 2 give only those configurations with nil or positive magnetization. However, the Hamiltonian above is symmetric under the simultaneous reversion of all spins, so both tables should be continued to the left to include the states with negative magnetization.

Table 1. Coefficients $p(n,k)$ for periodic boundary condition (PBC) configurations of a given energy E_n and magnetization $M_{n,k}$. The first column gives n; the second column gives the corresponding energy E_n; the following columns give the number of configurations for that energy (row) and magnetization (column). This table is symmetric and should be extended to the left by means of the symmetry properties so as to include the 8 negative values of the magnetization following Equations (10) and (12).

	k=	9	8	7	6	5	4	3	2	1
n	E_n	$M_k = 0$	2	4	6	8	10	12	14	16
15	32	2								
14	24		16							
13	20	64								
12	16	120	96	56						
11	12	576	512	64						
10	8	2112	1392	768	128					
9	4	3264	3136	1568	448					
8	0	4356	3680	2928	1248	228				
7	−4	1600	1920	1824	1664	576				
6	−8	768	624	704	688	736	208			
5	−12		64	96	192	256	256			
4	−16	8					24	96	88	
3	−20							32		
2	−24								16	
1	−32									1

Table 2. Coefficients $f(n,k)$ for free boundary condition (FBC) configurations of a given energy E_n and magnetization M_k. The first column gives n; the second column gives the corresponding energy E_n; the following columns give the number of configurations for that energy (row) and magnetization (column). This table is symmetric and should be extended to the left so as to include the 8 negative values of the magnetization following Equations (9) and (11).

	k=	9	8	7	6	5	4	3	2	1
n	E_n	$M_k = 0$	2	4	6	8	10	12	14	16
23	24	2								
22	20		4							
21	18	16	8							
20	16	36	16	2						
19	14	80	56	16						
18	12	192	148	48						
17	10	384	296	112	8					
16	8	814	596	272	44					
15	6	1360	1096	552	136					
14	4	1800	1608	864	288	12				
13	2	2256	1984	1296	512	72				
12	0	2320	2044	1604	740	188				
11	−2	1680	1616	1352	872	304	8			
10	−4	1008	1032	880	804	392	60			
9	−6	544	584	592	520	368	128			
8	−8	270	236	286	260	256	144	2		
7	−10	80	72	96	120	144	112	24		
6	−12	24	36	20	48	68	64	44		
5	−14		8	16	16	8	32	32		
4	−16	4				8	12	10	4	
3	−18							8	8	
2	−20								4	
1	−24									1

2.3. Numerical Simulations

In addition to some theoretical calculations, most of the work will deal with numerical calculations based on Monte Carlo (MC) simulations. A square lattice $L \times L$ is chosen; PBC or FBC are imposed; a site is randomly visited; and the energy cost δ of flipping the corresponding spin is calculated: if the energy is lowered, the change of orientation is accepted; otherwise, only when $\exp(-\delta/T) \leq r$, the spin flip is accepted, where r is a freshly-generated random number in the range [0,1]. This is the usual Metropolis algorithm. A Monte Carlo step (MCS) is reached after $N = L \times L$ spin-flip attempts. One of the main goals here is to report the sensitive temperatures (T^* or T^{**} analogues to critical temperatures in the TL) for different systems.

The energy is computed according to the Hamiltonian given by Equation (1). The instantaneous magnetization at time t and temperature T is obtained from Equation (2).

For each temperature, 2τ MCSs are performed: the first τ MCS is used to equilibrate at a fixed temperature T, while the second τ MCS is used to measure the magnetization every 20 MCSs, reaching a total of 50,000 measurements. Unless specified in a different way $\tau = 10^6$ in the rest of the paper; this τ value gives stable results and leads to coincidence with the analytic expressions obtained as described above.

2.4. The Clock Model

This is the discrete version of the famous 2D XY model, which is probably the most extensively studied example showing the Berezinskii–Kosterlitz–Thouless (BKT) transition [47,48]. It is often used as a reference of its peculiar critical behavior at the transition point and universal features [49–53]. Instead of the exclusion of an explicit continuous symmetry essential for the BKT transition, it can also emerge from a system without explicit continuous symmetry. The q-state clock model is the discrete version of the XY model, with discrete symmetry. The Hamiltonian of the clock model can be written as:

$$H = -J \sum_{i<j} \cos(\theta_i - \theta_j),$$ (13)

where $J > 0$ is the ferromagnetic coupling connecting pairs of nearest neighbors i and j; the discrete angle between the spin orientations is given by $\theta_{i,j}(\eta, q) = 2\pi\eta/q$, for $\eta = \{0, 1, .., q-1\}$. While the exact XY model is recovered only in the limit of infinite q, it has been found that the BKT characters appear in the clock models when $q \geq 5$ [54–57]. The nature of the phase transitions in the general clock model has been widely studied with different theoretical and numerical approaches, which however have given mixed results on the characterization of transitions at the lower bound of q (for instance, see the summary of the related debates in [58]).

In spite of all that, in this work, we want to use this model as a generalization of the Ising model presented above allowing for the increase in the number of internal states at each site. All of this is always aimed at the behavior of small systems. We will focus on a square lattice 8×8 with FBC, with the clock model ranging from $q = 2$ (Ising model) to $q = 20$ (one discrete version of the 2D finite XY model).

3. Results and Discussion

Let us begin by considering the theoretical approach for the Ising lattice with $L = 4$ introduced in the previous section. The magnetization per site can be calculated as:

$$m_P(T) = \frac{1}{N Z_P(T)} \sum_{n=1}^{15} \frac{\sum_{k=1}^{17} M_{n,k} p(n,k)}{\sum_{k=1}^{17} p(n,k)} c_{Pn} e^{-E_n/T}$$ (14)

and:

$$m_F(T) = \frac{1}{N Z_F(T)} \sum_{n=1}^{23} \frac{\sum_{k=1}^{17} M_{n,k} f(n,k)}{\sum_{k=1}^{17} f(n,k)} c_{Fn} e^{-E_n/T},$$ (15)

both of which give a nil result for any T since negative and positive contributions to the magnetization cancel out as they are weighted with the same probability due to the symmetry properties of Tables 1 and 2.

In macroscopic systems, ergodicity is spontaneously broken favoring any of the two halves of the configuration space. As we will see below, this is not always true for very small systems, but let us start from the ergodicity breaking approach so widely invoked to approach the TL. This can be achieved by either considering only the positive magnetization contributions of Tables 1 and 2 plus half of the coefficients for the $M_{n,k} = 0$ column of the corresponding table. Equivalently, we can force ergodicity breaking by considering all of the contributions to the magnetization, but using their absolute value, namely,

$$m_{Pe}(T) = \frac{1}{N Z_P(T)} \sum_{n=1}^{15} \frac{\sum_{k=1}^{17} |M_{n,k}| p(n,k)}{\sum_{k=1}^{17} p(n,k)} c_{Pn} e^{-E_n/T} \qquad (16)$$

and:

$$m_{Fe}(T) = \frac{1}{N Z_F(T)} \sum_{n=1}^{23} \frac{\sum_{k=1}^{17} |M_{n,k}| f(n,k)}{\sum_{k=1}^{17} f(n,k)} c_{Fn} e^{-E_n/T}, \qquad (17)$$

where e refers to the forced ergodicity breaking.

We invoke now the numeric simulations defined in the previous section to calculate the magnetization for the same system. Here, the same trick is done to break ergodicity, namely it is always $|m(T,t)|$ (obtained from Equation (2)) that is considered as the contribution to magnetization from every configuration visited. In this way, ergodicity breaking is forced to occur in any circumstance. We will call $\mu_{Pe}(T,t)$ (PBC) and $\mu_{Fe}(T,t)$ (FBC) the magnetizations obtained via numeric simulations for a temperature T and at an instant t.

We compare now previous approaches. Results from Equations (16) and (17) are shown by up and down solid triangles respectively in Figure 2. The results for the numeric simulations for PBC and FBC are shown here by empty diamonds and empty squares respectively. The values for $< \mu_{Pe}(T) >$ and $< \mu_{Fe}(T) >$ reported in Figure 2 correspond to positive averages over $\nu = 50{,}000$ measurements. The corresponding distribution of values allows us to report the error bars for each temperature.

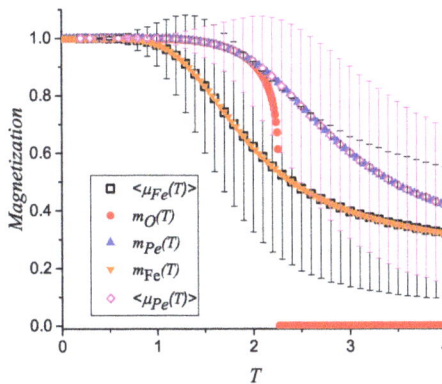

Figure 2. Magnetization for a lattice $L = 4$ obtained using the analytic expressions (16) (up blue triangles) and (17) (down yellow triangles) to compare with the results obtained from numeric simulations using PBC (empty magenta diamonds) and FBC (empty black squares); the Onsager solution is also plotted to mark the expected T_C in the thermodynamic limit (TL) (solid red circles).

For comparison and discussion, the Onsager solution is also added to this figure by means of solid circles.

Figure 2 deserves a careful analysis and discussion. The excellent agreement between theory and numeric simulation is very impressive. However, this should not be very surprising since the Metropolis algorithm is based on the same probability laws that appear in the partition function, hence, in the magnetization as well; both approaches consider ergodicity breaking, as already pointed out. The agreement is stressed in the sense that the computer algorithms used here are very reliable, and the equilibration times are enough. It can be noticed that both the theoretical and numeric results for PBC coincide with the Onsager solution up to almost $T = 2.0$. From there on, both treatments overestimate the magnetization values partly due to a small size effect, but also due to the forced ergodicity breaking, which is assumed in these calculations. We will come back to this point below.

Let us now pay attention to the different results obtained for FBC as compared to PBC regardless of the method employed in the calculation. It is PBC that can be used to compare with the results valid in the TL, which are illustrated in Figure 2, by the Onsager solution yielding $T_C \approx 2.27$. As can be seen, the theoretical results and magnetic simulations using PBC both agree with the Onsager solution up to $T = 2.0$ approximately; from there on, both solutions remain at high values, decreasing smoothly without an indication for a sharp phase transition. In the case of very small particles, PBC are unthinkable, since in this limit, surface states become very important. This is an indication that FBC should be used to model the properties of very small particles.

The light and smooth decrease in magnetization due to surface effects will determine that the nanomagnets involved in the simple systems of Figure 1 will have different responses as those made of the same basic material, but operating in the bulk sizes, namely in the TL. The sensitive temperatures T^* and T^{**} for which these devices will activate still depend on other features to be examined below, but this is a first result showing that properties characterized in the bulk need to be reexamined for very tiny particles.

Actually, we can define the corresponding sensitive temperature simply as the temperature for which $m(T)$ or $< \mu(T) >$ (with any subscripts) takes the value 0.5. This working definition is consistent with the TL, where a square step magnetization curve is expected at T_C (equivalent to our sensitive temperature), jumping from magnetization 1.0 to 0.0. Then, the sensitive temperatures can be read from Figure 2 as 2.4 and 3.4 for FBC and PBC, respectively. Both values are larger than the macroscopic expected result, which requires further discussion.

What is the real meaning of the average values we have just reported? Let us revisit the FBC data and do a different exercise: let us plot the instantaneous value of the magnetization $\mu_F(T, t)$ (no ergodicity breaking here) after the τ instants of measurement (last measured value for each temperature after 2τ total MCSs). This is plotted by solid squares (black) in Figure 3 using the same data already available from FBC on the 4×4 lattice. These results allow for just one interpretation: at least at $T = 0.8$ (or eventually at even lower T values), a magnetization reversal occurred, and ergodicity breaking is no longer valid. Therefore, the previous approach intended for the TL has to be reconsidered to analyze the results obtained in the small world. This is an indication that the ergodicity breaking assumption is valid only for systems over a minimum size and appropriate for the property under investigation.

All of the previous procedures are usual for TL analysis. However, the magnetic switches presented in the Introduction will respond instantaneously and without having the chance of performing derivatives, averages or even more sophisticated mathematical analyses, like the crossing of Binder cumulants [59] or the crossing of autocorrelation functions [60], to decide which is the appropriate value for the sensitive temperature. In these examples and in all nanosized devices, a careful analysis of the thermodynamics of the object in terms of the desired property should be appropriately done.

In the case of the magnetic switch illustrated by Figure 1a, the nanomagnet (nm) will stick to the ferromagnetic material (F) regardless of its polarity. This would be the case of systems whose magnetization reversals are much faster than any mechanical process (like elastic forces) usually involving atomic masses. Therefore, such a device would be insensitive to magnetization reversals and will respond effectively to the magnitude of the magnetization of the nm switching off the circuit at a certain sensitive temperature T^*.

On the other hand, in the case of the magnetic switch illustrated by Figure 1b, the magnetic nanorod will always point with its north pole to the fixed magnet on top, which coincides with the black portion of the rod in the upper part of this figure at low T values. As temperature rises, a magnetization reversal is possible in the softer magnetic material in the nanorod; the north pole is now on the white part of the rod, and the nanomagnet will rotate with respect to its fixed center. Therefore, this device is sensitive to the polarity of the nanomagnet and will disconnect the circuit at a certain sensitive temperature T^{**}.

Even if the nanomagnets in both switches are made of the same material, T^* and T^{**} will be different in general because they originate from different properties: the former is governed by the value of the absolute magnetization, and the latter responds to the reversal process depending on the instantaneous sign of the magnetization. As we will show below, they are different indeed, and for the systems to be studied, $T^* > T^{**}$.

Figure 3 also yields the average of the actual instantaneous magnetization values obtained for FBC $< \mu_F(T) >$ over the $\nu = 50{,}000$ instants, namely,

$$< \mu_F(T) >= \frac{1}{50{,}000} \sum_{t=1}^{50{,}000} \mu_F(T,t) \; ; \tag{18}$$

These results are reported by means of empty circles (red) with the corresponding error bars. If we look at these data, we should estimate $T_F^{**} \leq 0.8$.

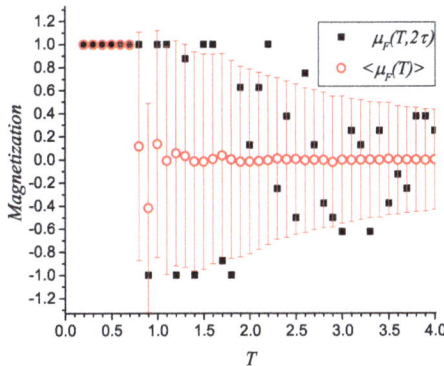

Figure 3. Solid (black) squares: last instantaneous magnetization $\mu(T, 2\tau)$ measured after 2×10^6 Monte Carlo steps (MCSs) for each temperature; empty circles (red) represent the average magnetization over $\nu = 50{,}000$ instantaneous measurements of magnetization at intervals of 20 MCSs after 10^6 MCSs of equilibration.

By enforcing ergodicity breaking in Figures 2 and 4, we are biasing the data in several senses: magnetization reversals are inhibited as if there is an infinite anisotropy; the instability of the systems is ignored; the curve is artificially smoothed; the average value of the magnetization is always positive; and it does not converge to 0.0 as $T \to \infty$. This last result is usually confused with the tails of the magnetization curves (and other order parameters) for $T > T^*$ due to finite size effects.

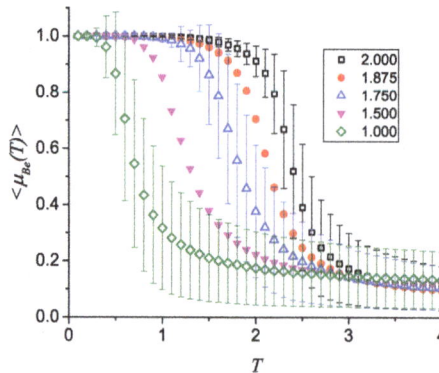

Figure 4. Simulations on 16×16 lattices for different effective coordination numbers κ whose values are given in the inset: 2.000 corresponds to PBC; 1.875 to FBC; 1.75 to the dilution of the lattice according to the decoration proposed in Figure 5 where four pieces $\ell \times \ell$ with $\ell = 4$ have been withdrawn; 1.500 corresponds to a similar decoration with $\ell = 6$; 1.00 corresponds to a similar decoration with $\ell = 7$. B means P when $\kappa = 2.000$, and it represents F for all of the other cases. For clarity, not all curves are shown with error bars.

From the previous discussion, it is clear that a unique definition for a Curie temperature in the nanoworld faces huge difficulties. At least the usual definition of a Curie temperature, where all magnetic properties change at one precise temperature, is not longer sustainable. Actually, in the TL, there would not be a difference between the two switches illustrated in Figure 1. However, we have just shown that for a magnetic systems of just very few interacting elements, different properties may have different sensitive temperatures.

The general message here is that when dealing with small particles, instantaneous responses may have a great importance and could determine completely the performance of the device. The details of this behavior will depend on the properties involved, geometry and other characteristics.

In any case, what is quite clear is that $< \mu(T) >$ is not the only property governing the response of the device at very small sizes. The large error bars in Figures 2 and 4 indicate that instantaneous measurements of magnetization can be erratic and at times could disconnect the device at unexpected temperatures under the desired one. As sizes get larger, error bars progressively decrease, and $< \mu(T) >$ becomes the right parameter to design devices as those shown in Figure 1.

Another important result shown by Figure 2 is the huge size of the error bars, which are likely associated with the small size of our system. To test this hypothesis, we will do numerical simulations similar to those of Figure 2 using both PBC and FBC for $L = 16$. Results (along with others to be defined immediately below) are presented in Figure 4 by empty squares and solid circles, respectively. In fact, error bars are now smaller; T^* has moved to lower values; the slope in the descent is more vertical (approaching the square step shape expected at the TL); and the remnant absolute magnetization towards high temperatures has decreased. Therefore, it is clear that we are moving in the direction where these simulations designed for the TL are effective.

We can make use of this opportunity to discuss another important consideration for small particle behavior: coordination number (or bonding, to put it in a more general context). Let us define effective coordination number κ as the ratio of the number of bonds B (nearest neighbors in our case) over the number of the N individuals (spins) interconnected by these B bonds. For any square lattice with PBC, $B = 2N$ and $\kappa_P^{(2)} = 2.0$; the upper index (2) refers to 2D. In the case of FBC, we have $B = 2L^2 - 2L$; $N = L^2$, so:

$$\kappa_F^{(2)} = \frac{2L^2 - 2L}{L^2} = 2 - \frac{2}{L}, \tag{19}$$

which for $L = 16$, yields $\kappa_F^{(2)} = 1.875$, as illustrated in the inset of Figure 4. It is clear that in the TL, $\kappa_F^{(2)} \to \kappa_P^{(2)}$, but for small particles, the difference might matter.

The effective coordination number defined above can be diminished upon diluting the original lattice. This can be achieved by human decoration on the original lattice (just removing atoms in a prescribed way) or could be due to some natural process that takes away magnetic atoms of our 2D sample (corrosion for instance). For simplicity, we will have the first alternative in mind, but the results for the second one would only add randomness to what we will discuss next.

Let us go back to the ferromagnetic lattice 16×16, and we will progressively dilute it upon removing four square sectors $\ell \times \ell$ each time in the way shown in Figure 5 for the case $\ell = 4$. To be more specific, the lattice is divided into four sectors $(L/2) \times (L/2)$, and the extraction of the ℓ^2 sites is done from the lower right corner of each sector. For the example under consideration, $N = L^2 - 4 \times \ell^2 = 192$ and $B = 336$, so $\kappa = 1.75$. Then, we remove four sectors with $\ell = 6$ and, finally, four sectors with $\ell = 7$, obtaining $\kappa = 1.5$ and $\kappa = 1.0$, respectively. In all of these cases, FBC are imposed. The last case is quite interesting because it is equivalent to the linear chain used by Ernest Ising in his original (and only) paper [38], for which the expected T_C is 0.0, namely no phase transition for finite temperatures.

Figure 5. Diluted 16×16 lattice upon decoration: four sectors 4×4 are removed to leave $N = 192$ spins with a total of $B = 336$ bonds; FBC are imposed to get $\kappa = 1.750$. Other similar decorations are defined in the text and in the previous figure.

Results for $< \mu(T) >$ are given in Figure 4, where the values of κ are given in the inset. Error bars were also calculated for all κ values, but they are shown for $\kappa = 2.000, 1.750$ and 1.000 only for the simplicity of the figure. Direct comparison shows that error bars for the lattice 16×16 are less pronounced than the corresponding ones for the lattice 4×4.

The main result of Figure 4 is to show the way T^* decreases as the effective coordination number decreases. This is even more so if ergodicity breaking is imposed, an effect that we will discuss again below.

Let us stress another shortcoming of the forced ergodicity breaking method mentioned before, namely the fact that $< \mu(T) >$ remains positive at high temperatures. In Figure 2, the last point of the simulation corresponds to $< \mu(4.0) > \approx 0.32$ with error bars of about 0.23 for FBC. For this lattice, we can actually make use of Equation (17) to obtain the exact value in the high T limit: $< \mu(T \to \infty) > \to 0.196$, which indicates that temperature has to be increased in the numeric simulations to attain a better comparison. We can follow this tendency simulating higher temperatures including larger lattices. Thus, Figure 6 shows results for $L = 4, 8$ and 16, where the positive asymptotic behavior is clearly shown; the values for $< \mu(40.0) >$ are $0.204, 0.104$ and 0.052, respectively. The first result is approaching the already quoted theoretical result 0.196. The set of values indicates that this sort of imposed remnant magnetization tends to be negligible for large lattices in the TL. A different way to put this is that magnetization reversals decrease as L increases, which makes the imposed ergodicity breaking coincide with the real ergodicity breaking. This tendency is shown in the inset of Figure 6 where in the logarithmic scale, we plot the number of magnetization reversals during the 50,000 measurements as a function of $1/L$ for $T = 1.5$. For higher temperatures, reversals are easier, but the amplitudes are lower; for lower temperatures, reversals are harder, but the amplitudes are higher (which makes the absolute value of $\mu(T, t)$ stay close to unity). When ergodicity breaking is imposed at low temperatures, magnetization always oscillates between values close to $+1$ or -1.

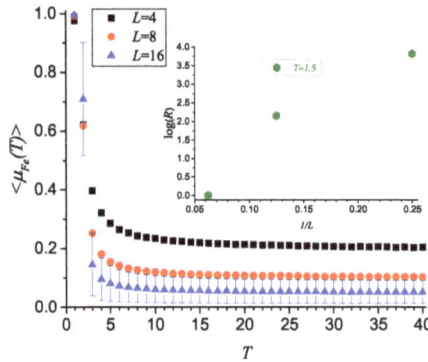

Figure 6. Average absolute magnetization for lattices of different sizes showing the tails for the artificial remnant magnetization induced by forced ergodicity breaking. The inset shows the decrease with the size of the number of magnetization reversal within 50,000 measurements after equilibration. For clarity, error bars are given for $L = 16$ only.

The curve for $\kappa = 1.000$ in Figure 4 deserves a special analysis since it is expected that the magnetization should go down as soon at $T > 0$. However, this figure is for $< \mu(T) >$, namely for forced ergodicity breaking, to compare with Figure 2, related to the analytic expressions given by Equations (16) and (17). If we go back to the data for this case, we can study each temperature to find that reversions are occurring at all temperatures. For instance, $< \mu(0.6) > \approx 0.7$ for $\kappa = 1.000$ in Figure 4; however, if we look at the instantaneous data for the 100 leading intervals of the measuring interval (immediately after equilibration), we find $\mu(0.6, t)$ oscillating as given in Figure 7.

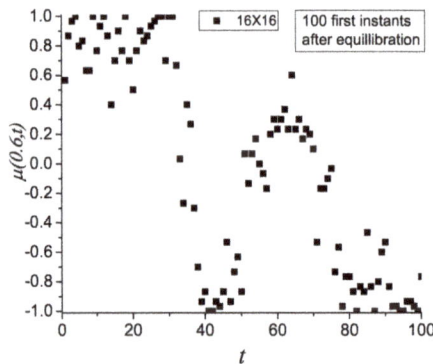

Figure 7. Instantaneous magnetization after $\tau = 10^6$ MCSs of equilibration for the lattice 16×16, $\kappa = 1.0$, with FBC at $T = 0.6$. Measurements are separated by 20 MCSs.

For lower temperatures, oscillations are progressively less and less frequent approaching alternations between $+1.0$ and -1.0 only at $T = 0.1$, where it is necessary to wait a huge amount of time to get the first reversal. In this way, we approach the result for the 1D Ising lattice: no phase transition is possible at a finite temperature.

Let us now investigate the role of the effective coordination number towards the 3D case with PBC ($\kappa_P^{(3)} = 3.0$). This is shown in Figure 8, where we begin by a single 4×4 layer with FBC ($\kappa_F^{(2)} = 1.875$); then, we continue the series: two layers 4×4 with FBC ($\kappa_{F1}^{(3)} = 2.0$); three layers 4×4 with FBC ($\kappa_{F2}^{(3)} = 2.167$); four layers 4×4 with FBC ($\kappa_{F3}^{(3)} = 2.25$); four layers 4×4, two directions with FBC and one direction with PBC ($\kappa_{FP}^{(3)} = 2.5$); four layers 4×4, one direction with FBC and two directions with PBC ($\kappa_{PF}^{(3)} = 2.75$); four layers 4×4, all directions with PBC ($\kappa_P^{(3)} = 3.0$). The result is eloquent: T^* moves to higher temperatures as the effective coordination number increases. Moreover, the dimensionality plays also an important role, which can be appreciated when we compare the second case here having $\kappa_{F1}^{(3)} = 2.0$, with the already seen case $\kappa_P^{(2)} = 2.0$ for the 4×4 PBC lattice (shown in Figure 4). It can be appreciated that for the 3D case, T^* is larger than for the 2D case.

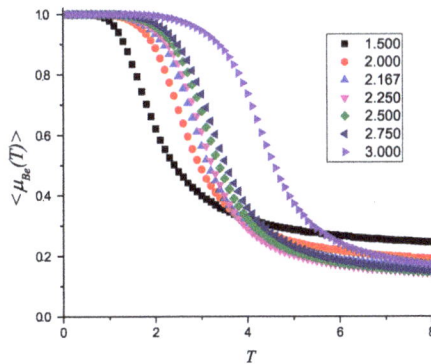

Figure 8. Magnetization for different systems based on 4×4 layers where the effective coordination number κ varies as given in the inset: 1.50 (one layer ; B = F); 2.0 (two layers 4×4; B = F); 2.167 (three layers 4×4; B = F); 2.25 (four layers 4×4; B = F); 2.50 (four layers 4×4; B = 2F + 1P); 2.75 (four layers 4×4; B = 1F + 2P); 3.0 (four layers 4×4; B = P) (the first curve is truly 2D, and part of it was already included in Figure 2; it is given here for completeness and comparison purposes).

Now, we show results for the q-states clock model simulated on an 8×8 square lattice. The density of magnetic states increases as q increases; therefore, we expect a q dependence of the sensitive temperature characterizing the transit from an ordered ferromagnetic initial regime to a magnetically-disordered phase.

Figure 9 shows the sensitive temperature T^* vs. $1/q$ ranging from $q = 2$ to $q = 20$. Here, we follow the same approximation as in Figure 2, namely we force ergodic separation starting always from a ferromagnetic state. T^* is defined as the temperature where the normalized magnetization is $< \mu_e(T^*) >= 0.5$, which is a practical way to define the transition to a disordered state. In order to sample with the same density the q-possible states for each spin in the 8×8 square lattice (there are q^{64} states for an 8×8 lattice), we increase the number of MCSs as q increases; higher q values need larger numbers of MCS to reach equilibrium. Therefore, we use $\tau q/2$ MCSs for the simulations of the q-clock model.

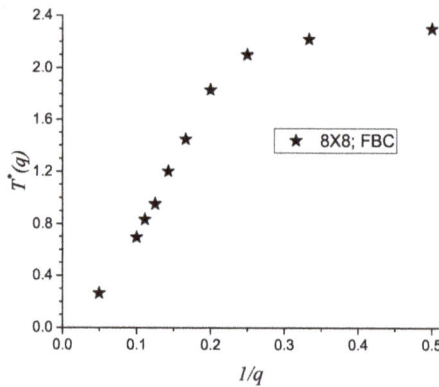

Figure 9. Sensitive temperature T^* from a magnetic ordered to a disordered phase for the q-states clock model as a function of $1/q$. A square lattice 8×8 with FBC is used. Measurements are obtained after 5×10^6 MCSs for averaging.

As we see in Figure 9, there is a monotonous decrease of T^* with increasing values of q. Larger values of q mean energy states separated by less energy differences, therefore moving the initial ferromagnetic state to other states with different magnetizations in an easier way: this implies that the disorder for higher q values will be reached at lower temperatures as compared with the starting $q = 2$ case (Ising model).

A very striking feature is shown by Figure 9: two regimes are clearly shown: for $q = 2$ to $q = 4$, and from $q = 5$ to $q = 20$ (and possible larger values). In the first regime, T^* decreases with a relatively low slope as a function of $1/q$, whereas for $q > 4$, the slope of the T^* vs. $1/q$ curve is larger. This change of slope could be associated with different types of transitions that occur for different q values. Namely, there is an order $\rightarrow$ disorder transition for $q <= 4$ and a double transition order $\rightarrow$ quasi-ordered $\rightarrow$ disordered for $q >= 5$ as discussed in [58,61].

It is clear from symmetry arguments that as q increases, lower temperatures are sufficient to originate new ferromagnetic states in any of the q directions irrespective of the starting configuration. Therefore, as q tends to infinity, T^* tends to zero. This is known as the Mermin–Wagner theorem [62]. This theorem establishes that there is no magnetic state at any finite temperature ($T > 0$) in the XY model. Moreover the formation of vortex-antivortex states is prohibited in an 8×8 lattice because of its small size. Therefore, the BKT transition is also absent in samples of such a small size.

4. Conclusions

As magnetic particles get very tiny, comprising a low number of elements, especially at the nanoscopic level, the ergodic separation is no longer possible in the way that it is handled in the TL. This means that as temperature progressively increases, very soon, all of the configuration space becomes available, with positive and negative components of the magnetization. While the transit time from a state with one magnetization sign to the opposite one is faster than other characteristic times of the system (elastic forces, vibrations, etc.), the particle will present a magnetization whose magnitude could make it stick to a ferromagnetic material, as is shown in Figure 1a. However, as the temperature continues to increase, such a magnitude will weaken, changing the static magnetic properties of the particle.

For tiny magnetic particles, statistical fluctuations are larger than the needed reliability of the particle, which could be reflected by a given experimental accuracy. This fact is reflected in the error bars found in previous measurements suggesting that at any given temperature, oscillations are large, and a unique value of magnetization cannot be found at any temperature. This behavior requires extreme care when using magnetic particles at the molecular level being part of a device since the performance of this tiny magnet could be under what could be needed. If the dimensions available allow it, it would be safer to use slightly larger particles that already behave as macroscopic ones. The sensitive temperature at which the particle changes according to its magnitude is T^*, which has to be thought of as a rather large range centered at a value. Towards the TL, such a range should disappear, and $T^* \to T_C$.

Such considerations are even more important when the response of the particle is based on the orientation of the nanomagnet, as is proposed in Figure 1b. Here, a magnetization reversal (even if the magnitude remains high) can trigger a reorientation of the particle. Such magnetization reversals could happen at almost any temperature for small particles. The main conditions are that they become less and less frequent as temperature is lowered and size increases. The sensitive temperature at which the particle changes according to its polarity is T^{**}, which again has to be thought of as a range centered at a value. From previous results, it follows that $T^{**} < T^*$, since reversals at low T tend to fluctuate between extreme magnetization values. Towards the TL, we can expect that $T^{**} \to T^* \to T_C$.

Previous results show that a definition for the Curie temperature for small systems, even if a defying and interesting task, does not add any new physical insight for what will be the real response of a single nanosized ferromagnet of any given material. The different characteristics of the tiny system contribute to the fluctuations of its magnetization that in the last run dominate its response. Alternatively, the Curie temperature is a thermodynamic parameter arising from ensemble averages; in this way, T_C finds a unique definition in the TL.

However, even some of the theoretical treatments intended for the TL need a critical reformulation. One of them is the fact that the high temperature sector of the magnetization curve is lifted since all magnetizations are taken as positive when ergodic separation is assumed and imposed. Then, critical temperatures tend to be overestimated. In addition, the zero magnetization value is not reachable for finite sizes and finite temperatures as presented in Figure 6 since the tails are forced to be always on the positive side of the magnetization; so not always are the long asymptotic positive tails exclusively due to finite size effects, as is frequently assumed.

In the case of very small particles, simulations cannot impose PBC since this would ignore the surface states, which become more important for these systems. The energy spectrum of surface elements means a more compact and homogeneously-distributed density of states for the same size (see Tables 1 and 2). In terms of the magnetization of the system, simulations with FBC lead to softer magnetism (lower T^*) than simulations for a system of the same size, but with PBC.

Dilution of the lattice upon remotion of magnetic sites (intended or accidental) will lead to an increase of the surface, thus lowering the effective coordination number κ. Materials get softer, namely sensitive temperatures move to lower values, as κ decreases. This phenomenon has still another implication since lower effective coordination numbers also mean higher chances of magnetization reversal. Thus, if magnetic tiny particles lose magnetic mass through corrosion, the sensitive temperatures could move to lower values due to aging. Previous effects were tested through several κ values ranging from 1.0 through 3.0 (namely from one to three dimensions). This allowed one to appreciate that for similar κ values, materials are harder in the higher dimensionality realization.

The instabilities due to magnetization reversal events decrease very rapidly with size. In the case of a 16×16 lattice, FBC, at $T = 1.5$, the ergodicity breaking was violated just once for 50,000 instantaneous measurements. This result seems to indicate that for lattices over 32×32 (about 10^3 magnetic sites), magnetization reversals can be progressively ignored for most of the practical applications.

When the density of states gets denser due to more internal degrees of freedom, the sensitive temperatures decrease. This was tested with the clock model varying q, the number of independent orientations of the spin on the 2D lattice. Actually, as q increases, T^* tends to approach 0.0 in the expected Mermin–Wagner theorem valid for $q \to \infty$. It is quite interesting that this result obtained for macroscopic systems is also obtained for a lattice as small as an 8×8 lattice with FBC.

Finally, we want to stress that in the present study, we have not found any evidence for failure of any of the basic thermodynamic principles, which seem to be very robust. However, we have presented evidence indicating that some approximations done in the TL cannot a priori be applied to magnetic particles with a thousand elements or less. Care has to be paid to the way averages are performed or even if taking averages to describe instant responses of the system makes any sense; actually, such a small magnetic particle could exhibit properties associated with its proper fluctuations.

Acknowledgments: Partial support from the following two Chilean sources is acknowledged: Fondecyt under Contract 1150019 and Financiamiento Basal para Centros Científicos y Tecnológicos de Excelencia (Chile) through the Center for Development of Nanoscience and Nanotechnology (CEDENNA). PV acknowledges DGIIPfrom Universidad Santa María under Contract PI-M-17-3. CONICET (Argentina) under Project Number PIP 112-201101-00615; Universidad Nacional de San Luis (Argentina) under Project 03-0816; and the National Agency of Scientific and Technological Promotion (Argentina) under project PICT-2013-1678.

Author Contributions: Eugenio E. Vogel coordinated the work of the different coauthors, did most of the writing, prepared most of the figures and did most of the calculations. Patricio Vargas wrote the code for the clock model and calculated the corresponding results including the last figure. He also collaborated to the writing of several passages of the manuscript. Gonzalo Saravia wrote the code and obtained some results for the simulations on 2D and 3D Ising systems. Julio Valdés obtained the partition functions and magnetization for small Ising systems with and without periodic boundary conditions. A.J. Ramirez-Pastor wrote the Introduction and other passages of the paper, performed the main bibliography search, reviewed the analytic results and helped with the discussions, comments and conclusions. Paulo Centres wrote the code and performed the calculations for the size variations of the 2D Ising model and elaborated the figures, providing the discussion and comments on them.

Conflicts of Interest: The authors declare no conflict of interest.

Glossary of Symbols

The following alphabetic list gives the main symbols used in this article, their basic meaning and the number of the equation (if any) where they were defined or used for the first time.

B	number of bonds in the lattice.
c_{Fn}	degeneracy of the n-th energy level for FBC, (8).
c_{Pn}	degeneracy of the n-th energy level for PBC, (7).
δ	energy cost due to the possible flip of the visited spin.
e	subindex to represent that ergodic separation has been enforced.
E_n	energy on the n-th level, (6).
$f_{n,k}$	degeneracy of the k-th configuration within the n-th energy level for FBC, (11).
F	denotes free boundary conditions (FBC).
H	Hamiltonian of a systems of N spins, (1).
i or j	position of spin S_i or S_j, (1).
J	exchange interaction ($J = 1$), (1).
κ	effective coordination number, (19).
L	side of the square lattice $L \times L$.
$m(T,t)$	magnetization for temperature T and at time t, (2).
$m_O(T)$	magnetization for Onsager solution at temperature T, (4).
$m_F(T)$	Magnetization at temperature T for FBC from analytic expressions.
$m_P(T,t)$	Magnetization at temperature T for PBC from analytic expressions.
$M_{n,k}$	magnetization for the k-th configuration within the n-th energy level, (9).
MCS	Monte Carlo step (plural: MCSs).
$\mu_F(T,t)$	Magnetization at instant t and temperature T for FBC from numeric calculations.
$\mu_P(T,t)$	Magnetization at instant t and temperature T for PBC from numeric calculations.
$< \mu(T) >$	average value of the numeric magnetization over 50,000 instants after equilibration, (18).
n	index to label energy levels, (6).
N	number of magnetic elements (spins), (1).
ν	number of measurements done within the interval of 10^6 MCCs.
$p_{n,k}$	degeneracy of the k-th configuration within the n-th energy level for PBC, (10).

P	periodic boundary conditions (PBC).
r	random number.
R	number of reversions in the nanomagnet during 50,000 measurements.
S	Ising spin: $+1$ or -1, ($\cdot$).
t	time instant.
τ	period of equilibration (10^6 MCSs).
T_C	Curie temperature, ($\cdot$).
T^*	Sensitive temperature depending on the magnitude of the magnetization.
T^{**}	Sensitive temperature depending on the sign of the magnetization.
$Z(T)$	Partition function at temperature T, ($\cdot$).
$Z_F(T)$	$Z(T)$ for free boundary conditions, ($\cdot$).
$Z_P(T)$	$Z(T)$ for periodic boundary conditions, ($\cdot$).

References

1. Mehta, A.; Rief, M.; Spudich, J.; Smith, D.; Simmons, R. Single-molecule biomechanics with optical methods. *Science* **1999**, *283*, 1689–1695.
2. Xu, X.; Yin, S.; Moro, R.; de Heer, W.A. Magnetic Moments and Adiabatic Magnetization of Free Cobalt Clusters. *Phys. Rev. Lett.* **2005**, *95*, 237209.
3. Ritort, F. Single molecule experiments in biological physics: Methods and applications. *J. Phys. C* **2006**, *18*, R531–R583.
4. Greenleaf, W.; Woodside, M.; Block, S. High-resolution, single-molecule measurements of biomolecular motion. *Annu. Rev. Biophys. Biomol. Struct.* **2007**, *36*, 171–190.
5. Roy, R.; Hohng, S.; Ha, T. A practical guide to single-molecule FRET. *Nat. Methods* **2008**, *5*, 507–516.
6. Sarkar, T.; Raychaudhuri, A.K.; Bera, A.K.; Yusuf, S.M. Effect of size reduction on the ferromagnetism of the manganite $La_{1-x}Ca_xMnO_3$ (x = 0.33). *New J. Phys.* **2010**, *12*, 123026.
7. Gross, D.H.E. *Microcanonical Thermodynamics. Phase Transitions in Small Systems*; World Scientific: Singapore, 2001.
8. Casetti, L.; Kastner, M. Nonanalyticities of Entropy Functions of Finite and Infinite Systems. *Phys. Rev. Lett.* **2006**, *97*, 100602.
9. Bertoldi, D.S.; Bringa, E.M.; Miranda, E.N. Exact solution of the two-level system and the Einstein solid in the microcanonical formalism. *Eur. J. Phys.* **2011**, *32*, 1485–1493.
10. Miranda, E.N.; Bertoldi, D.S. Thermostatistics of small systems: Exact results in the microcanonical formalism. *Eur. J. Phys.* **2013**, *34*, 1075–1087.
11. Baletto, F.; Ferrando, R. Structural properties of nanoclusters: Energetic, thermodynamic, and kinetic effects. *Rev. Mod. Phys.* **2005**, *77*, 371–423.
12. Berry, R.S.; Smirnov, B.M. Phase transitions in various kinds of clusters. *PHYS-USP* **2009**, *52*, 137–164.
13. Ohring, M. *Materials Science of Thin Films*; Academic: San Diego, CA, USA, 2002.
14. Yanagida, T. Fluctuation as a tool of biological molecular machines. *BioSystems* **2008**, *93*, 3–7.
15. Hill, T.L. Thermodynamics of small systems. *J. Chem. Phys.* **1962**, *36*, 3182–3197.
16. Hill, T.L. *Thermodynamics of Small Systems, Parts I & II*; Dover Publications: Mineola, NY, USA, 2013.
17. Hill, T.L.; Chamberlin, R.V. Extension of the thermodynamics of small systems to open metastable states: An example. *Proc. Natl. Acad. Sci. USA* **1998**, *951*, 12779–12782.
18. Chamberlin, R.V. Mean-field cluster model for the critical behaviour of ferromagnets. *Nature* **2000**, *408*, 337–339.
19. Hill, T.L. Perspective: Nanothermodynamics. *Nano Lett.* **2001**, *1*, 111–112.
20. Hill, T.L. A different approach to nanothermodynamics. *Nano Lett.* **2001**, *1*, 273–275.
21. Lu, A.-H.; Salabas, E.L.; Schüth, F. Magnetic Nanoparticles: Synthesis, Protection, Functionalization, and Application. *Angew. Chem. Int. Ed.* **2007**, *46*, 1222–1244.
22. Jun, Y.-W.; Seo, J.-W.; Cheon, J. Nanoscaling Laws of Magnetic Nanoparticles and Their Applicabilities in Biomedical Sciences. *Acc. Chem. Res.* **2008**, *41*, 179–189.
23. Rong, C.-B.; Li, D.; Nandwana, V.; Poudyal, N.; Ding, Y.; Wang, Z.L.; Zeng, H.; Liu, J.P. Size-Dependent Chemical and Magnetic Ordering in $L1_0$-FePt Nanoparticles. *Adv. Mater.* **2006**, *18*, 2984–2988.
24. Ziese, M.; Semmelhack, H.C.; Hans, K.H.; Sena, S.P.; Blythe, H.J. Thickness dependent magnetic and magnetotransport properties of strain-relaxed $La_{0.7}Ca_{0.3}MnO_3$ films. *J. Appl. Phys.* **2002**, *91*, 9930–9936.

25. Lu, H.M.; Li, P.Y.; Huang, Y.N.; Meng, X.K.; Zhang, X.Y.; Liu, Q. Size-dependent Curie transition of Ni nanocrystals. *J. Appl. Phys.* **2009**, *105*, 023516.

26. Palomares-Baez, J.-P.; Panizon, E.; Ferrando, R. Nanoscale effects on phase separation. *Nano Lett.* **2017**, doi:10.1021/acs.nanolett.7b01994.

27. Sun, Z.Z.; Wang, X.R. Theoretical Limit of the Minimal Magnetization Switching Field and the Optimal Field Pulse for Stoner Particles. *Phys. Rev. Lett.* **2006**, *97*, 077205.

28. Wernsdorfer, W. Magnetic Anisotropy and Magnetization Reversal Studied in Individual Particles. In *Surface Effects in Magnetic Nanoparticles*; Dino, F., Ed.; Springer: Berlin, Germany, 2005.

29. Schick, M.; Walker, J.S.; Wortis, M. Phase diagram of the triangular Ising model: Renormalization-group calculation with application to adsorbed monolayers. *Phys. Rev. B* **1977**, *16*, 2205–2219.

30. Binder, K.; Landau, D.P. Phase diagrams and critical behavior in Ising square lattices with nearest- and next-nearest-neighbor interactions. *Phys. Rev. B* **1980**, *21*, 1941–1962.

31. Kinzel, W.; Schick, M. Phenomenological scaling approach to the triangular Ising antiferromagnet. *Phys. Rev. B* **1981**, *23*, 3435–3441.

32. Landau, D.P. Critical and multicritical behavior in a triangular-lattice-gas Ising model: Repulsive nearest-neighbor and attractive next-nearest-neighbor coupling. *Phys. Rev. B* **1983**, *27*, 5604–5617.

33. Binder, K.; Landau, D.P. Finite-size scaling at first-order phase transitions. *Phys. Rev. B* **1984**, *30*, 1477–1485.

34. Landau, D.P.; Binder, K. Phase diagrams and critical behavior of Ising square lattices with nearest-, next-nearest-, and third-nearest-neighbor couplings. *Phys. Rev. B* **1985**, *31*, 5946–5953.

35. Chin, K.K.; Landau, D.P. Monte Carlo study of a triangular Ising lattice-gas model with two-body and three-body interactions. *Phys. Rev. B* **1987**, *36*, 275–284.

36. Landau, D.P.; Binder, K. Monte Carlo study of surface phase transitions in the three-dimensional Ising model. *Phys. Rev. B* **1990**, *41*, 4633–4645.

37. Kämmerer, S.; Dünweg, B.; Binder, K.; d'Onorio de Meo, M. Nearest-neighbor Ising antiferromagnet on the fcc lattice: Evidence for multicritical behavior. *Phys. Rev. B* **1996**, *53*, 2346–2351.

38. Ising, E. Beitrag zur Theorie des Ferromagnetismus. *Z. Phys.* **1925**, *31*, 253–258.

39. Domb, C. *Phase Transitions and Critical Phenomena*; Series Expansions for Lattice Models; Domb, C., Green, M.S., Eds.; Academic Press: New York, NY, USA, 1974; p. 1.

40. Fisher, M.E. The theory of equilibrium critical phenomena. *Rep. Prog. Phys.* **1967**, *30*, 615–730.

41. Onsager, L. Crystal Statistics. I. A Two-Dimensional Model with an Order-Disorder Transition. *Phys. Rev.* **1944**, *65*, 117–149.

42. Hill, T.L. *An Introduction to Statistical Thermodynamics*; Addison Wesley: Reading, MA, USA, 1960.

43. Bethe, H. Statistical theory of superlattices. *Proc. R. Soc. Lond. A* **1935**, *150*, 552–575.

44. Velásquez, E.A.; Mazo-Zuluaga, J.; Restrepo, J.; Iglesias, O. Pseudocritical behavior of ferromagnetic pure and random diluted nanoparticles with competing interactions: Variational and Monte Carlo approaches. *Phys. Rev. B* **2011**, *83*, 184432.

45. Bertoldi, D.S.; Bringa, E.M.; Miranda, E.N. Analytical solution of the mean field Ising model for finite systems. *J. Phys. Condens. Matter* **2012**, *24*, 226004.

46. Chamberlin, R.V. The Big World of Nanothermodynamics. *Entropy* **2015**, *17*, 52–73.

47. Berezinskii, V.L. Destruction of Long-range Order in One-dimensional and Two-dimensional Systems having a Continuous Symmetry Group I. Classical Systems. *Zh. Eksp. Teor. Fiz.* **1971**, *59*, 907–920.

48. Kosterlitz, J.M.; Thouless, D.J. Long range order and metastability in two dimensional solids and superfluids. (Application of dislocation theory). *J. Phys. C Solid State Phys.* **1972**, *5*, L124.

49. *40 Years of Berezinskii-Kosterlitz-Thouless Theory*; Jose, J.V., Ed.; World Scientific: London, UK, 2013.

50. Kosterlitz, J.M.; Thouless, D.J. Ordering, metastability and phase transitions in two-dimensional systems. *J. Phys. C Solid State Phys.* **1973**, *6*, 1181–1203.

51. Kosterlitz, J.M. The critical properties of the two-dimensional xy model. *J. Phys. C Solid State Phys.* **1974**, *7*, 1046–1060.

52. Jose, J.V.; Kadanoff, L.P.; Kirkpatrick, S.; Nelson, D.R. Renormalization, vortices, and symmetry-breaking perturbations in the two-dimensional planar model. *Phys. Rev. B* **1977**, *16*, 1217–1241.

53. Kenna, R. The XY Model and the Berezinskii-Kosterlitz-Thouless Phase Transition. *arXiv* **2005**, arXiv:cond-mat/0512356.

54. Elitzur, S.; Pearson, R.B.; Shigemitsu, J. Phase structure of discrete Abelian spin and gauge systems. *Phys. Rev. D* **1979**, *19*, 3698–3714.

55. Cardy, J.L. General discrete planar models in two dimensions: Duality properties and phase diagrams. *J. Phys. A Math. Gen.* **1980**, *13*, 1507–1515.

56. Fröhlich, J.; Spencer, T. The Kosterlitz-Thouless transition in two-dimensional Abelian spin systems and the Coulomb gas. *Commun. Math. Phys.* **1981**, *81*, 527–602.

57. Ortiz, G.; Cobanera, E.; Nussinov, Z. Dualities and the phase diagram of the p-clock model. *Nucl. Phys. B* **2012**, *854*, 780–814.

58. Borisenko, O.; Cortese, G.; Fiore, R.; Gravina, M.; Papa, A. Numerical study of the phase transitions in the two-dimensional Z(5) vector model. *Phys. Rev. E* **2011**, *83*, 041120.

59. Binder, K. Applications of Monte Carlo methods to statistical physics. *Rep. Prog. Phys.* **1997**, *60*, 487–559.

60. Vogel, E.E.; Saravia, G.; Bachmann, F.; Fierro, B.; Fischer, J. Phase transitions in Edwards-Anderson model by means of information theory. *Physica A* **2009**, *388*, 4075–4082.

61. Seung, K.B.; Petter, M. Non-Kosterlitz-Thouless transitions for the q-state clock models. *Phys. Rev. E* **2010**, *82*, 031102.

62. Mermin, N.D.; Wagner, H. Absence of Ferromagnetism or Antiferromagnetism in One- or Two-Dimensional Isotropic Heisenberg Models. *Phys. Rev. Lett.* **1966**, *17*, 1133–1136.

![entropy logo]

entropy

MDPI

Article

Exact Expressions of Spin-Spin Correlation Functions of the Two-Dimensional Rectangular Ising Model on a Finite Lattice

Tao Mei

Department of Journal, Central China Normal University, Wuhan 430079, China; meitao@mail.ccnu.edu.cn

Received: 8 February 2018; Accepted: 10 April 2018; Published: 12 April 2018

Abstract: We employ the spinor analysis method to evaluate exact expressions of spin-spin correlation functions of the two-dimensional rectangular Ising model on a finite lattice, special process enables us to actually carry out the calculation process. We first present some exact expressions of correlation functions of the model with periodic-periodic boundary conditions on a finite lattice. The corresponding forms in the thermodynamic limit are presented, which show the short-range order. Then, we present the exact expression of the correlation function of the two farthest pair of spins in a column of the model with periodic-free boundary conditions on a finite lattice. Again, the corresponding form in the thermodynamic limit is discussed, from which the long-range order clearly emerges as the temperature decreases.

Keywords: two-dimensional Ising model; spin-spin correlation functions; exact solution; short-range order; long-range order

---

1. Introduction

Since the exact solution of the partition function in the absence of a magnetic field of the two-dimensional rectangular Ising model with periodic-periodic boundary conditions is obtained in the thermodynamic limit [1] and in finite-size systems [2], many authors have contributed to the knowledge of various aspects of this model, such as different boundary conditions, the arrangement modes of the spin lattice, surfaces, or mathematical methods, etc. [3–7].

Besides the partition function of the model, the calculations of spin-spin correlation functions are an important subject in the research of the two-dimensional Ising model. Some expressions of correlation functions in the thermodynamic limit have been obtained [3–5,8,9], and the case in a finite lattice has been studied [10–12].

The determination of exact expressions of the partition function and spin-spin correlation functions of the model on a finite lattice is not only a theoretical subject; the results obtained can also be used in the research of finite-size scaling, finite-size corrections, and boundary effects [7,13–16].

In this paper, we present some exact expressions of spin-spin correlation functions of the two-dimensional rectangular Ising model on a finite lattice by employing the spinor analysis method [2].

In Section 2, for the model with L rows and N columns and periodic-periodic boundary conditions (Onsager's lattice), we calculate some exact expressions of the correlation function $\langle \sigma_{1,1}\sigma_{1,1+Q} \rangle$ and compare the corresponding forms in the thermodynamic limit obtained here with known results presented in Reference [9]. The investigation in Section 2 shows the main steps, key points, and problems of the approach used in this paper. Since the whole process are complex, here we outline the approach.

(1) Although any spin-spin correlation functions $\langle \sigma_{l,n}\sigma_{l',n'} \rangle$ can be expressed by matrices, and the matrices belong to spin representatives [8], here we only consider $\langle \sigma_{l,1}\sigma_{l,1+Q} \rangle$, i.e., the correlation functions of pairwise spins in one column, of which exact expressions can be obtained by the spinor analysis method.

(2) When we employ the spinor analysis method to evaluate $\langle \sigma_{l,1}\sigma_{l,1+Q} \rangle$, it is very difficult to find the exact eigenvalues of the corresponding rotation matrix; however, the operators of the first derivative and limit in the expressions of $\langle \sigma_{l,1}\sigma_{l,1+Q} \rangle$ (see (8)) allow us to obtain exact expressions of $\langle \sigma_{1,1}\sigma_{1,1+Q} \rangle$ by only finding approximate eigenvalues of the rotation matrix.

(3) We employ Rayleigh-Schrodinger Perturbation Theory (RSPT) in quantum mechanics and change "finding eigenvalue up to Q-th order" to "finding eigenvalue through Q times first-order approximation" (see the discussions in Section 2.5) to find approximate eigenvalues of the rotation matrix. On the one hand, this approximate method enables us to actually carry out the calculation process. On the other hand, since RSPT is irregular, the approximate method is in fact incapable when Q is very larger. Hence, by this approach we can only obtain exact expressions of correlation functions when Q is a small number, for example, that of $\langle \sigma_{1,1}\sigma_{1,2} \rangle$, $\langle \sigma_{1,1}\sigma_{1,3} \rangle$, $\langle \sigma_{1,1}\sigma_{1,4} \rangle$, $\cdots$, etc., which belong to the short-range order.

What is more interesting is the long-range order, as it is closely related to properties of the phase transformation of the system. To obtain the correlation function that can reveal the long-range order of the model on a finite lattice, we turn to the model with L rows and N columns and periodic boundary condition in the horizontal direction and free boundary condition in the vertical direction. For this model, because of the free boundary condition in a column, and $\sigma_{l,N}$ is the farthest spin of $\sigma_{l,1}$ in the column, we forecast that the correlation functions $\langle \sigma_{l,1}\sigma_{l,N} \rangle$, $\langle \sigma_{l,1}\sigma_{l,N-1} \rangle$, $\langle \sigma_{l,1}\sigma_{l,N-2} \rangle$, $\cdots$ will display the long-range order.

For the model with periodic-free boundary conditions, if we write $\langle \sigma_{l,1}\sigma_{l,1+Q} \rangle$ in matrix forms and use the method presented in Section 2, then we still can only obtain exact expressions of $\langle \sigma_{1,1}\sigma_{1,2} \rangle$, $\langle \sigma_{1,1}\sigma_{1,3} \rangle$, $\cdots$ but cannot obtain that of $\langle \sigma_{l,1}\sigma_{l,N} \rangle$, $\langle \sigma_{l,1}\sigma_{l,N-1} \rangle$, $\cdots$. On the other hand, if we write $\langle \sigma_{1,1}\sigma_{1,N-n} \rangle$ in matrix forms directly (see Formulas (52) and (53) in this paper), then the method presented in Section 2 is feasible to deal with the matrix forms of $\langle \sigma_{1,1}\sigma_{1,N-n} \rangle$, and we therefore can obtain some exact expressions of $\langle \sigma_{l,1}\sigma_{l,N} \rangle$, $\langle \sigma_{l,1}\sigma_{l,N-1} \rangle$, $\langle \sigma_{l,1}\sigma_{l,N-2} \rangle$, $\cdots$.

However, to save space, in this paper we no longer discuss $\langle \sigma_{l,1}\sigma_{l,N-1} \rangle$, $\langle \sigma_{l,1}\sigma_{l,N-2} \rangle$, $\cdots$ but only evaluate $\langle \sigma_{l,1}\sigma_{l,N} \rangle$, for which all of the matrix forms, the corresponding rotation matrices, and the eigenvalue equations have been given by Reference [12]. Therefore, we only need to derive the exact expression of $\langle \sigma_{l,1}\sigma_{l,N} \rangle$ by employing the method presented in Section 2. (The reason why the determination of the exact expression of $\langle \sigma_{l,1}\sigma_{l,N} \rangle$ fails in Reference [12] is explained in Section 3.3).

After obtaining the exact expression of $\langle \sigma_{l,1}\sigma_{l,N} \rangle$, we discuss the properties of the expression of $\langle \sigma_{l,1}\sigma_{l,N} \rangle$ in the thermodynamic limit, from which the long-range order emerges as the temperature decreases, as shown clearly.

2. Short-Range Order in Onsager's Lattice

2.1. The Definition of the Spin-Spin Correlation Functions and Their Basic Properties

According to the definition of the spin-spin correlation functions, $\langle \sigma_{l,n}\sigma_{l+P,n+Q} \rangle$ of pairwise spins $\sigma_{l,n}$ and $\sigma_{l+P,n+Q}$ of Onsager's lattice read:

$$\langle \sigma_{l,n}\sigma_{l+P,n+Q} \rangle = \frac{1}{Z} \sum_{\{\sigma_h,v=\pm1\}} \sigma_{l,n}\sigma_{l+P,n+Q} \exp\left(\frac{J'}{kT} \sum_{l'=1}^{L} \sum_{n'=1}^{N} \sigma_{l',n'}\sigma_{l'+1,n'} \right) \exp\left(\frac{J}{kT} \sum_{l'=1}^{L} \sum_{n'=1}^{N} \sigma_{l',n'}\sigma_{l',n'+1} \right), \quad (1)$$

where $\sigma_{L+1,n} = \sigma_{1,n}$, $\sigma_{l,N+1} = \sigma_{l,1}$; $J'(> 0)$, and $J(> 0)$ are the interaction constants for the horizontal and vertical directions, respectively;

$$Z = \sum_{\{\sigma_h,v=\pm1\}} \exp\left(\frac{J'}{kT} \sum_{l=1}^{L} \sum_{n=1}^{N} \sigma_{l,n}\sigma_{l+1,n} \right) \exp\left(\frac{J}{kT} \sum_{l=1}^{L} \sum_{n=1}^{N} \sigma_{l,n}\sigma_{l,n+1} \right) \quad (2)$$

is the partition function in absence of a magnetic field of the model.

According to the periodic-periodic boundary conditions of Onsager's lattice, it is easy to prove $\langle \sigma_{l,n}\sigma_{l+P,n+Q} \rangle = \langle \sigma_{1,1}\sigma_{1+P,1+Q} \rangle$; in this paper, we calculate $\langle \sigma_{l,n}\sigma_{l,n+Q} \rangle = \langle \sigma_{1,1}\sigma_{1,1+Q} \rangle$, i.e., we only calculate correlation functions of pairwise spins in one column.

Further, because of periodic-periodic boundary conditions, both $\sigma_{1,2}$ and $\sigma_{1,N}$ are the closest spins of $\sigma_{1,1}$, and we therefore have $\langle \sigma_{1,1}\sigma_{1,2} \rangle = \langle \sigma_{1,1}\sigma_{1,N} \rangle$. Generally speaking, it is easy to prove $\langle \sigma_{1,1}\sigma_{1,1+Q} \rangle = \langle \sigma_{1,1}\sigma_{1,1+N-Q} \rangle$ in terms of periodic-periodic boundary conditions. Hence, $\sigma_{1,1+[N/2]}$ is the farthest spin of $\sigma_{1,1}$, where $[x]$ denotes the greatest integer not exceeding x. Thus, we only need to calculate $\langle \sigma_{1,1}\sigma_{1,1+Q} \rangle$ for $Q = 1, 2, \cdots, \left[\dfrac{N}{2}\right]$.

2.2. Some Results Concerning the Partition Function Z

From (1), we see that to obtain $\langle \sigma_{1,1}\sigma_{1,1+Q} \rangle$, we need knowledge about the partition function Z given by (2), from which we summarize some results presented in Reference [2] as follows.

$$Z = \left(2\sinh\frac{2J'}{kT}\right)^{\frac{LN}{2}} \widetilde{Z}, \ \widetilde{Z} = \mathrm{Tr}\left(\frac{1+\Gamma_{2N+1}}{2}\mathbf{U}^{(\uparrow)} + \frac{1-\Gamma_{2N+1}}{2}\mathbf{U}^{(\downarrow)}\right). \tag{3}$$

$\widetilde{Z}$ can thus be obtained by finding the trace of a matrix, where Γ_{2N+1} is the matrix $\mathbf{U}$ defined by (15.68) in Reference [17], and the matrices $\dfrac{1+\Gamma_{2N+1}}{2}\mathbf{U}^{(\uparrow)}$ and $\dfrac{1-\Gamma_{2N+1}}{2}\mathbf{U}^{(\downarrow)}$ can be diagonalized at the same time.

On the other hand, both matrices $\mathbf{U}^{(\uparrow)}$ and $\mathbf{U}^{(\downarrow)}$ are spin representatives. By $\omega^{(\uparrow)}$ and $\omega^{(\downarrow)}$ we denote the corresponding rotation matrices of $\mathbf{U}^{(\uparrow)}$ and $\mathbf{U}^{(\downarrow)}$, respectively; both $\omega^{(\uparrow)}$ and $\omega^{(\downarrow)}$ can be diagonalized:

$$\begin{aligned}
&(\zeta^{(\uparrow\downarrow)})^{+}\omega^{(\uparrow\downarrow)}\zeta^{(\uparrow\downarrow)} = \lambda^{(\uparrow\downarrow)}; \\
&\lambda^{(\uparrow\downarrow)} = \mathrm{diag}\left[\ \lambda_1^{(\uparrow\downarrow)} \ \ \lambda_2^{(\uparrow\downarrow)} \ \ \cdots \ \ \lambda_N^{(\uparrow\downarrow)}\ \right], \ \lambda_n^{(\uparrow\downarrow)} = \mathrm{diag}\left[\ e^{\gamma_n^{(\uparrow\downarrow)}} \ \ e^{-\gamma_n^{(\uparrow\downarrow)}}\ \right](n = 1, 2, \cdots, N),
\end{aligned} \tag{4}$$

where $\mathbf{A}^{+}$ indicates a complex conjugate to a matrix $\mathbf{A}$,

$$\begin{aligned}
&\cosh\gamma_n = \cosh 2K'\cosh 2K - \cos\varphi_n\sinh 2K'\sinh 2K, \ e^{-2K'} = \tanh\frac{J'}{kT}, \ K = \frac{J}{kT}; \\
&\varphi_n^{(\uparrow)} = \frac{(2n-1)\pi}{N}, \ \varphi_n^{(\downarrow)} = \frac{2n\pi}{N} \ (n = 1, 2, \cdots, N = 2M).
\end{aligned} \tag{5}$$

In the above formulas, γ_n and φ_n are as the abbreviations for $\gamma_n^{(\uparrow\downarrow)}$ and $\varphi_n^{(\uparrow\downarrow)}$, respectively. To save space, we use these abbreviation as far as possible in this paper.

Since properties of the partition function Z vary between even and odd numbers N, we must calculate $\langle \sigma_{1,1}\sigma_{1,1+Q} \rangle$ separately in terms of whether N is an even or odd number. In this paper we only consider the case that $N = 2M$ as an even number (the case of $N = 2M + 1$ can be dealt with by the same approach). From Reference [2], we have:

$$\begin{aligned}
\widetilde{Z} = \frac{1}{2}&\left(\left(\prod_{m=1}^{M} 2\cosh\frac{L\gamma_m^{(\uparrow)}}{2}\right)^2 + \left(\prod_{m=1}^{M} 2\sinh\frac{L\gamma_m^{(\uparrow)}}{2}\right)^2\right. \\
&+ \left(2\cosh\frac{L\gamma_{2M}^{(\downarrow)}}{2}\right)\left(2\cosh\frac{L\gamma_M^{(\downarrow)}}{2}\right)\left(\prod_{m=1}^{M-1} 2\cosh\frac{L\gamma_m^{(\downarrow)}}{2}\right)^2 \\
&\left.+ \left(2\sinh\frac{L\gamma_{2M}^{(\downarrow)}}{2}\right)\left(2\sinh\frac{L\gamma_M^{(\downarrow)}}{2}\right)\left(\prod_{m=1}^{M-1} 2\sinh\frac{L\gamma_m^{(\downarrow)}}{2}\right)^2\right).
\end{aligned} \tag{6}$$

When $N = 2M$, γ_n and φ_n have properties:

$$\gamma_m = \gamma_{m'}, \; \varphi_m = 2\pi - \varphi_{m'}, \; m' = \begin{cases} 2M - (m-1), & \text{for } (\uparrow); \\ 2M - m, & \text{for } (\downarrow), \end{cases} \quad m = \begin{cases} 1, 2, \cdots, M, & \text{for } (\uparrow); \\ 1, 2, \cdots, M-1, & \text{for } (\downarrow), \end{cases}$$

$$\gamma_M^{(\downarrow)} = 2(K + K'), \; \gamma_{2M}^{(\downarrow)} = 2(K - K'), \; \varphi_M^{(\downarrow)} = \pi, \; \varphi_{2M}^{(\downarrow)} = 2\pi,$$

$$0 < \gamma_1^{(\uparrow)} < \gamma_2^{(\uparrow)} < \cdots < \gamma_M^{(\uparrow)}, \; 0 < \left| \gamma_{2M}^{(\downarrow)} \right| < \gamma_1^{(\downarrow)} < \gamma_2^{(\downarrow)} < \cdots < \gamma_{M-1}^{(\downarrow)} < \gamma_M^{(\downarrow)}. \tag{7}$$

2.3. Writing $\langle \sigma_{1,1} \sigma_{1,1+Q} \rangle$ in Matrix Form

Since $\sigma_{h,v}^2 = 1$, we have $\sigma_{1,1} \sigma_{1,1+Q} = \sigma_{1,1} \sigma_{1,2} \cdot \sigma_{1,2} \sigma_{1,3} \cdots \cdots \sigma_{1,Q-1} \sigma_{1,Q} \cdot \sigma_{1,Q} \sigma_{1,1+Q}$, and, further, considering $\sigma_{1,q} \sigma_{1,1+q} = \lim\limits_{\phi_q \to 0} \dfrac{\partial e^{\psi_q \sigma_{1,q} \sigma_{1,1+q}}}{\partial \phi_q}$, according to (1), $\langle \sigma_{1,1} \sigma_{1,1+Q} \rangle$ can be written in the form:

$$\langle \sigma_{1,1} \sigma_{1,1+Q} \rangle = \frac{1}{Z} \left(\prod_{q=1}^{Q} \lim_{\phi_q \to 0} \frac{\partial}{\partial \phi_q} \right) Y_Q,$$

$$Y_Q = \sum_{\{\sigma_{h,v} = \pm 1\}} \exp\left(\sum_{q=1}^{Q} \phi_q \sigma_{1,q} \sigma_{1,1+q} \right) \exp\left(\frac{J'}{kT} \sum_{l=1}^{L} \sum_{n=1}^{N} \sigma_{l,n} \sigma_{l+1,n} \right) \exp\left(\frac{J}{kT} \sum_{l=1}^{L} \sum_{n=1}^{N} \sigma_{l,n} \sigma_{l,n+1} \right), \tag{8}$$

Also, by a standard method [1,17] we can further write Y_Q in the form:

$$Y_Q = \left(2\sinh\frac{2J'}{kT} \right)^{\frac{LN}{2}} \tilde{Y}_Q, \; \tilde{Y}_Q = \mathrm{Tr}\left(\frac{1 + \Gamma_{2N+1}}{2} \mathbf{V}^{(\uparrow)} + \frac{1 - \Gamma_{2N+1}}{2} \mathbf{V}^{(\downarrow)} \right). \tag{9}$$

$\tilde{Y}_Q$ can thus be obtained by finding the trace of a matrix. Here we are not going to write out the explicit expressions of $\mathbf{V}^{(\uparrow)}$ and $\mathbf{V}^{(\downarrow)}$, but only point out that:

i.   The matrices $\dfrac{1 + \Gamma_{2N+1}}{2} \mathbf{V}^{(\uparrow)}$ and $\dfrac{1 - \Gamma_{2N+1}}{2} \mathbf{V}^{(\downarrow)}$ can be diagonalized at the same time.

ii.   Both $\mathbf{V}^{(\uparrow)}$ and $\mathbf{V}^{(\downarrow)}$ are spin representatives, whose corresponding rotation matrices are $\zeta^{(\uparrow)} \mathbf{H}^{(\uparrow)} (\zeta^{(\uparrow)})^+$ and $\zeta^{(\downarrow)} \mathbf{H}^{(\downarrow)} (\zeta^{(\downarrow)})^+$, respectively, where ζ are introduced by (4),

$$\mathbf{H} = \mathbf{H}^{(0)} + \sum_{q=1}^{Q} \sinh 2\phi_q \mathbf{H}^{(q)} + 2 \sum_{q=1}^{Q} \sinh^2 \phi_q \mathbf{H}'^{(q)}, \tag{10}$$

the forms of the matrices $\mathbf{H}^{(0)}$, $\mathbf{H}^{(q)}$, and $\mathbf{H}'^{(q)}$ can be expressed in terms of 2×2 blocks $\mathbf{H}_{lm}^{(0)}$, $\mathbf{H}_{lm}^{(q)}$ and $\mathbf{H}_{lm}'^{(q)}$, $1 \leq l, m \leq N(=2M)$, given by:

$$\mathbf{H}_{lm}^{(0)} = \begin{bmatrix} e^{L\gamma_l} & 0 \\ 0 & e^{-L\gamma_l} \end{bmatrix} \delta_{lm}, \tag{11}$$

$$\mathbf{H}_{lm}^{(q)} = e^{-iq\frac{(l-m)\pi}{M}} \mathbf{H}_{lm}, \; \mathbf{H}_{lm} = \frac{1}{N} e^{-i\frac{\theta_l - \theta_m}{2}} \begin{bmatrix} e^{\frac{L(\gamma_l + \gamma_m)}{2}} \cos\frac{\theta_l + \theta_m}{2} & -ie^{\frac{L(\gamma_l - \gamma_m)}{2}} \sin\frac{\theta_l + \theta_m}{2} \\ ie^{-\frac{L(\gamma_l - \gamma_m)}{2}} \sin\frac{\theta_l + \theta_m}{2} & -e^{-\frac{L(\gamma_l + \gamma_m)}{2}} \cos\frac{\theta_l + \theta_m}{2} \end{bmatrix}, \tag{12}$$

$$\mathbf{H}_{lm}'^{(q)} = e^{-iq\frac{(l-m)\pi}{M}} \mathbf{H}'_{lm}, \; \mathbf{H}'_{lm} = \frac{1}{N} e^{-i\frac{\theta_l - \theta_m}{2}} \begin{bmatrix} e^{\frac{L(\gamma_l + \gamma_m)}{2}} \cos\frac{\theta_l - \theta_m}{2} & ie^{\frac{L(\gamma_l - \gamma_m)}{2}} \sin\frac{\theta_l - \theta_m}{2} \\ ie^{-\frac{L(\gamma_l - \gamma_m)}{2}} \sin\frac{\theta_l - \theta_m}{2} & e^{-\frac{L(\gamma_l + \gamma_m)}{2}} \cos\frac{\theta_l - \theta_m}{2} \end{bmatrix}. \tag{13}$$

In (12) and (13), the quantities θ_n are defined by:

$$\sinh\gamma_n \cos\theta_n = \cosh 2K' \sinh 2K - \cos\varphi_n \sinh 2K' \cosh 2K,$$

$$\sinh\gamma_n \sin\theta_n = \sin\varphi_n \sinh 2K' \; (n = 1, 2, \cdots, N = 2M). \tag{14}$$

When $N = 2M$, θ_n have properties:

$$\theta_m = 2\pi - \theta_{m'}, \ \theta_M^{(\downarrow)} = \theta_{2M}^{(\downarrow)} = 0 \,. \tag{15}$$

In (15), the values of m' and m are exactly the same as those in (7).

We can first evaluate the eigenvalues of the rotation matrices $\zeta \mathbf{H} \zeta^+$, and then obtain $\widetilde{Y}_Q$ in terms of the spinor analysis method. Finally, we obtain $\langle \sigma_{1, 1} \sigma_{1, 1+Q} \rangle$ according to (8) and (9).

From (10) to (13), we see that $\mathbf{H}$ is a Hermitian conjugate matrix. Hence, the eigenvalues of both matrices $\zeta \mathbf{H} \zeta^+$ and $\mathbf{H}$ are the same, and we therefore can only evaluate the eigenvalues of the rotation matrix $\mathbf{H}$.

2.4. Basic Properties of the Eigenvalues and Eigenvectors of the Matrix $\mathbf{H}$

Any rotation matrix $\mathbf{A}$ has the following property: If τ is an eigenvalue of $\mathbf{A}$, then τ^{-1} is also an eigenvalue of $\mathbf{A}$ [2]. Since $\zeta \mathbf{H} \zeta^+$ is a rotation matrix and the eigenvalues of $\zeta \mathbf{H} \zeta^+$ and $\mathbf{H}$ are the same, $\mathbf{H}$ should have the same property. In this sub-section we prove this conclusion.

The eigen equation of $\mathbf{H}$ reads:

$$\mathbf{H}\mathbf{\Psi} = \tau\mathbf{\Psi}. \tag{16}$$

By $\mathbf{\Psi} = \begin{bmatrix} \psi_1 & \cdots & \psi_n & \cdots & \psi_N \end{bmatrix}^{\mathrm{T}}, \psi_n = \begin{bmatrix} \psi_n^\Delta \\ \psi_n^\nabla \end{bmatrix}$, where $\mathbf{A}^{\mathrm{T}}$ means the transpose of a matrix $\mathbf{A}$, we denote the eigenvector of $\mathbf{H}$, and, introducing C_q^Δ, C_q^∇ $(q = 1, 2, \cdots, Q)$ in terms of:

$$
\begin{bmatrix} \psi_n^\Delta \\ \psi_n^\nabla \end{bmatrix} = e^{-i\frac{\theta_n}{2}} \begin{bmatrix} e^{-\frac{L\gamma_n}{2}} & 0 \\ 0 & e^{\frac{L\gamma_n}{2}} \end{bmatrix} \begin{bmatrix} \cos\frac{\theta_n}{2} & i\sin\frac{\theta_n}{2} \\ i\sin\frac{\theta_n}{2} & \cos\frac{\theta_n}{2} \end{bmatrix}
$$
$$
\times \begin{bmatrix} \dfrac{\tau\cos h(L\gamma_n) - 1 + \tau\cos\theta_n\sin h(L\gamma_n)}{\tau^2 - 2\tau\cosh(L\gamma_n) + 1} & i\dfrac{\tau\sin\theta_n\sin h(L\gamma_n)}{\tau^2 - 2\tau\cosh(L\gamma_n) + 1} \\ -i\dfrac{\tau\sin\theta_n\sin h(L\gamma_n)}{\tau^2 - 2\tau\cosh(L\gamma_n) + 1} & \dfrac{\tau\cos h(L\gamma_n) - 1 + \tau\cos\theta_n\sin h(L\gamma_n)}{\tau^2 - 2\tau\cosh(L\gamma_n) + 1} \end{bmatrix} \begin{bmatrix} \sum\limits_{q=1}^{Q} e^{-iq\varphi_n} C_q^\Delta \\ \sum\limits_{q=1}^{Q} e^{-iq\varphi_n} C_q^\nabla \end{bmatrix},
$$

we can prove that the eigen Equation (16) is equivalent to:

$$\coth\phi_q \begin{bmatrix} C_q^\Delta \\ C_q^\nabla \end{bmatrix} = \sum_{q'=1}^{Q} \begin{bmatrix} A_+(q - q') & -B(q - q') \\ -B(q - q') & -A_-(q - q') \end{bmatrix} \begin{bmatrix} C_{q'}^\Delta \\ C_{q'}^\nabla \end{bmatrix} \ (q = 1, 2, \cdots, Q)\,; \tag{17}$$

where

$$
A_\pm(k) = \frac{1}{N}\sum_{n=1}^{N} e^{ik\frac{2n\pi}{N}} \frac{\tau^2 \pm 2\tau\cos\theta_n\sin h(L\gamma_n) - 1}{\tau^2 - 2\tau\cosh(L\gamma_n) + 1}
$$
$$
= \begin{cases} \dfrac{1}{M}\sum\limits_{m=1}^{M}\cos\dfrac{km\pi}{M}\dfrac{\tau^2 \pm 2\tau\cos\theta_m^{(\uparrow)}\sin h(L\gamma_m^{(\uparrow)}) - 1}{(e^{L\gamma_m^{(\uparrow)}}\tau - 1)(e^{-L\gamma_m^{(\uparrow)}}\tau - 1)}, & \text{for } (\uparrow)\,; \\[12pt] \dfrac{1}{M}\left(\dfrac{(-1)^k}{2}\dfrac{e^{\mp L\gamma_M^{(\downarrow)}}\tau + 1}{e^{\mp L\gamma_M^{(\downarrow)}}\tau - 1} + \dfrac{1}{2}\dfrac{e^{\mp L\gamma_{2M}^{(\downarrow)}}\tau + 1}{e^{\mp L\gamma_{2M}^{(\downarrow)}}\tau - 1} + \sum\limits_{m=1}^{M-1}\cos\dfrac{km\pi}{M}\dfrac{\tau^2 \pm 2\tau\cos\theta_m^{(\downarrow)}\sin h(L\gamma_m^{(\downarrow)}) - 1}{(e^{L\gamma_m^{(\downarrow)}}\tau - 1)(e^{-L\gamma_m^{(\downarrow)}}\tau - 1)}\right), & \text{for } (\downarrow), \end{cases}
$$

$$
B(k) = -i\frac{1}{N}\sum_{n=1}^{N} e^{ik\frac{2n\pi}{N}} \frac{2\tau\sin\theta_n\sin h(L\gamma_n)}{\tau^2 - 2\tau\cosh(L\gamma_n) + 1} = \begin{cases} \dfrac{1}{M}\sum\limits_{m=1}^{M}\sin\dfrac{km\pi}{M}\dfrac{2\tau\sin\theta_m^{(\uparrow)}\sin h(L\gamma_m^{(\uparrow)})}{\tau^2 - 2\tau\cosh(L\gamma_m^{(\uparrow)}) + 1}, & \text{for } (\uparrow)\,; \\[12pt] \dfrac{1}{M}\sum\limits_{m=1}^{M-1}\sin\dfrac{km\pi}{M}\dfrac{2\tau\sin\theta_m^{(\downarrow)}\sin h(L\gamma_m^{(\downarrow)})}{\tau^2 - 2\tau\cosh(L\gamma_m^{(\downarrow)}) + 1}, & \text{for } (\downarrow), \end{cases}
$$

where we have used (7) and (15). We see that all $A_\pm(k)$ and $B(k)$ are real functions and satisfy:

$$A_\pm(-k) = A_\pm(k), B(-k) = -B(k).$$

According to the above expressions and the properties of $A_{\pm}(k)$ and $B(k)$, we can conclude that if τ and $C_q^{\Delta}(\tau)$, $C_q^{\nabla}(\tau)$ $(q = 1, 2, \cdots, Q)$ satisfy (17), then $\tau' = \tau^{-1}$ and

$$C_q^{\Delta}(\tau') = C_0(\tau)C_q^{\nabla}(\tau) , \ C_q^{\nabla}(\tau') = C_0(\tau)C_q^{\Delta}(\tau)(q = 1, 2, \cdots, Q) \tag{18}$$

also satisfy (17), where $C_0(\tau)$ is an arbitrary function. From this discussion we not only prove the conclusion "If τ is an eigenvalue of the matrix $\mathbf{H}$, then τ^{-1} is also an eigenvalue", but also obtain the relation (18) between $\left\{ C_q^{\Delta}(\tau), C_q^{\nabla}(\tau) \right\}$ and $\left\{ C_q^{\Delta}(\tau^{-1}), C_q^{\nabla}(\tau^{-1}) \right\}$. The conclusion and the relation (18) are useful to determine the forms of approximate eigenvalues and the expressions of the normalized eigenvectors of $\mathbf{H}$, as well as to calculate the determinant of the matrix consisting of the eigenvectors in the actual calculation process.

2.5. Approximate Method for Solving the Eigen Equation (16)

It is very difficult to find the exact eigenvalues of $\mathbf{H}$ by solving the eigen Equation (16). On the other hand, the operator $\prod\limits_{q=1}^{Q} \lim\limits_{\phi_q \to 0} \dfrac{\partial}{\partial \phi_q}$ in (8) allows us to ignore all terms whose orders are higher than $\phi_q^1(= \phi_q)$ $(q = 1, 2, \cdots, Q)$ in all eigenvalues of $\mathbf{H}$. According to this key property, we can obtain the exact expressions of $\langle \sigma_{1, 1}\sigma_{1, 1+Q} \rangle$ by only finding approximate eigenvalues of $\mathbf{H}$.

Concretely, as the first step, the term $2\sum\limits_{q=1}^{Q} \sinh^2\phi_q \mathbf{H}'^{(q)}$ with the factors $\sinh^2\phi_q$ in (10) can be ignored, since $\sinh^2\phi_q \approx \phi_q^2$ have ϕ_q^2 order. Then, from (11) we see that $\mathbf{H}^{(0)}$ in (10) is a diagonal matrix, whose eigenvalues and eigenvectors are summarized in the following formulas:

$$\mathbf{H}^{(0)}\boldsymbol{\Psi}_{n,\pm}^{(0)} = e^{\pm L\gamma_n}\boldsymbol{\Psi}_{n,\pm}^{(0)} (n = 1, 2, \cdots, N = 2M);$$
$$\boldsymbol{\Psi}_{n,\pm}^{(0)} = \left[\ \boldsymbol{\psi}_{n,1,\pm}^{(0)} \ \cdots \ \boldsymbol{\psi}_{n,m,\pm}^{(0)} \ \cdots \ \boldsymbol{\psi}_{n,N,\pm}^{(0)} \ \right]^{\mathrm{T}}, \ \boldsymbol{\psi}_{n,m,\pm}^{(0)} = \begin{bmatrix} 0 \\ 0 \end{bmatrix} (m \neq n) ,$$
$$\boldsymbol{\psi}_{n,n,+}^{(0)} = \begin{cases} \begin{bmatrix} 1 \\ 0 \end{bmatrix}, \text{ for the eigenvalue } e^{L\gamma_n}; \\ \begin{bmatrix} 0 \\ 0 \end{bmatrix}, \text{ for the eigenvalue } e^{-L\gamma_n}, \end{cases} \quad \boldsymbol{\psi}_{n,n,-}^{(0)} = \begin{cases} \begin{bmatrix} 0 \\ 0 \end{bmatrix}, \text{ for the eigenvalue } e^{L\gamma_n}; \\ \begin{bmatrix} 0 \\ 1 \end{bmatrix}, \text{ for the eigenvalue } e^{-L\gamma_n}. \end{cases} \tag{19}$$

which are as the zeroth order approximation of the eigen Equation (16).

The term $\sum\limits_{q=1}^{Q} \sinh 2\phi_q \mathbf{H}^{(q)}$ with the factors $\sinh 2\phi_q \approx 2\phi_q$ $(q = 1, 2, \cdots, Q)$ in (10) can be regarded as a perturbation term. Then, by using RSPT, we can obtain the approximate eigenvalues of $\mathbf{H}$.

However, although what eigenvalues we need are only corrected to the $\phi_q^1(= \phi_q)$ order $(q = 1, 2, \cdots, Q)$, we must calculate the perturbation terms up to the Q-th order, not only for the first-order approximation, because all of the terms with the factor $\prod\limits_{q=1}^{Q} \sinh 2\phi_q$ appear in the Q-th order eigenvalues and are needed, which only include the ϕ_q^1 order for every ϕ_q.

However, if we calculate the eigenvalues up to the Q-th order by using RSPT, then not only the actual calculation process is very complex, but there are also many unwanted terms with factors ϕ_q^k $(k \geq 2)$, for example, the term with the factor $\sinh^3 2\phi_1 \prod\limits_{q=4}^{Q} \sinh 2\phi_q$, in the Q-th order eigenvalues.

To take out those terms with factors ϕ_q^k $(k \geq 2)$, we change "finding the eigenvalue up to the Q-th order" to "finding the eigenvalue through Q times first-order approximation".

Concretely, since now $\mathbf{H} = \mathbf{H}^{(0)} + \sum\limits_{q=1}^{Q} \sinh 2\phi_q \mathbf{H}^{(q)}$, we first consider the matrix $\mathbf{H}^{(0)} + \sinh 2\phi_1 \mathbf{H}^{(1)}$, in which the eigenvalues and eigenvectors $\left\{ \tau^{(0)}, \boldsymbol{\Psi}^{(0)} \right\}$ of $\mathbf{H}^{(0)}$ are given by (19) and $\sinh 2\phi_1 \mathbf{H}^{(1)}$ is as perturbation term. By only calculating first-order approximation we obtain all eigenvalues and eigenvectors $\left\{ \tau^{(1)},\ \mathbf{\Psi}^{(1)} \right\}$ of $\mathbf{H}^{(0)} + \sinh2\phi_1\mathbf{H}^{(1)}$; therefore, all terms in $\left\{ \tau^{(1)},\ \mathbf{\Psi}^{(1)} \right\}$ only correct to the ϕ_1^1 order.

Then, we consider the matrix $\mathbf{H}^{(0)} + \sinh2\phi_1\mathbf{H}^{(1)} + \sinh2\phi_2\mathbf{H}^{(2)}$. Since now all eigenvalues and eigenvectors $\left\{ \tau^{(1)},\ \mathbf{\Psi}^{(1)} \right\}$ of $\mathbf{H}^{(0)} + \sinh2\phi_1\mathbf{H}^{(1)}$ are known, we regard $\sinh2\phi_2\mathbf{H}^{(2)}$ as a perturbation term, and, by only calculating the first-order approximation, obtain all eigenvalues and eigenvectors $\left\{ \tau^{(2)},\ \mathbf{\Psi}^{(2)} \right\}$ of $\mathbf{H}^{(0)} + \sinh2\phi_1\mathbf{H}^{(1)} + \sinh2\phi_2\mathbf{H}^{(2)}$, in which all terms are only of the ϕ_1^1 and ϕ_2^1 orders. In particular, all of the terms with the factor $\sinh2\phi_1\sinh2\phi_2 \approx \phi_1\phi_2$ remain.

Then, we consider the matrix $\mathbf{H}^{(0)} + \sinh2\phi_1\mathbf{H}^{(1)} + \sinh2\phi_2\mathbf{H}^{(2)} + \sinh2\phi_3\mathbf{H}^{(3)}$. Since now all eigenvalues and eigenvectors $\left\{ \tau^{(2)},\ \mathbf{\Psi}^{(2)} \right\}$ of $\mathbf{H}^{(0)} + \sinh2\phi_1\mathbf{H}^{(1)} + \sinh2\phi_2\mathbf{H}^{(2)}$ are known, we regard $\sinh2\phi_3\mathbf{H}^{(3)}$ as a perturbation term, and, by only calculating the first-order approximation, obtain all eigenvalues and eigenvectors $\left\{ \tau^{(3)},\ \mathbf{\Psi}^{(3)} \right\}$ of $\mathbf{H}^{(0)} + \sinh2\phi_1\mathbf{H}^{(1)} + \sinh2\phi_2\mathbf{H}^{(2)} + \sinh2\phi_3\mathbf{H}^{(3)}$, in which all terms are only of the ϕ_1^1, ϕ_2^1 and ϕ_3^1 orders. In particular, all of the terms with the factor $\sinh2\phi_1\sinh2\phi_2\sinh2\phi_3 \approx \phi_1\phi_2\phi_3$ remain, and, many unwanted terms with factors $\sinh^2 2\phi_1\sinh2\phi_2$, $\sinh2\phi_2\sinh^2 2\phi_3$, etc., do not appear in the eigenvalues of $\tau^{(3)}$.

We follow this approach up to $\sinh2\phi_Q\mathbf{H}^{(Q)}$ and every time we only calculate thr first-order approximation, which leads to the eigenvalues and eigenvectors $\left\{ \tau^{(Q)},\ \mathbf{\Psi}^{(Q)} \right\}$ of all terms being only of the $\phi_q^1(= \phi_q)$ order. All of the terms with $\prod\limits_{q=1}^{Q} \sinh2\phi_q$ remain, and at the same time those unwanted terms with ϕ_q^k ($k \geq 2$) do not appear.

On the one hand, the above approximate method allows us to actually carry out the calculation process to find the eigenvalues and eigenvectors of $\mathbf{H}$. In particular, once we obtain $\left\{ \tau^{(1)},\ \mathbf{\Psi}^{(1)} \right\}$, we can obtain $\widetilde{Y}_1$ by the spinor analysis method, as well as obtain $\langle \sigma_{1,\,1}\sigma_{1,\,2} \rangle$ in terms of (8) and (9). Once we obtain $\left\{ \tau^{(2)},\ \mathbf{\Psi}^{(2)} \right\}$, we can obtain $\widetilde{Y}_2$ by the spinor analysis method, and, further, obtain $\langle \sigma_{1,\,1}\sigma_{1,\,3} \rangle$ in terms of (8) and (9), $\cdots$. Generally speaking, once we obtain $\left\{ \tau^{(q)},\ \mathbf{\Psi}^{(q)} \right\}$, we can obtain $\widetilde{Y}_q$, and, further, obtain $\langle \sigma_{1,\,1}\sigma_{1,\,1+q} \rangle$.

On the other hand, since RSPT is irregular, when Q is very large, e.g., $Q \approx \left[\dfrac{N}{2}\right]$, the above approach no longer functions. Hence, by this approach we can only obtain the exact expressions of correlation functions when Q is a small number, for example, $\langle \sigma_{1,\,1}\sigma_{1,\,2} \rangle$, $\langle \sigma_{1,\,1}\sigma_{1,\,3} \rangle$, $\langle \sigma_{1,\,1}\sigma_{1,\,4} \rangle$, $\cdots$, etc., which belong to the short-range order, but we cannot obtain the exact expressions of correlation functions when Q is larger, for example, $\left\langle \sigma_{1,\,1}\sigma_{1,\,[N/2]+1} \right\rangle$, $\left\langle \sigma_{1,\,1}\sigma_{1,\,[N/2]} \right\rangle$, $\left\langle \sigma_{1,\,1}\sigma_{1,\,[N/2]-1} \right\rangle$, $\cdots$, etc., which belong to the long-range order.

2.6. Recurrence Formulas of the Eigenvalues and Eigenvectors $\left\{ \tau^{(Q)},\ \mathbf{\Psi}^{(Q)} \right\}$

According to the discussions in the above sub-section, we first regard $\sinh2\phi_1\mathbf{H}^{(1)}$ as a perturbation term, and, by using RSPT, evaluate eigenvalues and eigenvectors $\left\{ \tau^{(1)},\ \mathbf{\Psi}^{(1)} \right\}$ of the matrix $\mathbf{H}^{(0)} + \sinh2\phi_1\mathbf{H}^{(1)}$ up to the first-order approximation. However, according to (7), all eigenvalues $e^{\pm L\gamma_n^{(\uparrow)}}$ are doubly-degenerate; and, except $e^{\pm L\gamma_{2M}^{(\downarrow)}}$ and $e^{\pm L\gamma_M^{(\downarrow)}}$, all the remaining eigenvalues $e^{\pm L\gamma_n^{(\downarrow)}}$ are also doubly-degenerate. Hence, for doubly-degenerate eigenvalues of $\mathbf{H}^{(0)}$, we must use the degenerate perturbation theory; the results obtained up to ϕ_1^1 order are as follows.

$$\alpha_{m,\,\uparrow\downarrow}^{(1)} = \frac{\sinh2\phi_1}{M}\cos^2\frac{\theta_m^{(\uparrow\downarrow)}}{2}\ ,\ \beta_{m,\,\uparrow\downarrow}^{(1)} = \frac{\sinh2\phi_1}{M}\sin^2\frac{\theta_m^{(\uparrow\downarrow)}}{2}\ ,\ \delta_M^{(1)} = \delta_{2M}^{(1)} = \frac{\sinh2\phi_1}{2M}, \tag{20}$$

$$\Delta\Psi^{(0)}_{m,\pm,\uparrow\downarrow,\mathrm{I}} = \frac{\cos\frac{\theta^{(\uparrow\downarrow)}_m}{2}}{\sqrt{2}}\left(\sum_{\substack{l=1 \\ l\neq m,\, l\neq m'}}^{2M} \frac{\cos\frac{\theta^{(\uparrow\downarrow)}_l}{2}}{\sinh\frac{L\left(\gamma^{(\uparrow\downarrow)}_l - \gamma^{(\uparrow\downarrow)}_m\right)}{2}}\Psi^{(0)}_{l,\pm} - \sum_{l=1}^{2M} \frac{i\sin\frac{\theta^{(\uparrow\downarrow)}_l}{2}}{\sinh\frac{L\left(\gamma^{(\uparrow\downarrow)}_l + \gamma^{(\uparrow\downarrow)}_m\right)}{2}}\Psi^{(0)}_{l,\mp} \right),$$

$$\Delta\Psi^{(0)}_{m,\pm,\uparrow\downarrow,\mathrm{II}} = \frac{i\sin\frac{\theta^{(\uparrow\downarrow)}_m}{2}}{\sqrt{2}}\left(\sum_{\substack{l=1 \\ l\neq m,\, l\neq m'}}^{2M} \frac{i\sin\frac{\theta^{(\uparrow\downarrow)}_l}{2}}{\sinh\frac{L\left(\gamma^{(\uparrow\downarrow)}_l - \gamma^{(\uparrow\downarrow)}_m\right)}{2}}\Psi^{(0)}_{l,\pm} - \sum_{l=1}^{2M} \frac{\cos\frac{\theta^{(\uparrow\downarrow)}_l}{2}}{\sinh\frac{L\left(\gamma^{(\uparrow\downarrow)}_l + \gamma^{(\uparrow\downarrow)}_m\right)}{2}}\Psi^{(0)}_{l,\mp} \right),$$

$$\Delta\Psi^{(0)}_{M,\pm,\downarrow} = \sum_{\substack{l=1 \\ l\neq M}}^{2M} \frac{\cos\frac{\theta^{(\downarrow)}_l}{2}}{\sinh\frac{L\left(\gamma^{(\downarrow)}_l - \gamma^{(\downarrow)}_M\right)}{2}}\Psi^{(0)}_{l,\pm} - \sum_{l=1}^{2M} \frac{i\sin\frac{\theta^{(\downarrow)}_l}{2}}{\sinh\frac{L\left(\gamma^{(\downarrow)}_l + \gamma^{(\downarrow)}_M\right)}{2}}\Psi^{(0)}_{l,\mp},$$

$$\Delta\Psi^{(0)}_{2M,\pm,\downarrow} = \sum_{\substack{l=1 \\ l\neq 2M}}^{2M} \frac{\cos\frac{\theta^{(\downarrow)}_l}{2}}{\sinh\frac{L\left(\gamma^{(\downarrow)}_l - \gamma^{(\downarrow)}_{2M}\right)}{2}}\Psi^{(0)}_{l,\pm} - \sum_{l=1}^{2M} \frac{i\sin\frac{\theta^{(\downarrow)}_l}{2}}{\sinh\frac{L\left(\gamma^{(\downarrow)}_l + \gamma^{(\downarrow)}_M\right)}{2}}\Psi^{(0)}_{l,\mp}. \tag{21}$$

In Table 1, (20) and (21), the values of m' and m are exactly the same as those in (7).

Table 1. The eigenvalues and eigenvectors of $\mathbf{H}^{(0)} + \sinh 2\phi_1 \mathbf{H}^{(1)}$ corrected to the $\phi_1^1 (= \phi_1)$ order.

Eigenvalue $\left\{ \tau^{(1)} \right\}$	Eigenvector $\left\{ \Psi^{(1)} \right\}$
$e^{\pm(L\gamma^{(\uparrow\downarrow)}_m + a^{(1)}_{m,\uparrow\downarrow})}$	$\frac{1}{\sqrt{2}}\left(\Psi^{(0)}_{m,\pm} - \Psi^{(0)}_{m',\pm}\right) - \frac{\sinh 2\phi_1}{2M}\Delta\Psi^{(0)}_{m,\pm,\uparrow\downarrow,\mathrm{I}}$
$e^{\pm(L\gamma^{(\uparrow\downarrow)}_m - \beta^{(1)}_{m,\uparrow\downarrow})}$	$\frac{1}{\sqrt{2}}\left(\Psi^{(0)}_{m,\pm} + \Psi^{(0)}_{m',\pm}\right) - \frac{\sinh 2\phi_1}{2M}\Delta\Psi^{(0)}_{m,\pm,\uparrow\downarrow,\mathrm{II}}$
$e^{\pm(L\gamma^{(\downarrow)}_M + \delta^{(1)}_M)}$	$\Psi^{(0)}_{M,\pm} - \frac{\sinh 2\phi_1}{4M}\Delta\Psi^{(0)}_{M,\pm,\downarrow}$
$e^{\pm(L\gamma^{(\downarrow)}_{2M} + \delta^{(1)}_{2M})}$	$\Psi^{(0)}_{2M,\pm} - \frac{\sinh 2\phi_1}{4M}\Delta\Psi^{(0)}_{2M,\pm,\downarrow}$

From Table 1, we see that all eigenvalues $\left\{ \tau^{(1)} \right\}$ of $\mathbf{H}^{(0)} + \sinh 2\phi_1 \mathbf{H}^{(1)}$ are nondegenerate. Hence, all degenerate eigenvalues of $\mathbf{H}^{(0)}$ are relieved by $\sinh 2\phi_1 \mathbf{H}^{(1)}$. Thus, when we calculate the eigenvalues and eigenvectors $\left\{ \tau^{(2)}, \Psi^{(2)} \right\}$ of $\mathbf{H}^{(0)} + \sinh 2\phi_1 \mathbf{H}^{(1)} + \sinh 2\phi_2 \mathbf{H}^{(2)}$, we only need use nondegenerate perturbation theory; this is applicable up to $\sinh 2\phi_Q \mathbf{H}^{(Q)}$. Further, since from $\left\{ \tau^{(q)}, \Psi^{(q)} \right\}$ to $\left\{ \tau^{(q+1)}, \Psi^{(q+1)} \right\}$ $(q = 1, 2, \cdots, Q-1)$, we need only to calculate the first-order approximation in terms of $\sinh 2\phi_{q+1} \approx 2\phi_{q+1}$, and the corresponding recurrence formulas are:

$$\tau^{(q+1)}_m = \tau^{(q)}_m + \frac{\sinh 2\phi_{q+1}}{2M}\left(\Psi^{(q)}_m\right)^+ \mathbf{H}^{(q+1)}\Psi^{(q)}_m,$$

$$\Psi^{(q+1)}_m = \Psi^{(q)}_m - \frac{\sinh 2\phi_{q+1}}{2M}\sum_{\substack{l=1 \\ l\neq m}}^{4M} \frac{\left(\Psi^{(q)}_l\right)^+ \mathbf{H}^{(q+1)}\Psi^{(q)}_m}{\tau^{(q)}_l - \tau^{(q)}_m}\Psi^{(q)}_l, \tag{22}$$

$$(m = 1, 2, \cdots, 4M;\ q = 1, 2, \cdots, Q-1),$$

In the calculation, all terms including the $\phi_q^k (k \geq 2)$ order are ignored.

In principle, by following the above approach we obtain the eigenvalues $\left\{ \tau^{(Q)} \right\}$. Furthermore, considering that up to the first-order approximation for ϕ_q, we have $1 + C\sinh 2\phi_q \approx e^{C\sinh 2\phi_q}$, the eigenvalues $\left\{ \tau^{(Q)} \right\}$ of **H** can be denoted by the forms:

$$e^{\pm(L\gamma_m^{(\uparrow\downarrow)} + \alpha_{m,\uparrow\downarrow}^{(Q)})}, \ e^{\pm(L\gamma_m^{(\uparrow\downarrow)} - \beta_{m,\uparrow\downarrow}^{(Q)})}, \ e^{\pm(L\gamma_M^{(\downarrow)} + \delta_M^{(Q)})}, \ e^{\pm(L\gamma_{2M}^{(\downarrow)} + \delta_{2M}^{(Q)})},$$

where the value of m is exactly the same as that in (7).

Based on the above forms of the eigenvalues $\left\{ \tau^{(Q)} \right\}$ and using the spinor analysis method, we obtain:

$$
\begin{aligned}
\tilde{Y}_Q = \ & \left(\prod_{l=1}^{M} 2\cosh\frac{L\gamma_l^{(\uparrow)} + \alpha_{l,\uparrow}^{(Q)}}{2} \right) \left(\prod_{m=1}^{M} 2\cosh\frac{L\gamma_m^{(\uparrow)} - \beta_{m,\uparrow}^{(Q)}}{2} \right) \\
& + \left(\prod_{l=1}^{M} 2\sinh\frac{L\gamma_l^{(\uparrow)} + \alpha_{l,\uparrow}^{(Q)}}{2} \right) \left(\prod_{m=1}^{M} 2\sinh\frac{L\gamma_m^{(\uparrow)} - \beta_{m,\uparrow}^{(Q)}}{2} \right) \\
& + \left(2\cosh\frac{L\gamma_{2M}^{(\downarrow)} + \delta_{2M}^{(Q)}}{2} \right) \left(2\cosh\frac{L\gamma_M^{(\downarrow)} + \delta_M^{(Q)}}{2} \right) \left(\prod_{l=1}^{M-1} 2\cosh\frac{L\gamma_l^{(\downarrow)} + \alpha_{l,\downarrow}^{(Q)}}{2} \right) \left(\prod_{m=1}^{M-1} 2\cosh\frac{L\gamma_m^{(\downarrow)} - \beta_{m,\downarrow}^{(Q)}}{2} \right) \\
& + \left(2\sinh\frac{L\gamma_{2M}^{(\downarrow)} + \delta_{2M}^{(Q)}}{2} \right) \left(2\sinh\frac{L\gamma_M^{(\downarrow)} + \delta_M^{(Q)}}{2} \right) \left(\prod_{l=1}^{M-1} 2\sinh\frac{L\gamma_l^{(\downarrow)} + \alpha_{l,\downarrow}^{(Q)}}{2} \right) \left(\prod_{m=1}^{M-1} 2\sinh\frac{L\gamma_m^{(\downarrow)} - \beta_{m,\downarrow}^{(Q)}}{2} \right).
\end{aligned}
\tag{23}
$$

Finally, according to (8) and (9), we obtain:

$$\langle \sigma_{1,1} \sigma_{1,1+Q} \rangle = \frac{1}{\tilde{Z}} \left(\prod_{q=1}^{Q} \lim_{\phi_q \to 0} \frac{\partial}{\partial \phi_q} \right) \tilde{Y}_Q, \tag{24}$$

where $\tilde{Z}$ and $\tilde{Y}_Q$ are given by (6) and (23), respectively.

2.7. The Exact Expressions of $\langle \sigma_{1,1} \sigma_{1,2} \rangle$ and $\langle \sigma_{1,1} \sigma_{1,3} \rangle$ on a Finite Lattice

Although in Section 2.5 we presented a simplified approximate method, the actual calculation process of $\langle \sigma_{1,1} \sigma_{1,1+Q} \rangle$ is still complex; here, we only present the expressions of $\langle \sigma_{1,1} \sigma_{1,2} \rangle$ and $\langle \sigma_{1,1} \sigma_{1,3} \rangle$ directly.

When $Q = 1$, substituting $\alpha_{m,\uparrow\downarrow}^{(1)}$, $\beta_{m,\uparrow\downarrow}^{(1)}$, $\delta_M^{(1)}$, and $\delta_{2M}^{(1)}$ given by (20) into (23), we obtain $\tilde{Y}_1$, and, further, we have:

$$
\begin{aligned}
\lim_{\phi_1 \to 0} \frac{\partial \tilde{Y}_1}{\partial \phi_1} = \ & \left(\sum_{m=1}^{M} \frac{\cos\theta_m^{(\uparrow)}}{M} \tanh\frac{L\gamma_m^{(\uparrow)}}{2} \right) \left(\prod_{l=1}^{M} 2\cosh\frac{L\gamma_l^{(\uparrow)}}{2} \right)^2 \\
& + \left(\sum_{m=1}^{M} \frac{\cos\theta_m^{(\uparrow)}}{M} \coth\frac{L\gamma_m^{(\uparrow)}}{2} \right) \left(\prod_{l=1}^{M} 2\sinh\frac{L\gamma_l^{(\uparrow)}}{2} \right)^2 \\
& + \left(\frac{1}{2M} \tanh\frac{L\gamma_{2M}^{(\downarrow)}}{2} + \frac{1}{2M} \tanh\frac{L\gamma_M^{(\downarrow)}}{2} + \sum_{m=1}^{M-1} \frac{\cos\theta_m^{(\downarrow)}}{M} \tanh\frac{L\gamma_m^{(\downarrow)}}{2} \right) \\
& \quad\times \left(2\cosh\frac{L\gamma_{2M}^{(\downarrow)}}{2} \right) \left(2\cosh\frac{L\gamma_M^{(\downarrow)}}{2} \right) \left(\prod_{l=1}^{M-1} 2\cosh\frac{L\gamma_l^{(\downarrow)}}{2} \right)^2 \\
& + \left(\frac{1}{2M} \coth\frac{L\gamma_{2M}^{(\downarrow)}}{2} + \frac{1}{2M} \coth\frac{L\gamma_M^{(\downarrow)}}{2} + \sum_{m=1}^{M-1} \frac{\cos\theta_m^{(\downarrow)}}{M} \coth\frac{L\gamma_m^{(\downarrow)}}{2} \right) \\
& \quad\times \left(2\sinh\frac{L\gamma_{2M}^{(\downarrow)}}{2} \right) \left(2\sinh\frac{L\gamma_M^{(\downarrow)}}{2} \right) \left(\prod_{l=1}^{M-1} 2\sinh\frac{L\gamma_l^{(\downarrow)}}{2} \right)^2.
\end{aligned}
\tag{25}
$$

Substituting $\widetilde{Z}$ given by (6) and the above expression into (24), we obtain the expressions of $\langle \sigma_{1,1}\sigma_{1,2} \rangle$ of the model on a finite lattice:

$$\langle \sigma_{1,1}\sigma_{1,2} \rangle = \frac{1}{\widetilde{Z}} \lim_{\phi_1 \to 0} \frac{\partial \widetilde{Y}_1}{\partial \phi_1}. \tag{26}$$

Then, using $\left\{ \tau^{(1)}, \mathbf{\Psi}^{(1)} \right\}$ presented in Table 1, (20) and (21), and according to (22), we obtain $\left\{ \tau^{(2)} \right\}$, which can be denoted by the forms:

$$e^{\pm (L\gamma_m^{(\uparrow\downarrow)} + \alpha_{m,\uparrow\downarrow}^{(2)})}, \ e^{\pm (L\gamma_m^{(\uparrow\downarrow)} - \beta_{m,\uparrow\downarrow}^{(2)})}, \ e^{\pm (L\gamma_M^{(\downarrow)} + \delta_M^{(2)})}, \ e^{\pm (L\gamma_{2M}^{(\downarrow)} + \delta_{2M}^{(2)})},$$

where

$$\alpha_{m,\uparrow\downarrow}^{(2)} = \frac{\sinh 2\phi_1 + \sinh 2\phi_2}{M} \cos^2 \frac{\theta_m}{2} - \frac{\sinh 2\phi_2}{M} \sin^2 \varphi_m$$
$$+ \frac{\sinh 2\phi_1 \sinh 2\phi_2}{4M^2} \sum_{\substack{k=1 \\ k \neq m,\, k \neq m'}}^{2M} \frac{4\cos^2 \frac{\theta_m}{2} \cos^2 \frac{\theta_k}{2} \cos \varphi_m \cos \varphi_k - \sin \theta_m \sin \theta_k \sin \varphi_m \sin \varphi_k}{e^{L(\gamma_m - \gamma_k)} - 1}$$
$$+ \frac{\sinh 2\phi_1 \sinh 2\phi_2}{4M^2} \sum_{k=1}^{2M} \frac{4\cos^2 \frac{\theta_m}{2} \sin^2 \frac{\theta_k}{2} \cos \varphi_m \cos \varphi_k + \sin \theta_m \sin \theta_k \sin \varphi_m \sin \varphi_k}{e^{L(\gamma_m + \gamma_k)} - 1},$$

$$\beta_{m,\uparrow\downarrow}^{(2)} = \frac{\sinh 2\phi_1 + \sinh 2\phi_2}{M} \sin^2 \frac{\theta_m}{2} - \frac{\sinh 2\phi_2}{M} \sin^2 \varphi_m$$
$$+ \frac{\sinh 2\phi_1 \sinh 2\phi_2}{4M^2} \sum_{\substack{k=1 \\ k \neq m,\, k \neq m'}}^{2M} \frac{-4\sin^2 \frac{\theta_m}{2} \sin^2 \frac{\theta_k}{2} \cos \varphi_m \cos \varphi_k + \sin \theta_m \sin \theta_k \sin \varphi_m \sin \varphi_k}{1 - e^{-L(\gamma_m - \gamma_k)}}$$
$$+ \frac{\sinh 2\phi_1 \sinh 2\phi_2}{4M^2} \sum_{k=1}^{2M} \frac{-4\sin^2 \frac{\theta_m}{2} \cos^2 \frac{\theta_k}{2} \cos \varphi_m \cos \varphi_k - \sin \theta_m \sin \theta_k \sin \varphi_m \sin \varphi_k}{1 - e^{-L(\gamma_m + \gamma_k)}},$$

$$\delta_M^{(2)} = \frac{\sinh 2\phi_1 + \sinh 2\phi_2}{2M} - \frac{\sinh 2\phi_1 \sinh 2\phi_2}{2M^2} \sum_{\substack{k=1 \\ k \neq M}}^{2M} \frac{1}{e^{L(\gamma_M^{(\downarrow)} - \gamma_k^{(\downarrow)})} - 1} \cos^2 \frac{\theta_k^{(\downarrow)}}{2} \cos \varphi_k^{(\downarrow)}$$
$$- \frac{\sinh 2\phi_1 \sinh 2\phi_2}{2M^2} \sum_{k=1}^{2M} \frac{1}{e^{L(\gamma_M^{(\downarrow)} + \gamma_k^{(\downarrow)})} - 1} \sin^2 \frac{\theta_k^{(\downarrow)}}{2} \cos \varphi_k^{(\downarrow)},$$

$$\delta_{2M}^{(2)} = \frac{\sinh 2\phi_1 + \sinh 2\phi_2}{2M} + \frac{\sinh 2\phi_1 \sinh 2\phi_2}{2M^2} \sum_{k=1}^{2M-1} \frac{1}{e^{L(\gamma_{2M}^{(\downarrow)} - \gamma_k^{(\downarrow)})} - 1} \cos^2 \frac{\theta_k^{(\downarrow)}}{2} \cos \varphi_k^{(\downarrow)}$$
$$+ \frac{\sinh 2\phi_1 \sinh 2\phi_2}{2M^2} \sum_{k=1}^{2M} \frac{1}{e^{L(\gamma_{2M}^{(\downarrow)} + \gamma_k^{(\downarrow)})} - 1} \sin^2 \frac{\theta_k^{(\downarrow)}}{2} \cos \varphi_k^{(\downarrow)}.$$

In the above expressions, the values of m' and m are exactly the same as those in (7). We see that all terms with the factor $\sinh 2\phi_1 \sinh 2\phi_2$ remain in the above expressions.

Substituting the above expressions of $\alpha_{m,\uparrow\downarrow}^{(2)}$, $\beta_{m,\uparrow\downarrow}^{(2)}$, $\delta_M^{(2)}$, and $\delta_{2M}^{(2)}$ into (23), we obtain $\widetilde{Y}_2$, and, further, we have:

$$
\lim_{\phi_2 \to 0} \frac{\partial}{\partial \phi_2} \lim_{\phi_1 \to 0} \frac{\partial \tilde{Y}_2}{\partial \phi_1}
$$

$$
= \left(\sum_{m=1}^{M} \frac{W_m^{(\uparrow)}}{M} \tanh \frac{L\gamma_m^{(\uparrow)}}{2} + \frac{1}{2M} \sum_{m=1}^{M} \frac{1}{M} \frac{2\cos^2 \varphi_m^{(\uparrow)} - \sin^2 \theta_m^{(\uparrow)}}{\cosh^2 \frac{L\gamma_m^{(+)}}{2}} + \left(\sum_{m=1}^{M} \frac{1}{M} \cos \theta_m^{(\uparrow)} \tanh \frac{L\gamma_m^{(\uparrow)}}{2} \right)^2 \right)
$$

$$
\times \left(\prod_{l=1}^{M} 2\cosh \frac{L\gamma_l^{(\uparrow)}}{2} \right)^2
$$

$$
+ \left(\sum_{m=1}^{M} \frac{W_m^{(\uparrow)}}{M} \coth \frac{L\gamma_m^{(\uparrow)}}{2} - \frac{1}{2M} \sum_{m=1}^{M} \frac{1}{M} \frac{2\cos^2 \varphi_m^{(\uparrow)} - \sin^2 \theta_m^{(\uparrow)}}{\sinh^2 \frac{L\gamma_m^{(+)}}{2}} + \left(\sum_{m=1}^{M} \frac{1}{M} \cos \theta_m^{(\uparrow)} \coth \frac{L\gamma_m^{(\uparrow)}}{2} \right)^2 \right)
$$

$$
\times \left(\prod_{l=1}^{M} 2\sinh \frac{L\gamma_l^{(\uparrow)}}{2} \right)^2
$$

$$
+ \left(\left(\frac{W_{2M}^{(\downarrow)}}{2M} \tanh \frac{L\gamma_{2M}^{(\downarrow)}}{2} + \frac{W_M^{(\downarrow)}}{2M} \tanh \frac{L\gamma_M^{(\downarrow)}}{2} + \sum_{m=1}^{M-1} \frac{W_m^{(\downarrow)}}{M} \tanh \frac{L\gamma_m^{(\downarrow)}}{2} \right) \right.
$$

$$
+ \frac{1}{2M} \left(\frac{1}{2M} \frac{1}{\cosh^2 \frac{L\gamma_{2M}^{(\downarrow)}}{2}} + \frac{1}{2M} \frac{1}{\cosh^2 \frac{L\gamma_M^{(\downarrow)}}{2}} + \sum_{m=1}^{M-1} \frac{1}{M} \frac{2\cos^2 \varphi_m^{(\downarrow)} - \sin^2 \theta_m^{(\downarrow)}}{\cosh^2 \frac{L\gamma_m^{(\downarrow)}}{2}} \right) \tag{27}
$$

$$
+ \left. \left(\frac{1}{2M} \tanh \frac{L\gamma_{2M}^{(\downarrow)}}{2} + \frac{1}{2M} \tanh \frac{L\gamma_M^{(\downarrow)}}{2} + \sum_{m=1}^{M-1} \frac{1}{M} \cos \theta_m^{(\downarrow)} \tanh \frac{L\gamma_m^{(\downarrow)}}{2} \right)^2 \right)
$$

$$
\times \left(2\cosh \frac{L\gamma_{2M}^{(\downarrow)}}{2} \right) \left(2\cosh \frac{L\gamma_M^{(\downarrow)}}{2} \right) \left(\prod_{l=1}^{M-1} 2\cosh \frac{L\gamma_l^{(\downarrow)}}{2} \right)^2
$$

$$
+ \left(\left(\frac{W_{2M}^{(\downarrow)}}{2M} \coth \frac{L\gamma_{2M}^{(\downarrow)}}{2} + \frac{W_M^{(\downarrow)}}{2M} \coth \frac{L\gamma_M^{(\downarrow)}}{2} + \sum_{m=1}^{M-1} \frac{W_m^{(\downarrow)}}{M} \coth \frac{L\gamma_m^{(\downarrow)}}{2} \right) \right.
$$

$$
- \frac{1}{2M} \left(\frac{1}{2M} \frac{1}{\sinh^2 \frac{L\gamma_{2M}^{(\downarrow)}}{2}} + \frac{1}{2M} \frac{1}{\sinh^2 \frac{L\gamma_M^{(\downarrow)}}{2}} + \sum_{m=1}^{M-1} \frac{1}{M} \frac{2\cos^2 \varphi_m^{(\downarrow)} - \sin^2 \theta_m^{(\downarrow)}}{\sinh^2 \frac{L\gamma_m^{(-)}}{2}} \right)
$$

$$
+ \left. \left(\frac{1}{2M} \coth \frac{L\gamma_{2M}^{(\downarrow)}}{2} + \frac{1}{2M} \coth \frac{L\gamma_M^{(\downarrow)}}{2} + \sum_{m=1}^{M-1} \frac{1}{M} \cos \theta_m^{(\downarrow)} \coth \frac{L\gamma_m^{(\downarrow)}}{2} \right)^2 \right)
$$

$$
\times \left(2\sinh \frac{L\gamma_{2M}^{(\downarrow)}}{2} \right) \left(2\sinh \frac{L\gamma_M^{(\downarrow)}}{2} \right) \left(\prod_{l=1}^{M-1} 2\sinh \frac{L\gamma_l^{(\downarrow)}}{2} \right)^2 ,
$$

where $W_m^{(\uparrow \downarrow)}$ is introduced by:

$$
W_m = \frac{1}{2M} \left(\sum_{\substack{k=1 \\ k \neq m, \, k \neq m'}}^{2M} \frac{4\cos^2 \frac{\theta_m}{2} \cos^2 \frac{\theta_k}{2} \cos \varphi_m \cos \varphi_k - \sin \theta_m \sin \theta_k \sin \varphi_m \sin \varphi_k}{e^{L(\gamma_m - \gamma_k)} - 1} \right.
$$

$$
+ \sum_{k=1}^{2M} \frac{4\cos^2 \frac{\theta_m}{2} \sin^2 \frac{\theta_k}{2} \cos \varphi_m \cos \varphi_k + \sin \theta_m \sin \theta_k \sin \varphi_m \sin \varphi_k}{e^{L(\gamma_m + \gamma_k)} - 1}
$$

$$
+ \sum_{\substack{k=1 \\ k \neq m, \, k \neq m'}}^{2M} \frac{4\sin^2 \frac{\theta_m}{2} \sin^2 \frac{\theta_k}{2} \cos \varphi_m \cos \varphi_k - \sin \theta_m \sin \theta_k \sin \varphi_m \sin \varphi_k}{1 - e^{-L(\gamma_m - \gamma_k)}} \tag{28}
$$

$$
\left. + \sum_{k=1}^{2M} \frac{4\sin^2 \frac{\theta_m}{2} \cos^2 \frac{\theta_k}{2} \cos \varphi_m \cos \varphi_k + \sin \theta_m \sin \theta_k \sin \varphi_m \sin \varphi_k}{1 - e^{-L(\gamma_m + \gamma_k)}} \right),
$$

where the value of m' is exactly the same as that in (7).

Substituting $\widetilde{Z}$ given by (6) and (27) into (24), we obtain the expressions of $\langle \sigma_{1,1}\sigma_{1,3}\rangle$ of the model on a finite lattice:

$$\langle \sigma_{1,1}\sigma_{1,3}\rangle = \frac{1}{\widetilde{Z}}\lim_{\phi_2\to 0}\frac{\partial}{\partial\phi_2}\lim_{\phi_1\to 0}\frac{\partial\widetilde{Y}_2}{\partial\phi_1} \tag{29}$$

2.8. The Expressions of $\langle \sigma_{1,1}\sigma_{1,2}\rangle$ and $\langle \sigma_{1,1}\sigma_{1,3}\rangle$ in the Thermodynamic Limit

We now consider the thermodynamic limit. First, if L is very large, then according to (7) we have:

$$2\cosh\frac{L\gamma_m^{(\uparrow\downarrow)}}{2}\approx 2\sinh\frac{L\gamma_m^{(\uparrow\downarrow)}}{2}\approx \exp\frac{L\gamma_m^{(\uparrow\downarrow)}}{2},\tanh\frac{L\gamma_m^{(\uparrow\downarrow)}}{2}\approx \coth\frac{L\gamma_m^{(\uparrow\downarrow)}}{2}\approx 1\,(m=1,2\cdots M);$$

However, when the system crosses its critical temperature, $\gamma_{2M}^{(\downarrow)} = 2(K-K')$ changes sign, following which we therefore have:

$$2\cosh\frac{L\gamma_{2M}^{(\downarrow)}}{2}\approx \exp\frac{L\left|\gamma_{2M}^{(\downarrow)}\right|}{2},\, 2\sinh\frac{L\gamma_{2M}^{(\downarrow)}}{2}\approx \frac{\gamma_{2M}^{(\downarrow)}}{\left|\gamma_{2M}^{(\downarrow)}\right|}\exp\frac{L\left|\gamma_{2M}^{(\downarrow)}\right|}{2},\tanh\frac{L\gamma_{2M}^{(\downarrow)}}{2}\approx \coth\frac{L\gamma_{2M}^{(\downarrow)}}{2}\approx \frac{\gamma_{2M}^{(\downarrow)}}{\left|\gamma_{2M}^{(-)}\right|}.$$

Hence, for $\widetilde{Z}$ and $\lim_{\phi_1\to 0}\frac{\partial\widetilde{Y}_1}{\partial\phi_1}$ given by (6) and (25), respectively, when L is very large, we obtain:

$$\widetilde{Z}\approx \frac{1}{2}\left(\left(\prod_{m=1}^{M}\exp\frac{L\gamma_m^{(\uparrow)}}{2}\right)^2 + \left(\prod_{m=1}^{M}\exp\frac{L\gamma_m^{(\uparrow)}}{2}\right)^2\right.$$
$$\left. + \exp\frac{L\left|\gamma_{2M}^{(\downarrow)}\right|}{2}\exp\frac{L\gamma_M^{(\downarrow)}}{2}\left(\prod_{m=1}^{M-1}\exp\frac{L\gamma_m^{(\downarrow)}}{2}\right)^2 + \frac{\gamma_{2M}^{(\downarrow)}}{\left|\gamma_{2M}^{(\downarrow)}\right|}\exp\frac{L\left|\gamma_{2M}^{(\downarrow)}\right|}{2}\exp\frac{L\gamma_M^{(\downarrow)}}{2}\left(\prod_{m=1}^{M-1}\exp\frac{L\gamma_m^{(\downarrow)}}{2}\right)^2\right) \tag{30}$$
$$\approx \left(\prod_{m=1}^{M}\exp\frac{L\gamma_m^{(\uparrow)}}{2}\right)^2 \times \begin{cases}1,\, K<K';\\ 2,\, K>K',\end{cases}$$

$$\lim_{\phi_1\to 0}\frac{\partial\widetilde{Y}_1}{\partial\phi_1}\approx \left(\sum_{n=1}^{M}\frac{\cos\theta_n^{(\uparrow)}}{M}\right)\left(\prod_{l=1}^{M}\exp\frac{L\gamma_l^{(\uparrow)}}{2}\right)^2$$
$$+\frac{1}{2}\left(\frac{1}{2M}\frac{\gamma_{2M}^{(\downarrow)}}{\left|\gamma_{2M}^{(-)}\right|}+\frac{1}{2M}\sum_{n=1}^{M-1}\frac{\cos\theta_n^{(\downarrow)}}{M}\right)\left(1+\frac{\gamma_{2M}^{(\downarrow)}}{\left|\gamma_{2M}^{(\downarrow)}\right|}\right)\left(\exp\frac{L\left|\gamma_{2M}^{(\downarrow)}\right|}{2}\right)\left(\exp\frac{L\gamma_M^{(\downarrow)}}{2}\right)\left(\prod_{l=1}^{M-1}\exp\frac{L\gamma_l^{(\downarrow)}}{2}\right)^2$$
$$\approx \left(\sum_{n=1}^{M}\frac{\cos\theta_n^{(\uparrow)}}{M}\right)\left(\prod_{m=1}^{M}\exp\frac{L\gamma_m^{(\uparrow)}}{2}\right)^2\times\begin{cases}1,\,K<K';\\2,\,K>K'.\end{cases}$$

Substituting the above two expressions into (26), we obtain:

$$\lim_{\substack{L\to\infty\\N\to\infty}}\langle\sigma_{1,1}\sigma_{1,2}\rangle = \lim_{M\to\infty}\sum_{m=1}^{M}\frac{\cos\theta_m^{(\uparrow)}}{M} = \lim_{M\to\infty}\left(\frac{1}{2}\sum_{m=1}^{M}\frac{\cos\theta_m^{(\uparrow)}}{M}+\frac{1}{2}\sum_{m=M+1}^{2M}\frac{\cos\theta_m^{(\uparrow)}}{M}\right)$$
$$= \lim_{N\to\infty}\sum_{n=1}^{N}\frac{\cos\theta_n^{(\uparrow)}}{N} = \int_0^1 dx\cos\theta(\pi x), \tag{31}$$

where the function $\theta(\pi x)$ in terms of (14) is defined by:

$$\cos\theta(\pi x) = \frac{\cosh 2K'\sinh 2K - \cos(\pi x)\sinh 2K'\cosh 2K}{\sqrt{(\cosh 2K'\cosh 2K - \cos(\pi x)\sinh 2K'\sinh 2K)^2 - 1}},$$
$$\sin\theta(\pi x) = \frac{\sin(\pi x)\sinh 2K'}{\sqrt{(\cosh 2K'\cosh 2K - \cos(\pi x)\sinh 2K'\sinh 2K)^2 - 1}}. \tag{32}$$

The result (31) is in accordance with that in Reference [8].

According to the similar discussions, for $\langle \sigma_{1,1}\sigma_{1,3}\rangle$ we first have:

$$\lim_{\substack{L \to \infty \\ N \to \infty}} \langle \sigma_{1,1}\sigma_{1,3}\rangle \approx \lim_{\substack{L \to \infty \\ M \to \infty}} \sum_{m=1}^{M} \frac{W_m^{(\uparrow)}}{M} + \left(\lim_{M\to\infty}\sum_{m=1}^{M}\frac{\cos\theta_m^{(\uparrow)}}{M}\right)^2. \tag{33}$$

We discuss the first term in (28) as an example to show how to calculate $\lim_{L\to\infty} W_m^{(+)}$. First, using (7) and (15), the first term in (28) can be written in the form:

$$\frac{1}{2M}\sum_{\substack{k=1 \\ k\neq m\,,\,k\neq m'}}^{2M}\frac{4\cos^2\frac{\theta_m^{(\uparrow)}}{2}\cos^2\frac{\theta_k^{(\uparrow)}}{2}\cos\varphi_m^{(\uparrow)}\cos\varphi_k^{(\uparrow)}-\sin\theta_m^{(\uparrow)}\sin\theta_k^{(\uparrow)}\sin\varphi_m^{(\uparrow)}\sin\varphi_k^{(\uparrow)}}{e^{L(\gamma_m^{(\uparrow)}-\gamma_k^{(\uparrow)})}-1}$$

$$=\frac{1}{M}\sum_{\substack{k=1 \\ k\neq m}}^{M}\frac{4\cos^2\frac{\theta_m^{(\uparrow)}}{2}\cos^2\frac{\theta_k^{(\uparrow)}}{2}\cos\varphi_m^{(\uparrow)}\cos\varphi_k^{(\uparrow)}-\sin\theta_m^{(\uparrow)}\sin\theta_k^{(\uparrow)}\sin\varphi_m^{(\uparrow)}\sin\varphi_k^{(\uparrow)}}{e^{L(\gamma_m^{(\uparrow)}-\gamma_k^{(\uparrow)})}-1}$$

$$=\frac{1}{M}\sum_{k=1}^{m-1}\frac{4\cos^2\frac{\theta_m^{(\uparrow)}}{2}\cos^2\frac{\theta_k^{(\uparrow)}}{2}\cos\varphi_m^{(\uparrow)}\cos\varphi_k^{(\uparrow)}-\sin\theta_m^{(\uparrow)}\sin\theta_k^{(\uparrow)}\sin\varphi_m^{(\uparrow)}\sin\varphi_k^{(\uparrow)}}{e^{L(\gamma_m^{(\uparrow)}-\gamma_k^{(\uparrow)})}-1}$$

$$+\frac{1}{M}\sum_{k=m+1}^{M}\frac{4\cos^2\frac{\theta_m^{(\uparrow)}}{2}\cos^2\frac{\theta_k^{(\uparrow)}}{2}\cos\varphi_m^{(\uparrow)}\cos\varphi_k^{(\uparrow)}-\sin\theta_m^{(\uparrow)}\sin\theta_k^{(\uparrow)}\sin\varphi_m^{(\uparrow)}\sin\varphi_k^{(\uparrow)}}{e^{L(\gamma_m^{(\uparrow)}-\gamma_k^{(\uparrow)})}-1}.$$

According to (7), when $k < m$, $\gamma_k^{(\uparrow)} < \gamma_m^{(\uparrow)}$, $\lim_{L\to\infty} e^{L(\gamma_m^{(\uparrow)}-\gamma_k^{(\uparrow)})}=\infty$, and, thus, the first term in the above expression vanishes; when $k > m$, $\gamma_k^{(\uparrow)} > \gamma_m^{(\uparrow)}$, $\lim_{L\to\infty} e^{L(\gamma_m^{(\uparrow)}-\gamma_k^{(\uparrow)})}=0$, we therefore obtain:

$$\lim_{L\to\infty}\frac{1}{2M}\sum_{\substack{k=1 \\ k\neq m\,,\,k\neq m'}}^{2M}\frac{4\cos^2\frac{\theta_m^{(\uparrow)}}{2}\cos^2\frac{\theta_k^{(\uparrow)}}{2}\cos\varphi_m^{(\uparrow)}\cos\varphi_k^{(\uparrow)}-\sin\theta_m^{(\uparrow)}\sin\theta_k^{(\uparrow)}\sin\varphi_m^{(\uparrow)}\sin\varphi_k^{(\uparrow)}}{e^{L(\gamma_m^{(\uparrow)}-\gamma_k^{(\uparrow)})}-1}$$

$$=\frac{1}{M}\sum_{k=m+1}^{M}\frac{4\cos^2\frac{\theta_m^{(\uparrow)}}{2}\cos^2\frac{\theta_k^{(\uparrow)}}{2}\cos\varphi_m^{(\uparrow)}\cos\varphi_k^{(\uparrow)}-\sin\theta_m^{(\uparrow)}\sin\theta_k^{(\uparrow)}\sin\varphi_m^{(\uparrow)}\sin\varphi_k^{(\uparrow)}}{-1}.$$

Using this method to deal with the remaining terms in $W_m^{(\uparrow)}$, we finally obtain:

$$\lim_{L\to\infty}W_m^{(\uparrow)}=\frac{2}{M}\left(\sum_{k=m+1}^{M}\sin\theta_m^{(\uparrow)}\sin\theta_k^{(\uparrow)}\sin\varphi_m^{(\uparrow)}\sin\varphi_k^{(\uparrow)}-\sum_{k=m+1}^{M}\cos\theta_m^{(\uparrow)}\cos\theta_k^{(\uparrow)}\cos\varphi_m^{(\uparrow)}\cos\varphi_k^{(\uparrow)}\right.$$

$$\left.+\sum_{k=1}^{m+1}\cos\varphi_m^{(\uparrow)}\cos\varphi_k^{(\uparrow)}\right)+\frac{2}{M}\left(2\sin^2\frac{\theta_m^{(\uparrow)}}{2}\left(\cos^2\frac{\theta_m^{(\uparrow)}}{2}-\cos^2\varphi_m^{(\uparrow)}\right)-\cos\varphi_m^{(\uparrow)}\cos\varphi_{m+1}^{(\uparrow)}\right).$$

As $M \to \infty$, the second term in the above expression vanishes, and, according to the definition of the Riemann integral, we have:

$$\lim_{\substack{L \to \infty \\ M \to \infty}} W_m^{(\uparrow)} = 2 \int_x^1 dy (\sin \theta(\pi x) \sin \theta(\pi y) \sin(\pi x) \sin(\pi y) - \cos \theta(\pi x) \cos \theta(\pi y) \cos(\pi x) \cos(\pi y))$$

$$+ 2 \int_0^x dy \, \cos(\pi x) \cos(\pi y) ,$$

where the function $\theta(\pi x)$ is introduced by (32), $x - \dfrac{m+1}{M}$. Further,

$$\lim_{\substack{L \to \infty \\ M \to \infty}} \sum_{m=1}^{M} \frac{W_m^{(\uparrow)}}{M} = 2 \int_0^1 dx \int_x^1 dy \sin \theta(\pi x) \sin \theta(\pi y) \sin(\pi x) \sin(\pi y) \tag{34}$$

$$- 2 \int_0^1 dx \int_x^1 dy \cos \theta(\pi x) \cos \theta(\pi y) \cos(\pi x) \cos(\pi y) + 2 \int_0^1 dx \int_0^x dy \, \cos(\pi x) \cos(\pi y) .$$

Generally speaking, for the function $f(u, v)$ and the domain D of the integration shown in Figure 1, we have:

$$\iint_D du dv f(u, v) = \int_a^b du \int_a^u dv f(u, v) = \int_a^b dv \int_v^b du f(u, v) . \tag{35}$$

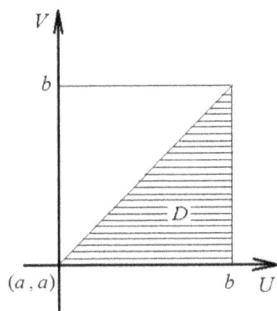

Figure 1. The domain of the integration in (35).

Using (35), for the first term in (34) we obtain:

$$2 \int_0^1 dx \int_x^1 dy \, \sin \theta(\pi x) \sin \theta(\pi y) \sin(\pi x) \sin(\pi y) = 2 \int_0^1 dx \sin \theta(\pi x) \sin(\pi x) \int_x^1 dy \, \sin \theta(\pi y) \sin(\pi y)$$
$$= \int_0^1 dx \, \sin \theta(\pi x) \sin(\pi x) \int_x^1 dy \, \sin \theta(\pi y) \sin(\pi y) + \int_0^1 dy \, \sin \theta(\pi y) \sin(\pi y) \int_0^y dx \, \sin \theta(\pi x) \sin(\pi x)$$
$$= \int_0^1 dx \, \sin \theta(\pi x) \sin(\pi x) \int_0^1 dy \, \sin \theta(\pi y) \sin(\pi y) .$$

Likewise, the second term in (34) becomes:

$$-2 \int_0^1 dx \int_x^1 dy \, \cos \theta(\pi x) \cos \theta(\pi y) \cos(\pi x) \cos(\pi y) = -\int_0^1 dx \, \cos \theta(\pi x) \cos(\pi x) \int_0^1 dy \, \cos \theta(\pi y) \cos(\pi y) .$$

Further, the third term in (34) vanishes due to:

$$\int_0^1 dx \int_0^x dy \, \cos(\pi x) \cos(\pi y) = \frac{1}{\pi} \int_0^1 dx \cos(\pi x) \sin(\pi x) = 0$$

Therefore, (34) becomes:

$$
\lim_{\substack{L \to \infty \\ M \to \infty}} \sum_{m=1}^{M} \frac{W_m^{(\uparrow)}}{M} = \int_0^1 dx \, \sin\theta(\pi x)\sin(\pi x) \int_0^1 dy \, \sin\theta(\pi y)\sin(\pi y)
$$

$$
- \int_0^1 dx \, \cos\theta(\pi x)\cos(\pi x) \int_0^1 dy \, \cos\theta(\pi y)\cos(\pi y) \qquad (36)
$$

$$
= - \int_0^1 dx \, \cos(\theta(\pi x) - \pi x) \int_0^1 dy \, \cos(\theta(\pi y) + \pi y) \, .
$$

Substituting (31) and (36) into (33), we obtain the form of $\langle \sigma_{1,\,1}\sigma_{1,\,3} \rangle$ in the thermodynamic limit:

$$
\lim_{\substack{L \to \infty \\ N \to \infty}} \langle \sigma_{1,\,1}\sigma_{1,\,3} \rangle = -\int_0^1 dx \, \cos(\theta(\pi x) - \pi x) \int_0^1 dy \, \cos(\theta(\pi y) + \pi y) + \left(\int_0^1 dx \, \cos\theta(\pi x) \right)^2 . \quad (37)
$$

On the other hand, the expressions of $\langle \sigma_{1,\,1}\sigma_{1,\,1+Q} \rangle$ in the thermodynamic limit have been obtained [3,5,9]. Thus, we here cite the formulas (B6) and (B7) in Reference [9] for comparison. According to those two formulas:

$$
\lim_{\substack{L \to \infty \\ N \to \infty}} \langle \sigma_{1,\,1}\sigma_{1,\,2} \rangle = a_0, \quad \lim_{\substack{L \to \infty \\ N \to \infty}} \langle \sigma_{1,\,1}\sigma_{1,\,3} \rangle = \begin{vmatrix} a_0 & a_1 \\ a_{-1} & a_0 \end{vmatrix} = a_0^2 - a_1 a_{-1}; a_r = \frac{1}{\pi}\int_0^\pi d\omega \cos(\theta(\omega) - r\omega). \quad (38)
$$

where $\theta(\omega)$ is the function $\delta * (\omega)$ in Reference [9]. We see that (31) and (37) obtained here are exactly the same as (38).

3. Long Range-Order in the Model with Periodic-Free Boundary Conditions

For the model with L rows and N columns and periodic boundary condition in the horizontal direction and free boundary condition in the vertical direction, we consider $\langle \sigma_{l,\,n}\sigma_{l,\,n'} \rangle$, i.e., correlation functions of pairwise spins in one column, periodic boundary condition in the horizontal direction leads to:

$$
\langle \sigma_{l,\,n}\sigma_{l,\,n'} \rangle = \langle \sigma_{1,\,n}\sigma_{1,\,n'} \rangle = \frac{1}{Z_0} \sum_{\{\sigma_h,\,v=\pm1\}} \sigma_{1,\,1}\sigma_{1,\,n'} \exp\left(\frac{J'}{kT}\sum_{l=1}^{L}\sum_{m=1}^{N}\sigma_{l,\,m}\sigma_{l+1,\,m} \right) \exp\left(\frac{J}{kT}\sum_{l=1}^{L}\sum_{m=1}^{N-1}\sigma_{l,\,m}\sigma_{l,\,m+1} \right), \quad (39)
$$

where

$$
Z_0 = \sum_{\{\sigma_h,\,v=\pm1\}} \exp\left(\frac{J'}{kT}\sum_{l=1}^{L}\sum_{m=1}^{N}\sigma_{l,\,m}\sigma_{l+1,\,m} \right) \exp\left(\frac{J}{kT}\sum_{l=1}^{L}\sum_{m=1}^{N-1}\sigma_{l,\,m}\sigma_{l,\,m+1} \right) \quad (40)
$$

is the partition function of the system in absence of a magnetic field, where $\sigma_{L+1,\,m} = \sigma_{1,\,m}$, $J'(>0)$ and $J(>0)$ are the interaction constants for the horizontal and vertical directions, respectively.

3.1. Some Results Concerning the Partition Function Z_0

We summarize some results concerning Z_0 given by (40), some of which are obtained in Reference [12]. However, the approximate values of some quantities presented here show improvement over those given by Reference [12].

By using the spinor analysis method, Z_0 is obtained in Reference [12]:

$$
Z_0 = \left(2\sinh\frac{2J'}{kT} \right)^{\frac{LN}{2}} \prod_{n=1}^{N} \left(2\cosh\frac{L\gamma_{n-1}}{2} \right), \quad (41)
$$

where γ_{n-1} $(n = 1, 2, \cdots, N)$ are determined by:

$$\cosh\gamma_{n-1} = \cosh 2K' \cosh 2K - x_{n-1}\sinh 2K'\sinh 2K, e^{-2K'} = \tanh\frac{J'}{kT}, \ K = \frac{J}{kT}, \tag{42}$$

where $(n = 2, 3, \cdots, N)$ $(n = 1, 2, \cdots, N)$ are N roots of the N-th order algebraic equation in x:

$$g_N(x) - 2g_{N-1}(x)\coth 2K'\tanh K + g_{N-2}(x)\tanh^2 K = 0, \tag{43}$$

where

$$g_n(x) = \sum_{k=0}^{[n/2]} \frac{(n+1)!}{(2k+1)!(n-2k)!} x^{n-2k}(x^2-1)^k \tag{44}$$

is an n-th degree polynomial in x. If by $x \equiv \dfrac{d+d^{-1}}{2}$ we introduce the quantity d, then $g_n(x)$ can be written in the form:

$$g_n(x) = \frac{d^{n+1} - d^{-(n+1)}}{d - d^{-1}}. \tag{45}$$

The expression in (45) is not only simple but also convenient for investigating the properties of $g_n(x)$, especially if we assume $x = \cos\varphi$, then $g_n(x) = \dfrac{\sin(n+1)\varphi}{\sin\varphi}$. Substituting these forms of $g_n(x)$ into (43), for the $N-1$ roots of the N-th order algebraic Equation (43) we obtain:

$$x_{n-1} = \cos\varphi_{n-1}, \varphi_{n-1} = \frac{(n-1)\pi + \theta_{n-1}}{N}, 0 < \theta_{n-1} < \pi (n = 2, 3, \cdots, N), \tag{46}$$

Further, θ_{n-1} can be determined by solving a transcendental equation about θ; if N is finite, then the evaluation of θ_{n-1} is complex because of the so-called "finite size effect"; for the limit case $\langle\sigma_{l,1}\sigma_{l,N}\rangle$, we can assume $\theta_{n-1} = \sum\limits_{k=0}^{\infty} \dfrac{\theta_{n-1}^{(k)}}{N^k}$ and obtain θ_{n-1} by the iterative method. Further, we obtain $\gamma_1, \gamma_2, \cdots, \gamma_{N-1}$ in terms of (42). Concretely, correcting to $\dfrac{1}{N}$ order, we have:

$$x_{n-1} = \cos\frac{(n-1)\pi + \theta_{n-1}^{(0)}}{N}, \ \gamma_{n-1} \approx \gamma_{n-1}^{(0)} + 2\sin\frac{(n-1)\pi + \theta_{n-1}^{(0)}}{N}\sin\frac{\theta_{n-1}^{(0)}}{2N}\frac{\sinh 2K'\sinh 2K}{\sinh\gamma_{n-1}^{(0)}}, \tag{47}$$
$$(n = 2, 3, \cdots, N)$$

where $\gamma_{n-1}^{(0)}$ and $\theta_{n-1}^{(0)}$ $(n = 2, 3, \cdots, N)$ are introduced by:

$$\cosh\gamma_{n-1}^{(0)} = \cosh 2K' \cosh 2K - \cos\frac{(n-1)\pi}{N}\sinh 2K'\sinh 2K, \ \gamma_{n-1}^{(0)} > 0;$$
$$\cosh 2K'\sinh 2K - \cos\frac{(n-1)\pi}{N}\sinh 2K'\cosh 2K = \sinh\gamma_{n-1}^{(0)}\cos\theta_{n-1}^{(0)}, \tag{48}$$
$$\sin\frac{(n-1)\pi}{N}\sinh 2K' = \sinh\gamma_{n-1}^{(0)}\sin\theta_k^{(0)}, \ 0 < \theta_{n-1}^{(0)} < \pi.$$

More important are the values of x_0 and γ_0; to present x_0 and γ_0, we first introduce a temperature T_c in terms of:

$$N = \frac{\sinh 2K'_c}{\sinh 2(K_c - K'_c)}, \tanh K'_c = e^{-\frac{2J'}{kT_c}}, K_c = \frac{J}{kT_c}. \tag{49}$$

When $T \geq T_c$, $0 < x_0 \leq 1$; however, when $T < T_c$, $K' < K$ and $1 < x_0 < \dfrac{\tanh 2K}{\tanh 2K'}$. For the limit case $N \to \infty$, we can obtain the approximate values of x_0 and γ_0, whose low-order approximations are:

$$
x_0 \approx \begin{cases}
\cos \dfrac{\pi}{N}, & T \geq T_c, \ K' > K ; \\[2mm]
\cos \dfrac{\pi}{2N}, & T \geq T_c, \ K' = K ; \\[2mm]
\dfrac{1}{2}\left(\dfrac{\tanh K}{\tanh K'} + \dfrac{\tanh K'}{\tanh K} \right) - 2\left(\dfrac{\tanh K'}{\tanh K} \right)^{2N} \dfrac{(\cosh 2K - \cosh 2K')^2}{\sinh^2 2K' \sinh^3 2K}, & T < T_c,
\end{cases} \tag{50}
$$

$$
\gamma_0 \approx \begin{cases}
2(K' - K) + 2\sin^2 \dfrac{\pi}{2N} \dfrac{\sinh 2K' \sinh 2K}{\sinh 2(K' - K)}, & T \geq T_c, \ K' > K ; \\[2mm]
2\sin \dfrac{\pi}{4N} \sinh 2K, & T \geq T_c, \ K' = K ; \\[2mm]
2\left(\dfrac{\tanh K'}{\tanh K} \right)^{N} \dfrac{\cosh 2K - \cosh 2K'}{\sinh 2K}, & T < T_c .
\end{cases} \tag{51}
$$

We can thus make a comparison between Onsager's lattice and the model with periodic-free boundary conditions. For Onsager's lattice, when the system crosses its critical temperature, $\gamma_N^{(-)} = \gamma_{2M}^{(-)} = 2(K - K')$ given by (7) changes sign; however, from (51) we see that, for the model with periodic-free boundary conditions, when $T \geq T_c$, $\gamma_0 \approx 2(K' - K)$. Once the system crosses its critical temperature T_c, γ_0 becomes exponentially smaller and then vanishes rapidly as $N \to \infty$. This property of γ_0 plays a key role for the correlation function $\langle \sigma_{1,\,1} \sigma_{1,\,N} \rangle$.

3.2. The Matrix Forms of $\langle \sigma_{1,\,1} \sigma_{1,\,N-Q} \rangle$ and Some Results Concerning $\langle \sigma_{1,\,1} \sigma_{1,\,N} \rangle$ Obtained in Reference [12]

If we write (39) in forms similar to (8), then by employing the method presented in Section 2, the exact expressions we can obtain are still $\langle \sigma_{1,\,1} \sigma_{1,\,2} \rangle$, $\langle \sigma_{1,\,1} \sigma_{1,\,3} \rangle$, $\cdots$, which belong to the short-range order. We still cannot obtain the exact expressions of $\langle \sigma_{l,\,1} \sigma_{l,\,N} \rangle$, $\langle \sigma_{l,\,1} \sigma_{l,\,N-1} \rangle$, $\cdots$.

To obtain the exact expressions of $\langle \sigma_{l,\,1} \sigma_{l,\,N} \rangle$, $\langle \sigma_{l,\,1} \sigma_{l,\,N-1} \rangle$, $\cdots$, we consider the forms:

$$
\langle \sigma_{1,\,1} \sigma_{1,\,N-n} \rangle = \frac{1}{Z_0} \sum_{\{ \sigma_{h,\,v} = \pm 1 \}} \sigma_{1,\,1} \sigma_{1,\,N-n} \exp\left(\frac{J'}{kT} \sum_{l=1}^{L} \sum_{m=1}^{N} \sigma_{l,\,m} \sigma_{l+1,\,m} \right) \exp\left(\frac{J}{kT} \sum_{l=1}^{L} \sum_{m=1}^{N-1} \sigma_{l,\,m} \sigma_{l,\,m+1} \right). \tag{52}
$$

Taking advantage of $\sigma_{l,\,m}^2 = 1$ and $\sigma_{1,\,l} \sigma_{1,\,m} = \lim\limits_{\alpha \to 0} \dfrac{\partial e^{\alpha \sigma_{1,\,l} \sigma_{1,\,m}}}{\partial \alpha}$, we have:

$$
\sigma_{1,\,1} \sigma_{1,\,N-n} = \sigma_{1,\,1} \sigma_{1,\,N} \sigma_{1,\,N-1} \cdots \sigma_{1,\,N-(n-2)} \sigma_{1,\,N-(n-1)} \sigma_{1,\,N-(n-1)} \sigma_{1,\,N-n}
$$
$$
= \left(\prod_{k=1}^{n} \lim_{\beta_k \to 0} \frac{\partial}{\partial \beta_k} \right) \lim_{\beta_N \to 0} \frac{\partial}{\partial \beta_N} \left(e^{\beta_N \sigma_{1,\,1} \sigma_{1,\,N}} \prod_{k=0}^{n-1} e^{\beta_{k+1} \sigma_{1,\,N-k} \sigma_{1,\,N-(k+1)}} \right),
$$

Further, (52) can be written in the form:

$$
\langle \sigma_{1,\,1} \sigma_{1,\,N-n} \rangle = \frac{1}{Z_0} \left(2\sinh \frac{2J'}{kT} \right)^{\frac{LN}{2}} \left(\prod_{k=1}^{n} \lim_{\beta_k \to 0} \frac{\partial}{\partial \beta_k} \right) \lim_{\beta_N \to 0} \frac{\partial}{\partial \beta_N} \mathrm{Tr}(\mathbf{W}), \tag{53}
$$

where the matrix $\mathbf{W}$ belongs to the spin representative, and, by employing the method presented in Section 2, we can obtain the exact expressions of $\langle \sigma_{l,\,1} \sigma_{l,\,N} \rangle$, $\langle \sigma_{l,\,1} \sigma_{l,\,N-1} \rangle$, $\langle \sigma_{l,\,1} \sigma_{l,\,N-2} \rangle$, $\cdots$.

However, to save space, here we no longer discuss $\langle \sigma_{l,\,1} \sigma_{l,\,N-1} \rangle$, $\langle \sigma_{l,\,1} \sigma_{l,\,N-2} \rangle$, $\cdots$, but only consider $\langle \sigma_{l,\,1} \sigma_{l,\,N} \rangle$, for which a closed formula was given by Reference [12]:

$$
\langle \sigma_{1,\,1} \sigma_{1,\,N} \rangle = \frac{1}{Z_0} \left(2\sinh \frac{2J'}{kT} \right)^{\frac{LN}{2}} \lim_{\phi \to 0} \frac{\partial Y}{\partial \phi}, \tag{54}
$$

$$Y = \frac{1}{2}\left(\prod_{l=1}^{\lfloor (N+1)/2 \rfloor} 2\cosh\frac{\chi_{2(l-1)}^{(+)}}{2}\right)\left(\prod_{m=1}^{\lfloor N/2 \rfloor} 2\cosh\frac{-\chi_{2m-1}^{(+)}}{2}\right) + \frac{1}{2}\left(\prod_{l=1}^{\lfloor (N+1)/2 \rfloor} 2\sinh\frac{\chi_{2(l-1)}^{(+)}}{2}\right)\left(\prod_{m=1}^{\lfloor N/2 \rfloor} 2\sinh\frac{-\chi_{2m-1}^{(+)}}{2}\right)$$
$$+ \frac{1}{2}\left(\prod_{l=1}^{\lfloor (N+1)/2 \rfloor} 2\cosh\frac{\chi_{2(l-1)}^{(-)}}{2}\right)\left(\prod_{m=1}^{\lfloor N/2 \rfloor} 2\cosh\frac{-\chi_{2m-1}^{(-)}}{2}\right) - \frac{1}{2}\left(\prod_{l=1}^{\lfloor (N+1)/2 \rfloor} 2\sinh\frac{\chi_{2(l-1)}^{(-)}}{2}\right)\left(\prod_{m=1}^{\lfloor N/2 \rfloor} 2\sinh\frac{-\chi_{2m-1}^{(-)}}{2}\right), \tag{55}$$

where $\chi_{2(l-1)}^{(\pm)}$ and $\chi_{2m-1}^{(\pm)}$ are determined by:

$$e^{\chi_{2(l-1)}^{(\pm)}} = \tau_{2(l-1)}^{(\pm)}\left(l = 1, 2, \quad , \left[\frac{N+1}{2}\right]\right); \; e^{\chi_{2m-1}^{(\pm)}} = \tau_{2m-1}^{(\pm)}\left(m - 1, 2, \cdots, \left[\frac{N}{2}\right]\right), \tag{56}$$

$\tau_n^{(\pm)}$ $(n = 1, 2, \cdots, N)$ are N roots of the N-th order algebraic equation $F_\pm(\tau) = 0$, where

$$F_\pm(\tau) = \left(\prod_{l=1}^{\lfloor (N+1)/2 \rfloor}\left(\tau\,e^{-L\gamma_{2(l-1)}} - 1\right)\right)\left(\prod_{m=1}^{\lfloor N/2 \rfloor}\left(\tau\,e^{L\gamma_{2m-1}} - 1\right)\right)$$
$$\mp\tanh\phi\left(\prod_{l=1}^{\lfloor (N+1)/2 \rfloor}\left(\tau\,e^{-L\gamma_{2(l-1)}} - 1\right)\right)\left(\prod_{m=1}^{\lfloor N/2 \rfloor}\left(\tau\,e^{L\gamma_{2m-1}} - 1\right)\right)$$
$$\mp 4\tanh\phi\sum_{n=1}^{\lfloor (N+1)/2 \rfloor}\left(\Omega_{2(n-1)}^2\left(\prod_{\substack{l=1\\l\neq n}}^{\lfloor (N+1)/2 \rfloor}\left(\tau\,e^{-L\gamma_{2(l-1)}} - 1\right)\right)\left(\prod_{m=1}^{\lfloor N/2 \rfloor}\left(\tau\,e^{L\gamma_{2m-1}} - 1\right)\right)\right) \tag{57}$$
$$\mp 4\tanh\phi\sum_{n=1}^{\lfloor N/2 \rfloor}\left(\Omega_{2n-1}^2\left(\prod_{l=1}^{\lfloor (N+1)/2 \rfloor}\left(\tau\,e^{-L\gamma_{2(l-1)}} - 1\right)\right)\left(\prod_{\substack{m=1\\m\neq n}}^{\lfloor N/2 \rfloor}\left(\tau\,e^{L\gamma_{2m-1}} - 1\right)\right)\right),$$

where

$$\Omega_{n-1} = \sinh 2K'\cosh K\sqrt{\frac{1 - x_{n-1}^2}{N\sinh^2\gamma_{n-1} + \cosh\gamma_{r-1}\cosh 2K' - \cosh 2K}} \quad (n = 1, 2, \cdots, N) \tag{58}$$

are the normalization constants of the eigenvectors of a rotation matrix [12], as $N \to \infty$. Thus, according to (47)~(51), we obtain:

$$\lim_{N\to\infty}\Omega_{n-1} \approx \frac{1}{\sqrt{N}}\sin\frac{(n-1)\pi}{N}\frac{\sinh 2K'\cosh K}{\sinh\gamma_{n-1}} \quad (n = 2, 3, \cdots, N),$$

$$\lim_{N\to\infty}\Omega_0^2 \approx \begin{cases} \dfrac{1}{N}\sin^2\dfrac{\pi}{N}\dfrac{\sinh^2 2K'\cosh^2 K}{\sinh^2 2(K' - K)} &, T \geq T_c\,, K' > K\,; \\[2mm] \dfrac{1}{N}\dfrac{\sinh^2 2K'\cosh^2 K}{\sinh^2 2K} &, T \geq T_c\,, K' = K\,; \\[2mm] \dfrac{\cosh 2K - \cosh 2K'}{4\sinh^2 K} &.T < T_c\,. \end{cases} \tag{59}$$

3.3. An Exact Expression of $\langle \sigma_{1,1}\sigma_{1,N} \rangle$ on a Finite Lattice

In Reference [12], all roots of the equation $F_\pm(\tau) = 0$ are obtained by correcting to the e^{-LC_0} order of magnitude (C_0 is a positive constant). These approximate roots can lead to the exact expression of $\langle \sigma_{1,1}\sigma_{1,N} \rangle$ in the thermodynamic limit, since $\lim_{L\to\infty} e^{-LC_0} = 0$, but cannot lead to the exact expression of $\langle \sigma_{1,1}\sigma_{1,N} \rangle$ on a finite lattice. Hence, the expression of $\langle \sigma_{1,1}\sigma_{1,N} \rangle$ presented in Reference [12] is only an approximate result.

On the other hand, similar to the analysis in Section 2.5, the operator $\lim_{\phi\to 0}\frac{\partial}{\partial\phi}$ in (54) allows us to ignore all terms whose order is higher than $\phi^1(= \phi)$ in all roots of the equation $F_\pm(\tau) = 0$. Hence, to obtain the exact expression of $\lim_{\phi\to 0}\dfrac{\partial Y}{\partial \phi}$, we need only to find all roots of the equation $F_{\pm}(\tau) = 0$ corrected to the $\phi^1(=\phi)$ order, The corresponding calculations are in fact simpler than those required of find the roots corrected to the e^{-LC_0} order of magnitude in Reference [12]; concretely, we obtain:

$$\chi^{(\pm)}_{2(l-1)} \approx L\gamma_{2(l-1)} \pm 4\Omega^2_{2(l-1)}\tanh\phi, \quad \chi^{(\pm)}_{2m-1} \approx -L\gamma_{2m-1} \pm 4\Omega^2_{2m-1}\tanh\phi.$$

Substituting the above results into (55), we obtain Y correcting to $\tanh\phi(\approx \phi^1)$ order:

$$
\begin{aligned}
Y \approx \ & \frac{1}{2}\left(\prod_{l=1}^{[(N+1)/2]} 2\cosh\frac{L\gamma_{2(l-1)} + 4\Omega^2_{2(l-1)}\tanh\phi}{2}\right)\left(\prod_{m=1}^{[N/2]} 2\cosh\frac{L\gamma_{2m-1} - 4\Omega^2_{2m-1}\tanh\phi}{2}\right) \\
& + \frac{1}{2}\left(\prod_{l=1}^{[(N+1)/2]} 2\sinh\frac{L\gamma_{2(l-1)} + 4\Omega^2_{2(l-1)}\tanh\phi}{2}\right)\left(\prod_{m=1}^{[N/2]} 2\sinh\frac{L\gamma_{2m-1} - 4\Omega^2_{2m-1}\tanh\phi}{2}\right) \\
& + \frac{1}{2}\left(\prod_{l=1}^{[(N+1)/2]} 2\cosh\frac{L\gamma_{2(l-1)} - 4\Omega^2_{2(l-1)}\tanh\phi}{2}\right)\left(\prod_{m=1}^{[N/2]} 2\cosh\frac{L\gamma_{2m-1} + 4\Omega^2_{2m-1}\tanh\phi}{2}\right) \\
& - \frac{1}{2}\left(\prod_{l=1}^{[(N+1)/2]} 2\sinh\frac{L\gamma_{2(l-1)} - 4\Omega^2_{2(l-1)}\tanh\phi}{2}\right)\left(\prod_{m=1}^{[N/2]} 2\sinh\frac{L\gamma_{2m-1} + 4\Omega^2_{2m-1}\tanh\phi}{2}\right).
\end{aligned}
$$

Substituting the above result and (41) into (54), we obtain the exact expression of $\langle \sigma_{1,1}\sigma_{1,N}\rangle$ of the model on a finite lattice:

$$\langle \sigma_{1,1}\sigma_{1,N}\rangle = 2\left(\sum_{l=1}^{[(N+1)/2]} \Omega^2_{2(l-1)}\coth\frac{L\gamma_{2(l-1)}}{2} - \sum_{m=1}^{[N/2]} \Omega^2_{2m-1}\coth\frac{L\gamma_{2m-1}}{2}\right)\left(\prod_{n=1}^{N}\tanh\frac{L\gamma_{n-1}}{2}\right) \quad (60)$$

Although the whole calculation process is complex, the final result (60) is simple.

3.4. The Expression of $\langle \sigma_{1,1}\sigma_{1,N}\rangle$ in the Thermodynamic Limit

To derive the expression of $\langle \sigma_{1,1}\sigma_{1,N}\rangle$ in the thermodynamic limit, we first discuss some properties of γ_{n-1} $(n = 1, 2, \cdots, N)$.

For γ_{n-1} $(n = 2, 3, \cdots, N)$ given by (47), we have:

$$\lim_{L\to\infty} L\gamma_{n-1} = \infty, \ \lim_{L\to\infty}\tanh\frac{L\gamma_{n-1}}{2} = \lim_{L\to\infty}\coth\frac{L\gamma_{n-1}}{2} = 1 (n = 2, 3, \cdots, N). \quad (61)$$

As for γ_0, when $T \geq T_c$, from (51) we see that (61) still holds for γ_0; however, when $T < T_c$, from the last expression in (51) we see that maybe $\lim_{L\to\infty} L\gamma_0 = \infty$ does not hold. For example, if $L = N^a$, where a is a positive integer, then:

$$\lim_{L\to\infty} L\gamma_0 = \lim_{N\to\infty} N^a \cdot 2\left(\frac{\tanh K'}{\tanh K}\right)^N \csc h2K(\cosh 2K - \cosh 2K') = 0,$$

since now $0 < \dfrac{\tanh K'}{\tanh K} < 1$, for $0 < b < 1$, $\lim_{N\to\infty} N^a b^N = 0$. Hence, for γ_0 we have:

$$\lim_{L\to\infty, N\to\infty} L\gamma_0 = \begin{cases} \infty, & T \geq T_c; \\ 0, & T < T_c, \end{cases} \quad \lim_{L\to\infty, N\to\infty}\tanh\frac{L\gamma_0}{2} = \begin{cases} 1, & T \geq T_c; \\ 0, & T < T_c, \end{cases} (L = N^a) \quad (62)$$

According to the above discussions, to obtain the expression of $\langle \sigma_{1,\,1} \sigma_{1,\,N} \rangle$ in the thermodynamic limit, we first write (60) in the form:

$$
\begin{aligned}
\langle \sigma_{1,\,1} \sigma_{1,\,N} \rangle = \ & 2\Omega_0^2 \left(\prod_{n=2}^{N} \tanh \frac{L\gamma_{n-1}}{2} \right) \\
& + 2\tanh \frac{L\gamma_0}{2} \left(\sum_{l=2}^{[(N+1)/2]} \Omega_{2(l-1)}^2 \coth \frac{L\gamma_{2(l-1)}}{2} - \sum_{m=1}^{[N/2]} \Omega_{2m-1}^2 \coth \frac{L\gamma_{2m-1}}{2} \right) \left(\prod_{n=2}^{N} \tanh \frac{L\gamma_{n-1}}{2} \right),
\end{aligned}
$$

as $L \to \infty$, according to (61), the above expression becomes:

$$
\lim_{L \to \infty} \langle \sigma_{1,\,1} \sigma_{1,\,N} \rangle \approx 2\Omega_0^2 + 2\tanh \frac{L\gamma_0}{2} \left(\sum_{l=2}^{[(N+1)/2]} \Omega_{2(l-1)}^2 - \sum_{m=1}^{[N/2]} \Omega_{2m-1}^2 \right). \tag{63}
$$

When $T \geq T_c$, according to (59) and (62), Equation (63) becomes:

$$
\lim_{L \to \infty} \langle \sigma_{1,\,1} \sigma_{1,\,N} \rangle \approx 2 \left(\sum_{l=1}^{[(N+1)/2]} \Omega_{2(l-1)}^2 - \sum_{m=1}^{[N/2]} \Omega_{2m-1}^2 \right),
$$

Further, as $N \to \infty$,

$$
\sum_{l=1}^{[(N+1)/2]} \Omega_{2(l-1)}^2 \approx \sum_{m=1}^{[N/2]} \Omega_{2m-1}^2 \approx \frac{1}{2} \sum_{n=1}^{N} \Omega_{n-1}^2, \tag{64}
$$

hence, for this case $\lim\limits_{L \to \infty,\, N \to \infty} \langle \sigma_{1,\,1} \sigma_{1,\,N} \rangle = 0$.

When $T < T_c$, according to (62) and (64), the second term in (63) vanishes, and (63) thus becomes $\lim\limits_{L \to \infty} \langle \sigma_{1,\,1} \sigma_{1,\,N} \rangle \approx 2\Omega_0^2$, where Ω_0^2 is given by the last expression in (59) as $N \to \infty$.

Summarizing the above results, in the thermodynamic limit, if $L = N^a$, then (60) becomes:

$$
\lim_{L \to \infty,\, N \to \infty} \langle \sigma_{1,\,1} \sigma_{1,\,N} \rangle = \begin{cases} 0, & T \geq T_c ; \\ \dfrac{\cosh 2K - \cosh 2K'}{2\sinh^2 K}, & T < T_c . \end{cases}
$$

The above result was obtained in Reference [12] in terms of an approximate result of $\langle \sigma_{1,\,1} \sigma_{1,\,N} \rangle$. Some further discussions about the above result can be found in Reference [12].

From the above discussions, it is revealed how the changes of the values of γ_{n-1} ($n = 1, 2, \cdots, N$), especially the change of the value of γ_0, lead to the change of $\lim\limits_{L \to \infty,\, N \to \infty} \langle \sigma_{1,\,1} \sigma_{1,\,N} \rangle$ when the system crosses its critical temperature T_c, as well as how the long-range order emerges as the temperature decreases.

Conflicts of Interest: The authors declare no conflict of interest.

References

1. Onsager, L. Crystal statistics. I. A two-dimensional model with an order-disorder transition. *Phys. Rev.* **1944**, *65*, 117–149. [CrossRef]
2. Kaufman, B. Crystal statistics. II. Partition function evaluated by spinor analysis. *Phys. Rev.* **1949**, *76*, 1232–1243. [CrossRef]
3. McCoy, B.M.; Wu, T.T. *The Two-Dimensional Ising Model*; Harvard University Press: Cambridge, MA, USA, 1973.
4. Baxter, B.J. *Exactly Solved Models in Statistical Mechanics*; Academic Press: London, UK, 1982.
5. McCoy, B.M. *Advanced Statistical Mechanics*; Oxford University Press: Oxford, UK, 2010.
6. Lu, W.T.; Wu, F.Y. Ising model on nonorientable surfaces: Exact solution for the Möbius strip and the Klein bottle. *Phys. Rev. E* **2001**, *63*, 026107. [CrossRef] [PubMed]
7. Hucht, A. The square lattice Ising model on the rectangle I: Finite systems. *J. Phys. A Math. Theor.* **2017**, *50*, 065201. [CrossRef]

8. Kaufman, B.; Onsager, L. Crystal statistics. III. Short-range order in a binary Ising lattice. *Phys. Rev.* **1949**, *76*, 1244–1252. [CrossRef]

9. Montroll, E.W.; Potts, R.B.; Ward, J.C. Correlations and spontaneous magnetization of a two-dimensional Ising model. *J. Math. Phys.* **1963**, *4*, 308–322. [CrossRef]

10. Bugrij, A.I. Correlation function of the two-dimensional Ising model on a finite lattice. I. *Theor. Math. Phys.* **2001**, *127*, 528–548. [CrossRef]

11. Bugrij, A.I.; Lisovyy, O.O. Correlation function of the two-dimensional Ising model on a finite lattice. II. *Theor. Math. Phys.* **2004**, *140*, 987–1000. [CrossRef]

12. Mei, T. An exact closed formula of a spin-spin correlation function of the two-dimensional Ising model with finite size. *Int. J. Theor. Phys.* **2015**, *54*, 3462–3489. [CrossRef]

13. Ferdinand, A.E.; Fisher, M.E. Bounded and inhomogeneous Ising models. I. Specific-heat anomaly of a finite lattice. *Phys. Rev.* **1969**, *185*, 832–846. [CrossRef]

14. O'Brien, D.L.; Pearce, P.A.; Warnaar, S.O. Finitized conformal spectrum of the Ising model on the cylinder and torus. *Phys. A Stat. Mech. Appl.* **1996**, *228*, 63–77. [CrossRef]

15. Izmailian, N.S. Finite-size effects for anisotropic 2D Ising model with various boundary conditions. *J. Phys. A Math. Theor.* **2012**, *45*, 494009. [CrossRef]

16. Hucht, A. The square lattice Ising model on the rectangle II: Finite-size scaling limit. *J. Phys. A Math. Theor.* **2017**, *50*, 265205. [CrossRef]

17. Huang, K. *Statistical Mechanics*, 2nd ed.; Wiley: New York, NY, USA, 1987.

MDPI

St. Alban-Anlage 66

4052 Basel

Switzerland

Tel. +41 61 683 77 34

Fax +41 61 302 89 18

www.mdpi.com

Entropy Editorial Office

E-mail: entropy@mdpi.com

www.mdpi.com/journal/entropy